Handbook of Experimental Pharmacology

Volume 87/I

Pharmacology of the Skin I

Pharmacology of Skin Systems
Autocoids in Normal and Inflamed Skin

Contributors

D. I. Abramson, J. C. Allen, H. P. Baden, R. C. Benyon, A. G. Bird, S. D. Brain, K. A. Brown, E. Christophers, M. K. Church, L. S. Clegg, K. J. Collins, W. I. Cranston, R. P. R. Dawber, D. C. Dumonde, C. J. Dunn, J. A. Edwardson, V. Eisen, B. A. Ellis, J. C. Foreman, M. J. Forrest, M. Goos, M. W. Greaves, S. T. Holgate, V. K. Hopsu-Havu, A. B. Kay, C. M. Lapière, F. Lawlor, B. Lynn, P. Mauvais-Jarvis, J. Morley, B. V. Nusgens, E. M. Saihan, C. Schubert, S. Shuster, A. J. Thody, G. Volden, A. J. Wardlaw, S. I. Wasserman, T. J. Williams, D. A. Willoughby, F. Wright, M. A. Zar, V. A. Ziboh

Editors

Malcolm W. Greaves and Sam Shuster

Springer-Verlag
Berlin Heidelberg New York
London Paris Tokyo

Professor MALCOLM W. GREAVES, M.D., Ph.D., F.R.C.P.

The Institute of Dermatology
St. Thomas's Hospital
Lambeth Palace Road
London SE1 7EH
Great Britain

Professor SAM SHUSTER, M.D., F.R.C.P.

The University of Newcastle upon Tyne
Department of Dermatology
The Royal Victoria Infirmary
Newcastle upon Tyne NE1 4LP
Great Britain

With 78 Figures

ISBN-13: 978-3-642-73799-2 e-ISBN-13: 978-3-642-73797-8
DOI: 10.1007/978-3-642-73797-8

Library of Congress Cataloging in Publication Data. Pharmacology of the skin/contributors, D. I. Abramson ... [et al.]; editors, M. W. Greaves and S. Shuster. p. cm. – (Handbook of experimental pharmacology; v. 87/1–2) Includes index. (v. 1: alk. paper). 1. Dermatopharmacology. 2. Skin – Physiology. I. Abramson, David I. (David Irvin), 1905– . II. Greaves, M. W. III. Shuster, Sam. IV. Series. [DNLM: 1. Dermatologic Agents – pharmacology. 2. Skin – drug effects. 3. Skin Diseases – drug therapy. W1 HA51L v. 87 pt. 1–2/QV 60 P536] QP905.H3 vol. 87/1–2 [RM303] 615'.1 s – dc19 [615'.778] DNLM/DLC for Library of Congress

Softcover reprint of the hardcover 1st edition 1989

2122/3130-543210 – Printed on acid-free paper

List of Contributors

D. I. Abramson, 916 North Oak Avenue, Oak Park, IL 60302, USA

J. C. Allen, Research Division, The North East Wales Institute, Kelsterton College, Connah's Quay, Deeside, Clwyd CH5 4BR, Great Britain

H. P. Baden, Harvard Medical School, Department of Dermatology, Massachusetts General Hospital, 32, Fruit Street, Boston, MA 02114, USA

R. C. Benyon, Clinical Pharmacology, University of Southampton, Southampton General Hospital, Tremona Road, Southampton SO9 4XY, Great Britain

A. G. Bird, Regional Immunology Department, Newcastle General Hospital, Westgate Road, Newcastle upon Tyne NE1 6BE, Great Britain

S. D. Brain, Clinical Research Centre, Section of Vascular Biology, Watford Road, Harrow, Middlesex HA1 3UJ, Great Britain

K. A. Brown, Department of Immunology, St. Thomas's Hospital, Lambeth Palace Road, London SE1 7EH, Great Britain

E. Christophers, Department of Dermatology, Hautklinik, Christian-Albrechts-Universität, Schittenhelmstr. 7, D-2300 Kiel

M. K. Church, Clinical Pharmacology, University of Southampton, Southampton General Hospital, Tremona Road, Southampton SO9 4XY, Great Britain

L. S. Clegg, Clinical Pharmacology, University of Southampton, Southampton General Hospital, Tremona Road, Southampton SO9 4XY, Great Britain

K. J. Collins, Medical Research Council, Environmental Physiology Unit, (Annexe), 242, Pentonville Road, London NE1 9LB, Great Britain

W. I. Cranston, Department of Medicine, St. Thomas's Hospital, Medical School, London SE1 7EH, Great Britain

R. P. R. Dawber, The Department of Dermatology, The Slade Hospital, Headington, Oxford OX3 7JH, Great Britain

D. C. Dumonde, Department of Immunology, St. Thomas's Hospital, Lambeth Palace Road, London SE1 7EH, Great Britain

C. J. Dunn, Department of Hypersensitivity Diseases Research, The Upjohn Company, Kalamazoo, MI 49001, USA

J. A. Edwardson, Medical Research Council, Neuroendocrinology Unit, Newcastle General Hospital, Westgate Road, Newcastle upon Tyne NE4 6BE, Great Britain

V. Eisen, Department of Pharmacology and Therapeutics, Cobbold Laboratories, Thorn Institute of Clinical Sciences, The Middlesex Hospital Medical School, London W1N 8AA, Great Britain

B. A. Ellis, Department of Immunology, St. Thomas's Hospital, Lambeth Palace Road, London SE1 7EH, Great Britain

J. C. Foreman, The Department of Pharmacology, University College London, Gower Street, London WC1E 6BT, Great Britain

M. J. Forrest, Merck Sharp & Dohme Research Laboratories, P.O. Box 2000, Rahway, NJ 07065, USA

M. Goos, Medizinische Einrichtungen der Universität, Hautklinik, Hufelandstr. 55, D-4300 Essen 1

M. W. Greaves, Institute of Dermatology, St. Thomas's Hospital, Lambeth Palace Road, London SE1 7EH, Great Britain

S. T. Holgate, Medicine I, Southampton General Hospital, Tremona Road, Southampton SO9 4XY, Great Britain

V. K. Hopsu-Havu, Department of Dermatology, University of Turku, SF-20520 Turku 52

A. B. Kay, Department of Allergy and Clinical Immunology, Cardiothoracic Institute, Dovehouse Street, London SW3 6LY, Great Britain

C. M. Lapière, Université de Liège, Clinique Dermatologique, CHU de Bavière, Bld. de la Constitution, 66, B-4020 Liège

F. Lawlor, Institute of Dermatology, St. Thomas's Hospital, Lambeth Palace Road, London SE1 7EH, Great Britain

B. Lynn, Department of Physiology, University College London, Gower Street, London WC1E 6BT, Great Britain

P. Mauvais-Jarvis, Reproductive Endocrinology Department, Hôpital Necker, 149, rue de Sèvres, F-75730 Paris Cedex 15

J. Morley, Preclinical Pharmacology, Sandoz Ltd., CH-4002 Basle

B. V. Nusgens, Université de Liège, Clinique Dermatologique, Hôpital de Bavière, Bld. de la Constitution, 66, B-4020 Liège

E. M. Saihan, Department of Dermatology, University Hospital, Queen's Medical Centre, Nottingham NG7 2UH, Great Britain

C. Schubert, Department of Dermatology, Hautklinik, Christian-Albrechts-Universität, Schittenhelmstr. 7, D-2300 Kiel

S. Shuster, The University of Newcastle upon Tyne, Department of Dermatology, The Royal Victoria Infirmary, Newcastle upon Tyne NE1 4LP, Great Britain

A. J. Thody, Department of Dermatology, The University of Newcastle upon Tyne, The Royal Victoria Infirmary, Newcastle upon Tyne NE1 4LP, Great Britain

G. Volden, Department of Dermatology, Regional Hospital, University of Trondheim, N-7006 Trondheim

A. J. Wardlaw, Department of Allergy and Clinical Immunology, Cardiothoracic Institute, Brompton Hospital, Fulham Road, London SW3 6HP, Great Britain

S. I. Wasserman, Division of Allergy/Rheumatology, Department of Medicine, School of Medicine, University of California Medical Center, 225 Dickinson Street, San Diego, CA 92103-9981, USA

T. J. Williams, Department of Applied Pharmacology, Cardiothoracic Institute, Fulham Road, Brompton, London SW3 6HP, Great Britain

D. A. Willoughby, Department of Experimental Pathology, St. Bartholomews Hospital Medical College, (University of London), West Smithfield, London EC1A 7BE, Great Britain

F. Wright, Faculté de Medecine, Pitié Salpétrière, Service de Biochimie Medicale, 91 Bld. Hôpital, F-75634 Paris Cedex 13

M. A. Zar, Department of Pharmacological Sciences, Medical School, University of Newcastle upon Tyne, Newcastle upon Tyne NE1 4HH, Great Britain

V. A. Ziboh, Department of Dermatology, TB192-School of Medicine, University of California Davis, Davis, CA 95616, USA

Preface

The recent interest in the pharmacology of the skin and the treatment of its diseases has come about for two reasons. The first is a realisation that many aspects of pharmacology can be studied as easily in human skin, where they may be more relevant to human physiology and diseases, as in animal models. Examples of this are the action of various vasoactive agents and the isolation of mediators of inflammation after UV irradiation and antigen-induced dermatitis. The second reason is the fortuitous realisation that a pharmacological approach to the treatment of skin disease need not always await the full elucidation of etiology and mechanism. For example, whilst the argument continued unresolved as to whether the pilo-sebaceous infection which constitutes acne was duc to a blocked duct or to a simple increase in sebum production, 13-*cis*-retinoic acid was found quite by chance totally to ablate the disease; again, whilst cyclosporin, fresh from its triumphs in organ transplantation, has been found able to suppress the rash of psoriasis, it has resuscitated the debate on etiology.

We are therefore entering a new era in which the pharmacology and clinical pharmacology of skin are being studied as a fascinating new way of exploring questions of human physiology and pharmacology as well as an important step in the development and study of new drugs, use of which will improve disease control and at the same time help to define pathological mechanisms.

It was because of this burgeoning interest in pharmacology of skin and its diseases that this book came about. Indeed, it is long overdue and was planned several years ago; we console ourselves with the thought that the delay may have served to help define certain principles which were then only just emerging.

The book is divided into two volumes, which are independent but complementary, the first being an account of the general pharmacology of skin, and the second being more concerned with disease and drugs. The first volume is divided into two parts, the one dealing with the pharmacology of skin systems and their control and the other with autocoids in normal and inflamed skin. The second volume has three parts: the first part deals with the methods of measurement which are becoming of increasing importance in studying both the pharmacology and clinical pharmacology of skin; the second deals with toxicology in its widest sense – including metabolism and percutaneous absorption; and the third is an account of both specific drugs and the drugs used for specific diseases.

The aim has been to give an up-to-date review in which there was sufficient background detail for an understanding of the subject without having to refer beyond the two volumes – but, of course, with sufficient referencing to serve as a guide to deeper reading. Authors were encouraged to present both consensus and personal views so that both possibilities and what appear today to be probabilities are presented. In this way it is hoped that the two extremes of dogma and fantasy have been avoided. In presenting this account we were well aware of the problems of a rapidly advancing field, but we hope that the chapters are so written that newer knowledge, which is now almost continuously becoming available, can be more easily understood and incorporated into a body of knowledge. Finally but inevitably, we have had to make certain omissions – for example in the field of anti-viral drugs, AIDS and one or two other areas – if only for reasons of size.

The volumes are written for pharmacologists, clinical and non-clinical, and will be pertinent to pharmacists and to many with an interest in the skin now working in the pharmaceutical industry, as well as to physiologists. The work too will be of major interest to dermatologists. Although the book is directed primarily at a postgraduate audience, there is much in it that will be of helpful to Honours and Ph.D. students. We hope that by including as much background as is relevant to an understanding within the two volumes, we will have reduced the likelihood of the reader having to look elsewhere for primary explanations. This plan may have led to some duplication and over-simplification, but we hope readers will agree that this was justified by the overriding objective. Finally we would like to thank our secretaries, Miss Angela Dell and Mrs. Madeline Young, for their inexplicably angelic assistance.

Malcolm W. Greaves
Sam Shuster

Contents

Section A: Pharmacology of Skin Systems

CHAPTER 5

Fibroblasts, Collagen, Elastin, Proteoglycans and Glycoproteins
C. M. LAPIÈRE and B. V. NUSGENS 69

CHAPTER 6

Dermal Blood Vessels and Lymphatics
D. I. ABRAMSON. With 9 Figures 89

CHAPTER 8

Immunopharmacology of Mast Cells

CHAPTER 9

Lymphocytes

CHAPTER 10

Structure, Function and Control: Afferent Nerve Endings in the Skin
B. LYNN. With 1 Figure . . . 175

CHAPTER 11

Sweat Glands: Eccrine and Apocrine

CHAPTER 12

Thermoregulation and the Skin

CHAPTER 13

Hair and Nail

CHAPTER 21

Prostaglandins, Leukotrienes, Related Compounds and Their Inhibitors

CHAPTER 24

Neutrophil and Eosinophil Chemotaxis and Cutaneous Inflammatory Reactions

CHAPTER 25

Neuropeptides and the Skin

CHAPTER 26

Polyamines

Contents of Companion Volume 87, Part II

Section A: Methods

Section B: Absorption, Metabolism, and Toxicity

Section C: Drugs and Diseases

Section A: Pharmacology of Skin Systems

CHAPTER 1

The Epidermis

E. Christophers, C. Schubert, and M. Goos

Mammalian epidermis is a renewing tissue which provides the outer surface coat of the body. The high impermeability of the stratum corneum membrane and its physical strength protect the body against external influences and maintain the milieu intérieur. Besides representing the continuous supply of cells forming the external barrier (the stratum corneum), epidermal cells have been shown to possess a number of additional functional capabilities as listed in Table 1.

In the following the main aspects of epidermal structure and function will be reviewed. Since the literature on this subject has become quite extensive, this overview is written knowing that it may not be complete.

Table 1. Functional capacities of epidermal cells

Main functions	Additional functions upon need
Cell division	Adhesion
Keratin synthesis	Migration
Synthesis of basal lamina	Phagocytosis
Synthesis of glycocalyx	
Degranulation	

A. The Structure of Epidermis

The mammalian integument is composed of an ectodermal epithelium (the epidermis) which covers the external surface of the body and an underlying mesodermal connective tissue layer (the corium or dermis). The subcutis is a layer underneath the corium containing mostly adipose tissue (panniculus adiposus). Epidermis, dermis and subcutis form a functional unit and are called the skin.

I. The External Surface of Skin

The surface of the skin is not smooth but – with the exception of palmae and plantae – divided into fields (surface ridges: Ackerman 1978) by furrows of varying depths. These fields are of polygonal, often rhombic shape. At the intersections of the furrows, hair follicles emerge (Fig. 1 a). The shape and distribution of the epidermal fields and the depth of the furrows exhibit characteristic regional dif-

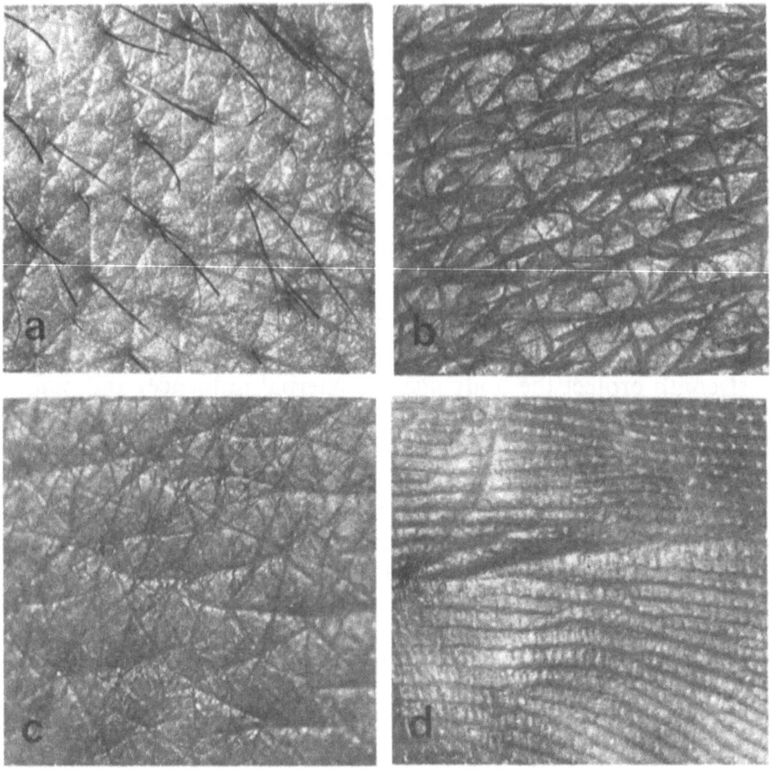

Fig. 1 a–d. Normal skin, surface ridges: **a** back of hand, **b** knee, **c** elbow, **d** palma

ferences. Furthermore the pattern of the fields, at least of certain body regions, e.g. the female breast, is individually unique (MONTAGNA and PARAKKAL 1974; BANDMANN 1980). The orientation of surface ridges is thought to result from:

1. Epidermal rete ridges plus dermal papillae configuration
2. Arrangement of underlying collagen bundles
3. Fusion of muscles and fasciae in the dermis (ACKERMAN 1978)

Because of this it is plausible that different skin regions which are subject to comparably strong mechanical stress show similar patterns of surface ridges, e.g. skin of the knee or elbow (Fig. 1 b, c).

The distensibility of skin is dependent on its thickness, elasticity, number of folds, the degree of fixation to the subcutis and the age of the individual. Abdominal skin is notable because of its unusual capacity for distension, e.g. during pregnancy (MONTAGNA and PARAKKAL 1974). A different surface view can be seen at the palmae and plantae. Here about 0.5 mm wide ridges which are separated by parallel sulci form an individual pattern (Fig. 1 d). On top of the surface ridges the openings of the sweat glands are seen. The patterns of ridges and sulci are known as dermatoglyphics (CUMMINS 1942, 1946, 1964). Their important role in anthropology, genetics and criminology is well known. Peculiar dermatoglyphic patterns are typical for a number of diseases and genetic abnormalities, e.g. ichthyosis vulgaris, psoriasis vulgaris, Down's syndrome and Klinefelter's syndrome (LEUTGEB et al. 1972; ACKERMAN 1978; EBLING 1979).

II. Histology of Epidermis

1. Introduction

The epidermis, i.e. the ectodermal part of the integument, is a stratified, orthokeratotic, squamous epithelium. It is composed of a living stratum malpighii and a dead superficial stratum corneum. The stratum malpighii is divided into a stratum basale, a stratum spinosum and a stratum granulosum (see HAM 1974; BUCHER 1980). BLOOM and FAWCETT (1975) confine the term stratum malpighii to the stratum basale and stratum spinosum (Fig. 2). Controversy exists about the presence of a separate stratum lucidum between the stratum granulosum and the stratum corneum (see KLIGMAN 1964; MONTAGNA and PARAKKAL 1974; BLOOM and FAWCETT 1975; RUPEC 1980).

This distinction into several horizontal layers is schematic and more of didactic value since it is actually only one cell type in different phases of maturation which builds up the epidermis. The epidermal cell (keratinocyte) originates by mitotic division in the basal cell layer, migrates to the surface and becomes fully cornified (see below). Besides the keratinocytes, three other cell types are present in epidermis: melanocytes, Langerhans cells and Merkel cells. These cell types are dealt with in detail elsewhere in this volume.

The junction between epidermis and dermis has been carefully analysed after separation of the two integumental components (GREB 1940; HORSTMANN 1957). The undersurface of the epidermis exhibits a complicated relief of ridges and

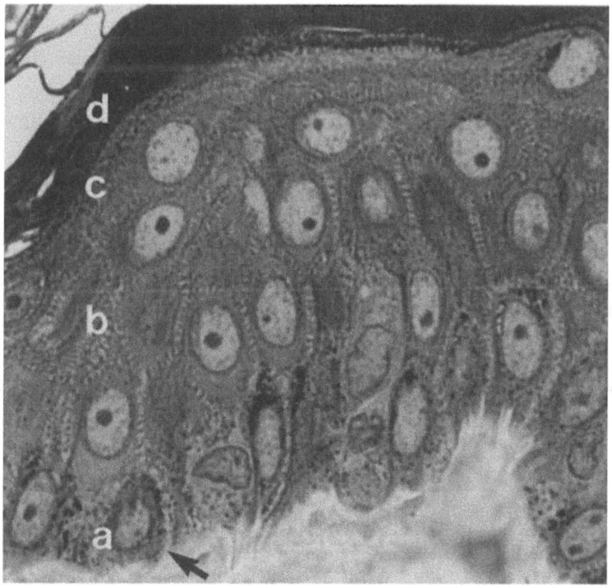

Fig. 2. Cross-section of the normal epidermis of the forearm. *a*, stratum basale, *b*, stratum spinosum, *c*, stratum granulosum, *d*, stratum corneum. *Arrow:* keratinocyte with melanin granules. (× 1400)

grooves which fits into the corresponding papillary body of the corium (GREB 1940; HORSTMANN 1957). The papillary body shows marked regional differences (MONTAGNA and PARAKKAL 1974).

HORSTMANN (1957) distinguished five types of dermo-epidermal interfaces. Part of his classification is based on the assumption that the epidermal ridges, the dermal papillae and the skin appendages are in a specific mutual arrangement. This, however, is in contrast to the suggestion by MONTAGNA and PARAKKAL (1974) that the organisation of these structures is random.

Wide variations of the dermo-epidermal interdigitations can be observed, ranging from a flattened plane in an atrophic (senile) epidermis to hypertrophy which causes an enlargement of the corresponding surfaces and facilitates the metabolic exchange between the blood capillaries and the epithelium. According to SCHUMACHER (1931) the height of the dermal papillae is directly related to the thickness of the epidermis (see also CHRISTOPHERS and LAURENCE (1976).

2. The Basement Membrane

At the light microscopic level the dermo-epidermal junction area is marked by a PAS-positive basement membrane about 1 µm thick, which has been described as a network of fibres embedded in an amorphous matrix (BRIGGAMAN 1982; BRIGGAMAN and WHEELER 1975; EBLING and ROOK 1979). According to ultrastructural investigations the dermo-epidermal junction consists of four components (BRIGGAMAN and WHEELER 1975; BRIGGAMAN 1981, 1982). The plasma membrane of the basal epidermal cells presents with varying specialized attachment devices (hemidesmosomes) (Fig. 3 b); this membrane is about 7–9 nm thick. Beneath it are first the lamina lucida, an electron-lucent zone about 30 nm thick, and then the basal lamina, which is about 40 nm thick. Underneath the basal lamina are bundles of filaments consisting of three morphological components (BRIGGAMAN 1981; Fig. 3 a):

1. As a fibrous element, anchoring fibrils are recognized on the basis of their asymmetrically banded central zone which with one end extends into the basal lamina and with the other into the dermis. The dermal ends of several adjacent anchoring fibrils form a network of interconnecting arcades.
2. Between the anchoring fibrils individual collagen fibres are present. Direct connection between the two fibre types has not been found.
3. In addition, bundles of fibrous elements have been described which seem to be identical with microfibrils associated with elastic fibres. One end of these fibrils fuses with the basal lamina, while the other end extends into the dermis.

The basal lamina is composed of an amorphous matrix-like structure approximately 40 nm thick into which fibrillar components are embedded. Chemically it contains collagen, non-collagenous glycoprotein and glycosaminoglycans. Filamentous and granular material representing pro-collagen and collagen is embedded into the matrix of glycoproteins (DARÓCZY and FELDMANN 1981). The basal lamina collagen is different from other types of collagen and belongs to type IV, type V (AB) and 7S collagen (JAOITA et al. 1978; Daróczy and FELDMANN

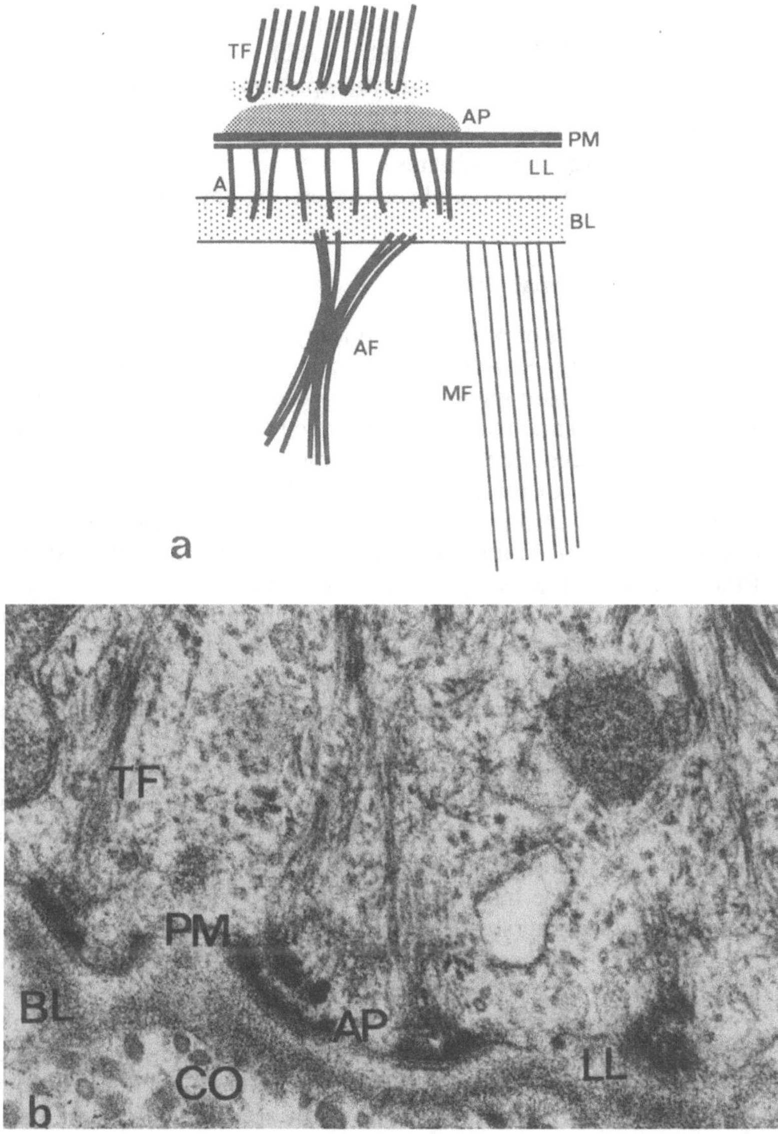

Fig. 3 a, b. Dermo-epidermal junction. **a** Simplified drawing of the dermo-epidermal junction; *top:* basal epidermal cell. **b** High magnification electron micrograph of the dermo-epidermal junction (× 40 000). *A*, anchoring filaments, *AF*, anchoring fibril, *AP*, attachment plaque (hemidesmosome), *BL*, basal lamina, *CO*, collagen fibres, *LL*, lamina lucida, *MF*, dermal microfibril bundle, *PM*, plasma membrane, *TF*, tonofilaments

1981; BRIGGAMAN 1982). The basal lamina is a product of the epithelial cells (for references see RUPEC 1980; BRIGGAMAN 1981, 1982).

Three functions are ascribed to the dermo-epidermal junction:

1. Dermo-epidermal adherence
2. Mechanical support for the epidermis
3. A barrier function for material and cells across the junction.

3. Fine Structure of Epidermal Cells

The stratum basale consists of one continuous layer of cuboidal or low columnar cells with the long axis aligned vertical to the skin surface. The cells contain large oval nuclei with one or more nucleoli (Fig. 2). Their cytoplasm is characterized by mitochondria, smooth and rough endoplasmic reticulum (ER) a poorly developed Golgi apparatus and numerous ribosomes; the latter render these cells basophilic, an indication for marked protein synthesis.

The cytoplasm is pervaded by abundant fine filaments (tonofilaments), the diameter of which is 4–5 nm. These form bundles of tonofibrils. The plasma membranes of the basal and spinous layers are about 8 nm thick and show the typical unit membrane structure of most mammalian cell membranes. They are covered by a glycocalyx (WOLFF and SCHREINER 1968; FAWCETT 1981).

As the post-mitotic keratinocyte leaves the basal cell layer, it moves into the stratum spinosum, a layer of variable thickness with polygonal or increasingly flattened cells. All of the cells bear short processes or spines (Fig. 2) by which they are attached to similar projections from adjacent cells. At the point of convergence of these spines desmosome–tonofilament complexes are present. This contact area of neighbouring cells is thickened and forms a nodular structure at the light microscopic level [nodes of BIZZOZERO (1870)].

The cytoplasmic volume has increased almost fourfold in comparison with that of the basal cells without significant changes of the nuclear volume (ACKERMAN 1978; ELIAS 1981). Furthermore, an increase of tonofilaments and a decrease of ribosomes can be observed from the basal to the spinous area. In the upper layers of the stratum spinosum ovoid cellular organelles termed lamellated bodies, (Odland bodies, keratinosomes or membrane coating granules) become apparent (LANDMANN 1980). Their diameter is about 0.1–0.3 μm. These organelles are always most numerous in the cells of the stratum spinosum, and in the stratum granulosum they are located toward the upper portion of the cells (EBLING and ROOK 1979).

The stratum granulosum (granular layer, stratum intermedium, BRODY 1977) is composed of two or three layers of flattened living cells (Fig. 2). Their surface area is increased by a factor of 2–3 compared with the stratum spinosum (ELIAS 1981). The cytoplasm contains keratohyalin granules which are basophilic inclusion bodies of irregular shape. In the electron microscope these granules are electron-dense bodies not lined by a membrane. They are surrounded by ribosomes and are in close contact with tonofilaments. Morphological, histochemical, radioisotopic, biochemical and immunological investigations have shown that keratohyalin granules consist of two components (DELECLUSE 1981):

1. A histidine-rich fibrillar protein (see also KAKIMI et al. 1980)
2. A dense homogeneous protein(s), rich in cystine and proline

During the transformation of granular cells to horny cells, keratohyalin granules are dissolved and a dense homogeneous protein(s) is seen in close relation with desmosomes. This is believed to contribute to the thickening of the cellular membrane of the horny cells. At this time the histidine-rich fibrillar protein is no longer visible.

The relation between keratohyalin and the development of the keratin pattern needs further investigation, since there is indication of a keratin pattern without keratohyalin (ANTON-LAMPRECHT 1973; DE BERSAQUES 1975; BLOOM and FAWCETT 1975; BRODY 1977).

Major changes of granular cells during keratinisation are as follows: At first the inner leaflet of the plasma membrane is thickened while the trilaminar structure is maintained. In connection with this WATT and GREEN (1981) as well as BANKS-SCHLEGEL and GREEN (1981) demonstrated a soluble protein precursor, called involucrin, the synthesis of which begins in the stratum spinosum. In vivo as well as in vitro the amount of this protein related to the increase in cell size as the keratinocytes mature. Involucrin is thought to be the precursor of the insoluble proteinaceous envelope on the cytoplasmic side of the plasma membrane, which may be partly responsible for the thickening of the inner membrane leaflet.

As the cellular organelles disappear from the cytoplasm, lamellar granules which are located close to the apical plasma membrane deplete their contents into the intercellular space. Histochemically there is a strikingly high content of acid phosphatase located in lamellar granules (BRAUN-FALCO 1965). HADLER and SILVEIRA (1981) have demonstrated the presence of tocopherol in the granular layer.

Between the stratum granulosum and the stratum corneum a discontinuous layer of transitional cells (T cells) is described by BRODY (1959). These transitional cells are caught within the brief process of dissolution of cellular organelles and the establishment of a stratum corneum cell pattern. During transformation into corneocytes the nuclei and virtually all of the cytoplasmic organelles are lost. In addition the water content of corneocytes is low (approx. 8% of dry weight) and the cells are extremely flattened. They are completely filled with keratin, which appears as bundles of filaments embedded in an opaque interfilamentous substance.

Investigations of the plasma membrane have shown (GRAY 1981) that the major proportion of the surface glycoprotein is degraded and any glycoprotein remaining is probably associated with desmosomes. On the other hand the trilaminar structure of the plasma membrane is preserved with a much thickened inner leaflet (about 10–16 nm thick) in which the protein is stabilised by γ-(ε-glutamyl) lysine cross-linking and a lipid composition which is unique by virtue of the absence of phospholipids (for lit. see GRAY 1981).

In summary, during differentiation, among other changes, keratinocytes alter shape (flatten), increase in diameter [from 6 μm in the basal cell (PINKUS 1952) to 34–44 μm in the horny cell (PLEWIG and MARPLES 1970)], lose organelles, form fibrous proteins, become dehydrated and acquire thickened cell membranes.

In the following two particular structures of the epidermis are dealt with in more detail: the lamellar granules and the intercellular junctions. They have received considerable attention in recent years.

4. Lamellar Granules

The structure and the function of lamellar granules (LGs) have been reviewed recently by ALLEN and POTTEN (1975), SQUIER et al. (1978), LANDMANN (1980) and ELIAS (1981). In contrast to older views (MONTAGNA and PARAKKAL 1974), LGs can be detected not only in mammalian epidermis but also in the epidermis of reptiles and birds (LANDMANN 1980). LGs appear in the cells of the upper stratum spinosum and in the stratum granulosum. According to MATOLTSY and PARAKKAL (1965) they originate in the Golgi apparatus, whereas HAYWARD (1979) and ELIAS (1981) support the view that they are derived from the smooth ER. In mammals they exhibit a uniform ultrastructure (LANDMANN 1980). They are surrounded by a membrane 6–8 nm thick and contain several stacks of lamellar material embedded in an amorphous matrix. The lamellar pattern is brought about by electron-lucent discs with a central interrupted striation, separated by a somewhat thinner electron-dense band (Fig. 4a, b). Like coins, these discs are stacked one above each other. The LGs seem to contain a high amount of polysaccharides, hydrolytic enzymes and polar lipids, especially glycolipids and phospholipids (for lit. see LANDMANN 1980; ELIAS 1981). In the mid-to-upper stratum granulosum and between granular and horny cells, LGs discharge their contents into the intercellular space by exocytosis.

Lamellar granules contain a variety of active substances (ELIAS et al. 1979). These are most likely involved in various functions of epidermis, such as lipid-rich intercellular material, which plays a role in epidermal permeability (ELIAS and FRIEND 1975; ELIAS et al. 1977, 1979; ELIAS 1981; LANDMANN 1980; SQUIER et al. 1978). Tracer perfusion studies have shown that the permeability barrier for water in part functions already at the level of the stratum granulosum (ELIAS 1981).

The formation of the LGs and the discharge of their contents may affect the volume of the stratified cells as well as the intercellular mass. According to ELIAS (1981) LGs occupy as much as 15% of the granular cells' cytoplasm. As the keratinocyte surface area is increased by a factor of 2–3, this is presumably due to the membranous material added during LG secretion. In the lower stratum granulosum, before the extrusion of LGs, the intercellular space constitutes less than 1% of the total tissue volume. After LG secretion the intercellular volume fraction increases to 5%–30% of the entire epidermal volume. Furthermore ELIAS (1981) suggests that LGs secrete glycolipids and free sterols into the intercellular

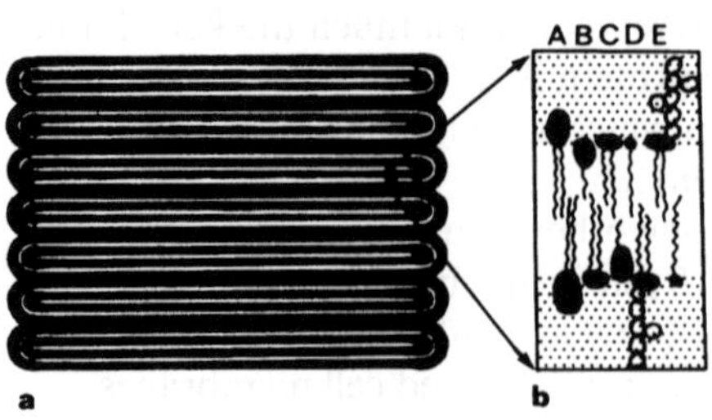

Fig. 4a, b. Model of a lamellar granule. **a** Cross-section through a stack of lamellar material in a lamellar granule. **b** Model of the chemical composition of a bilayer of a lamellar granule. *A*, phospholipids, *B*, cholesterol, *C*, ceramides, *D*, free fatty acids, *E*, glucosylceramides. (Redrawn from LANDMANN 1980)

space of the stratum granulosum, where these are metabolized to sphingolipids, free fatty acids and cholesterol esters, all of which participate in the formation of bilayers of the stratum corneum.

Further aspects of LGs are discussed by ALLEN and POTTEN (1975). According to these authors discharged lysosomal enzymes may contribute to the dissolution of desmosomal contacts and to the subsequent peeling of corneocytes.

5. Intercellular Junctions

Adjacent keratinocytes, as well as Merkel cells and keratinocytes, are interconnected by intercellular junctions. Such specific junctions include macula adherens (= desmosome) (Fig. 5a, b) or zonula adherens, macula or zonula occludens (= tight junction) and macula communicans (= nexus = gap junction). These membrane specifications are absent between keratinocytes and melanocytes or Langerhans cells. For their visualisation various electron microscopic techniques are necessary, e.g. freeze fracture, negative stain or tracer methods (BREATHNACH 1981; FAWCETT 1981).

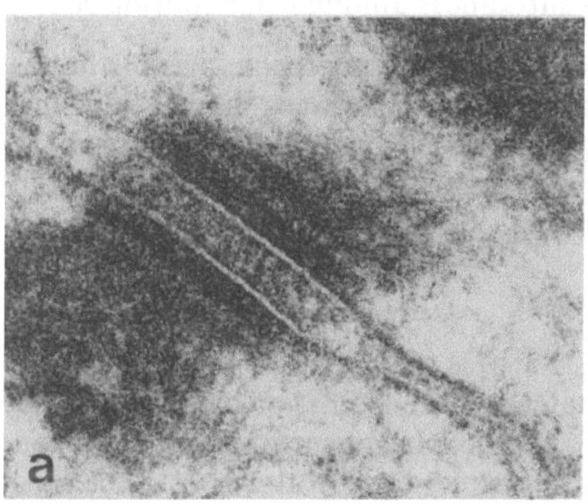

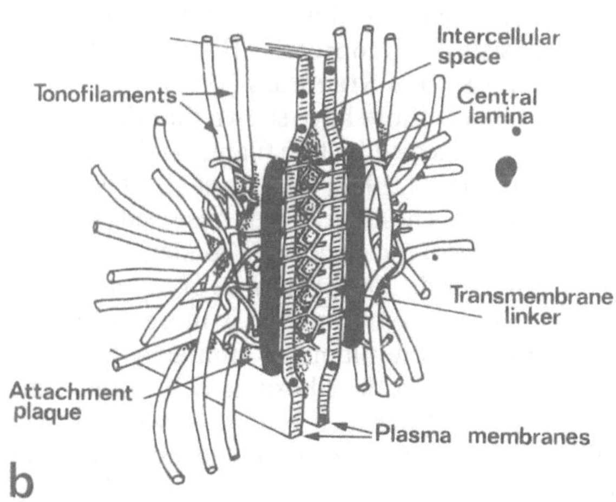

Fig. 5a, b. Desmosome. **a** High magnification electron micrograph of a desmosome between adjacent keratinocytes in the human epidermis (× 90000). **b** Structure of a desmosome. (Redrawn from FAWCETT 1981)

a) Gap Junctions

Using transmission electron microscopy, gap junctions appear as septilaminar structures. The plasma membranes of neighbouring cells, each of which measures 7.5 nm in thickness (consisting of the three laminae), closely approach each other and are separated by an intercellular cleft of 2–4 nm (STAEHELIN 1974). In freeze fracture replicas of epidermal cells the gap junctions can be recognized by groups of more or less regularly arranged 10-nm particles (PERACCHIA 1980). These appear to be composed of subunits of a structural protein (connexin) which spans the intercellular space. There are aqueous channels about 2 nm in diameter, through which the cytoplasms of the two neighbouring cells are supposed to communicate (for references see PITTS 1981).

Gap junctions are relatively frequent in the basal layer and slowly decrease toward the periphery. They usually occur in close connection with the desmosomes. So far, gap junctions have not been found in the stratum corneum (CAPUTO and PELUCHETTI 1977; LELOUP et al. 1979; ELIAS 1981). Two main functions are ascribed to this type of junction (for lit. see STAEHELIN 1974):

1. Transmission of action potentials (electrical communication)
2. Distribution of low molecular weight substances among neighbouring cells

Molecules up to 1000 daltons spread through the microchannels of the gap junctions almost as fast as if there were no membranes between the cells. For such substances groups of cells connected by gap junctions form a syncytium (see FAWCETT 1981; PITTS 1981). Hypertrophy and hyperplasia of gap junctions are stimulated by treatment with retinoic acid (ELIAS 1981).

b) Desmosomes and Hemidesmosomes

The desmosome (macula adherens) (Fig. 5a, b) is the most prominent intercellular junction in epidermis. In sectioned material desmosomes are characterised by a widened intercellular space (25–35 nm), parallel plasma membranes, and a zone of fine filamentous material (desmosomal plaque) closely attached to the inner aspect of the plasma membrane and in which intracellular microfilaments (tonofilaments) seem to be anchored (Fig. 5a). The intercellular space frequently contains electron-dense material which consists of glycoproteins (PAPPAS 1975) and has been analysed in more detail by GORBSKY and STEINBERG (1981).

Within the desmosomal plaques various authors have shown that the plasma membranes of the neighbouring cells are connected by fine filaments (KELLY and LUFT 1966; STAEHELIN 1974; STAEHLIN and HULL 1978; LELOUP et al. 1979; FAWCETT 1981) (Fig. 5b). Various stages of desmogenesis have been observed (LELOUP et al. 1979): In a first step, the coat substance of the glycocalyx condenses in three parallel and equidistant dotted lines. Afterwards the appearance of asymmetrical desmosomes is observed, in which one desmosomal plaque is more developed than the facing one. Later on, newly formed tonofilaments intimately linked to now symmetrical osmiophilic plaques develop in the ribosome-rich cytoplasmic microvilli bordering the intercellular space. It has also been observed that gap junctions develop in the basal cell layers in places where desmogenesis occurs. In contrast to gap junctions, desmosomes are more numerous in the upper layers of the epidermis than in the basal cell layer. They persist into the stratum corneum, where the attachment plaque disappears and increased opacity of the intercellular component can be seen (CAPUTO and PELUCHETTI 1977).

Hemidesmosomes are half-desmosomes (Fig. 3b) which occur at the basal half of the basally located keratinocytes opposite the basal lamina. CHRISTOPHERS and WOLFF (1975) observed that in vitro the number of hemidesmosomes increases twofold after treatment with retinoic acid.

c) Tight Junctions

In addition to the contact apparatuses described above, the freeze fracture technique has revealed the presence of epidermal tight junctions (maculae occludentes). Here the plasma membranes of neighbouring cells fuse, measuring 14–15 nm (WELSCH and BUCHHEIM 1978; PINTO DA SILVA and KACHAR 1982). Tight junctions are mainly observed in the stratum granulosum, where their function remains unknown and is still debated (ELIAS 1981).

A further membrane structure has been described in the horny cells (where probably the main permeability barrier of the epidermis is located) (CAPUTO et al. 1981; ELIAS 1981). These membrane specialisations superficially resemble tight junctions. In freeze fracture preparations the plasma membranes exhibit an extensive network of interconnecting ridges on the E-face and grooves on the P-face, which are continuous with desmosomes. Their exact function is unknown (ELIAS 1981).

6. Regional Differences in Epidermal Structure

The epidermal thickness varies dramatically with body site. Most of the body surface has an average epidermal thickness of 40–50 μm. Mean epidermal thickness values for different body sites are given by WHITTON and EVERALL (1973): face: 49.8 μm; trunk: 42.4 μm; arms and legs: 60.1 μm; back for hand: 84.5 μm; wrist: 80.7 μm; fingertips: 36.9 μm; side of finger: 222.7 μm; back of finger: 138.1 μm. These authors found no significant correlation between epidermal thickness, sex and age.

The malphighian layer of normal human epidermis rarely exceeds seven or eight cells in thickness: one layer of basal cells, about four layers of stratum spinosum and about three layers of granular cells (KLIGMAN 1964; BLOOM and FAWCETT 1975). In the stratum corneum at least twice as many cell layers occur as in the living epidermis. Normally, about 15–20 cell layers can be observed in the horny layer of the back, which has an average thickness of about 12–15 μm (KLIGMAN 1964; CHRISTOPHERS and KLIGMAN 1964). For most of the body surface the stratum corneum has a maximum thickness of about 10–15 μm; the stratum corneum of the scrotum has an average thickness of about 9 μm (KLIGMAN 1964).

Epidermis in man is composed of nearly 5 million living cells per cm^2. As the area of one corneocyte measures 937 μm^2 (PLEWIG and MARPLES 1970; BERGSTRESSER and TAYLOR 1977), the entire stratum corneum consisting of 17 cell layers (CHRISTOPHERS and KLIGMAN 1964), calculation of the total number of corneocytes included, reveals 6.6×10^6 cells per cm^2. Thus nearly a quarter of the total epidermal cell mass consists of corneocytes.

Although the stratum corneum is composed of many more cellular layers than the rest of the epidermis, it contributes approximately one-third of the latter's total thickness.

B. Epidermal Replacement

I. Epidermal Renewal Rates

Counting mitotic figures was among the earliest attempts to quantitate epidermal cell production. Since then various attempts have been made to determine more precisely the rates at which cell birth and cell loss at the surface of skin take place. Several methods have been applied for the measurement of rates at which epidermis or portions of it are being produced. These include:

1. Counting mitosis with or without mitotic arrest (PINKUS 1952; FISHER 1968; FISHER and WELLS 1968; ROWE et al. 1978)
2. Determining the number of labelled nuclei in DNA synthesis after uptake of radioactive precursors in vitro (CHRISTOPHERS and LAURENCE 1973), with measurement of cell cycle length (CHOPRA and FLAXMAN 1972)
3. Quantitation of transit times of labelled cells by multiple biopsies (PORTER and SHUSTER 1968; ROTHBERG et al. 1961)
4. Microspectrophotometric measurements of Feulgen-stained nuclei to determine the distribution of proliferative cells in the cell cycle (BAUER et al. 1980)
5. In vivo staining of the horny layers (e.g. dansyl chloride, BAKER and KLIGMAN 1967) and determining the rate of disappearance of stained cells

Precise methods of determining the proliferation activity of epidermis include the administration of an S phase marker, tritiated thymidine and subsequent autoradiography. As for mitotic counts, the number of labelled cells depends upon the time the cells spend in DNA synthesis (S phase, T_S); therefore the length of T_S is vital. The most accurate method for measuring the length of the S phase consists in determining the fraction of labelled mitoses (FLM curves) (QUASTLER and SHERMAN 1959; for discussion see WRIGHT 1977, 1981).

Under strict steady state conditions a double labelling method can be applied (PILGRIM and MAURER 1965; MAURER et al. 1965; PULLMANN et al. 1974) which considers the cells under study being randomly distributed throughout the cell cycle. Partial synchronisation (due to circadian rhythms, external influences etc.) will affect such data and as a result information concerning the presence of circadian rhythms in human epidermis is conflicting (KAHN et al. 1968; SCHELL et al. 1981).

With data given for T_S (or even T_M or T_C) the turnover or a tissue compartment of the entire organ may be calculated provided the number of cells present in this compartment is known. Therefore rigorous quantification of epidermal cellularity applying Abercombie's correction for tissue geometry is needed (LAURENCE and CHRISTOPHERS 1976).

As shown by PINKUS (1952), there are approximately 0.1–1.6 basal cells in mitosis per thousand cells. The mitotic process lasts around 90 min. Between 3% and 5% of cells are in DNA synthesis and the length of the S phase tends to be around 7 h. This figure corresponds to the T_S of other renewing tissues and is widely used for calculating the total cell cycle time (T_c). Since the number of cells present at any phase of the cell cycle is proportional to the time the cells spend in that phase, T_c can be calculated from the labelling index and the duration of T_S (QUASTLER and SHERMAN 1959). As has been pointed out before, data concerning the site of the proliferate pore of the epidermis under study are mandatory in addition to the conditions of steady state with a regular age distribution. Often both parameters are not well known and as a result turnover data calculated on the basis of labelling indices and S phase durations appear to differ considerably.

More direct observations on epidermal turnover are obtained by following the movement of ^{3}H-TdR labelled post-mitotic cells through the differentiating compartment. This transit time (for the living portion) was observed to last 12–14 days (WEINSTEIN and VAN SCOTT 1965; EPSTEIN and MAIBACH 1965).

Assessment of the turnover times of the germinative cell population has produced variable results. Using tritiated thymidine two methods, namely continuous labelling and FLM curves, were applied. The results are summarized in a recent review by HILL (1980).

WEINSTEIN and FROST (1968) repeatedly administered ^{3}H-TdR intradermally and obtained a preliminary FLM curve. With a T_S lasting 16 h, a turnover time of 19 days was calculated for the germinative layer.

GELFANT (1978) recently indicated non-unity of the epidermal growth fraction. He emphasized the presence of a significant shift in the ratio of cycling to non-cycling germinative cells and suggested that possibly more than 70% of the germinative cells are in a non-cycling state. Indeed, intravenous infusion of radiolabelled thymidine over more than 2 weeks in a cancer patient treated with chemotherapy revealed that only 17% of the epidermal basal cells had become labelled (Frindle et al. 1968). This indicates a large proportion of non-cycling cells if artifacts due to decreased bioavailability of the tracer can be excluded.

By calculating the epidermal replacement rate in a three-dimensional view of the epidermal cell mass, BERGSTRESSER and TAYLOR (1977) arrived at a turnover time of 31 days for the living portion and an additional 14 days for the stratum corneum, totalling 45 days. This corresponds to the data provided by WEINSTEIN and FROST (1969): basal cell replacement 19 days plus an epidermal transit time of 14 days makes 33 days with 14 days for the corneocyte compartment. As a result the total epidermal turnover time lasts 47 days. Calculated data published by HALPRIN (1972) range between 40 and 56 days for the non-keratinized portion. Taken together these data are considerably longer than those originally provided by ROTHBERG and CROUNSE (1961), i.e. approximately 4 weeks.

II. Effects of External Influences

In presenting such information it should be kept in mind that in human skin the replacement rates of epidermis may vary depending upon the body region, environmental, nutritional and hormonal conditions, temperature and humidity, all of which may profoundly affect the rate of cell proliferation. Unfortunately avoiding external influences on skin is nearly impossible and some of the discrepancies present in the literature on epidermal replacement rates may in fact be due to uncontrolled conditions. Mitotic activity, although low when compared with other renewing tissues such as the intestinal lining, the hair follicle matrix or the haematopoietic system, can easily be triggered in epidermis and reaches maximum values within 12–24 h, depending upon the stimulating dose (HUNTER et al. 1956; BROPHY and LOBITZ 1959; CHRISTOPHERS 1972a). Thus, one principal mechanism by which the epidermis adapts to environmental influences is via changes in the rate of cell production. This leads to changes in tissue mass (hyperplasia), the extent of which is regulated by functional demand (GROSS 1965). Any insult, physical or chemical, is quickly answered by the initiation of cell proliferation. In epidermis susceptibility to damage is highest in the differentiating portion, the prime site of epithelial breakdown. This again emphasises the importance of germinative cell production as one of the most important functions of skin.

Since normally the rates of epidermal proliferation and maturation are closely linked, increased new cell production is paralleled by an increased rate of maturation. Separation of these two metabolic rates can rarely be seen. ETOH et al. (1975) observed active migration and maturation of epidermal cells after radiosuppression of mitosis. LAURENCE and CHRISTOPHERS (1976) showed that after systemic application of hydrocortisone there was an increased rate of cell maturation (keratinisation) whereas proliferation remained unaltered. These two examples are the only ones demonstrating dissociability of the rates and indicate how difficult it may be to experimentally change rates of synthesis in one epidermal compartment without concomitant alterations of the metabolic rate in the other.

III. Migration Out of the Basal Layer

Under the conditions of strict tissue homeostasis, epidermal cell loss is equal to the dividing activity in the germinative layer. The question has been raised as to which cell after mitotic division is destined to leave the basal layer and move into the differentiative compartment. In the squamous epithelium of young rats LEBLOND et al. (1964) studied the fate of post-mitotic daughter cells after ^{3}H-TdR labelling. They found that cells ready to divide or in division are unlikely to leave the basement membrane. However, they are transferable during the period between mitosis and the beginning of the next DNA synthesis. IVERSEN et al. (1968) studied the fate of post-divisional daughter cells in hairless mouse epidermis and reached similar conclusions. BULLOUGH and MITRANI (1976) reviewed evidence to indicate that not all basal cells are equally liable to lose their "baseline grip". In accordance with above-mentioned observations, they imply that a cell can only leave the basal layer during a defined intermitotic period, the G_{1b} phase.

In normal mouse ear epidermis, BULLOUGH and MITRANI found more than 90% of mitotic figures to be horizontally oriented, which is indicative of both daughter cells remaining basal. In comparison, in hyperplastic ear epidermis, in which mitotic activity spreads from the basal layer into the suprabasal layer, there was a considerable increase in vertical mitosis. Similarly, in the thicker epidermis of man, PINKUS and HUNTER (1966) found that more than half of the mitotic figures were vertical and there was not much change under forced regeneration (stripping). DUFFIL et al. (1976) saw 24% vertical mitoses in a psoriatic plaque. In addition DNA-synthesising cells located suprabasally have been shown to be quite numerous in normal human epidermis (PENNEYS et al. 1970). Therefore one may hypothesise that in thicker epidermis, where the basal layer is more densely populated, a daughter cell may move directly out of the basal layer, so that retention of daughter cells at the basement membrane after mitosis may be restricted to thin stratifying epithelia.

Unfortunately the mechanisms regulating detachment of cells from the basal layer are not well understood. Attachment appears to vary with the proliferative activity of the germinative layer and, as pointed out by BULLOUGH and MITRANI (1976), is an active process which lasts only very briefly after division. The cohesive strength of the dermo-epidermal junction, essential for the exposure to environmental stress, needs to be maintained while simultaneously cells are dividing and becoming detached in order to migrate distally.

IV. The Formation of Epidermal Cell Columns

It has long been well known that the epidermis exhibits a horizontally oriented stratification. By means of special techniques it has been shown that particular regions of the epidermis of birds and mammals (including man) display a vertical organisation in the form of columns (MACKENZIE 1969, 1972; CHRISTOPHERS 1970; CHRISTOPHERS et al. 1974; CHRISTOPHERS and LAURENCE 1980; MENTON 1976).

The cornified cells form regularly arranged stacks in which the peripheral parts of the cells (comprising about 10% of the cellular surface) of neighbouring stacks interdigitate (Figs. 6, 7a). The firmness of the stratum corneum is largely due to these zipper-like interdigitations of the cornified cells. In surface view preparations of body areas showing a columnar structure the cornified cells are usually of hexagonal outline. This architectural principle is of wide distribution in the biosphere (principle of the honeycomb).

A three-dimensional construction of the stratum corneum can be drawn by combining the hexagonal outline of the cells in the surface view preparation with the columnar structure of the sectioned preparation. It turns out that each cornified cell forms a body with 14 facets, which is in contact with eight neighbouring cells. A consequence of the shape of the cells is that neighbouring cells are not at the same level but stepwise displaced (CHRISTOPHERS 1971 a, b). Furthermore it was found that stacking of cells is most regular in very thin epidermis with slow cell turnover, where it can also be recognized in the malpighian layer. This structural organisation therefore was suggested to be a function of epidermal thickness and cell turnover rather than of animal species (CHRISTOPHERS 1980 a; CHRISTOPHERS and LAURENCE 1976) (Fig. 7 a–c).

By the use of a fluorescent staining technique, CHRISTOPHERS (1971 a, b) observed post-mitotic cells in the state of leaving the basal layer. These cells showed

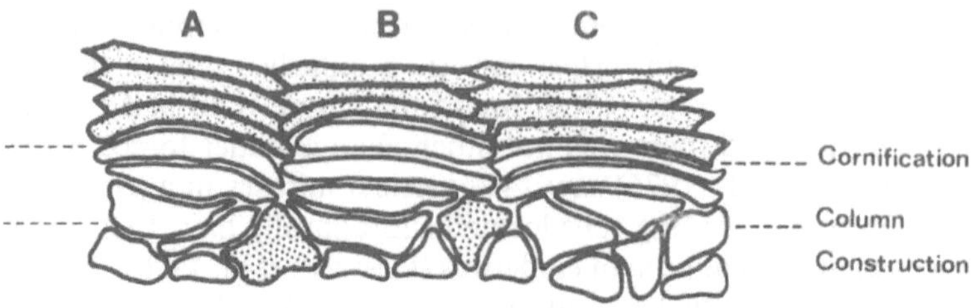

Fig. 6. Mouse ear epidermis. Beneath the stacked horny layer cells (*shaded*), two or three layers of stacked malpighian cells are present. Undetermined "intercolumnar" basal cells (*stippled*) are presumed to move into their respective columns according to the number of cell layers present (see text). Column A: lowermost cornified cell "down" position in relation to lowermost cornified cell in column B, which is in the "up" position

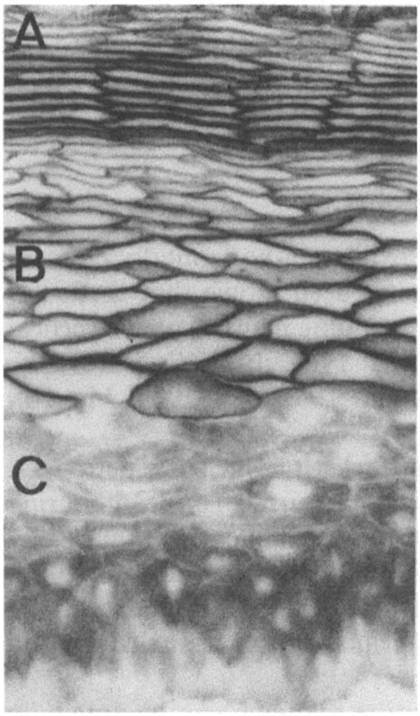

Fig. 7. Guinea-pig ear epidermis. *A* Zone of normally stacked stratum corneum, *B* zone of stratum corneum cells cornified in an unordered way (after irritation), *C* zone of living cells (stratum malpighii)

strong staining, indicating differentiation (Christophers and Braun-Falco 1971), and were found mainly between two adjacent columns. They could be identified migrating upward and becoming positioned beneath the overlying stacks. These findings indicate that cells located beneath the column produce the overlying cells of that column and that there is no exchange between adjacent stacks (Christophers et al. 1974). These observations have recently been confirmed and extended by Potten (1981). According to Potten, in the epidermis of the mouse (back and ears) there are on average 10.6 basal cells located beneath each column. In the centre of these a subgroup is formed by three or five tightly associated basal cells. One of them is located centrally and is a Langerhans cell. The entire group of approximately ten basal cells and their maturing progeny is called the "epidermal proliferative unit". Cells located in the periphery of a column are more mature than those at the centre and they would appear to be cycling at a faster rate than the central subgroups. The central portion of cells is supposed to represent immature stem cells which respond differently to stimulation.

V. The "Zipper Mechanism" Leading to Column Formation

Unfortunately the mechanism of column formation cannot be elucidated by direct in vivo observation and a different interpretation of epidermal column formation has been given by CHRISTOPHERS (1971 a, b, 1980 a, b), CHRISTOPHERS et al. (1974) and CHRISTOPHERS and LAURENCE (1976). Their concept is based upon various features of stacked epidermis which include cell proliferation, movement of differentiating cells off the basal layer and the subsequent transformation of keratinocytes into corneocytes.

An essential feature of an epidermal cell column is the constancy of the stacked living cell population. Approximately three spinous cells, positioned in the upper layers of the living epidermis, are stacked precisely and symmetrically beneath the overlying cornified column. This suggests that the formation of columns takes place beneath the stratum corneum. In the very thin mouse ear epidermis, the column is constructed by the positioning of the youngest differentiating (spinous) cell as it leaves the basal layer (Fig. 6), i.e. the organisation occurs in the suprabasal layer. Clearly the problem then is reduced to the question of why these cells should become stacked and not remain in a disorganized arrangement.

In the epidermis of the guinea-pig ear it was found that column formation was present in some animals while in others no order could be detected (CHRISTOPHERS 1972 b). Changes from order to disorder occasionally could be found even within one histological specimen, indicating that the regulation of column formation in this tissue is at threshold. Thus, the guinea-pig ear was ideally suited to correlate factors possibly involved, such as cell proliferation thickness and cellular order, since here order and disorder are not related to differences in body site (CHRISTOPHERS 1972 b).

Two characteristic parameters of epidermal growth, namely the proliferative activity and the tissue mass (e.g. thickness), were determined in a group of guinea-pig ears and correlated to the presence and/or absence of cell columns (Fig. 7). A highly ordered stratum corneum was consistently present where the living epidermis was less than 38 μm thick, whereas the horny layer became disordered whenever the epidermis was more than 45 μm thick. On the other hand, between 14 and 28 ^{3}H-thymidine labelled basal cells per millimeter surface length were found still to be coexistent with the presence of epidermal cell columns. Where there were more than 30 labelled cells per millimeter surface length no order could be detected. These data indicate that a relatively slight change in new cell production and thickness is paralleled by a change in structure (Fig. 7).

The fact that a low rate of cell proliferation is paralleled by the formation of epidermal cell columns makes it necessary to consider the other factor upon which functional homeostasis is dependent, namely the rate of maturation of post-mitotic cells. BULLOUGH (1972 a, b) has previously indicated that the rate of mitosis and the rate of maturation must be precisely matched so that various thicknesses of the epidermis are maintained. As has been shown before, the rate of cell maturation is low in the thin epidermis of mouse ear, lasting approximately 20 days (from cell birth to cell death) (CHRISTOPHERS and LAURENCE 1973). Accordingly, mitotic figures are rare in this tissue. Thus, the production of daughter cells in this long-lasting epidermal cell proliferation will cause considerable differences in the age of neighbouring cells in adjacent columns, i.e. any newly producted post-divisional daughter cell will find itself surrounded by cells which are much older. As pointed out by LEBLOND et al. (1964), GREULICH (1964), IVERSEN et al. (1968) and subsequently CHRISTOPHERS and LAURENCE (1973) and POTTEN (1981), considerable differences can therefore be postulated to exist among the basal cell population of thin epidermis. It seems to be composed of several subpopulations which are separated by their respective ages.

Differential susceptibility of cells to staining with fluorescein isothiocyanate (FITC) has demonstrated differences in maturation among the basal cell population alone (CHRISTOPHERS 1971 b; CHRISTOPHERS and LAURENCE 1973). At the ultrastructural level, CHRISTOPHERS et al. (1974) demonstrated that the differentiating cells of adjacent columns, although located at the same horizontal level, also differed in their degree of maturity. Using keratohyalin, keratinosomes and the plasma membrane as markers for cellular maturation, differences were found between adjacent cells of two columns. In addition, differences in synthesizing activity of adjacent keratinocytes situated at the same subcorneal cell row were demonstrated by the uptake of tritiated proline (CHRISTOPHERS et al. 1974). Such differences in degree of maturation of adjacent cells were found to be absent in nonstacked epidermis, indicating that their existence plays a role in the formation of epidermal columns.

Additional information was gained by quantitating the number of flattened cells present in adjacent columns. It had been noticed that, due to the constant overlap of epidermal corneocytes, a stepped position of the lowermost horny cells is produced. One cell is found in the "up" position adjacent to a cell in the "down" position (Fig. 6). By counting the number of non-keratinised cells in relation to such corneocyte positions, it was found that below a "down" position significantly less maturing stacked keratinocytes were present (CHRISTOPHERS et al. 1974).

Taking together these observations on the maturation process, it is indicative that the transition of a keratinocyte into a corneocyte takes place alternately in cells in adjacent columns. This results in an ordered complex of cornified cells consisting of interlocking stacks. The alternating way in which cells leave the living portion of the epidermis (keratinocyte transforming to corneocyte) is an important feature of the mechanism which sorts the cells into stacks as they leave the basal layer. Only those cells which find adequate "space" will leave the basement membrane, flatten and enter the appropriate columns. Transformation of an uppermost keratinocyte into a corneocyte (cornification) in a given stack will produce a corneocyte in the "down" position. In consequence, the number of stacked living cells will be less than in the adjacent columns, thus providing "space" for an undetermined subcolumnar cell to become sorted into this column (Fig. 6). In other words, this "space" (or reduced pressure) in the respective columns is probably created by the progressive flattening of the maturing cell due to the loss of turgidity and cellular contents during maturation and the transition from keratinocyte to corneocyte, alternating from one column to the next. Cells in adjacent columns are already locked into position and entry of the basal cell into these columns is barred until the cornification of another keratinocyte takes place. This individual method of sorting cells into columns caused by the cornification of the upper living epidermal cells provides positional differences between cells in two neighbouring columns. Therefore, entry into a column is predetermined by maturational differences between the cells of adjacent stacks. Thus, architectural organisation of the epidermis is a consequence of "differential" maturation.

In man the epidermal columnar structure in general is a feature of the skin of the trunk and nearby areas (CHRISTOPHERS 1980a). According to MACKENZIE et al. (1981) considerable variation exists in the precision of cell alignment among different human individuals. On the whole the degree of cell alignment is less precise than in experimental animals and in the rhesus monkey. Certainly, additional studies are required in order to elucidate the reasons for these differences.

So far no specific types of organisation have been observed in the epidermis of palmae and plantae, peripheral extremities and the head (CHRISTOPHERS 1980a). Here the epidermis cornifies in an "unordered" way. The corneocytes of these areas are characterised by their variable shape and by a higher degree of overlap between neighbouring cells. These features are possibly the basis for relatively high tolerance against mechanical stress, e.g. of the planta. Similar structural findings have been observed in diseases with an increased cellular proliferation (CHRISTOPHERS 1980a, b).

Another type of horny cell arrangement is described in the mouse tail epidermis (ALLEN and POTTEN 1976) and at the edge of the guinea-pig ear (BULLOUGH and STOLZE 1978). Here the cornified cells show an imbricated (roof tile) arrangement. BULLOUGH and STOLZE (1978) suggest that the roof tile pattern is typical of slight to moderate hyperplasia.

C. The Langerhans Cell

As one of the principal epidermal symbionts (besides melanocytes and Merkel cells), Langerhans cells have become a major field of interest recently. Langerhans cells were first described in 1868 by PAUL LANGERHANS after staining with gold chloride. He described them as dendritic cells in the suprabasal layer of the epidermis.

I. Morphological Features

For a long time, the method for adenosine triphosphatase (ATPase) served as a very useful tool for the study of epidermal Langerhans cells in man, guinea-pig and mouse. The enzyme histochemical markers are different in different species (see WOLFF 1972).

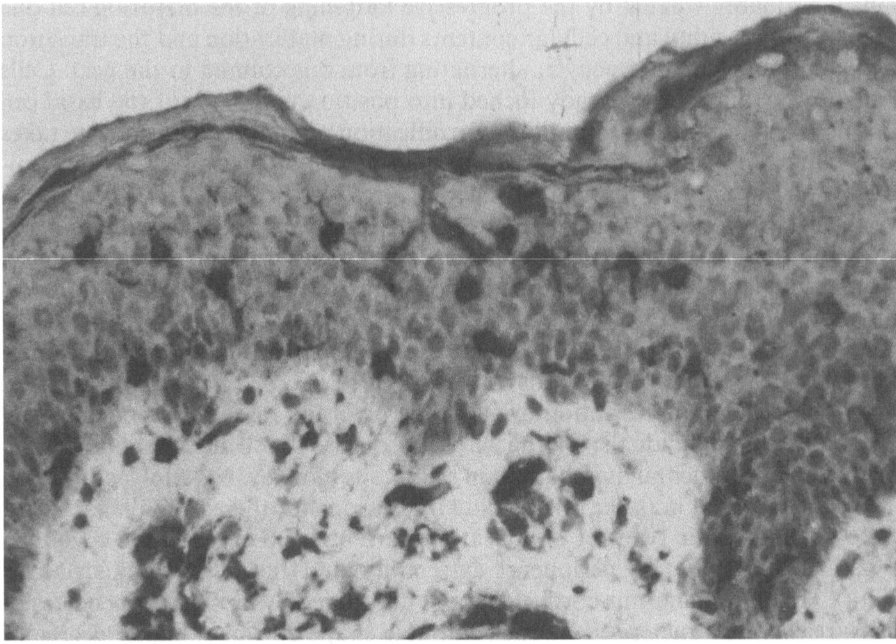

Fig. 8. Langerhans cells in human epidermis as revealed by immunoperoxidase technique using anti-T6 monoclonal murine antibody. × 120

Equally good or even better results are achieved with recently developed immunohistochemical methods using monoclonal antibodies on frozen sections. The indirect immunoperoxidase techniques are superior to immunofluorescence methods since they yield better morphological details (Fig. 8).

Both methods, ATPase and monoclonal antibody techniques, have their limitations. Negative staining results do not exclude the existence of Langerhans cells. Therefore, the most reliable marker for Langerhans cells is their ultrastructural feature of specific granules and a characteristic overall appearance. These criteria were first established by BIRBECK et al. in 1961 and consist of:

1. A clear cytoplasm
2. A mostly lobulated nucleus
3. The absence of desmosomes and tonofilaments
4. Characteristic rod-shaped organelles with a central lamella (Fig. 9)

Besides these granule-containing Langerhans cells there are so-called indeterminate cells in the epidermis. For a long time they were thought to be a different dendritic cell type (for review see WOLFF 1972). Probably these cells are related to Langerhans cells and to interdigitating cells of the lymphatic tissue. Except for the Birbeck granule, these cells are isomorphous and form part of a universal system including thymus, spleen, lymph node, tonsil and stratified epithelia. It may well be that indeterminate cells of the epidermis and interdigitating cells of lymphoid organs turn out to be identical cell types.

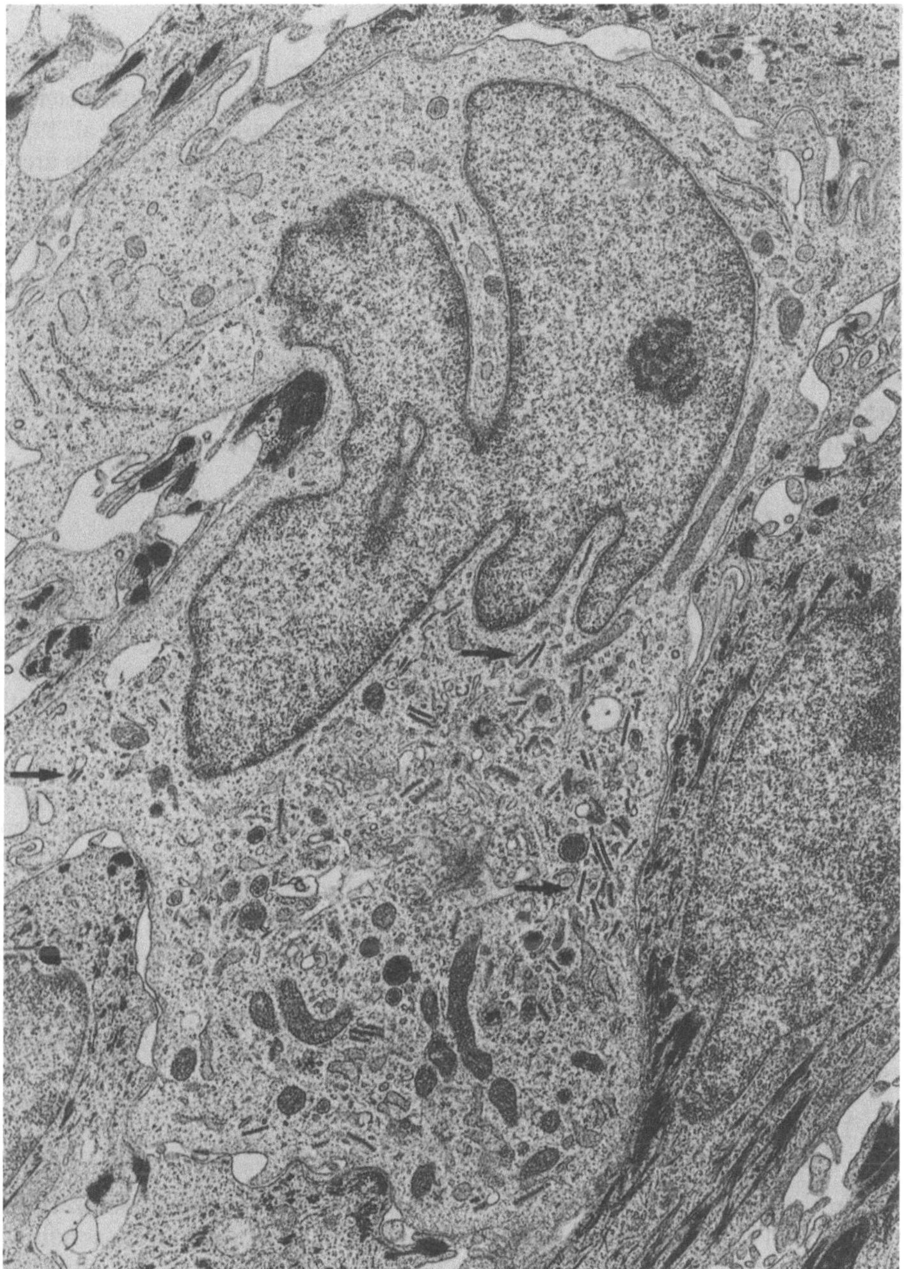

Fig. 9. Langerhans cell of human epidermis. *Arrows* indicate Langerhans cell granules. × 144000

II. Origin

Although Langerhans cells are predominantly located in the epidermis, it has long been speculated that they are a migrating cell system with a characteristic kinetic behaviour (for review see WOLFF 1972; SILBERBERG-SINAKIN et al. 1978). Recently it was shown by immunological methods that Langerhans cells are of mesenchymal origin. In 1979, KATZ et al. conducted experiments with chimeric mice using Ia antigens as a marker for the identification of cells of different origin (Ia antigens for immune response associated antigens). Chimerisation was induced by irradiation of mice A with 750–800 R and subsequent reconstitution with semi-allogeneic (A × B) or allogeneic (B) bone marrow cells. In the first week after chimerisation, the Ia haplotype specificities of Langerhans cells were still of recipient origin. Three months later, however, the Langerhans cells expressed Ia specificities of donor type. These findings suggest that Langerhans cells are derived from a mobile pool of precursor cells in the bone marrow.

III. Functional Properties

In 1977, STINGL et al. were able to show that Langerhans cells are the only epidermal cells which possess Fc-IgG and C_3-receptors. In addition, Langerhans cells were shown to synthesise Ia antigens (ROWDEN et al. 1977). Since these are features Langerhans cells have in common with macrophages, it was postulated that Langerhans cells might have an antigen-presenting function. Subsequently STINGL et al. (1978) proved that Langerhans cells indeed have the capacity of presenting to T cells soluble protein or haptenised antigens and stimulating both sygeneic and allogeneic T cells. This provided the basis for the current concept of the induction of T cell dependent immune responses in the skin in that antigens introduced into the epidermis are processed by Ia-bearing Langerhans cells, which, in turn, deliver a sensitizing signal to the immune apparatus. Thus the generation of effector mechanisms is triggered for the elimination of the respective antigens. Furthermore, the facts that Langerhans cells are potent stimulators of allogeneic T cell activation and that tissue allografts with no or only a few Langerhans cells have a prolonged survival on I-region disparate recipients suggests that Langerhans cells are a principal sensitizing structure (STINGL and WOLFF 1982).

IV. Role in Disease

Regarding the functional properties of Langerhans cells, it seems most likely that they play an essential role in immunologically mediated inflammatory conditions of the skin. Obviously, normally functioning Langerhans cells and Langerhans cell density are a prerequisite for optimal induction of contact hypersensitivity. Using C57BL/6J mice, TOEWS et al. (1980) have investigated the induction of contact hypersensitivity to DNFB through normal skin and through skin deficient in Langerhans cells. Tail skin epidermis is relatively deficient in Langerhans cells. The normal number of Langerhans cells in abdominal body wall skin can be reduced or even depleted by exposure to short course UV light irradiation. When sensitisation with DNFB was attempted through normal body wall skin, UV-irradiated body wall skin and normal tail skin, it was only through normal body wall skin that induction of sensitisation to DNFB was achieved. More importantly, animals whose first exposure to DNFB occurred through Langerhans cell deficient skin were unable subsequently to become specifically sensitized when immunisation was attempted through the conventional route using normal body wall skin. Thus it appears not only that Langerhans cells are essential for induction of contact hypersensitivity to chemical allergens, but that in their absence, exposure to these agents leads to unresponsiveness. In further experiments with Syrian hamsters, STREILEIN and BERGSTRESSER (1981) have shown that viable lymphoid cells from unresponsive animals can transfer the unresponsiveness to naive hamsters.

D. Conclusion

The vast collection of data on mammalian epidermis, as has been put forward in excellent monographs (MONTAGNA and PARAKKAL 1974) and tentatively outlined in this review, places emphasis on the structural uniqueness of epidermis: it consists of a living and a dead portion, constantly being replaced, with dramatic rate changes upon demand and the ability to construct its own architecture. Besides such functional adaptation toward the need for protection, epidermis serves as an immunologically competent organ, harbouring a macrophage-like cell and, as recently shown, being able to produce lymphocyte activating factor(s) and in addition change the spectrum of cell surface immune receptors. Epidermis may be subject to modulating influences from dermal origin, as has been known for many years, yet it may itself exert modulating effects. With these observations epidermis may be viewed not as an isolated tissue but as an integral part of the body.

References

Ackerman AB (1978) Histologic diagnosis of inflammatory skin diseases. Lea and Febiger, Philadelphia

Allen T, Potten C (1975) Desmosomal form, fate, and functions in mammalian epidermis. J Ultrastruct Res 51:94–105

Allen TD, Potten C (1976) Ultrastructural site variations in mouse epidermal organization. J Cell Sci 21:341–359

Anton-Lamprecht I (1973) Zur Ultrastruktur hereditärer Verhornungsstörungen: III. Autosomal dominante Ichthyosis vulgaris. Arch Dermatol Forsch 248:149–156

Baker H, Kligman AM (1967) Technique for estimating turnover time of human stratum corneum. Arch Dermatol 95:408

Bandmann HJ (1980) Makroskopische Anatomie der Haut. In: Korting GW (ed) Dermatologie in Praxis und Klinik, vol 1. Thieme, Stuttgart, pp 1.1–1.11

Banks-Schlegel S, Green H (1981) Involucrin synthesis and tissue assembly by keratinocytes in natural and cultured human epithelia. J Cell Biol 90:732–737

Bauer FW, Crombar NHCMN, De Grood RM, De Jongh GJ (1980) Flow cytometry as a tool for the study of cell kinetics in epidermis. Br J Dermatol 102:629–639

Bergstresser PR, Taylor JR (1977) Epidermal "turnover time" – a new examination. Br J Dermatol 96:503–509

Birbeck MS, Breathnach AS, Everall JD (1961) An electron microscopic study of basal melanocytes and high level clear cell (Langerhans cell) in vitiligo. J Invest Dermatol 37:51–64

Bizzozero G (1870) Osservazione sulla structura degli epiteli pavimentosi stratificrate. Rend r 1st Lombardo 3:675
Bloom W, Fawcett OW (1975) A textbook of histology, 10th edn. Saunders, Philadelphia
Braun-Falco O (1965) Die Histochemie der Barriere. In: Lejhanec G, Hybasek P (eds) De structura et functione stratorum epidermis s.d. barrierae. Lékarská Fakulta University U.E. Parkyne, Brünn, pp 49–61
Breathnach AS (1981) Application of freeze-fracture replication to study of hyperkeratosis. In: Marks R, Christophers E (eds) The epidermis in disease. MTP Press, Lancaster, pp 31–43
Briggaman RA (1981) Basement membrane formation and origin with special reference to skin. Front Matrix Biol 9:142–155
Briggaman RA (1982) Biochemical composition of the epidermal-dermal junction and other basement membranes. J Invest Dermatol 78:1–6
Briggaman RA, Wheeler CE (1975) The epidermal-dermal junction. J Invest Dermatol 65:71–84
Brody I (1959) The keratinization of epidermal cells of normal guinea pig skin as revealed by electron microscopy. J Ultrastruct Res 2:482–493
Brody I (1977) Ultrastructure of the stratum corneum. Int J Dermatol 16:245–257
Brophy D, Lobitz WC (1959) Injury and reinjury to human epidermis. J Invest Dermatol 32:495
Bucher O (1980) Cytologie, Histologie und Mikroskopische Anatomie des Menschen, 10th edn. Huber, Bern
Bullough WS (1972a) The control of epidermal thickness. Br J Dermatol 87:187–199
Bullough WS (1976b) The control of epidermis thickness. Br J Dermatol 87:347–354
Bullough WS, Mitrani E (1976) An analysis of the epidermal chalone control mechanism. In: Chalones. Houck JC (ed) Chalones. North-Holland, New York, pp 7–36
Bullough WS, Stolze J (1978) A new form of cellular arrangement in guinea-pig ear epidermis. Br J Dermatol 99:519–525
Caputo R, Peluchetti D (1977) The junctions of normal human epidermis. J Ultrastruct Res 61:44–61
Caputo R, Gasparini G, Innocenti M (1981) Normal and abnormal keratinization processes. A freeze-fracture study. In: Marks R, Christophers E (eds) The epidermis in disease. MTP Press, Lancaster, pp 43–61
Chopra DP, Flaxman BA (1972) Human epidermal cell cycle in vitro. Br J Dermatol 87:13–17
Christophers E (1970) Eine neue Methode zur Darstellung des stratum corneum. Arch Klin Exp Derm 237:717–721
Christophers E (1971a) Die epidermale Columnärstruktur. Z Zellforsch 114:441–450
Christophers E (1971b) Cellular architecture of the stratum corneum. J Invest Dermatol 56:165–169
Christophers E (1972a) Correlation between columns formation, thickness and rate of new cell production in guinea pig epidermis. Virchows Arch [B] 10:286–292
Christophers E (1972b) Kinetic aspects of epidermal healing. In: Maibach HI, Rovee D (eds) Epidermal wound healing. Year Book Publishers, Chicago, pp 53–69
Christophers E (1980a) Epidermopoese und Keratinisation. In: Korting GW (Hrsg) Dermatologie in Praxis und Klinik, vol 1. Thieme, Stuttgart, pp 164–167
Christophers E (1980b) Structural aspects of mammalian epidermal growth. In: Spearman RIC, Riley PA (eds) The skin of vertebrates. Academic Press, London, pp 137–139
Christophers E, Braun-Falco O (1971) Fluorescein-isothiocyanate as a stain for keratinizing epithelia. Arch Derm Forsch 241:199–209
Christophers E, Kligman AM (1964) Percutaneous absorption in aged skin. In: Montagna W (ed) Advances in biology of skin, vol VI. Pergamon, London, pp 163–175
Christophers E, Laurence EB (1973) Regional variations in mouse skin: quantitation of epidermal compartments in two different body sites. Virchows Arch [B] 12:212–222
Christophers E, Laurence EB (1976) Kinetics and structural organisation of the epidermis. Curr Probl Dermatol 6:87–106
Christophers E, Laurence EB (1980) Architectural considerations of epidermal growth. Pharmacol Ther 10:329–336

Christophers E, Wolff HH (1975) Differential formation of desmosomes and hemidesmosomes in epidermal cell cultures treated with retinoic acid. Nature 256:209–210
Christophers E, Wolff HH, Laurence EB (1974) The formation of epidermal cell columns. J Invest Dermatol 62:555–559
Cummins H (1942) The skin and mammary glands. In: Morris' human anatomy, 10th edn. Blakeston, Philadelphia
Cummins H (1946) Dermatoglyphics: significant patternings of the body surface. Yale J Biol Med 18:551–565
Cummins H (1964) Dermatoglyphics: a brief review. In: Montagna W, Lobitz CW (eds) The epidermis. Academic Press, New York, pp 375–385
Daróczy J, Feldmann J (1981) Microfilaments of the human epidermal-dermal junction. Front Matrix Biol 9:155–174
De Bersaques J (1975) Keratohyalin und Epidermisverhornung. Hautarzt 26:177–189
Delecluse C (1981) Keratohyaline synthesis. Front Matrix Biol 9:102–112
Duffil M, Wright N, Shuster S (1976) The cell proliferation kinetics of psoriasis examined by three in vivo techniques. Br J Dermatol 94:355–362
Ebling FJG (1979) The normal skin. In: Rock A, Wilkinson DS, Ebling FJG (eds) Textbook of dermatology, vol 1, 3rd edn. Blackwell, Oxford, pp 5–31
Ebling FJG, Rook A (1979) Disorders of keratinization. In: Rook A, Wilkinson DS, Ebling FJG (eds) Textbook of dermatology, vol 2, 3rd edn. Blackwell, Oxford, pp 1253–1315
Elias PM (1981) Epidermal lipids membranes and keratinization. Int J Dermatol 20:1–19
Elias P, Friend D (1975) The permeability barrier in mammalian epidermis. J Cell Biol 65:185–197
Elias P, Goerke J, Friend D (1977) Permeability barrier lipids: composition and influence on epidermal structure. J Invest Dermatol 69:535–547
Elias P, Brown B, Goerke J (1979) Localization and composition of lipids in neonatal mouse stratum granulosum and stratum corneum. J Invest Dermatol 73:339–348
Epstein WL, Maibach HT (1965) Cell renewal in human epidermis. Arch Dermatol 92:462–468
Etoh H, Taguchi YH, Tabachnik J (1975) Movement of beta-irradiated epidermal basal cells to the spinous-granula layers in the absence of cell division. J Invest Dermatol 64:431–435
Fawcett D (1981) The cell, 2nd edn. Saunders, Philadelphia
Fisher LB (1968) Determination of the normal rate and duration of mitosis in human epidermis. Br J Dermatol 80:24–28
Fisher LB, Wells GC (1968) The mitotic rate and duration in lesions of psoriasis and ichthyosis. Br J Dermatol 80:235–240
Frindle E, Malaise E, Tubiana M (1968) Cell proliferation kinetics in five human solid tumors. Cancer 22:611–620
Gelfant S (1978) Understanding chemotherapy of psoriasis in terms of epidermal cell proliferation. Cell Tissue Kinet 11:577–579
Gorbsky G, Steinberg M (1981) Isolation of the intercellular glycoproteins of desmosomes. J Cell Biol 90:243–248
Gray G (1981) Keratinization and the plasma membrane of the stratum corneum. Cell Front Matrix Biol 9:83–102
Greb W (1940) Untersuchungen über die Gestalt des Papillarkörpers der menschlichen Haut. Z Anat 110:245–263
Greulich RC (1964) Aspects of cell individuality in the renewal of stratified squamous epithelium. In: Montagna W, Lobitz W (eds) The epidermis. Academic Press, New York, pp 117–133
Gross RJ (1965) Kinetics of compensatory growth. Q Rev Biol 40:126–146
Hadler W, Silveira S (1981) Identification of tocopherol (vitamin E) in the skin and its histochemical detection. Acta Histochem 68:11–21
Halprin KM (1972) Epidermal "turnover time" – a re-examination. Br J Dermatol 86:14–19
Ham AW (1974) Histology, 7th edn. Lippincott, Philadelphia

Hayward AF (1979) Membrane-coating granules. Int Rev Cytol 59:97–124
Hill MW (1980) Cell kinetics in premalignancy. In: Mackenzie IC, Dabelsteen E, Squier Chr (eds) Oral premalignancy. University of Iowa Press, Iowa, pp 191–219
Horstmann E (1957) Die Haut. In: Bargmann W (ed) Handbuch der mikroskopischen Anatomie des Menschen, vol 3/1. Springer, Berlin, Göttingen, Heidelberg, pp 1–276
Hunter R, Pinkus H, Steele Ch (1956) Examination of the epidermis by the strip method. J Invest Dermatol 27:31–34
Iversen OH, Bjerknes R, Devik F (1968) Kinetics of cell renewal, cell migration and cell loss in the hairless mouse dorsal epidermis. Cell tissue Kinet 1:351–367
Jaoita H, Foidart J, Katz S (1978) Localization of the collagenous component in skin basement membrane. J Invest Dermatol 70:191–193
Kahn G, Weinstein GD, Frost P (1968) Kinetics of human epidermal cell proliferation: diurnal variation. J Invest Dermatol 50:459–462
Kakimi S, Fukuyama K, Epstein W (1980) A study of ultrathin frozen sections of granular cells in newborn rat epidermis. J Ultrastruct Res 70:8–14
Katz SI, Tamaki K, Sachs DH (1979) Epidermal Langerhans cells are derived from cells originating in the bone marrow. Nature 282:324–326
Kelly D, Luft J (1966) Fine structure development and classification of desmosomes and related attachment mechanism. In: Electron microscopy. 6th edn. International Congress on Electron Microscopy, Tokyo 2:401–402
Kligman AM (1964) The biology of the stratum corneum. In: Montagna W, Lobitz GW (eds) The epidermis. Academic Press, New York, pp 387–430
Landmann L (1980) Lamellar granules in mammalian, avian and reptilian epidermis. J Ultrastruct Res 72:245–263
Langerhans P (1868) Über die Nerven der menschlichen Haut. Virchows Arch Pathol Anat 44:325–337
Laurence EB, Christophers E (1976) Selective action of hydrocortisone on postmitotic epidermal cells in vivo. J Invest Dermatol 66:201–281
Leblond CP, Greulich RC, Peirera JP (1964) Relationship of cell formation and cell migration in the renewal of stratified squamous epithelia. In: Montagna W, Billingham RB (eds) Advances in biology of skin, vol 5. Pergamon, Oxford, pp 39–66
Leloup R, Laurent L, Ronveaux M, Doochmans P, Wanson JC (1979) Desmosomes and desmogenesis in the epidermis of calf muzzle. Biologie Cellulaire 34:137–152
Leutgeb CH, Bandmann HJ, Breit R (1972) Handlinienmuster, Ichthyosis vulgaris und Dermatitis atopica. Arch Dermatol Forsch 244:354–363
Mackenzie IC (1969) Ordered structure of the stratum corneum of mammalian skin. Nature 222:881–882
Mackenzie IC (1972) The ordered structure of the stratum corneum of mammalian skin. In: Maibach H, Rovee D (eds) Epidermal wound healing. Year Book Medical Publishers, Chicago, pp 5–26
Mackenzie IC, Zimmermann K, Peterson L (1981) The pattern of cellular organization of human epidermis. J Invest Dermatol 76:459–461
Matoltsy AG, Parakkal PE (1965) Membrane-coating granules of keratinizing epithelia. J Cell Biol 24:297–307
Maurer W, Pilgrim Ch, Wegener K, Hellweg S (1965) Messung der Dauer der DNS-Verdoppelung und der Generationszeit bei verschiedenen Zellarten von Maus und Ratte durch Doppelmarkierung mit H-3- und C-14-Thymidin. Radioaktive Isotopen in Klinik und Forschung 6:96–107
Menton DN (1976) A minimum-surface mechanism to account for the organization of cells into columns in the mammalian epidermis. Am J Anat 145:1–22
Montagna W, Parakkal PF (1974) The structure and function of skin, 3rd edn. Academic Press, New York
Pappas G (1975) Junctions between cells. In: Weismann G, Claiborne R (eds) Cell membranes. Hospital Practice, New York, pp 87–94
Penneys NS, Fulton JE, Weinstein GD, Frost P (1970) Location of proliferating cells in human epidermis. Arch Dermatol 101:323–327
Peracchia C (1980) Structural correlates of gap junction permeation. Int Rev Cytol 66:81–146

Pilgrim C, Maurer W (1965) Autoradiographische Untersuchung über die Konstanz der DNS-Verdoppelungsdauer bei Zellarten von Maus und Ratte durch Doppelmarkierung mit ^{3}H- und ^{14}C-Thymidin. Exp Cell Res 37:187–199
Pinkus H (1952) Examination of the epidermis by the strip method: II. Biometric data on regeneration of the human epidermis. J Invest Dermatol 19:431–444
Pinkus H, Hunter R (1966) The direction of the mitotic axis in human epidermis. Arch Dermatol 94:351–354
Pinto da Silva P, Kachar B (1982) On tight junction structure. Cell 28:441–450
Pitts JD (1981) How do animal cells communicate? Verh Dtsch Zool Ges 74:134–137
Plewig G, Marples R (1970) Regional differences of cell sizes in the human stratum corneum: part I. J Invest Dermatol 54:13–21
Porter D, Shuster S (1968) Epidermal renewal and amino acids in psoriasis and pityriasis rubra pilaris. Arch Dermatol 98:339–343
Potten CS (1981) Cell replacement in epidermis (keratopoiesis) via discrete units of proliferation. Int Rev Cytol 69:271–318
Pullmann H, Lennartz KJ, Steigleder GK (1974) In vitro examination of cell proliferation in normal and psoriatic epidermis, with special regard to diurnal variations. Arch Dermatol Forsch 250:177
Quastler H, Sherman FG (1959) Cell population kinetics in the intestinal epithelium of the mouse. Exp Cell Res 17:420–438
Rothberg S, Crounse RG, Lee JL (1961) Glycin-^{14}L incorporation into the proteins of normal stratum corneum and the abnormal stratum corneum of psoriasis. J Invest Dermatol 37:497–505
Rowden G, Lewig MG, Sullivan AK (1977) Ia antigen expression on human epidermal Langerhans cells. Nature 268:247–248
Rowe L, Dixon WJ, Forsythe A (1978) Mitosis in normal and psoriatic epidermis. Br J Dermatol 98:293–299
Rupec M (1980) Mikroskopische und elektronenmikroskopische Anatomie der Haut. In: Korting GW (ed) Dermatologie in Praxis und Klinik, vol 1. Thieme, Stuttgart, pp 1.14–1.52
Schell H, Schwarz W, Hornstein OP, Bernlochner W, Weghorn C (1981) Evidence of diurnal variation of human epidermal cell proliferation. Arch Dermatol Res 271:41–47
Schumacher S (1931) Integument der Mammalier. In: Bolk L, Goeppert E, Kallus E, Lubosch W (eds) Handbuch der vergleichenden Anatomie der Wirbeltiere, vol 1. Urban & Schwarzenberg, Berlin, pp 449–504
Silberberg-Sinakin I, Baer RL, Thorbecke GJ (1978) Langerhans cells: a review of their nature with emphasis on their immunologic functions. Prog Allergy 24:268–294
Squier C, Fejerskov O, Jepsen A (1978) The permeability of a keratinizing squamous epithelium in culture. J Anat 126:103–109
Staehelin LA (1974) Structure and function of intercellular junctions. Int Rev Cytol 39:191–283
Staehelin LA, Hull BE (1978) Cell junctions between living cells. Sci Am 238:141–152
Stingl G, Wolff K (1982) Origin and function of Langerhans cells and their role in disease. In: Goos M, Christophers E (eds) Lymphoproliferative diseases of the skin. Springer, Berlin Heidelberg New York, pp 34–40
Stingl G, Wolff-Schreiner EC, Pichler W, Gschnait F, Knapp W, Wolff K (1977) Epidermal Langerhans cells bear Fc and C3 receptors. Nature 258:245–246
Stingl G, Katz SI, Clement L, Green I, Shevach EM (1978) Immunologic functions of Ia-bearing epidermal Langerhans cells. J Immunol 121:2005–2013
Streilein JW, Bergstresser PR (1981) Langerhans cell function dictates induction of contact hypersensitivity or unresponsiveness to DNFB in Syrian hamsters. J Invest Dermatol 77:272–277
Toews GB, Bergstresser PR, Streilein JW (1980) Epidermal Langerhans cell density determines whether contact hypersensitivity or unresponsiveness follows skin painting with DNFB. J Immunol 124:445–453
Watt F, Green H (1981) Involucrin synthesis is correlated with cell size in human epidermal cultures. J Cell Biol 90:738–742

Weinstein GD, Frost P (1968) Abnormal cell proliferation in psoriasis. J Invest Dermatol 50:254–259
Weinstein GD, Frost P (1969) Cell proliferation in benign and malignant skin diseases in humans. Natl Cancer Inst Monogr 30:225–246
Weinstein GD, Van Scott EJ (1965) Autoradiographic analysis of turnover times in normal and psoriatic epidermis. J Invest Dermatol 45:257–262
Welsch U, Buchheim WC (1978) Zelljunktionen. Verh Anat Ges 72:199–215
Whitton J, Everall J (1973) The thickness of the epidermis. Br J Dermatol 89:467–476
Wolff K (1972) The Langerhans cell. Curr Probl Dermatol 4:79–145
Wolff K, Schreiner E (1968) An electron microscopic study on the extraneous coat of keratinocytes and the intercellular space of the epidermis. J Invest Dermatol 51:418–432
Wright NA (1977) Cell population kinetics in human epidermis. Int J Dermatol 16:499
Wright NA (1981) A methodological approach to epidermopoiesis. In: Marks R, Christophers E (eds) The epidermis in disease. MTP Press, Lancaster, pp 139–153

CHAPTER 2

Keratin

H. P. BADEN

A. Introduction

Keratinisation is a complex orderly process by which viable epidermal cells are converted into dead cornified cells. In the skin this occurs in the epidermis, hair and nails, although there are significant differences in the products formed in the three tissues. A structure common to all of them, however, is the filamentous fibrous protein, keratin (FRASER et al. 1972), which is synthesised and remains intracellularly. Keratin has a coiled coil structure in all three tissues as revealed by X-ray diffraction studies but the amino acid composition of epidermal keratin differs from that of nail and hair, which are very similar (Table 1) (BADEN et al. 1973).

Epidermis and nail grow continuously but hair undergoes growth and resting phases. Nail and hair appear to grow at their maximal rate while epidermis can turn over at much higher rates in response to trauma and in certain diseases such as psoriasis (WEINSTEIN and VAN SCOTT 1965). This may in part be due to shortening of the germinative cell cycle (WEINSTEIN and FROST 1968), but some of the

Table 1. Amino acid analysis of fibrous proteins: residue per 100 residues

Amino acid	Hair	Nail	Stratum corneum
Cystine (half)	7.1	5.5	1.0
Lysine	2.9	2.8	4.3
Histidine	0.6	0.7	1.0
Arginine	7.0	7.1	4.8
Aspartic acid	8.4	10.4	9.5
Threonine	5.9	3.9	3.9
Serine	10.5	9.4	11.7
Glutamic acid	17.1	14.8	13.7
Proline	5.6	5.3	2.3
Glycine	5.3	8.8	18.6
Alanine	6.0	8.3	5.6
Valine	5.3	3.9	3.7
Methionine	0.6	Trace	1.5
Isoleucine	3.0	2.9	3.2
Leucine	9.8	10.7	8.5
Tyrosine	2.8	3.0	3.2
Phenylalanine	2.1	2.5	3.4

basal cells are normally blocked in G_1 and with proper stimulation they can be recruited into the germinative cell population (GELFANT 1976). The stimulus that causes epidermal cells to leave the germinative cell population and differentiate is not known, but mitotic pressure is not involved (ETOH et al. 1975).

A number of metabolic activities of epidermal cells are similar to those of other tissue, such as energy production and synthesis of macromolecules, but there are several which are peculiar to the cornification process. This review will emphasise certain biochemical events in differentiation of epidermal cells and relate these to the structural features of the cell. The major components to be considered are tonofilaments, keratohyalin granules, desmosomes, membrane coating granules and cornified cell envelopes.

B. Fibrous Proteins

The fibrous protein of the epidermis appears as 80-Å filaments within the cells which attach to a modified region of the desmosome (Fig. 1). The protein was identified by its characteristic X-ray diffraction pattern, which showed a pitch periodicity of 5.1 Å and a distance between peptide chains of 9.8 Å (Fig. 2). This pitch periodicity is smaller than that of the usual α helix, suggesting that the molecule has a super coiled structure (CRICK 1952). RUDALL (1952) was the first to isolate epidermal fibrous protein using cow snout epidermis and extracting with a urea buffer, but MATOLTSY (1965) later showed that the fibrous protein of the malpighian layer, prekeratin, could be extracted with citrate buffer, pH 2.65. This protein was insoluble in the pH range 3–10 in the absence of denaturing agents. A similar protein could also be isolated from the epidermis surrounding the cow hoof (STEINERT and IDLER 1975). Other investigators using sodium dodecyl sulphate (SDS) polyacrylamide gel electrophoresis (PAGE) showed that the protein consisted of as many as seven different polypeptides in the molecular weight range of 40000–70000 depending on the electrophoretic technique being used (BADEN 1980; STEINERT and IDLER 1975; LEE et al. 1975; SKERROW 1974) (Fig. 3).

By mixing different combinations of the isolated polypeptides it was found that at least two different polypeptides were necessary to form the helical molecule as judged by X-ray diffraction techniques (BADEN and LEE 1977; STEINERT and GULLINO 1976). Using two-dimensional electrophoresis, it has been found that there are acid and basic groups of polypeptides (COOPER et al. 1985) and one of each of these is present in the helical molecule. It was first thought that the monomer subunit consisted of three chains (LEE and BADEN 1976; STEINERT 1978), but recent structural studies indicate that the most basic subunit consists of a two chain coiled coil which is organised into a four chain complex (STEINERT et al. 1985). These results indicate that the fibrous protein is heterogeneous, consisting of a family of molecules with different combinations of peptide chains.

Only about half of the keratin molecule has a helical structure so it is clear that both helical and non-helical regions must exist. Amino acid sequence analysis indicates that there is a central helical core with random regions present at the ends of the peptide chains (STEINERT et al. 1985). Association of these molecules into the larger macromolecular structures most likely occurs by side to side and end to end alignment.

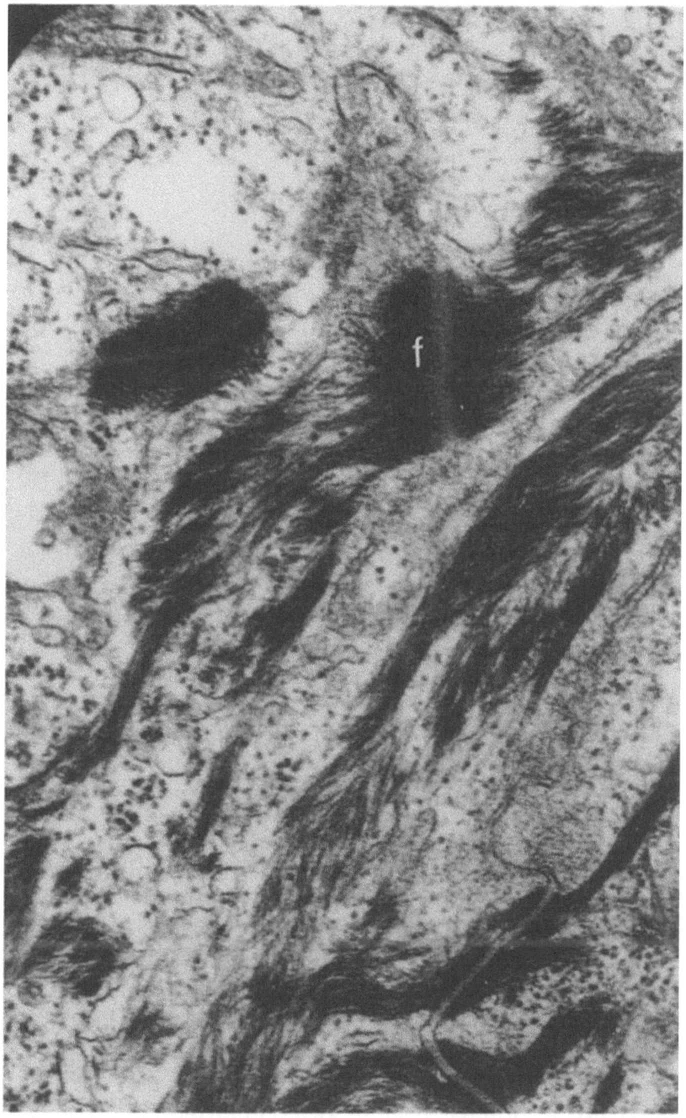

Fig. 1. Electron micrograph showing bundles of filaments (*f*) streaming out of desmosomes

In the change from viable malpighian layer cells to dead stratum corneum cells several alterations in the fibrous proteins occur. Prekeratin contains no cystine but only cysteine, while progressive oxidation to cystine occurs in the stratum corneum fibrous protein. Extraction of the stratum corneum protein from cow snout epidermis requires a denaturing agent such as urea at alkaline pH, but most of the protein can be solubilised without a reducing agent (BADEN et al. 1976). However, the solubilised protein can be shown to contain cystine, suggesting that there

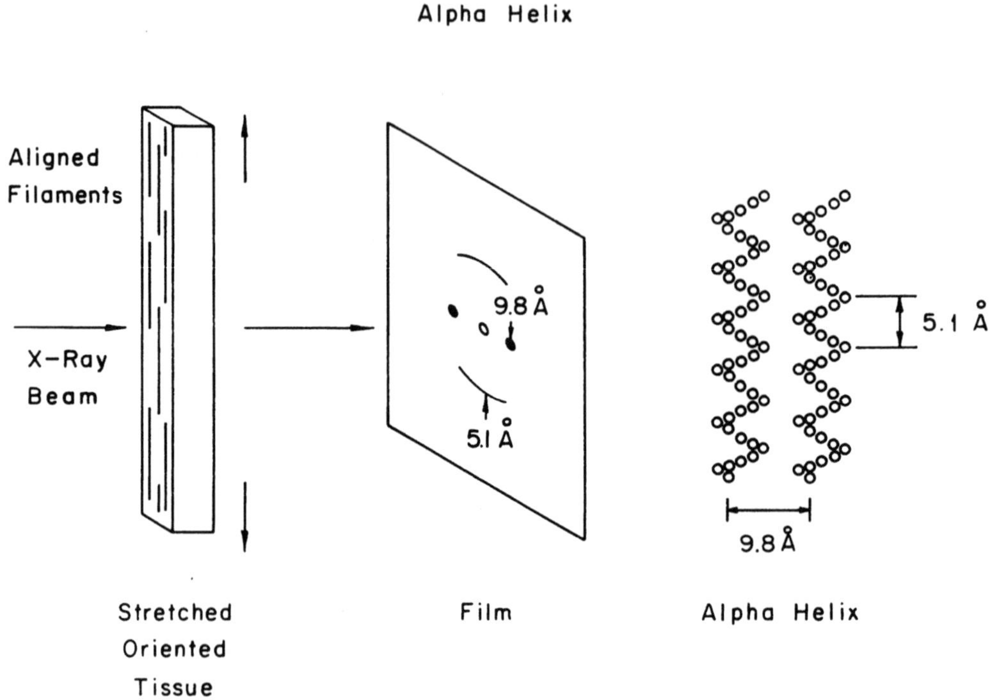

Fig. 2. The X-ray diffraction pattern and its relationship to the molecular structure of keratin

is not extensive cross-linking between keratin molecules. In stratum corneum of other animals such as man this is not the case and a reducing agent is necessary for extractions of protein (BADEN et al. 1976; STEINERT and IDLER 1979). The SDS PAGE patterns of the fibrous proteins are very similar to prekeratin, indicating that little alteration in the size of the polypeptides has occurred.

Human prekeratin also consists of a number of polypeptides which are in the same molecular weight range as those reported for bovine epidermis (BADEN and LEE 1978; BAYNES et al. 1978; SKERROW 1977; COOPER et al. 1985). Some variability in the relative proportion has been reported in different individuals and in different areas of the body. Isolation of fibrous proteins from human stratum corneum requires the presence of a reducing agent in alkaline denaturing buffers, suggesting that there is extensive disulphide cross-linking. Although the SDS PAGE pattern of the isolated stratum corneum polypeptides is similar to that of keratin of the living layer, the pattern of protein isolated from the most superficial layers is different, indicating that post-translational modification has occurred (BADEN et al. 1976; STEINERT and IDLER 1979). This may be the result of proteolytic breakdown, non-disulphide cross-linking, or addition or loss of components that cause an apparent change in molecular weight.

Several post-translational modifications of the fibrous proteins have been found in the stratum corneum fibrous keratins. The ε-(γ-glutamyl)lysine cross-

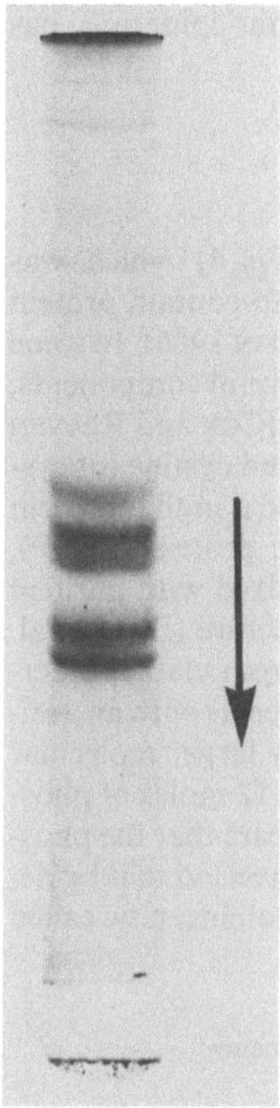

Fig. 3. SDS PAGE pattern of prekeratin from bovine snout epidermis

link has been definitively identified in the fibrous protein of human stratum corneum although the cornified envelope is the major site where this cross-linking occurs (ABERNATHY et al. 1977; SUGAWARA 1977). Phosphorylation of the fibrous polypeptides has been observed in both whole epidermis and in cultured epidermal cells (GILMARTIN et al. 1980; GREEN and SUN 1978). Finally, peptide-bound citrulline has been found in stratum corneum fibrous proteins, and an enzyme which converts peptide-bound arginine to peptide-bound citrulline has been found in neutral buffer extracts of epidermis (KUBILUS et al. 1979, 1980). What special properties this gives the proteins has not yet been determined. Methylation of the basic amino acids has not been observed.

Evidence has been presented that different polypeptides are formed in the different epidermal cell layers (VIAC et al. 1980). Similar results were obtained in studies of snake epidermis, where it was observed that the scale (β) and epidermal (α) fibrous proteins were synthesised at different levels of the regenerating epidermis prior to shedding. Although it would appear that the heaviest polypeptides are synthesised in the higher layers and thick epidermis contains the largest amount of heaviest polypeptides, this has not proved to be a general rule. Perihoof bovine epidermis, for example, which is thicker than snout epidermis, has smaller polypeptides (LEE et al. 1979).

C. Keratohyalin

Keratohyalin is a unique epidermal subcellular particle (Fig. 4) which was thought to represent cellular debris but has now been shown to contain protein which is synthesised in the granular layer (FUKUYAMA and EPSTEIN 1966). In some animals it has been shown to be heterogeneous, consisting of several components. The major component appears to be a protein rich in histidine (COX and REAVEN 1967; BADEN et al. 1974) and arginine and to lack methionine and cystine (or cysteine) (Table 2). DALE (1977) and BALL et al. (1978) have isolated a similar protein from newborn rat stratum corneum, stratum corneum basic protein (SCBP), which has a molecular weight of 50000. This protein, when mixed with purified filamentous protein, causes it to associate into a rope-like structure (DALE et al. 1978). LONSDALE-ECCLES et al. (1980) have isolated a protein from viable epidermis which has a similar amino acid composition to SCBP and reacts with an antibody prepared to SCBP. This protein, however, has a slightly larger molecular weight of 53000 and contains phosphate groups. There are 10–12 moles of phosphate per mole of SCBP and these are on serine groups. It appears that the phosphorylated protein is a precursor of the SCBP and becomes converted to it by dephosphorylation. Most recently it has been found that these proteins, now called

Table 2. Amino acid composition of rat stratum corneum proteins: residue per 100 residues (DALE 1977)

	Basic protein	Fibrous protein	
		Heavy chain	Light chain
Cystine (half)	0	0.8	0.8
Lysine	0	7.1	4.6
Histidine	7.9	0.8	0.7
Arginine	13.9	7.4	5.8
Aspartic acid	3.6	8.9	8.2
Threonine	5.7	2.8	3.2
Serine	17.3	9.5	11.9
Glutamic acid	20.5	14.3	17.9
Proline	2.8	2.0	0.6
Glycine	14.4	11.5	15.1
Alanine	11.8	4.0	2.8
Valine	0.6	3.2	1.7
Methionine	0	4.9	2.8
Isoleucine	1.4	3.6	2.8
Leucine	0	5.9	7.7
Tyrosine	Trace	2.2	5.5
Phenylalanine	0	4.8	3.6
Tryptophan		1.5	2.3

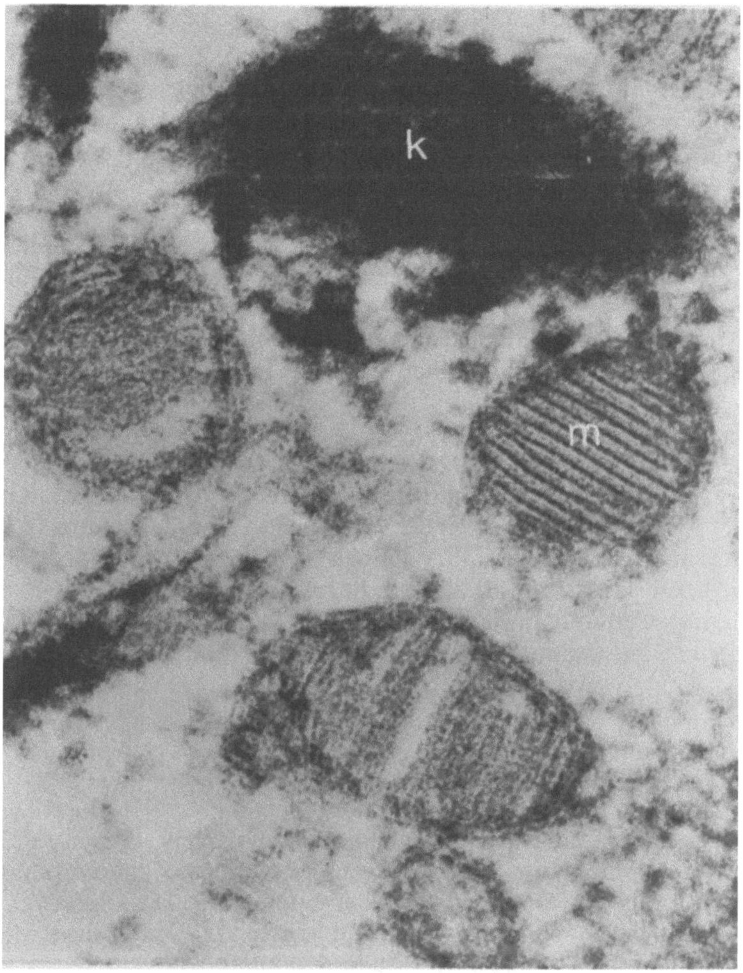

Fig. 4. Electron micrograph showing keratohyalin granule (*k*) and membrane coating granules (*m*) (ODLAND and REED 1967)

filaggrin, are derived from a precursor of molecular weight >200000 (HARDING and SCOTT 1983; MEEK et al. 1983).

The keratohyalin basic proteins have been identified in human epidermis and partially characterised (LYNLEY and DALE 1983) and also appear to be derived from a higher molecular weight precursor (FLECKMAN et al. 1985; KUBILUS et al. 1985). The basic protein appears to act as a matrix protein in which the filaments are imbedded and is responsible for the electron microscopic keratin pattern seen in stratum corneum. However, aggregation of filamentous protein and formation of cornified cells can occur in the absence of SCBP, this being observed in some areas of normal epidermis and in the epidermis of ichthyosis vulgaris (ANTON-LAMPRECHT and SCHNYDER 1974). It is not known what substitutes for basic protein but it has been observed that in the presence of calcium ion and other metal ions the filaments aggregate in a manner similar to that induced by SCBP (FUKUYAMA et al. 1978). An important new role for the basic protein has been found. The arginine groups in peptide linkage are converted to citrulline and the protein separates from the filaments. It is then broken down by proteolytic enzymes to free amino acids and other compounds which are important for binding water in the stratum corneum.

A second component of keratohyalin has been identified, called dense homogeneous deposits (DHDs), in rat epidermis (FUKUYAMA et al. 1972) and mucous membranes (JEPSEN 1970) by electron microscopy using special staining techniques. Morphological and labelled amino acids incorporation studies suggested it was rich in cystine and represented a precursor of the cornified cell envelope (FUKUYAMA and EPSTEIN 1975). MATOLTSY and MATOLTSY (1970) isolated a protein from rat epidermis which they called keratohyalin that had a high content of cystine, suggesting it was the DHD component. However, definitive identification using immunological and morphological studies has not yet been reported. The DHD particles have not been demonstrated in human epidermis and extensive studies of epidermis from other animals have not yet been reported to determine whether this finding is peculiar to the rat or is characteristic of all mammalian epidermis. Certainly the cornified envelope is a constant feature of mammalian epidermis, but the mechanism of its formation could vary in different species.

D. Cornified Envelope

The finding that stratum corneum could not be completely solubilised with denaturing solvents containing a reducing agent led to the recognition that the thick layer which forms beneath the plasma membrane in the granular layer contains highly resistant cross-links (MATOLTSY and MATOLTSY 1966). This structure, the cornified cell envelope (Fig. 5), was found to contain ε-(γ-glutamyl)lysine cross-links (Fig. 6) in the amount of 5.15 mol/10^5 g in newborn rat epidermis (SUGAWARA 1977). This was at least 5 times the amount of cross-link reported in the fibrous protein of stratum corneum from patients with lamellar ichthyosis (ABERNATHY et al. 1977). The cross-link has also been found in the cornified envelopes of human callus and keratinising epidermal cells in culture (RICE and GREEN 1977).

Although it has been speculated that the DHDs are the precursors of the envelope in newborn rat epidermis, this has not been determined in other species. A precursor for the envelope of about 90000 molecular weight, involucrin, has been found in cultured human epidermis using the incorporation of labelled amines (RICE and GREEN 1979). An antibody to involucrin shows peripheral staining of the granular cells by immunofluorescence, as would be expected for a precursor of the cornified envelope which is being laid down. Amino acid analysis of the protein shows about 50% of the residues to be glutamic acid, which is much higher than the glutamic acid content of the cornified envelopes isolated from keratinised cultured cells. Additional precursors of the cornified envelope

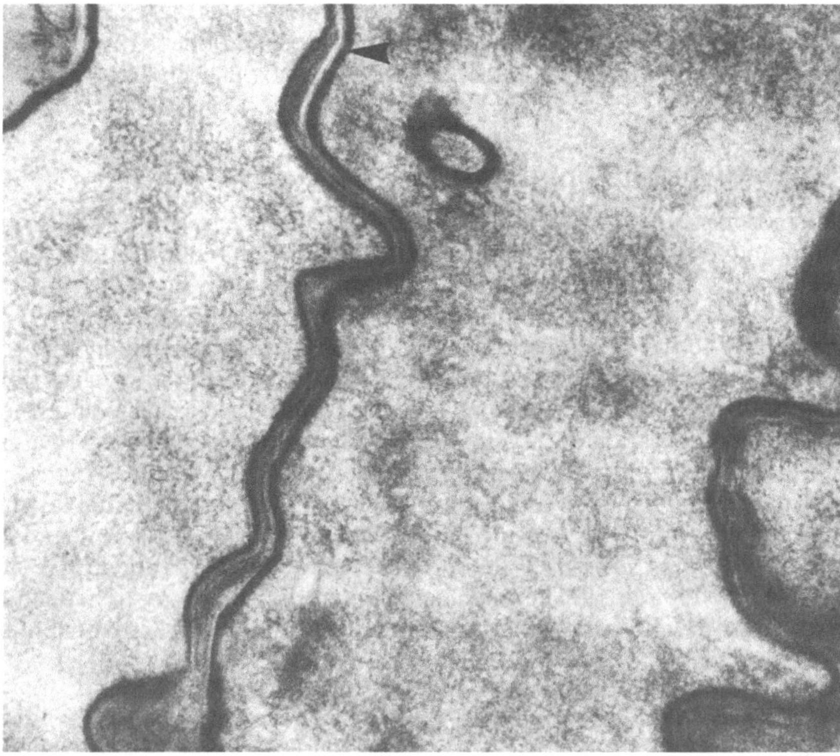

Fig. 5. A thickened cornified envelope (*arrowhead*) is seen surrounding the cells of the stratum corneum (MATOLTSY and PARAKKAL 1967)

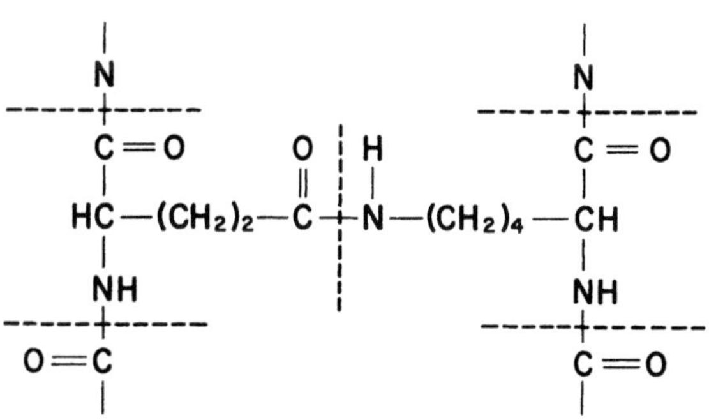

Fig. 6. The ε-(γ-glutamyl)lysine cross-link

are now being reported (SIMON and GREEN 1984; MA and SUN 1986), which probably explains the discrepancy. This fits well with the known heterogeneity of structural proteins such as the fibrous and matrix ones in epidermis and hair.

The enzyme which is responsible for this cross-linking has been identified in the epidermis of a large number of animals and has been purified from bovine (BUXMAN et al. 1980) and human epidermis (OGAWA and GOLDSMITH 1976). Although these enzymes have a similar molecular weight they are immunologically, at least, distinct. Several different enzymes have now been identified in human keratinocytes and it appears that a membrane-associated one is responsible for envelope formation (THACHER and RICE 1985; SCHMIDT et al. 1985).

E. Desmosomes

The desmosomes (BIZZOZERO 1871) have been studied extensively by electron microscopy and shown to consist of a number of components (BRODY 1968) (Fig. 7). Several of these are modified regions of the cell membrane while others are truly intercellular. The desmosomes must be broken and reformed as cells move from the basal layer to the stratum corneum. Although it seems most likely that some intracellular attachment is reversibly broken and reformed, the mechanism for this has not been identified in mammalian epidermis. At present the only known mechanism for separation of epidermal cells and retention of viability is treatment with proteolytic enzymes for short periods.

Desmosomes attached to fragments of cell membranes have been isolated free of other cellular components using bovine snout epidermis (SKERROW and MATOLTSY 1974a). They have been found to contain about 76% protein, 17% carbohydrate and 10% lipid. The sialic acid content is 5 nmol/mg of protein and 40% of the lipid is cholesterol and the remainder phospholipid. A number of pro-

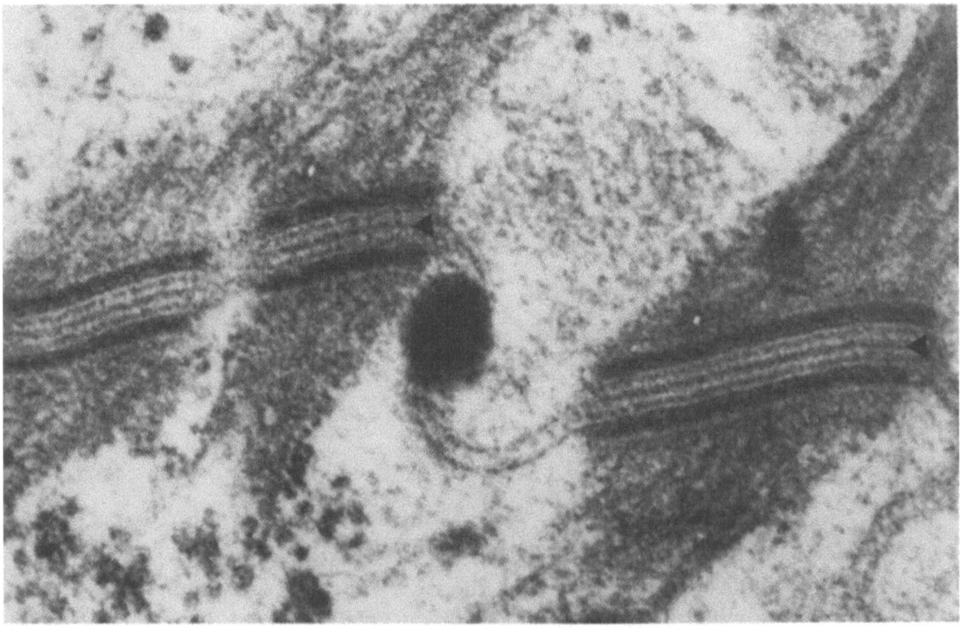

Fig. 7. Electron micrograph of a desmosome showing the multiple layers. The central dark band (*arrowhead*) and the adjacent light areas are in the intercellular space (ODLAND and REED 1967)

teins have been identified using SDS PAGE, and several of these have been purified and assigned to different components of the desmosome (Cowin et al. 1985).

F. Membrane Coating Granules

Attention has been focused in the past several years on the intercellular region as an important component of the barrier of the stratum corneum. Ovoid bodies, the membrane coating granules, are first observed in the upper spinous layer and are extruded into the intercellular space in the granular layer. They are 0.1–0.5 μm in size, surrounded by a 30 Å thick membrane, and their internal structure contains lamellae with alternating density (Fig. 4). The membrane coating granules which have been extruded partially disintegrate to give rise to lamellar sheets of alternating electron density at the junction of the stratum corneum (Elias et al. 1977).

The stratum corneum lipids are essential for barrier function and the location of these lipids in the stratum corneum membrane has been under intense investigation. Phospholipids are present in the granular layer but these are replaced by neutral lipids in the stratum corneum (Elias et al. 1977; Gray and Yardley 1975). At present the preponderance of evidence indicates that the lipid is not within the cell, but rather is located at the membrane or intercellular space. This is supported by freeze fracture studies of stratum corneum which indicate that the plane is through the intercellular space in intact tissue. This has been interpreted as signifying that the lipids are localised in the lamellar sheets between the cells which were derived from the membrane coating granules. X-ray diffraction studies of membrane couplets have more definitively localised the stratum corneum lipids intercellularly (Elias et al. 1983).

Acknowledgement. This work was supported by NIH Grant AM 06838.

References

Abernathy JL, Hill RL, Goldsmith LA (1977) The ε-(γ-glutamyl)lysine cross-links in human stratum corneum. J Biol Chem 252:1837–1839

Anton-Lamprecht I, Schnyder UW (1974) Ultrastructure of inborn errors of keratinization. VI. Inherited ichthyoses – a model system for heterogeneities in keratinization disturbances. Arch Dermatol Forsch 250:207–227

Baden HP (1980) Structure of epidermal keratin and variations in its polypeptide composition. In: Bernstein IA, Seiji M (eds) Biochemistry of abnormal epidermal differentiation. University of Tokyo Press, Tokyo, pp 345–363

Baden HP, Lee LD (1977) The structure of epidermal keratin. In: Seiji M, Bernstein IA (eds) Biochemistry of cutaneous epidermal differentiation. University of Tokyo Press, Tokyo, pp 478–492

Baden HP, Lee LD (1978) Fibrous protein of human epidermis. J Invest Dermatol 71:148–151

Baden HP, Goldsmith LA, Fleming B (1973) A comparative study of the physicochemical properties of human keratinized tissues. Biochim Biophys Acta 322:269–278

Baden HP, Goldsmith LA, Baden SB, Roth SI, Lee LD (1974) Keratohyalin protein in disorders of keratinization. J Invest Dermatol 62:411–414

Baden HP, Lee LD, Kubilus J (1976) The fibrous proteins of stratum corneum. J Invest Dermatol 67:573–576
Ball RD, Walker GK, Bernstein IA (1978) Histidine-rich proteins as molecular markers of epidermal differentiation. J Biol Chem 253(16):5861–5868
Baynes J, Levine M, McLeod A, Wilkinson A (1978) Precursor of keratin protein from human epidermis. Br J Dermatol 98:165–173
Bizzozero G (1871) Sulla struttura degli epiteli pavimentosi stratifcati (Abstract). Zentralmed Wochenschr 9:482–483
Brody I (1968) An electron-microscopic study of the junctional and regular desmosomes in normal human epidermis. Acta Derm Venereol 48:290–302
Buxman MM, Lobitz CJ, Wuepper KD (1980) Epidermal transglutaminase. J Biol Chem 255:1200–1203
Cooper D, Schermer A, Sun T-T (1985) Biology of disease. Classification of human epithelia and their neoplasms using monoclonal antibodies to keratins: strategies, applications, and limitations. Lab Invest 52:243–256
Cowin P, Kapprell H-P, Franke WW (1985) The complement of desmosomal plaque proteins in different cell types. J Cell Biol 101:1442–1454
Cox AJ, Reaven EP (1967) Histidine and keratohyalin granules. J Invest Dermatol 49:31–34
Crick FHC (1952) Is α-keratin a coiled-coil? Nature 170:882–883
Dale BA (1977) Purification and characterization of a basic protein from stratum of mammalian epidermis. Biochim Biophys Acta 491:193–204
Dale BA, Holbrook KA, Steinert PM (1978) Assembly of stratum corneum basic protein and keratin filaments in macrofibrils. Nature 276:729–731
Elias PM, Goerke J, Frend DS (1977) Mammalian epidermal barrier layer lipids: composition and influence on structure. J Invest Dermatol 69:535–546
Elias PM, Bonar L, Grayson S, Baden HP (1983) X-ray diffraction analysis of stratum corneum membrane couplets. J Invest Dermatol 80:213–214
Etoh H, Taguchi YH, Tabachnick J (1975) Movement of beta-irradiated epidermal basal cells to the spinous-granular layers in the absence of cell division. J Invest Dermatol 64:431–435
Fleckman P, Dale BA, Holbrook KA (1985) Profilaggrin, a high-molecular-weight precursor of filaggrin in human epidermis. J Invest Dermatol 85:507–512
Fraser RDB, MacRae TP, Rogers GE (1972) In: Kugelmass IN (ed) Keratins, their composition, structure and biosynthesis. Thomas, Springfield, pp 1–247
Fukuyama K, Epstein WL (1966) Epidermal keratinization: localization of isotopically labelled amino acids. J Invest Dermatol 47:551–560
Fukuyama K, Epstein WL (1975) A comparative autoradiographic study of keratohyalin granules containing cysteine and histidine. J Ultrastruct Res 51:314–325
Fukuyama K, Wier KA, Epstein WL (1972) Dense homogeneous deposits of keratohyalin granules in newborn rat epidermis. J Ultrastruct Res 38:16–26
Fukuyama K, Murozuka T, Caldwell R, Epstein WL (1978) Divalent cation stimulation of in vitro fibre assembly from epidermal keratin protein. J Cell Sci 33:255–263
Gelfant S (1976) The cell cycle in psoriasis. A reppraisal. Br J Dermatol 95:577–590
Gilmartin ME, Culbertson VB, Friedberg IM (1980) Phosphorylation of epidermal keratins. J Invest Dermatol 75:211–216
Gray GM, Yardley HJ (1975) Different populations of pig epidermal cells: isolation and lipid composition. J Lipid Res 16:441–450
Green H, Sun TT (1978) Keratin filaments of cultured human epidermal cells. J Biol Chem 253:2053–2060
Harding CR, Scott IR (1983) Histidine-rich proteins (filaggrins): structural and functional heterogeneity during epidermal differentiation. J Mol Biol 170:651–673
Jepsen H (1970) Two types of keratohyalin granules. J Ultrastruct Res 33:95–115
Kubilus J, Waitkus R, Baden HP (1979) The presence of citrulline in epidermal proteins. Biochim Biophys Acta 581:114–121
Kubilus J, Waitkus R, Baden HP (1980) Partial purification and specificity of an arginine-converting enzyme from bovine epidermis. Biochim Biophys Acta 615:246–251

Kubilus J, Scott I, Harding CR, Yendle J, Kvedar J, Baden HP (1985) The occurrence of profilaggrin and its processing in cultured keratinocytes. J Invest Dermatol 85:513–517
Lee LD, Baden HP (1976) Organisation of the polypeptide chains in mammalian keratin. Nature 264:377–378
Lee LD, Fleming BC, Waitkus RF, Baden HP (1975) Isolation of the polypeptide chains of prekeratin. Biochim Biophys Acta 412:82–90
Lee LD, Kubilus J, Baden HP (1979) Intraspecies heterogeneity of epidermal keratins isolated from bovine hoof and snout. Biochem J 177:187–196
Lonsdale-Eccles JD, Hangen JA, Dale BA (1980) A phosphorylated keratohyalin derived precursor of epidermal stratum corneum basic protein. J Biol Chem 255:2235–2238
Lynley AM, Dale BA (1983) The characterization of human epidermal filaggrin, a histidine-rich, keratin filament-aggregating protein. Biochim Biophys Acta 744:28–35
Ma AS-P, Sun T-T (1986) Differentiation-dependent changes in the solubility of a 195 kd protein in human epidermal keratinocytes. J Cell Biol 103:41–48
Matoltsy AG (1965) Soluble prekeratin. In: Lyne AG, Short BF (eds) Biology of the skin and hair growth. Elsevier, New York, pp 291–305
Matoltsy AG, Matoltsy MN (1966) The membrane protein of horny cells. J Invest Dermatol 46:127–129
Matoltsy AG, Matoltsy MN (1970) Chemical nature of keratohyalin granules of the epidermis. J Cell Biol 47:593–603
Matoltsy AG, Parakkal PF (1967) Keratinization. In: Zelickson AS (ed) Ultrastructure of normal and abnormal skin. Lea and Febiger, Philadelphia, p 97
Meek RL, Lonsdale-Eccles JD, Dale BA (1983) Epidermal filaggrin is synthesized on a large messenger ribonucleic acid as a high-molecular weight precursor. Biochemistry 22:4867–4871
Odland G, Reed T (1967) Epidermis. In: Zelickson AS (ed) Ultrastructure of normal and abnormal skin. Lea and Febiger, Philadelphia, pp 62–68
Ogawa H, Goldsmith LA (1976) Human epidermal transglutaminase preparation and properties. J Biol Chem 251:7281–7288
Rice RH, Green H (1977) The cornified envelope of terminally differentiated human epidermal keratinocytes consist of cross-linked protein. Cell 11:417–422
Rice RH, Green H (1979) Presence in human epidermal cells of a soluble protein precursor of the cross-linked envelope: activation of the cross-linking by calcium ions. Cell 18:681–694
Rudall KM (1952) The protein of mammalian epidermis. In: Anson ML, Bailey K, Edsall JT (eds) Advances in protein chemistry, vol 7. Academic, New York, pp 253–290
Schmidt R, Reichert U, Michel S, Shroot B, Bouclier M (1985) Plasma membrane transglutaminase and cornified envelope competence in cultured human keratinocytes. FEBS Lett 186:201–204
Simon M, Green H (1984) Participation of membrane-associated proteins in the formation of the cross-linked envelope of the keratinocyte. Cell 36:827–834
Skerrow D (1974) The structure of prekeratin. Biochem Biophys Res Commun 59:1311–1316
Skerrow D (1977) The isolation and preliminary characterization of human prekeratin. Biochim Biophys Acta 494:477–451
Skerrow CJ, Matoltsy AG (1974a) Isolation of epidermal desmosomes. J Cell Biol 63:515–523
Skerrow CJ, Matoltsy AG (1974b) Chemical characterization of isolated epidermal desmosomes. J Cell Biol 63:524–530
Steinert PM (1978) Structure of three-chain unit and bovine epidermal keratin filament. J Mol Biol 123:49–70
Steinert PM, Gullino MI (1976) Bovine epidermal keratin filament assembly in vitro. Biochem Biophys Res Commun 70:221–227
Steinert PM, Idler WW (1975) Polypeptide composition of bovine epidermal α-keratin. Biochem J 151:603–614

Steinert PM, Idler WW (1979) Postsynthethic modifications of mammalian epidermal α-keratin. Biochemistry 18:5664–5669
Steinert PM, Steven AC, Roop DR (1985) The molecular biology of intermediate filaments. Cell 42:411–419
Sugawara K (1977) Intermolecular cross-links in epidermal differentiation. In: Seiji M, Bernstein IA (eds) Biochemistry of cutaneous epidermal differentiation. University of Tokyo Press, Tokyo, pp 387–397
Thacher SM, Rice RH (1985) Keratinocyte-specific transglutaminase of cultured human epidermal cells: relation to cross-linked envelope formation and terminal differentiation. Cell 40:685–695
Viac J, Staquet MJ, Thivolet J, Goujon C (1980) Experimental production of antibodies against stratum corneum keratin polypeptides. Arch Dermatol Res 267(12):179–188
Weinstein GD, Frost P (1968) Abnormal cell proliferation in psoriasis. J Invest Dermatol 50:254–259
Weinstein GD, Van Scott EJ (1965) Autoradiographic analysis of turnover times of normal and psoriatic epidermis. J Invest Dermatol 45:257–262

CHAPTER 3

Regulation of Epidermal Growth

E. M. SAIHAN

The epidermis is a classical example of a renewing tissue. There is constant shedding of horny cells, these being replaced by new cells. Continuous replacement is of major importance for the proper maintenance of the epidermis. Since skin is the largest organ in the body, this is one of the major cell renewal systems. Moreover, epidermal growth can be increased temporarily over a short period by local injury or over a longer term as in psoriasis. This control of epidermal proliferation and differentiation is a highly complex and integrated process, and it depends on a delicate balance between factors promoting proliferation and factors favouring differentiation. Various chemical regulators have been postulated, including epidermal growth factor, cyclic nucleotides, chalones, prostaglandins, histamines and calcium and calmodulin. In this chapter I shall briefly review these substances.

A. Cyclic Nucleotides

Cyclic AMP and cyclic GMP are regarded as the "second" or intracellular messengers of hormones and chemical agents. They are involved in cellular functions as diverse as cell growth, differentiation, locomotion and membrane transport, but experimental data in one system are not always reproduced in another system because of the diverse action of nucleotides in different biological systems. Hence any statements have to be qualified with details of tissue and species specificity. In 1971 VOORHEES and DUELL suggested that in psoriasis there is a possible defect of the adenyl cyclase and cyclic AMP cascade. Their hypothesis was based on the fact that in psoriasis there is increased cellular proliferation, increased epidermal glycogen concentration and decreased cellular differentiation. In non-epidermal cells, cyclic AMP was known to decrease proliferation and to increase glycogen and differentiation. Since then many studies have been published on cyclic nucleotides in skin, particularly in psoriasis, and at times with contradictory results.

I. Effects of Cyclic AMP on Different Epidermal Cells

Cultured guinea-pig epidermal cells were used to study the effect of agents which increase intracellular cyclic AMP, including dibutyryl cyclic AMP, papaverine, theophylline, isoprenaline and prostaglandin E_2 methyl ester. Treatment for 24 h with dibutyryl cyclic AMP inhibited cell growth by 50%–95%, whereas butyrate showed essentially no effect. Papaverine, theophylline, isoprenaline and prostaglandin E_2 methyl ester all produced a dose-dependent growth inhibition (DELESCLUSE et al. 1974). Cyclic AMP added to briefly cultured epidermis of the mouse was found to inhibit cell division (MARKS and REBIEN 1972a; VOORHEES et al. 1972). β-Adrenergic stimulation resulted in an increase in cyclic AMP content of the isolated epidermis along with a reduction in the number of mitoses (POWELL et al. 1971).

GREEN (1978) looked at the effect of four agents known to increase the level of cellular cyclic AMP by different means (cholera toxin, dibutyryl cyclic AMP, methyl isobutyl xanthine and isoprenaline) and found that in the presence of supporting 3T3 cells, they increased the growth of colonies of cultured human epidermal cells and keratinocytes. With the addition of epidermal growth factor (EGF) most of the agents exerted an effect of considerable magnitude. Cholera toxin could exert an effect in the absence of supporting 3T3 cells and EGF. Using monolayer culture of neonatal mouse epidermal cells, MARCELO (1979) showed that the addition of dibutyryl cyclic AMP for 4 days increased the proliferative rate of cells in culture.

Using primary culture of adult human keratinocytes as described by LIU and KARASEK (1978), MARCELO and DUELL (1979) investigated the effect of 8-bromo cyclic AMP on proliferation and differentiation. 8-Bromo cyclic AMP decreased the incorporation of tritiated thymidine into DNA, whereas addition of cholera toxin to the culture resulted in an early stimulation of incorporation of tritiated thymidine into DNA, the effect of which was lost after 6 days. 8-Bromo cyclic GMP had no effect on the proliferative rate of culture. Thus it seems that there is a divergent response of adult culture to two different cyclic AMP elevating agents. Working on outgrowths of epidermal cells from explants, HARPER et al. (1974a) found that cyclic AMP, dibutyryl cyclic AMP and adenine nucleotides inhibited mitosis by 50%–60% whereas adenine, guanine, cytosine and uridine nucleotides did not inhibit mitosis. Epinephrine inhibited mitosis by 50%–70% in the epidermal outgrowth obtained from either normal skin or psoriatic lesions (HARPER et al. 1974b). MARCELO and TOMICH (1983) studied the effect of cyclic AMP analogues (8-bromo cyclic AMP and dibutyryl cyclic AMP) and cholera toxin on epidermal basal cells trypsinised from neonatal mouse and adult neonatal human skin, grown on plastic or on gelled collagen surfaces. Greatly increased intracellular cyclic AMP levels, that is 60- to 70-fold, stimulated neonatal mouse keratinocyte proliferation and differentiation, but the same dose was cytotoxic to both neonatal and adult human cells. However, a modest increase in intracellular cyclic AMP did stimulate adult human keratinocyte proliferation. Triamcinolone acetonide inhibited 8-bromo cyclic AMP stimulated proliferation in a dose-dependent manner. Likewise, etretinate and 13-*cis*-retinoic acid inhibited cyclic AMP stimulated proliferation and cholera toxin induced cyclic AMP synthesis.

OKADA et al. (1982) found that in the culture of epidermal keratinocytes, when a small number (10^5) of cells were inoculated in a 60×15 mm culture dish, cholera toxin strongly stimulated colony growth, while when a relatively larger number (8×10^5) of cells were inoculated in a dish, cholera toxin moderately accelerated cell division and increased DNA and protein level of the culture during early days of cultivation. However, after about 20 days of cultivation, when the culture reached confluence, cholera toxin decreased both DNA and protein content in the culture dish. The number of cells in the culture media and the age of the cultured cells may account for some of the contradictory results. It is possible that in a crowded culture or in a sheet in which keratinocytes are connected tightly and stratified, cyclic AMP and cyclic AMP elevating agents suppress the proliferation of keratinocytes, whereas in the sparse culture, cyclic AMP stimulates their proliferation.

II. Cyclic GMP

In dermatology cyclic AMP has received more attention than cyclic GMP. The major difficulties faced in the study of cyclic GMP are (a) the extremely small amounts present in epidermis and (b) the so-called ischaemic effect, which lowers the endogenous cyclic GMP level further. When tissue is traumatised, for example by injection or removal, there are immediate marked changes in the cyclic nucleotide level (STEINER et al. 1972). In contrast to the rapid decline in the cyclic GMP level, ischaemia causes a rapid and transient increase in epidermal cyclic AMP (YOSHIKAWA et al. 1975a; ADACHI et al. 1980; IIZUKA et al. 1979). The decrease in cyclic GMP is probably due to leakage of cyclic GMP extracellularly and sudden activation of epidermal cyclic GMP phosphodiesterase.

There are very few studies of the cyclic GMP system in skin. Three studies (VOORHEES et al. 1973; MARCELO et al. 1979; SAIHAN et al. 1980) have shown an increased level of cyclic GMP in psoriasis, but another study showed that the addition of cyclic GMP or dibutyryl cyclic GMP to culture media did not significantly influence the mitotic rate of epidermal outgrowth (FLAXMAN and HARPER 1975).

III. Receptors

In the cyclic AMP system in the epidermis, four different classes of receptor have been identified. The catecholamine receptor is a β-adrenergic receptor. In mouse and pig epidermis, both in slice and in homogenate, the increase in cyclic AMP content could be blocked by the addition to the incubation mixture of a β-antagonist, propranolol, but not by the addition of an α-antagonist, phentolamine (MARK and REBIEN 1972b; YOSHIKAWA et al. 1975b; DUELL 1980). The β-adrenergic receptors present are β_2 in nature (DUELL 1980), since the β_2-agonist salbutamol produced a dose-dependent increase in cyclic AMP content of the tissue. The other receptors include histamine, prostaglandin and adenine nucleotides. The histamine receptor is of the H_2 type, since its stimulation is blocked by specific antagonists such as cimetidine and metiamide (IIZUKA et al. 1976a, b). ADACHI et al. (1975) have shown that in human epidermis in keratotome sheets, addition of PGE_1 and PGE_2 (but not PGA and $PGF_{2\alpha}$) produced accumulation of cyclic AMP. Similar findings were noted by ASO et al. (1975) in guinea-pig epidermis. WILKINSON and ORENBERG (1979), in primary culture of guinea-pig epidermal cells, noted that maximum stimulation of cyclic AMP levels was associated with PGD_2; a smaller increase occurred with PGE_2 and a relatively transient rise with $PGF_{2\alpha}$, PGD_1, PGE_1, and $PGF_{1\alpha}$. Significant increases in cyclic GMP were immediately observed with PGD_2 and PGE_2. With $PGF_{2\alpha}$, maximum levels were noted after some delay, but unlike in the pig, guinea-pig and human, PGE_1 had no effect on the cyclic AMP level of mouse epidermis (GARTE and BELMAN 1983). The adenosine receptor is also responsive to adenosine 5′-mono-,-di- and -triphosphate but adenine is ineffective (IIZUKA et al. 1976b). Guanylate cyclase system is activated by both H_1- and H_2-receptors (IIZUKA et al. 1979; AOYAGI et al. 1980). It was also noted that addition of 0.5 μg/ml of EGF to slices of pig epidermis produced a significant increase in cyclic GMP levels after 40–60 min of incubation; however, the peak stimulation occurred after 6 h of treatment (AOYAGI et al. 1980).

B. Prostaglandins

Studies on epidermal regulation by prostaglandins are limited, although there have been a few studies on the changes of prostaglandins in psoriatic epidermis. In the preceding section the prostaglandins, particularly E_1 and E_2, have already been mentioned in connection with cyclic AMP and membrane receptors. KISCHER (1967) first observed that the prostaglandins of the E series stimulated epidermal proliferation and keratinisation in chick embryo cultures. BEM and GREAVES (1974) noted that addition of 0.1–20 μg PGE_1 per ml to the culture media of mouse embryo epidermis produced stimulation of DNA synthesis after 1–4 h. EAGLSTEIN and WEINSTEIN (1975), by employing an autoradiographic technique, showed that PGE_2 increased the number of DNA-synthesising cells in human forearm epidermis, whereas HARPER (1976) found that addition of prostaglandin E and F series to the culture medium of human epidermal cells resulted in 35%–87% inhibition of ^{3}H-thymidine uptake. Topical application of synthetic analogue of PGE_2 to the epidermis of hairless mice was studied by LOWE and STOUGHTON (1977): 1 μg of PGE analogue increased the DNA synthesis significantly by 5 h and a maximum increase of 360% was reached by 12 h, returning to the control level at 24 h. Whilst 20 μg of PGE_2 analogue reduced the epidermal DNA synthesis for 12 h after application, DNA synthesis was increased at 24 h, returning to control level at 48 h. No significant effect on epidermal DNA synthesis over 48 h was produced by 100 μg of topical PGE_2, but 1 μg of intradermal PGE_2 increased DNA synthesis by 160% at 24 h. These results suggest that PGE_2 analogues are biologically active compared with PGE_2.

BENTLEY-PHILLIPS et al. (1977) looked at the effects of intradermal injection of PGE_1 and $PGF_{2\alpha}$ on the growth of guinea-pig epidermis and found that 5–20 μg of PGE_1 caused a significant increase in autoradiographic labelling indices at 48 h. This increase was still present at 72 h and returned to normal at 96 h, but $PGF_{2\alpha}$ had no consistent effect. In an earlier study ZIBOH and HSIA (1972) found that in the essential fatty acid deficient mouse, topical application of PGE_2 led to clearing of scaling lesions. Linoleic acid, which increases PGE_2, also produced decreased epidermal proliferation in the essential fatty acid deficient mouse (LOWE and DEQUOY 1978). KRAGBALLE et al. (1985) studied the capacity of leukotrienes to stimulate the DNA synthesis of cultured human epidermal keratinocytes and noted that at a concentration ranging from 10^{-12} to 10^{-8} M, LTB_4

produced a 100% increase in DNA synthesis. PENTLAND and NEEDLEMAN (1986) thought that the conflicting reports of prostaglandins and epidermal proliferation could be attributed to several factors: (a) the wide variety of doses and types of prostaglandin selected for studies, (b) variations in the time interval of treatment and (c) variations in animal models and culture conditions. In their study they found that there was a fourfold higher increase in the synthesis of PGE_2 in non-confluent than in confluent keratinocyte cultures.

Moreover, confluent, unstratified cultures proliferate as fast as non-confluent cultures but synthesise much less PGE_2, indicating that whilst confluence modulates PGE_2 synthesis, it does not directly modulate proliferation.

C. Epidermal Growth Factor

Cellular proliferation in vivo and in vitro is controlled by several hormones and growth factors which are present in serum and in tissue fluids. In 1960, during the course of a study on the nerve growth promoting protein of the submaxillary gland of the mouse, COHEN noted that when this protein was injected into new-born animals, it produced developmental changes which could not be related to nerve growth factor. These changes included precocious opening of the eyelids (7 days instead of the usual 12–14 days) and precocious eruption of the teeth (at 6–7 days instead of the normal 8–10 days). In 1962 COHEN reported the isolation of a heat stable, non-dialysable, antigenic protein which was termed epidermal growth factor.

I. Chemical Composition and Properties

It was found that the substance was an antigenic protein, non-dialysable and heat stable. The biological activity was destroyed by incubation with chymotrypsin or bacterial proteinase. Distinctive chemical characteristics included the absence of lysine and phenylalanine (COHEN 1962).

TAYLOR et al. (1972) reported the major physical and chemical properties of the mouse EGF. Molecular weight determined by sedimentation equilibrium was 6400 and by amino acid composition, 6045.

Epidermal growth factor was found to be composed of a single polypeptide chain comprising 53 amino acid residues with the absence of three specific amino acid residues, lysine, alanine and phenylalanine. There were no detectable free sulphydryl groups, hexosamines or neutral sugar. EGF had an isoelectric point of pH 4.6. However, in crude homogenates of the submaxillary gland of male mouse (TAYLOR et al. 1970, 1974) EGF activity was found almost entirely in a high molecular weight complex. This complex had a molecular weight of 74000 and was composed of two molecules of EGF (6045 molecular weight) and two molecules of binding protein (29300 molecular weight) and showed arginyl esterase activity. The complex was stable at the pH range of 5–8. The biological activity of the high molecular weight complex EGF was found to be approximately one-sixth that of an equal weight of low molecular weight EGF. EGF binding protein did not affect the biological activity of the low molecular weight EGF, suggesting that EGF binding protein is not directly involved in the regulation of the activity of the biologically active low molecular weight EGF. In contrast to the apparent lack of relationship between salivary gland EGF levels and basal plasma EGF concentrations, there was a direct correlation with the response of plasma RIA-EGF to adrenergic stimulation. The increase in plasma EGF in response to α-adrenergic stimulation was abolished by excision of the submaxillary glands, but basal plasma and urinary levels were unchanged (BYYNY et al. 1974).

II. Human EGF

In 1975 STARKEY et al. detected EGF in the urine of pregnant and non-pregnant women and in men. GREGORY (1975) showed the similarity in the amino acid composition of EGF and urogastrone, and suggested that they were probably the same substance. Subsequently the presence of EGF was also reported by GREGORY et al. (1977) and DAILEY et al. (1978). HIRATA et al. (1980) reported the human plasma level of EGF; they did not find any significant difference between the sexes and there was no circadian periodicity.

III. Level of EGF

In the mouse, in plasma the level of EGF is between 1 and 300 ng/ml in milk and about 1000 ng/ml in saliva and urine (BYYNY et al. 1974; HIRATA and ORTHO 1979 a). In adult male mice, EGF levels are much higher in the submaxillary gland but surgical ablation of the submaxillary gland does not reduce the plasma levels of EGF (BYYNY et al. 1974).

In man the EGF level in saliva is between 5 and 17 ng/ml and in milk about 80 ng/ml; plasma IRh EGF was found to be 163.4 ± 13.7 (male) and 138.5 ± 6.9 (female) pg/ml (HIRATA et al. 1980). Twenty-four hour urinary excretion of hEGF using radioimmunoassay was found to be between 52–63 μg/g creatinine (STARKEY and ORTH 1977) and 28–40 μg/g creatinine (DAILEY et al. 1978). The linear correlation with urinary creatinine concentrations in each sample suggests that the body mass is more important than the sex difference. However, excretion by females taking an oral contraceptive was significantly greater (DAILEY et al. 1978). The site or sites of synthesis or storage of hEGF are not known. HIRATA and ORTHO (1979 b) showed low levels of EGF in submandibular gland, duodenum, pancreas, jejunum, thyroid and kidney.

IV. EGF Receptor

Evidence indicates that the initial interaction of EGF with target cells occurs on specific membrane receptors (HOLLENBERG and CUATRECASAS 1973; COHEN et al. 1975). A study with ^{125}I-labelled EGF showed the presence of specific and high affinity receptor molecules on the surface of responsive cell types (CARPENTER and COHEN 1979). The fate of EGF after binding the cell surface receptors was investigated by several laboratories using different techniques. These included fluorescence microscopy and fluorescence photobleaching recovery (SCHLESSINGER et al.

1978a, b; HAIGLER et al. 1978), electron microscopy to follow the fate of ferritin EGF (HAIGLER et al. 1979), colloidal gold avidin EGF (HOPKINS et al. 1981) and electron microscopic autoradiography (GORDON et al. 1978). From these studies it seems that the EGF becomes internalised by a process called receptor-mediated endocytosis. EGF binds to diffusely distributed, laterally mobile membrane receptors which rapidly cluster in a temperature-dependent process and become endocytosed and degraded by lysosomal enzymes.

V. EGF in Cell Proliferation and Differentiation

In vivo EGF has been shown to increase the number of mitotic activity in the epidermis of new-born mice and rats (BIRBAUM et al. 1976). EGF can also increase the dry weight, DNA and RNA content of the epidermis, (ANGELETTI et al. 1964), the disulphide content of the mouse epidermis (FRATI et al. 1972a) and the activity of the epidermal enzyme ornithine decarboxylase (STASNY and COHEN 1970; BLOSSE et al. 1974); in addition it stimulates release of arachidonic acid in pig epidermis (AOYAGI et al. 1985). COVELLI et al. (1972) showed that within 3 h of injecting ^{131}I-labelled EGF there was 300% concentration of EGF in the epidermis of rat and rabbits compared with blood. Topical application of EGF on the wounded cornea in rabbit produced epithelial hyperplasia (FRATI et al. 1972a; SAVAGE and COHEN 1973). DANIELE et al. (1979) also reported that there was accelerated healing of wounded human cornea provided the damage did not disrupt the integrity of the corneal stroma. However, GREAVES (1980) noted lack of effect of topically applied EGF on epidermal growth in man. NANNEY et al. (1984) showed an inverse relationship between the number of EGF receptors and the degree of epidermal differentiation and/or keratinisation, which may also suggest a physiological role of EGF in human epidermis.

D. Chalones of the Skin

Under normal conditions, the epidermis maintains a constant thickness. Local injury leads to an increased proliferation of the cells in the basal layer, followed by a transient period of hyperplasia. It is likely that there is a negative feedback mechanism to control this proliferation. Following injury, the healing process starts locally, suggesting that the signal of regeneration is local rather than blood borne (IVERSEN 1969). In 1962 BULLOUGH adopted the word "chalone" (derived from the Greek word *chalao*, meaning to slacken or to lower) to describe tissue-specific antimitotic substances extracted from the epidermis. In 1964 BULLOUGH and LAURENCE partially purified a water-soluble thermolabile and non-dialysable factor that inhibited the epidermal mitotic rate both in vitro and in vivo. Chalones have been shown to a greater or lesser extent in several tissues other than the epidermis, for example melanocytes, granulocytes, erythrocytes, lymphocytes, fibroblasts, liver, kidney and intestinal epithelium (for reference see IVERSON 1976). Chalones have also been demonstrated by CHOPRA and colleagues (1972). Both G1 inhibitor (DNA synthesis) and G2 inhibitor (mitosis) have been partially purified (ELGJO et al. 1971; MARKS 1971, 1973). Iverson's conclusions (1976) are now accepted by many workers in the field:

1. Chalones are produced in and present in the tissues on which they selectively act.
2. Other properties of chalone include water solubility, tissue specificity but species non-specificity, and reversibility of action.

Chalone action is short-lived, which may be due to the presence of chalone antagonist. Whether chalones work on cell membranes or intercellularly is not yet known. It was suggested that the action of chalone could be related to cyclic AMP.

The idea of an epidermal chalone is attractive, but the failure to develop a totally purified chalone has led to difficulty in establishing the credibility of the chalone concept. This was further impaired when MOHR et al. (1972) discovered that the extracts of melanoma, which had demonstrated a melanocytic effect in vivo, were contaminated by bacterial spores. The biological effect of these clostridia spores in vivo could more than account for the results obtained on the regression of melanoma in experimental animals. Even though DEWEY (1973) showed that sterile melanoma extracts were able to inhibit the proliferation of melanoma cell in culture, this did not restore confidence in the chalone concept and as a result the work on chalones has not been vigorously pursued.

E. Calcium and Calmodulin

Calcium is involved in the regulation of several different enzyme systems and cell motility (RASMUSSEN et al. 1976; DEDMAN et al. 1979). Moreover, Ca^{2+} might not act in its free ionic form but rather require the presence of a binding protein. In 1972 WOLF and SIEGEL purified a Ca^{2+} binding protein from bovine brain. The calcium binding protein is now commonly known as calmodulin. Calmodulin is a 16700 molecular weight acidic protein and has been identified in the skin (IIZUKA et al. 1982; MURRAY and ROGERS 1978; PETERSON and WUEPPER 1983). The calcium–calmodulin complex influences enzymes such as adenylcyclase, phosphodiesterase, protein kinases and phospholipase A_2 (MEANS and DEDMAN 1980; WIGHTMAN et al. 1982) and physiological processes such as endocytosis (SALISBURY et al. 1981).

VAN DE KERKHOF and VAN ERP (1983) found calmodulin content of the psoriatic lesions to be almost 30 times higher than the normal controlled and uninvolved skin in psoriatic patients. However, the level of the uninvolved psoriatic skin was not significantly different from the normal controls. Raised levels of calmodulin in psoriatic skin were also noted by TUCKER et al. (1984), MIZUMOTO et al. (1985), and FAIRLEY et al. (1985). But like many other chemicals, calmodulin most likely plays an indirect role in normal epidermal regulations.

F. Histamine

IIZUKA et al. (1979) studied the effect of histamine in pig epidermis and found that there was a rapid increase in the intracellular cyclic GMP level; maximum accumulation occurred at 1 min after the addition of histamine and the level remained relatively high for 5 min. The level of cyclic AMP, by contrast, continuously increased and reached a peak in 5 min. The other difference that was noted was the fact that the histamine caused only a two- to fourfold accumulation of cyclic GMP, compared with a 20-fold accumulation of cyclic AMP. Iizuka et al. also noted that the histamine-adenylate cyclase system has strictly H_2 type receptors, whereas the histamine guanylate cyclase system had both H_1 and H_2, but predominantly H_1 receptors. FLAXMAN and HARPER (1975), using human epidermal outgrowths, showed that histamine (10^{-2}–10^{-6} *M*) has a definite inhibitory effect on mitosis, whereas VOORHEES et al. (1972) found stimulation of mitosis in rat skin.

The bulk of the histamine in the skin resides in the mast cell granules in the dermis (RILEY and WEST 1953). The histamine content of the dermis is also substantial, at 10–20 µg/g wet weight (ROTHMAN 1954). However, the lack of epidermal changes at the site of previous weals in an urticarial rash, where there is a dramatic release of histamine, raises doubt about the effect of histamine on epidermal growth.

G. Conclusion

Despite many efforts, mechanisms regulating epidermal cell proliferation and differentiation are yet to be elucidated. We are still far from a precise understanding of the physiological role played by hormones like glucocorticoids, catecholamines, EGF and other related compounds like calcium and secondary signals like cyclic nucleotides. Lack of a proper animal model for the human skin has led to the study of different species which at times has produced contradictory results. Still, there is ample evidence that both local and systemic factors act in controlling epidermal growth.

Although EGF has been recognised for two decades and is found in both human blood and urine, very little is known about its changes in different keratinisation disorders. Moreover, studies on the effect of EGF on human keratinocytes have been very limited. A fair amount of work has been done on cyclic nucleotides, particularly cyclic AMP, but as the studies were carried out in different species and tissues, no clear picture has emerged. From the data available it is likely that cyclic nucleotides are indirectly involved in disorders of keratinisation. Elucidation of the controlling factors of the nucleotides should give a better understanding of epidermal cellular regulation. So-called chalones have been in the field of epidermal regulation for about 25 years, but it seems they have been completely forgotten in recent years. A negative feedback mechanism has to exist but whether it is carried out by a specific chemical is not known. It is possible that cyclic nucleotides working as a second messenger act as both a positive and a negative controlling factor. Prostaglandins have gained importance in many fields and their involvement in inflammatory processes has led to the introduction of several anti-prostaglandin compounds.

Prostaglandin most likely plays an important role in inflammatory disorders. Its role in controlling epidermal growth regulation is probably indirect, through epidermal receptors.

A consolidated effort by the different groups interested in the control mechanism of epidermal growth is long overdue. Without this basic understanding of epidermal growth regulation, progress in disorders of keratinisation will be very slow.

References

Adachi K, Yoshikawa K, Halprin K, Levine V (1975) Prostaglandins and cyclic AMP in epidermis. Br J Dermatol 92:381–388

Adachi K, Iizuka H, Halprin K, Levine V (1980) Epidermal cyclic AMP is not decreased in psoriatic lesions. J Invest Dermatol 74:74–76

Angeletti PU, Salvi ML, Chesanow RL, Cohen S (1964) Azione dell epidermal growth factor – sulla sintesti di acidi nucleicic apriteine dell'epithelio cutaneo. Experientia 20:1–6

Aoyagi T, Adachi K, Halprin K, Levine V (1980) The effects of epidermal growth factor on the cyclic nucleotide system in pig epidermis. J Invest Dermatol 74:238–241

Aoyagi T, Suya H, Kato N, Nemoto O, Kobayashi H, Miura Y (1985) Epidermal growth factor stimulates release of arachidinic acid in pig epidermis. J Invest Dermatol 84:168–171

Aso K, Orenberg EK, Farber EM (1975) Reduced epidermal cyclic AMP accumulation following prostaglandin stimulation: its possible role in the pathology of psoriasis. J Invest Dermatol 65:375–378

Bem JL, Greaves MW (1974) Prostaglandin E_1 effects on epidermal cell growth "in vitro". Arch Dermatol Forsch 251:35–41

Bentley-Phillips CB, Paulli-Jogensen H, Marks R (1977) The effects of prostaglandin E_1 and $F_{2\alpha}$ on epidermal growth. Arch Dermatol Res 257:233–237

Birbaum JE, Sapp TM, Moore JB (1976) Effects of reserpine, epidermal growth factor and cyclic nucleotide modulators in epidermal mitosis. J Invest Dermatol 66:313–318

Blosse PT, Fenton EL, Henningsson S, Kahlson G, Rogengren E (1974) Activities of decarboxylase of histidine and ornithine in young male mice after injection of epidermal growth factor. Experientia 30:22–23

Bullough WS (1962) The control of mitotic activity in adult mammalian tissues. Biol Rev 37:307–342

Bullough WS, Lawrence EB (1964) Mitotic control by internal secretion: the role of chalone–adrenalin complex. Exp Cell Res 33:176–194

Byyny RL, Orth DN, Cohen S, Doyne ES (1974) Epidermal growth factor: effects on androgen and andrenergic agents. Endocrinology 95:776–782

Carpenter G, Cohen S (1979) Epidermal growth factor. Annu Rev Biochem 48:193–216

Chopra DP, Yu RJ, Flaxman BA (1972) Demonstration of a tissue-specific inhibitor of mitosis of human epidermal cells in vitro. J Invest Dermatol 59:207–210

Cohen S (1960) Purification of a nerve growth promoting protein from the mouse salivary gland and its neuro-cytotoxic antiserum. Proc Natl Acad Sci USA 46:302–311

Cohen S (1962) Isolation of a mouse submaxillary gland protein accelerating incisor eruption and eyelid opening in the new-born animal. J Biol Chem 237:1555–1562

Cohen S, Carpenter G, Lembach KJ (1975) Interaction of epidermal growth factor with cultured fibroblasts. Adv Metab Disord 8:265–284

Covelli I, Rossi R, Mozzi R, Frati L (1972) Synthesis of bioactive ^{131}I-labelled epidermal growth factor and its distribution in rat tissues. Eur J Biochem 27:225–230

Dailey GE, Krana JW, Orth DN (1978) Homologans radioimmunoassay for human epidermal growth factor (urogastrone). J Clin Endocrinol Metab 46:929–936

Daniele S, Frati L, Fiore C, Santoni G (1979) The effect of the epidermal growth factor (EGF) on the corneal epithelium. Graefes Arch Clin Exp Opthalmol 210:159–165
Dedman JR, Brinkley BR, Means AR (1979) Regulation of microfilaments and microtubules by calcium and cyclic AMP. Adv Cyclic Nucleotide Res 11:131–174
Delescluse C, Colburn NH, Duell EA, Voorhees JJ (1974) Cyclic AMP elevating agents inhibit proliferation of keratinizing guine pig epidermal cells. Differentiation 2:343–350
Dewey DL (1973) The melanocytic chalone. In: Forscher BK, Houck JC (eds) Chalones: concepts and current researches. Nat Cancer Inst Monogr 38:213–216
Duell E (1980) Identification of a beta$_2$-adrenergic receptor in mammalian epidermis. Biochem Pharmacol 29:97–101
Eaglstein WH, Weinstein GD (1975) Prostaglandin and DNA synthesis in human skin. Possible relationship to ultraviolet light effects. J Invest Dermatol 64:386–389
Elgjo K, Henning SH, Edgehill W (1971) Epidermal mitotic rate and DNA synthesis after infection of water extracts made from mouse skin treated with actinomycin D: two or more growth-regulating substances? Virchows Arch [B] 7:342–347
Fairley JA, Marcelo CL, Hogan VA, Voorhees JJ (1985) Increased calmodulin levels in psoriasis and low Ca^{++} regulated mouse epidermal keratinocyte cultures. J Invest Dermatol 84:195–198
Flaxman BA, Harper RA (1975) In vitro analysis of the control of keratinocyte proliferation in human epidermis by physiologic and pharmacologic agents. J Invest Dermatol 65:52–59
Frati C, Covelli I, Mozzi R, Frati L (1972a) Mechanism of action of epidermal growth factor: effect on the sulfhydryl and disulfide group content of mouse epidermis during keratinization. Cell Differ 1:239–244
Frati L, Daniele S, Delogu A, Covelli I (1972b) Selective binding of the epidermal growth factor and its specific effects on the epithelial cells of the cornea. Exp Eye Res 14:135–141
Garte SJ, Belman S (1983) Prostaglandins fail to elevate cyclic AMP levels in mouse epidermis in vivo and in vitro. J Invest Dermatol 81:422–423
Gordon P, Carpenter J, Cohen S, Orci L (1978) Epidermal growth factor: morphological demonstration of binding internalization and lysosomal association in a human fibroblast. Proc Natl Acad Sci USA 75:5025–5029
Greaves MW (1980) Lack of effect of topically applied epidermal growth factor (EGF) on epidermal growth in man in vivo. Clin Exp Dermatol 5:101–103
Green H (1978) Cyclic AMP in relation to proliferation of the epidermal cell: a new view. Cell 15:801–811
Gregory H (1975) Isolation and structure of urogastrone and its relationship to epidermal growth factor. Nature 257:325–327
Gregory H, Bower JM, Willshire IR (1977) Urogastone and epidermal growth factor. In: Kastrup KW, Nielsen JH (eds) Growth factors. Pergamon, Elmsford, pp 75–84
Haigler HT, Ash JF, Singer SJ, Cohen S (1978) Visualization by fluorescence of the binding and internalization of epidermal growth factor in human carcinoma cell A-431. Proc Natl Acad Sci USA 75:3317–3321
Haigler HT, McKanna JA, Cohen S (1979) Direct visualization of the binding and internalization of a ferritin conjugate of epidermal growth factor in a human carcinoma cell A-431. J Cell Biol 81:382–395
Harper RA (1976) Effect of prostaglandins on ^{3}H-thymidine uptake into human epidermal cells in vitro. Prostaglandins 12:1019–1025
Harper RA, Flaxman A, Chopra D (1974a) Effect of pharmacological agents on human keratinocyte mitosis in vitro. 1. Inhibition by adenine nucleotides. Proc Soc Exp Biol Med 146:1032–1036
Harper RA, Flaxman B, Chopra D (1974b) Mitotic response of normal and psoriatic keratinocytes in vitro to compounds known to effect intracellular cyclic AMP. J Invest Dermatol 62:384–387
Hirata Y, Orth DN (1979a) Concentrations of epidermal growth factor, nerve growth factor, and submandibular gland renin in male and female mouse tissue and fluids. Endocrinology 105:1382–1387

Hirata Y, Orth DN (1979b) Epidermal growth factor (urogastrone) in human tissues. J Clin Endocrinol Metab 48:667–672
Hirata Y, Moore GM, Bertagna C, Orth DN (1980) Plasma concentration of immunoreactive human epidermal growth factor (urogastrone) in man. J Clin Endocrinol Metab 50:440–444
Hollenberg MD, Cuatrecasas P (1973) Epidermal growth factor: receptors in human fibroblasts and modulation by cholera toxin. Proc Natl Acad Sci USA 70:2964–2968
Hopkins CR, Boothroyd B, Gregory H (1981) Early events following the binding of epidermal growth factor to surface receptors on ovarian granulosa cells. Eur J Cell Biol 24:259–265
Iizuka H, Adachi K, Halprin K, Levine V (1976a) Histamine H_2 receptor-adenylate cyclase system in pig skin (epidermis). Biochim Biophys Acta 437:150–157
Iizuka A, Adachi K, Halprin K, Levine V (1976b) Adenosine and adenine nucleotides stimulation of skin (epidermis) adenylate cyclase. Biochim Biophys Acta 444:685–693
Iizuka H, Adachi K, Aoyagi T, Halprin K, Levine V (1979) Cyclic GMP system in epidermis II histamine stimulates cyclic GMP formation. J Invest Dermatol 73:313–316
Iizuka H, Ischizawa H, Koizumi H, Aoyagi T, Miura Y (1982) Pig skin epidermal calmodulin: effect of calmodulin deficient phosphodiesterase. J Invest Dermatol 78:230–233
Iversen OH (1969) Chalones of the skin in homeostatis regulators. In: Wolstenholme GE, Knight J (eds) Ciba Foundation symposium on homeostatic regulators. Churchill, London, pp 29–53
Iversen OH (1976) The history of chalones in chalones. In: Houck JC (ed) North-Holland, pp 37–69
Kischer CW (1967) Effects of specific prostaglandins on development of chick embryo skin and down feather organ in vitro. Dev Biol 16:203–215
Kragballe K, Desjarlais L, Voorhees JJ (1985) Leukotrienes B_4, C_4, and D_4 stimulate DNA synthesis in cultured human epidermal keratinocytes. Br J Dermatol 113:43–52
Liu S, Karasek M (1978) Isolation and growth of adult human epidermal keratinocytes in cell culture. J Invest Dermatol 71:157–162
Lowe NJ, DeQuoy P (1978) Linoleic acid effects on epidermal DNA synthesis and cutaneous prostaglandin levels in essential fatty acid deficiency. J Invest Dermatol 70:200–203
Lowe NJ, Stoughton R (1977) Effects of topical prostaglandins E_2 analogue on normal hairless mouse epidermal DNA synthesis. J Invest Dermatol 68:134–137
Marcelo C (1979) Differential effects of cyclic AMP and cyclic GMP on in vitro epidermal cell growth. Exp Cell Res 120:201–210
Marcelo C, Duell EA (1979) Cyclic AMP stimulates and inhibits adult human epidermal cell growth. J Invest Dermatol 72:279
Marcelo CL, Tomich J (1983) Cyclic AMP, glucocorticoid and retinoid modulation of in vitro keratinocyte growth. J Invest Dermatol 81:64s–68s
Marcelo C, Duell E, Stawiski M, Anderson T, Voorhees J (1979) Cyclic nucleotide levels in psoriatic and normal keratomed epidermis. J Invest Dermatol 72:20–24
Marks F (1971) Direct evidence of two tissue-specific chalone-like factors regulating mitosis and DNA synthesis in mouse epidermis. Hoppe-Seylers Z Physiol Chem 353:1273–1274
Marks F (1973) A tissue specific factor inhibiting DNA synthesis in mouse epidermis. Natl Cancer Inst Monogr 38:79–90
Marks F, Rebien W (1972a) Cyclic 3-5 AMP and theophylline inhibit epidermal mitosis in G_2 phase. Naturwissenschaften 59:41–42
Marks F, Rebien W (1972b) The second messenger system of mouse epidermis. 1. Properties and beta adrenergic activation of adenylate cyclase in vitro. Biochim Biophys Acta 284:556–567
Means AR, Dedman JR (1980) Calmodulin and intracellular calcium receptor. Nature 285:73–77
Mizumoto T, Hashimoto Y, Hirokawa M, Ohkuma N, Iizuka H, Ohkawara A (1985) Calmodulin activities are significantly increased in both uninvolved and involved epidermis in psoriasis. J Invest Dermatol 85:450–452

Mohr U, Hondius Boldingh W, Althoff J (1972) Identification of contaminating clostridium spores as the oncholytic agent in some chalone preparations. Cancer Res 32:1117–1121
Murray AW, Rogers A (1978) Calcium dependent protein modulator of cyclic nucleotide phosphodiasterases from mouse epidermis. Biochem J 176:727–732
Nanney LB, McKanna JA, Stoscheck CM, Carpenter G, King LE (1984) Visualization of epidermal growth factor receptors in human epidermis. J Invest Dermatol 82:165–169
Okada N, Kitano Y, Ischihara K (1982) Effect of cholera toxin on proliferation of cultured human keratinocytes in relation to intracellular cyclic AMP levels. J Invest Dermatol 79:42–47
Pentland A, Needleman P (1986) Modulation of keratinocyte proliferation in vitro by endogenous prostaglandin synthesis. J Clin Invest 77:246–251
Peterson LL, Wuepper KD (1983) Purification of human epidermal calmodulin. J Invest Dermatol 81:68–70
Powell JA, Duell EA, Voorhees JJ (1971) Beta adrenergic stimulation of endogenous epidermal cyclic AMP formation. Arch Dermatol 104:359–365
Rasmussen H, Jensen P, Goodman DB (1976) Interactions between calcium and cyclic nucleotides in control of secretion. In: Case RM, Goebell H (eds) Stimulus-secretion coupling in the gastrointestinal tract. Baltimore University Park Press, Baltimore, pp 33–47
Riley J, West GB (1953) The presence of histamine in tissue mast cells. J Physiol (Lond) 120:528–537
Rothman S (1954) Physiology and biochemistry of skin. The University of Chicago Press, Chicago
Saihan EM, Albano J, Burton J (1980) The effect of steroid and dithranol therapy on cyclic nucleotides in psoriatic epidermis. Br J Dermatol 102:565–569
Salisbury JL, Condelis JS, Maihle NJ, Satir P (1981) Calmodulin localisation during capping and receptor-mediated endocytosis. Nature 294:163–166
Savage CR Jr, Cohen S (1973) Proliferation of corneal epithelium induced by epidermal growth factor. Exp Eye Res 15:361–366
Schlessinger J, Schechter Y, Willingham MC, Pastan I (1978 a) Direct visualisation of binding aggegation and internalization of insulin and epidermal growth factor on living fibroblastic cells. Proc Natl Acad Sci USA 75:2659–2663
Schlessinger J, Schechter Y, Cuatrecasas P, Willingham MC, Pastan I (1978 b) Quantitative determination of the lateral diffusion coefficients of the hormone-receptor complexes of insulin and epidermal growth factor on the plasma membrane of cultured fibroblasts. Proc Natl Acad Sci USA 75:5353–5357
Starkey RH, Orth DN (1977) Radioimmunoassay of human epidermal growth factor (urogastrone). J Clin Endocrinol Metab 45:1144–1153
Starkey RH, Cohen S, Orth DN (1975) Epidermal growth factor: identification of a new hormone in human urine. Science 189:800–802
Stastny M, Cohen S (1970) Epidermal growth factor IV. The induction of ornithine decarboxylase. Biochim Biophys Acta 204:578–589
Steiner AL, Ferrendelli JA, Kipuis DM (1972) Radioimmunoassay for cyclic nucleotides III. Effect of ischaemia, changes during development and regional distribution of adenosine 3′,5′-monophosphate and guanosine 3′,5′-monophosphate in mouse brain. J Biol Chem 247:1121–1124
Taylor JM, Cohen S, Michell WM (1970) Epidermal growth factor: high and low molecular weight forms. Proc Natl Acad Sci USA 67:164–171
Taylor JM, Mitchell WM, Cohen S (1972) Epidermal growth factor: physical and chemical properties. J Biol Chem 247:5928–5934
Taylor JM, Mitchell WM, Cohen S (1974) Characterization of the high molecular weight form of epidermal growth factor. J Biol Chem 249:3198–3203
Tucker WFG, MacNeil S, Bleehen SS, Tomlinson S (1984) Biological active calmodulin levels are elevated in both involved and uninvolved epidermis in psoriasis. J Invest Dermatol 82:298–299

Van de Kerkhof PCM, Van Erp PEJ (1983) Calmodulin levels are grossly elevated in psoriatic lesions. Br J Dermatol 108:217–218
Voorhees JJ, Duell EA (1971) Psoriasis as a possible defect of the adenyl cyclase–cyclic AMP cascade. Arch Dermatol 104:352–358
Voorhees JJ, Duell EA, Kelsey WH (1972) Dibutyryl cyclic AMP inhibition of epidermal cell division. Arch Dermatol 105:384–386
Voorhees J, Stawiski M, Duell E (1973) Increased cyclic GMP and decreased cyclic AMP levels in the hyperplastic, abnormally differentiated epidermis of psoriasis. Life Sci 13:639–653
Wightman PD, Dahlgren ME, Bonney RJ (1982) Protein kinase activation of phospholipase A_2 in sonicates of mouse peritoneal macrophages. J Biol Chem 257:6650–6652
Wilkinson DI, Orenberg EK (1979) Effect of prostaglandins on cyclic nucleotide levels in cultured keratinocytes. Prostaglandins 17:419–429
Wolf DJ, Siegel F (1972) Purification of a calcium-binding phosphoprotein from pig brain. J Biol Chem 247:4180–4185
Yoshikawa K, Adachi K, Halprin KM, Levine V (1975a) Cyclic AMP in skin: effects of acute ischaemia. Br J Dermatol 92:249–254
Yoshikawa K, Adachi K, Halprin K, Levine V (1975b) The effect of catecholamine and related compounds on the adenyl cyclase system in epidermis. Br J Dermatol 93:29–36
Ziboh V, Hsia S (1972) Effects of prostaglandin E_2 on rat skin: inhibition of sterol ester-biosynthesis and clearing of scaly lesions in essential fatty acid deficiency. J Lipid Res 13:458–467

CHAPTER 4

Epidermal Lipogenesis (Essential Fatty Acids and Lipid Inhibitors)

V. A. ZIBOH

A. Introduction and Historical Considerations

The fact that skin is an active lipid-synthesising organ was first demonstrated by NICOLAIDES et al. (1955), who incubated human scalp skin with [^{14}]acetate and found the incorporation of ^{14}C into squalene, sterols and fatty acids. Further evidence to demonstrate the lipogenic activity of the skin was provided by the studies of PETTERSON and GRIESEMER (1959), who reported variations of lipogenic activity of the skin at various body sites. VROMAN et al. (1969) incubated preputial skin from new-born and abdominal skin of adult humans in vitro with [^{14}C]acetate and showed the incorporation of ^{14}C into all lipid classes of the skin. Of particular interest was the lack of ^{14}C incorporation into arachidonic acid even though this fatty acid constituted approximately 9% of total skin fatty acids. A similar observation was made by WILKINSON (1970), who incubated trypsinised epidermal cells with ^{14}C acetate in culture medium and found negligible incorporation of ^{14}C into 20:4,n6 acid, although smaller amounts of ^{14}C were incorporated into other polyunsaturated fatty acids (PUFAs). It is unlikely from these studies that essential fatty acids (EFAs) are biosynthesised de novo by skin in sufficient quantity to sustain the normal functioning of the epidermis. Thus epidermal EFAs must be derived from an exogenous source such as dietary linoleate.

Enthusiasm for the nutritional significance of certain (PUFAs) was heightened after BURR and BURR (1929, 1930) reported that rats maintained on a fat-free diet over a long period developed external abnormalities characterised by growth retardation, severe scaly dermatosis of the dorsal skin and the feet, caudal necrosis and extensive water loss through the skin. Also described were internal abnormalities such as fatty liver, kidney lesions, lung changes and impaired reproduction. Because these deficiency symptoms could be reversed by certain dietary PUFAs, notably linoleic acid, and because these fatty acids are not biosynthesised in quantities sufficient for proper functioning of physiological processes, the expression "essential fatty acids" was coined. Since growth retardation is a notable feature of EFA deficiency, it has been employed as a parameter for the comparison of individual fats or fatty acids for their respective nutritive values. Arachidonic acid was shown to have a greater growth-promoting activity than linoleate (MOHRHAUER and HOLMAN 1963). This suggests that arachidonic acid may exert a more direct physiological role than its precursor linoleate.

B. Essential Fatty Acids

I. Biosynthesis and Metabolism

It is now recognised that there are three major essential fatty acids, linoleic acid (9,12-octadecadienoic acid), linolenic acid (9,12,15-octadecatrienoic acid) and arachidonic acid (5,8,11,14-eicosatetraenoic acid). Recent investigations have shown that although linoleic and linolenic acids are related chemically, they are of two metabolically unrelated families in animal metabolism. For instance, linoleic acid and arachidonic acid are directly related as precursor and product respectively in animal metabolism, and they belong to the same family of n6 series of fatty acids, while linolenic acid (18 : 3) belongs to the n3 series. Figure 1 illustrates the metabolism of these two dietary fatty acids into their longer chain metabolites. Linoleic acid, representing the n6 series, is metabolised into arachidonic acid (20 : 4,n6) and 22 : 5,n6. An interesting observation by VERDINO et al. (1964) is that the 22 : 5,n6 acid is transformed back to 20 : 4,n6. Linolenic acid, representing the n3 series, is metabolised into the 20 : 5,n3, 22 : 5,n3 and 22 : 6,n3 acids.

II. Physiological Functions in the Skin

Assessments of the physiological role of the EFAs have for several years been associated with the latter's ability to maintain membrane fluidity. Although exhaus-

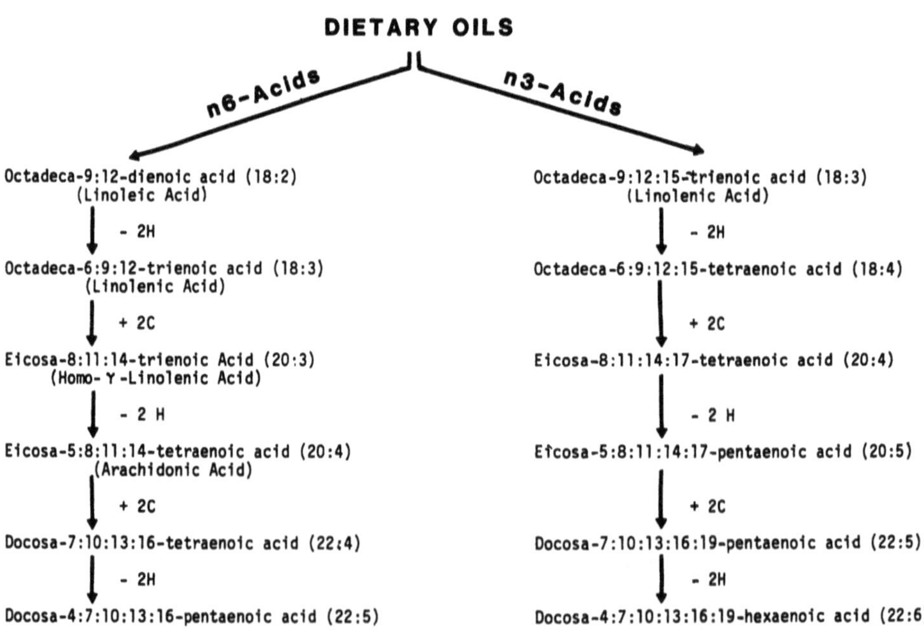

Fig. 1. The in vivo desaturation and elongation of dietary EFAs. n3 and n6 notations signify the position of the first double bond, numbered from the methyl terminal end of the fatty acid. The numbers in parentheses designate the number of carbons in the chain length and the number of unsaturation respectively

tive studies to demonstrate the phenomenon of membrane fluidity in epidermal cells are not available, it is likely that EFAs function to sustain the membrane fluidity of the epidermis, as in other tissues, since they are predominantly esterified to epidermal phospholipids (VROMAN et al. 1969; GRAY and YARDLEY 1975).

III. Role as Precursors of Prostaglandins and Related Lipids

Since the demonstration that the EFAs are precursors of the prostaglandins (VAN DORP et al. 1964), the possibility that they are biosynthesised in the skin has been investigated. A report by JESSUP et al. (1970) first indicated the release of prostaglandins from frog skin after hormonal challenge. Later VAN DORP (1971), using a gas liquid chromatographic system, showed that PGE_2 is a major prostaglandin in the epidermis. The direct in vitro transformations of arachidonic acid into prostaglandins soon followed with skin homogenates (GREAVES and MCDONALD-GIBSON 1973; JONSSON and ANGAARD 1972) and by microsomal fraction (ZIBOH 1973). More recently, transformations of arachidonic acid into prostaglandins and lipoxygenase products (HETEs) by skin preparations from humans (HAMMERSTROM et al. 1975) and keratinocytes (HAMMERSTROM et al. 1979; ZIBOH et al. 1981) have been demonstrated. These results imply that apart from their function in the maintenance of membrane fluidity the EFAs, do serve as precursors for the biosynthesis of these local hormone-like eicosanoids in the epidermis. This latter function of the EFAs may yet prove to be one of their most significant.

C. Essential Fatty Acid Deficiency

I. Macroscopic and Microscopic Appearance of the Skin During Deficiency

Visible evidence of EFA deficiency in rats is provided by a severe decrease in growth (Fig. 2a). The figure shows a general matting and disorientation of the fur, loss of hair and skin elasticity as indicated by a fold in the skin, and an extensive scaling of the dorsal skin and the tail. A yellow-brown, waxy and flaky pigmented substance normally found layered on the skin of normal-fed rats disappears during deficiency. The composition of this substance and the significance of its loss during EFA deficiency remain unknown.

Microscopically, the thickness of the epidermis increases to approximately threefold that of normal (Fig. 2b). The increase is evident in the stratum malpighii, stratum granulosum and stratum corneum. Extensive reviews have been provided of morphological and histological changes in the skin during EFA deficiency (for further information see MENTON 1968; KINGERY and KELLUM 1965).

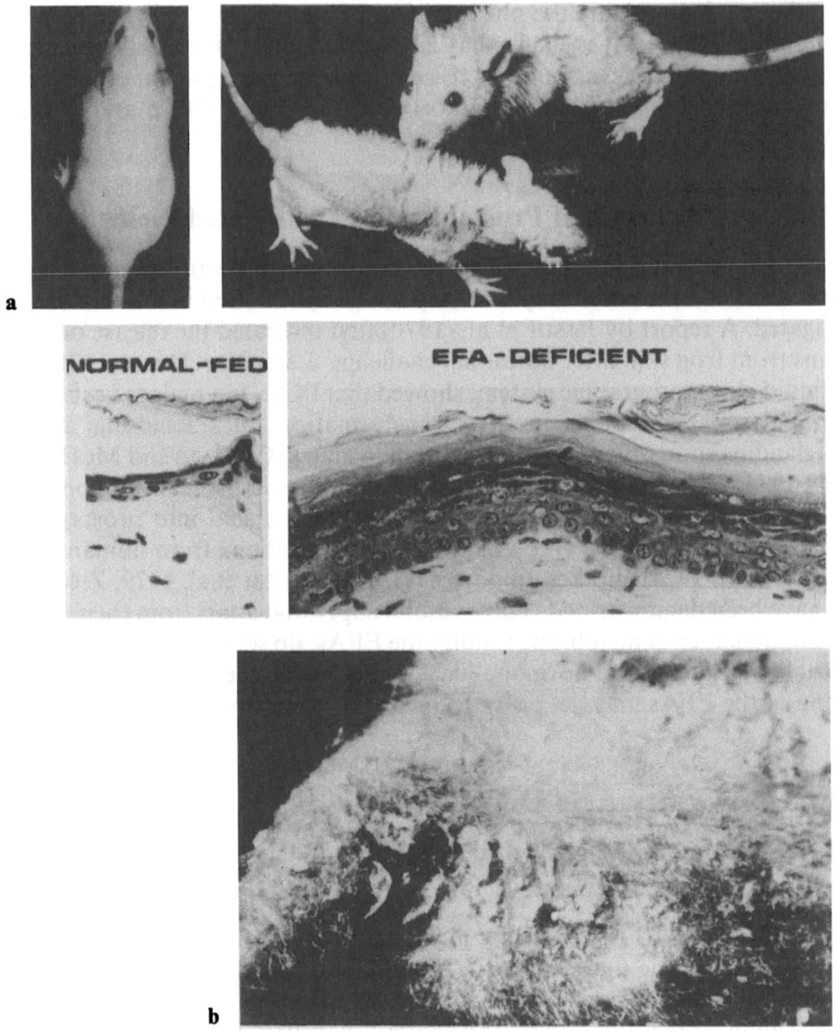

Fig. 2. a Macroscopic and **b** microscopic illustrations of rats on EFA-deficient diet

II. Altered Patterns of Polyunsaturated Fatty Acids

A notable alteration during EFA deficiency is the accumulation of PUFAs in tissues, particularly the 20:3,n9 fatty acid. An analysis of fatty acids in epidermal phospholipid isolated from normal and EFA-deficient rats is shown in Table 1. In the absence of dietary EFA, endogenous linoleic acid (18:2,n6) and arachidonic acid (20:4,n6) are decreased while 20:3,n9 fatty acid is increased markedly. Presumably, the tissue, sensing a decrease in dietary EFA, initiates an increased endogenous biosynthesis of monoenes and PUFAs, particularly the 20:3,n9 fatty acid. The metabolic transformation of oleic acid (18:1,n9) into the 20:3,n9 fatty acid has been demonstrated by liver preparations from EFA-deficient rats (Mead

Table 1. Composition of phospholipid fatty acids in skin of normal and EFA-deficient rats[a]

Fatty acids	Control	EFA-deficient
16.0	30.5±0.8	31.7±2.9
16.1	8.4±1.3	18.1±2.4
18.0	4.4±0.9	3.4±0.5
18.1 (n9)	29.3±1.2	39.1±4.2
18.2 (n6)	24.2±1.2	0.8±0.1
18.3	0.4±0.1	0.9±0.1
20.3 (n9)	Trace	5.2±1.1
20.4 (n6)	2.4±0.4	0.9±0.1

[a] Values are means±SEM. Tissue values are compared with authentic standard mixtures of fatty acid methyl esters

and SLATON 1956). It is thought that this enhanced formation of the trienoic acid during EFA deficiency is to sustain the overall degree of unsaturation which bestows fluidity to the cell membrane. In studies from the author's laboratory (ZIBOH et al. 1974) the 20:3,n9 fatty acid which was isolated from skin of EFA-deficient rats was shown to inhibit in vitro the oxygenation of arachidonic acid by the sheep vesicular gland cyclo-oxygenase. More recently, we have demonstrated that the topical application of this 20:3,n9 acid to the skin of normal EFA-fed rats induced scaly lesions similar to the characteristic dermatoses of EFA-deficient rats (NGUYEN et al. 1981).

III. Increased Metabolic Activity During Deficiency

A general increase in energy metabolism had previously been reported in EFA-deficient rats (PANOS et al. 1958; WESSON and BURR 1931). Incubation of [I-^{14}C]glucose and [U-^{14}C]glucose with skin from EFA-deficient rats results in increased production of $^{14}CO_2$ and ^{14}C-labelled lipids (ZIBOH and HSIA 1972). The increased $^{14}CO_2$ reflects an increased pentose cycle activity and generation of NADPH, which is essential for lipogenesis. A similar increase in metabolic activity was demonstrated when EFA skin was incubated with [^{3}H]PGE$_2$ (ZIBOH et al. 1977), which resulted in increased formation of [^{3}H]PGF$_{2\alpha}$ (a reaction catalysed by the NADPH-dependent PGE$_2$-9-ketoreductase). Thus the increased NADPH generated during EFA deficiency is also used for the reductive transformation of PGE$_2$ into PGF$_{2\alpha}$. Another example of increased metabolic activity of the EFA-deficient rat skin was revealed after incubation with tritiated thymidine, which resulted in increased labelling and mitotic indices (MCCULLOUGH et al. 1978). These data suggest that the EFAs or their metabolites function to maintain normal metabolism in the skin.

IV. Deficiency in Human Skin

Although the development of EFA deficiency in man was once thought unlikely (HANSEN and BURR 1946), numerous reports now establish that this deficiency can be induced secondarily in man after surgical or dietary interventions. For instance, human infants with chylous ascites who were maintained on a low-fat diet (WARWICK et al. 1959) and others with eczema (HANSEN 1933; FINNERUD et al. 1941) were shown to have a diminished iodine value in their serum fatty acids. These values were found to correspond with diminished contents of linoleic acid and arachidonic acid (BROWN and HANSEN 1937). The adaptation of the triene/-tetraene ratio, an index of the severity of EFA deficiency (HOLMAN 1960), has helped to demonstrate more clearly infantile EFA deficiency with associated scaly skin (PAULSTRUD et al. 1972). Similarly, the development of dietary deficiency in adult humans has also been described. COLLINS et al. (1971) and PRESS et al. (1974) described patients with gastrointestinal resections who developed EFA deficiency with associated skin scaliness. PROTTEY and HARTOP (1975) demonstrated increased 20:3,n9 fatty acid and decreased linoleic acid in the epidermal phospholipids of patients with chronic malabsorption syndrome levels. Topical application of sunflower seed oil on these patients cleared the scaly lesions and restored the elevated levels of the 20:3,n9 fatty acid to normal. A similar observation was reported by SKOLNIK et al. (1977) in a 19-year-old man who was maintained on a long-term regimen of fat-free, intravenous hyperalimentation fluids. The EFA deficiency and the scaly skin were reversed after topical application of linoleic acid. Thus, an alteration of the levels of EFAs in the epidermal phospholipids or their substitution by other non-essential PUFAs may contribute to the development of scaly dermatoses.

D. Epidermal Lipogenesis and Its Regulation

I. Interrelationships of Metabolic Pathways

Lipogenesis at large can be modified or inhibited by a variety of factors and at several steps in the metabolic pathways. Figure 3 illustrates the interrelationships between glycolysis, gluconeogenesis, citric acid cycle and fatty acid synthesis. Under normal conditions and in the presence of glucose, active glycolysis in tissues and in skin occurs, resulting in the formation of pyruvate. The latter diffuses into mitochondria where it is oxidised to acetyl-CoA and/or carboxylated to oxaloacetate. The acetyl-CoA and oxaloacetate condense to form citrate, which diffuses from the mitochondria into the cytosol, where it is cleaved into acetyl-CoA and oxaloacetate. The acetyl-CoA is converted into malonyl-CoA, which is then rapidly converted into palmitic acid with the concomitant utilisation of NADPH, which is generated from glycolysis and from the pentose cycle. The activities of the acetyl-CoA carboxylase and the fatty acid synthetase can be modified by the presence of various metabolites (WAITE and WAKIL 1962; MARTIN and VAGELOS 1962; SPENCER and LOWENSTEIN 1962). The end product of fatty acid synthesis in the form of long-chain acetyl-CoA also inhibits the acetyl-CoA carboxylase (the rate-limiting enzyme in fatty acid biosynthesis).

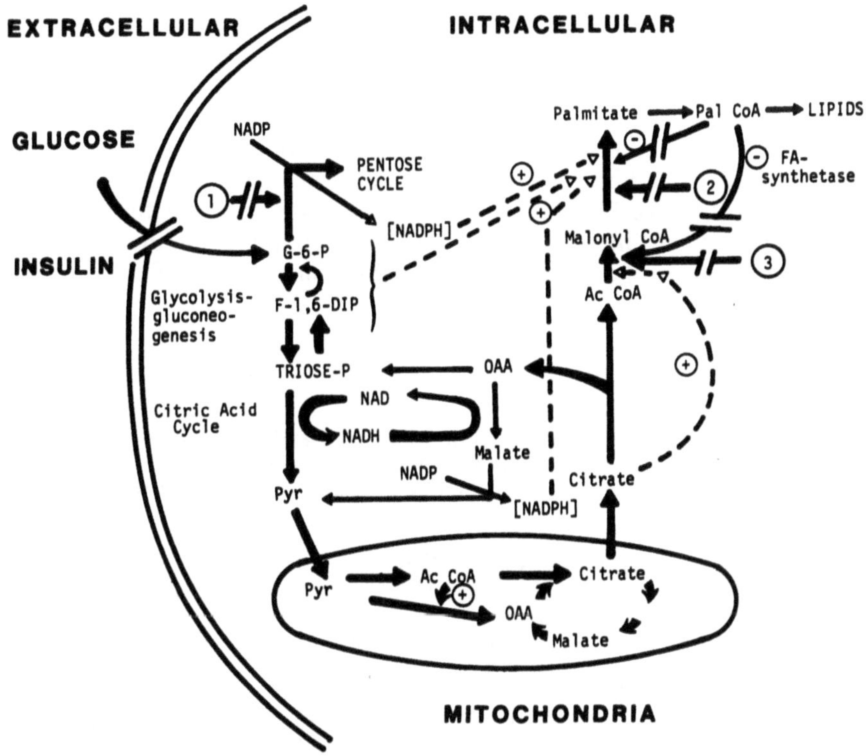

Fig. 3. Relationship between carbohydrate metabolism and lipogenesis. The numbers ①, ②, and ③ shown on the chart indicate possible sites of inhibition of lipogenesis by DHA, tetrolyl-Pn and clofibrate respectively

II. Regulation of Epidermal Lipogenesis

Metabolism of glucose via the glycolytic and pentose cycle has been reported to occur in skin of both human and experimental animals. A number of studies have attempted to regulate lipogenesis via these pathways by known metabolic modifiers. For instance, incubation of rat skin specimens with [I-^{14}C]acetate and tetrolyl-panthetheine (tetrolyl-Pn) inhibited non-competitively fatty acid and sterol biosynthesis (Ziboh and Brady 1971). This acetylenic analogue of butyryl-CoA most likely inhibits the condensation of acetyl-CoA with malonyl-CoA to form a thioester derivative (acetoacetyl-S-enzyme) which is common to fatty acid and sterol biosynthesis. Interestingly, a variety of acetylenic substances have been reported to inhibit fatty acid in other systems. These compounds are therefore useful tools for exploring epidermal lipogenesis. Eicosatetraenoic acid, an acetylenic analog of arachidonic acid, has also been reported to inhibit competitively the transformation of arachidonic acid into prostaglandins (Ahern and Downing 1970). Dehydroepiandrosterone (DHA), a powerful inhibitor of glucose-6-phosphate dehydrogenase activity, has been shown to inhibit epidermal lipogenesis (Ziboh et al. 1970). The data showed that DHA inhibited the pentose cycle activity which generates 41%–57% of the NADPH needed for lipid synthesis in the skin. Clofibrate (ethyl *p*-chlorophenoxy-isobutyrate), a drug previously employed to lower serum triglycerides and cholesterol (Thorp and Waring 1962; Best and Duncan 1964), was shown to inhibit lipogenesis in human and rat skin specimens (Fulton and Hsia 1972). The mechanism of its action remains speculative, although its inhibition of acetyl-CoA carboxylase has been suggested (Maragoudakis 1969). Thus, chemical agents which interfere with pathways of carbohydrate and lipid metabolism can be used to inhibit epidermal lipogenesis. This is indeed a fruitful area for more investigations because of their potential clinical implications.

References

Ahern DG, Downing DT (1970) Inhibition of prostaglandin biosynthesis by eicosa-5-8,11,14-tetraenoic acid. Biochim Biophys Acta 210:456–461

Best MM, Duncan DH (1964) Effects of cholesterol-lowering drugs on serum triglycerides. JAMA 187:37–40

Brown WR, Hansen AE (1937) Arachidonic and linoleic acid of serum in normal and eczematous human subjects. Proc Soc Exp Biol Med 36:113–117

Burr GO, Burr MM (1929) A new deficiency disease produced by the rigid exclusion of fat from the diet. J Biol Chem 82:345–367

Burr GO, Burr MM (1930) On the nature and role of fatty acids essential in nutrition. J Biol Chem 86:587–621

Collins FD, Sinclair AF, Royle JP, Coats DA, Maynard AT, Lenard RF (1971) Plasma lipids in human linoleic acid deficiency. Nutr Metab 13:150–155

Finnerud CW, Kesler RL, Wiese HF (1941) Ingestion of lard in the treatment of eczema and allied dermatoses: a clinical and biochemical study. Arch Dermatol Syphilol 44:849–861

Fulton JE, Hsia SL (1972) Inhibition of lipid synthesis by clofibrate: comparable study of human skin, rat skin, and rat liver in vitro. J Lipid Res 13:78–85

Gray GM, Yardley HJ (1975) Lipid compositions of cells isolated from pig, human and rat epidermis. J Lipid Res 16:434–437

Greaves MW, McDonald-Gibson W (1973) Effect of non-steroid anti-inflammatory drugs on prostaglandin biosynthesis by skin. Br J Dermatol 88:47–50

Hammerstrom S, Hamberg M, Samuelsson B, Duell EA, Strawiski M, Voorhees JJ (1975) Increased concentration of non-esterified arachidonic acid. 12L-hydroxy-5,8,10,14-eicosatetraenoic acid, prostaglandin E_2 and prostaglandin $F_{2\alpha}$ in epidermis of psoriasis. Proc Natl Acad Sci USA 72:5130–5132

Hammerstrom S, Lingren JA, Marcelo C, Duell EA, Anderson RF, Voorhees JJ (1979) Arachidonic acid transformations in normal and psoriatic skin. Dermatologica 73:180–183

Hansen AE (1933) Serum lipid changes and therapeutic effects of various oils in infantile eczema. Proc Soc Exp Biol Med 31:160–161

Hansen AE, Burr GO (1946) Essential fatty acids and human nutrition. JAMA 132:855–859

Holman RT (1960) The ratio of trienoic: tetraenoic acids in tissue lipids as a measure of essential fatty acid requirement. J Nutr 70:405–410

Jessup SJ, McDonald-Gibson WJ, Ramwell PJ, Shaw JE (1970) Biosynthesis and release of prostaglandins on hormonal treatment of frog skin and their effect on ion transport. Fed Proc 29:387

Jonsson DC, Angaard E (1972) Biosynthesis and metabolism of prostaglandin E_2 in human skin. Scand J Clin Lab Invest 29:289–296

Kingery FA, Kellum RE (1965) Essential fatty acids. Histochemical changes the skin of rats. Arch Dermatol 91:272–279

Margoudakis ME (1969) Inhibition of hepatic acetyl coenzyme A carboxylase by hypolipidemic agents. J Biol Chem 244:5005–5013
Martin DB, Vagelos PR (1962) The mechanism of tricarboxylic acid cycle regulation of fatty acid synthesis. J Biol Chem 237:1787–1792
McCullough JL, Schreiber SJ, Ziboh VA (1978) Cell proliferation kinetics of epidermis in the essential fatty acid deficient rat. J Invest Dermatol 70:318–320
Mead JF, Slaton WJ Jr (1956) Metabolism of essential fatty acids. III. Isolation of 5,8,11-eicosatraenoic acid from fat deficient rats. J Biol Chem 219:705–709
Menton D (1968) The effects of essential fatty acid deficiency on the skin of the mouse. Am J Anat 122:337–356
Mohrhauer H, Holman RT (1963) The effect of dietary essential fatty acids upon the composition of polyunsaturated fatty acids in depot fat and erythrocytes of the rat. J Lipid Res 4:151–159
Nguyen TT, Ziboh VA, Uematsu S, McCullough JL, Weinstein G (1981) New model of a sealing dermatosis: induction of hyperproliferation in hairless mice with eicosa-5,8,11-trienoic acid. J Invest Dermatol 76:381–387
Nicolaides N, Reiss OK, Langdon RG (1955) Studies of the in vitro lipid metabolism of human skin. I Biosynthesis in scalp skin. J Am Chem Soc 77:1535–1538
Panos TC, Finerty JC, Klein GF, Wall RL (1958) Essential fatty acids. Academic Press, New York, Butterworth, London, pp 205–220
Paulsrud JR, Whitten DF, Stewart SE, Holman RT (1972) Essential fatty acid deficiency in infants induced by fat-free intravenous feeding. Am J Clin Nutr 25:897–904
Petterson JF, Griesemer RD (1959) Lipogenesis in human skin. J Invest Dermatol 33:281–285
Press M, Hartop PJ, Prottey C (1974) The correction of essential fatty acid deficiency in man by the cutaneous administration of sunflower seed oil. Lancet I:597
Prottey C, Hartop PJ, Press M (1975) Correction of the cutaneous manifestations of essential fatty acid deficiency in man by the application of sunflower seed oil to the skin. J Invest Dermatol 64:228–234
Skolnik P, Eaglstein WH, Ziboh VA (1977) Human essential fatty acid deficiency: treatment of topical application of linoleic acid. Arch Dermatol 113:939–941
Spencer AF, Lowenstein JM (1962) The supply of precursors for the synthesis of fatty acids. J Biol Chem 237:3640–3648
Thorp JM, Waring WS (1962) Modification of metabolism and distribution of lipids by ethyl chlorophenoxyisobutyrate. Nature 194:948–949
Van Dorp DA (1971) Recent developments in the biosynthesis and the analysis of prostaglandins. Ann NY Acad Sci 180:181–199
Van Dorp DA, Beerthuis RW, Nugteren DH, Von Keman H (1964) Enzymatic conversion of all *cis*-polyunsaturated fatty acids into prostaglandins. Nature 203:839–841
Verdino B, Blank ML, Privett OS, Lundberg WO (1964) Metabolism of 4,7,10,13,16-docosapentaenoic acid in essential fatty acid-deficient rat. J Nutr 83:234–238
Vroman HE, Nemecek RA, Hsia SL (1969) Synthesis of lipids from acetate by human preputial and abnominal skin in vitro. J Lipid Res 10:507–514
Waite BM, Wakil SJ (1962) Studies on the mechanism of fatty acid synthesis XII. Acetyl coenzyme-A carboxylase. J Biol Chem 237:2750–2757
Warwick WJ, Holman RT, Quie PG, Good RA (1959) Chylous ascites and lymphedema. Am J Dis Child 98:317–329
Wesson LG, Burr GO (1931) The metabolic rate and respiratory quotients of rats on a fat-deficient diet. J Biol Chem 91:525–528
Wilkinson DI (1970) Incorporation of acetate-1-^{14}C into fatty acids of isolated epidermal cells. J Invest Dermatol 54:132–138
Ziboh VA (1973) Biosynthesis of prostaglandin E_2 in human skin: subcellular localization and inhibition of unsaturated fatty acids and anti-inflammatory drugs. J Lipid Res 14:377–384
Ziboh VA, Brady RO (1971) Inhibition of lipid synthesis in rat skin by tetrolyl-pantetheine. Proc Soc Exp Biol Med 136:15–18

Ziboh VA, Hsia SL (1971) Prostaglandin E_2; biosynthesis and effects on glucose and lipid metabolism in rat skin. Arch Biochem Biophys 146:100–109
Ziboh VA, Dreize MA, Hsia SL (1970) Inhibition of lipid synthesis and glucose-6-phosphate dehydrogenase in rat skin by dehydroepiandrosterone. J Lipid Res 11:346–354
Ziboh VA, Vanderhoek JY, Lands EM (1974) Inhibition of sheep vesicular gland oxygenase by unsaturated fatty acids from skin of essential fatty acid deficient rats. Prostaglandins 5:233–239
Ziboh VA, Lord JT, Penneys NS (1977) Alterations of prostaglandin E_2-9-ketoreductase activity in proliferating skin. J Lipid Res 18:37–43
Ziboh VA, Marcelo C, Voorhees JJ (1981) Induced lipoxygenation of arachidonic acid in mouse epidermal keratinocytes by calcium ionophore A23182. J Invest Dermatol 76:307

CHAPTER 5

Fibroblasts, Collagen, Elastin, Proteoglycans and Glycoproteins

C. M. LAPIÈRE and B. V. NUSGENS

A. Introduction

In vertebrates the specific morphological and physiological features of skin result predominantly from the function of the epidermis. It is, however, well established that the organogenesis of the epidermal cover and its functional properties are intimately related to precisely timed and spatially defined interactions with the dermis.

The dermis is a differentiated connective tissue which forms a supple and resistant support to the epidermis and allows the diffusion of nutrients from the bloodstream to the ectodermal cells. It is composed of three layers of different quality. The peri-adnexial dermis, also called the papillary dermis, supports the epidermis and surrounds the blood vessels, the nerves and all ectodermal adnexa. It is separated from them by a basement membrane. The reticular dermis represents the middle layer of the dermis and the most resistant part of the connective support of the skin. The hypodermis consists of fibrous sheets of connective tissue which surround the lobules of adipose cells of the deepest part of the skin and extend to the underlying fascia or to the periosteum.

The various sections of the dermis display obvious differences in composition and architectural organisation of their components, reflecting the differentiated state of the cells responsible for their synthesis, the fibroblasts.

B. Fibroblasts Are Differentiated Cells

At the age of 60 days in the human embryo the dermis is composed of a dense population of mesenchymal cells surrounded by little intercellular substance. Blood vessels are already formed but the complex vascular system differentiates later. At 120 days the two layers are individualised and reticulin fibres are present. Fibrogenesis increases and at the end of gestation the dermis is similar to its adult state although the density of the fibrous framework is not as important. The differentiation of the hypodermis starts during the second half of foetal life. The histogenesis of the dermis (for review and details, see SENGEL 1976) indicates that differentiation occurs early in the fibroblasts. It will keep developing until the established pattern observed in adult life is achieved. Such differentiation is most significant in the interpretation of the pharmacology of skin since drugs are known to act upon cells in defined stages of differentiation.

The several facets of connective tissue cells are expressed by a vast repertoire of biosynthetic products: the various types of collagen, elastin, the microfibrillar glycoproteins, fibronectin and other glycoproteins, glycosaminoglycans and proteoglycans. All these components are integral parts of the connective tissue and their synthesis is highly regulated to ensure the correct amount of macromolecules and organisation of the latter into adapted architectural structures.

The differentiation of the fibroblast can be defined by its biosynthetic production and other functional properties. The papillary dermal fibroblast displays a doubling time in culture lower than the reticular dermal fibroblast, resulting in a larger number of generation that can be achieved in culture (HARPER and GROVE 1979). Connective tissue naevi are composed of fibroblasts displaying abnormal biosynthetic capacities. Myofibroblasts are specialised connective tissue cells present in granulation tissue (MAJNO et al. 1971). Beside the indentations of the nucleus and the presence in the cytoplasm of a large number of microfilament bundles, their surface is covered by a material similar to basement membranes. They might derive from blood vessels, and they display contractile properties responsible for wound contraction and specialised biosynthetic activities such as deposition of collagen type III in the early phase of healing. This biosynthetic activity shifts progressively to production of collagen type I (GAY et al. 1978), perhaps by replacement of these cells by local fibroblasts.

Differentiation of fibroblasts is an even more striking trait when considering specialised connective tissues, e.g. cartilage and bone. Chondrocytes produce collagen type II while osteoblasts produce collagen type I and a calcifying matrix in vivo. This differentiation is not, however, irreversible since cultured chondrocytes in suspension produce collagen type II while they synthesise collagen type I in monolayer when attached to a rigid substrate. In vivo, the fibroblasts display differentiated characteristics under the influence of various mediators. Chemical factors can act as inducers since bone differentiation and calcification can be produced by implanting demineralised bone powder in the dermis (REDDI et al. 1977). An additional example is provided by fibroblasts from the genital area that acquire at puberty the receptors to sex hormones. Local differentiation and receptivity to hormones or mediators might explain the restricted occurrence of conditions such as pretibial myoedema and exophthalmos in Graves' disease.

C. Collagen

I. Molecular Structure and Distribution

The "collagen family" is a complex group of macromolecules sharing a common structural feature, a protease-resistant triple-helical domain made of repetitive triplets (Gly-X-Y) where one-third of the X and Y positions are occupied by proline and hydroxyproline. This repetitive sequence allows a self-assembly of molecules into a triple helix stabilised by hydrogen bonds between hydroxyproline residues. However, these related molecular species present a wide diversity in structure and chemical composition which is defined by the different genes coding for the various collagen polypeptides but also by intra- and extracellular post-translational modifications of the nascent chains. A tentative classification according to MILLER (1985) based on the latest available structural data and the known tissular distribution is shown in Table 1. Additional information and ref-

Table 1. The collagen family

	Type	Molecular composition	Polymers	Distribution
Group I: Continuous helical domain $M_r \geqq 95000$	I	$(\alpha_1)2\text{-}\alpha_2$	D-staggered fibres	Skin, bone, tendon, ubiquitous except in cartilage
	I trimer	$(\alpha_1)3$	D-staggered fibres	Tumours, skin, liver
	II	$(\alpha_1)3$	D-staggered fibres	Hyaline cartilage, invertebral disks, vitreous body
	III	$(\alpha_1)3$	D-staggered fibres	Skin, vessel walls, liver placenta, intestine, embryonic tissue
	V	$\alpha_1\text{-}\alpha_2\text{-}\alpha_3$ $(\alpha_1)2\text{-}\alpha_2$ $(\alpha_1)3$ $(\alpha_3)3$	Fibres	Placenta, amniotic membrane, skin, pericellular, cyto-skeletal
	XI	1α, 2α, 3α		Cartilage
Group II: Interrupted helical domain by non-helical sections $\geqq 95000$	IV	$(\alpha_1)2\text{-}\alpha_2$	Network	Basement membranes
	VI	$\alpha_1\text{-}\alpha_2\text{-}\alpha_3$	Beaded filaments	Placenta, uterus, skin, intima of aorta, cornea, muscle
	VII	$(\alpha_1)3$	Unstaggered fibrils	Anchoring fibrils
	VIII	$(\alpha_1)3$	?	Endothelial cells, human astrocytoma, Descemet's membrane
Group III: $M_r < 95000$	IX	$\alpha_1\text{-}\alpha_2\text{-}\alpha_3$	?	Cartilage
	X	$(\alpha_1)3$	?	Hypertrophic and calcifying cartilage
	FCL-1			Foetal fibroblasts

erences can be found in recent reviews (BORNSTEIN and SAGE 1980; MILLER 1985).

Up to now, 11 different types of collagen have been described in vertebrates tissues. They can be defined by their structure, size, supramolecular aggregation and extracellular location. A first group of collagen molecules characterised by a single continuous triple-helical domain of 300 nm comprises the interstitial collagens forming, by D-staggered lateral association, fibrous polymers displaying the characteristic 67 nm striation. The three major types, collagen I, II and III, are the products of four distinct genes. Type I collagen is a heterotrimer containing two α_1 (I) chains and one α_2 (I) chain. It forms large fibrils associated in thick bundles which are responsible for most of the tensile strength of the dermis and various other connective tissues. A homotrimer of α_1 (I) chain is present in traces in normal skin and occurs in some pathological conditions. Homotrimeric type II and type III collagen are composed of α_1 (II) and α_1 (III) chains respectively. Type II collagen is the main type of collagen found in hyaline cartilage and vitreous humour. Type III collagen is found in many connective tissues under the form of thin fibrils associated in small bundles. It is abundant in the foetal dermis and represents ca. 25% of collagen in the adult dermis. It is also a major component of all hollow and mobile organs, the blood vessels, the intestinal and genital tracts etc.

Type V collagen can be found under various associations. Both heterotrimeric forms [α_1 (V)$_2$ α_2 (V)] and (α_1 (V) α_2 (V) and α_3 (V)] and homotrimeric forms [α_1 (V)$_3$] and [α_3 (V)$_3$] have been described. It seems that the triple-helical domain of 300 nm is not interrupted and could therefore form fibrous polymers as observed under certain conditions in vitro. Type V collagen is present in skin, placenta and amniotic membranes. Its function is not yet clear. It seems to be pericellular and associated with the exoskeleton of the cells.

The minor collagens 1α, 2α and 3α (type XI) are present in normal cartilage. A great similarity exists between 1α, 2α and α_1 (V), α_2 (V) while the 3α chain is very close to the α_1 (II) chain.

The second group of molecules are characterised by their high molecular weight and interruptions in the triple helix by globular domains of different sizes that are sensitive to protease digestion. They form end-to-end aggregates.

Type IV collagen is composed of two different chains: α_1 (IV)$_2$ and α_2 (IV). Type IV molecules are highly flexible and interact at both the amino and carboxyl terminal ends to form a network of polymers constituting the lamina densa of basement membranes in association with laminin, proteoglycans and other glycoproteins. Type VI collagen contains a large globular domain and a relatively short helical part. They form beaded filaments by antiparallel association of tetrameric units. Although this type of collagen is ubiquitously distributed, its function is not yet clear. Type VII collagen is a very high molecular weight molecule with a 424-nm triple-helical domain. It is found as centrosymmetrically SLS-banded dimeric polymers in anchoring fibrils beneath the epithelial basement membranes. Type VIII collagen is a homotrimer of α_1 (VIII) chain. The triple helix is interrupted by pepsin-sensitive sequences yielding three nearly identical fragments. Its polymeric form is unknown. It is synthesised by endothelial cells and is present in Descemet's membrane.

The third group of molecules is characterised by their short chains. Type IX is found in cartilage under a heterotrimeric form of three different gene products [α_1 (IX), α_2 (IX), α_3 (IX)] and type X is a homotrimer of α_1 present in hypertrophic cartilage. The α_2 chain of type IX collagen has recently been shown to be linked to a chondroitin sulphate chain and to be involved in a covalent cross-link with the type II collagen. A very short chain collagen (FCL 1) is synthesised by foetal skin and ligamentum nuchae fibroblasts. Its expression seems to be developmentally regulated.

Thus in skin, at least seven different types of collagen are present. Type I and type III striated fibrils form the bulk of the connective tissue; type IV collagen is found in the basement membrane and type VII in anchoring fibrils. The roles of type V and type VI collagen have not yet been clearly defined.

II. Biosynthesis

The biosynthesis of the various types of collagen involves the same basic principles that are common to all proteins plus several particular enzyme-mediated post-translational modifications of the side chain of some amino acids. Specialised proteolytic processing of the precursor sequences occurs as in other proteins designed for secretion and transport in the extracellular space.

The structure of collagen genes is remarkable in the sense that they comprise at least 50 exons separated by large intervening sequences (VOGELI et al. 1980). The structure of the genes for interstitial collagens is now well characterised (for review, see RAMIREZ et al. 1985 and DE CROMBRUGGHE et al. 1985). The entire gene [38 Kb for α_2 (I) and α_1 (III), 18 Kb for α_1 (I)] is transcribed into precursor mRNA, which is processed to a functional mRNA of ± 5000 by splicing of the intervening sequences. All the exons covering the triple-helical domain are multiples of triplets and many of them encode six triplets (54 bp), always beginning with a glycine codon and ending with a Y codon. The other exons are 108 bp (2×54), 99 bp (108-9), 45 bp (54-9), and one 162 bp (3×54), suggesting that 54-bp exon perhaps represents the ancestral collagen gene. The arrangement of the exons for the triple helix of the three main interstitial collagens is remarkably conserved. Exons at the 5′ and 3′ end coding for the amino- and the carboxy-peptide are not so well conserved. The promoter region of the collagen genes is located in the 5′-flanking sequence at a rather long distance upstream of the start site of transcription. It is worth noting that the promoter of α_2 (I) collagen contains a binding site for a peptidic activator which is absent in the α_1 (III) gene. This suggests that the two genes, which are often coordinately expressed in many tissues, are not necessarily regulated by the same factor (HATAMOCHI et al. 1986).

The functional mRNA is translated on membrane-bound polysomes and the nascent polypeptides, beginning with a signal sequence, are introduced into the channels of the endoplasmic reticulum. During its transmembrane transport the signal sequence is cleaved and the precursor polypeptide of $\pm$ 1500 amino acids, pro-α-chains, is progressively completed. During this process, defined prolyl and lysyl residues are hydroxylated through the catalytic activity of two membrane-bound specific hydroxylases using as co-factors and co-substrates molecular oxygen, ferrous ions, ascorbate and α-ketoglutarate. Hydroxylation of proline is required to ensure the thermal stability of the triple-helical structure of the polypeptides.

Lysyl hydroxylation is performed by a different hydroxylase requiring the same co-factors as polylhydroxylase. Hydroxylysine is involved in covalent cross-kinking and some hydroxylysyl residues are glycosylated. First a galactosyl residue is added to hydrolysine through galactosyltransferase activity. In some instance a glucosyl residue is transferred to galactosyl-hydroxylysine. The function of the sugar residues is not known. These post-translational modifications vary in extent with the type of collagen and also from tissue to tissue. Details and references regarding these processes can be found in a recent review (KIVIRIKKO and MYLLYLA 1984).

During synthesis the three pro-α-chains assemble in triple helix and their association is stabilised by interchain disulphide bonds in the carboxy propeptide.

This step is enzyme mediated by a disulphide isomerase. Further addition of an N-linked oligosaccharide occurs in the carboxy propeptide to an asparaginyl residue through the dolichol pathway. Pro-collagen is stored in Golgi-derived vesicles.

Secretion is an active process requiring energy. It is blocked by colchicine and other poisons of the microfilament–microtubule system. After fusion with the cell membrane, the collagen precursor is released in the extracellular space, possibly in a confined environment (folds of the membrane), where further processing and polymerisation take place.

The proteolytic excision of the precursor sequences is performed by a series of specific endopeptidases – the pro-collagen peptidases (see review by PELTONEN et al. 1985). While most of the co-translational reactions seem to be catalysed by enzymes displaying little specificity in terms of collagen type, there exist different pro-collagen peptidases. The p-N-pro-collagen peptidases are type-specific for type I and III collagens (NUSGENS et al. 1980) and it seems that two different enzymes are also required to convert type I and type III p-C-collagen to collagen (PELTONEN et al. 1985). Pro-collagen peptidases share common characteristics: they are neutral endopeptidases requiring a native substrate, Ca^{2+}, and are inhibited by EDTA and some divalent ions.

III. Polymerisation

Fibrillogenesis is a multistep event involving both intracellular and extracellular assembly reactions (TRELSTAD 1982) probably controlled by fibroblasts. Collagen polymerisation can be reproduced in vitro using extracted collagen at 4 °C, neutral pH and isotonic ionic strength by bringing the solution to body temperature. It proceeds in two steps that both require thermal energy; the first is nucleation and the second is the assembly of the nuclei to form the polymers. The first step is stable and irreversible. The second step can be reversed by cooling (WILLIAMS et al. 1978). The ordered aggregation of the collagen monomers in the fibrils creates a periodicity of 67 nm that can be seen by electron microscopy. It depends on the quarter staggering between adjacent molecules plus a regular 20 nm spacing between successive molecules (HODGE and PETRUSKA 1963). Laterally, the quasi-hexagonal molecular packing fits most of the physical measurements (HULMES and MILLER 1979). These associative properties depend on the structure of the collagen molecules (PIEZ and TRUS 1978), probably the telopeptides (HELSETH et al. 1979) and the propeptides (LAPIÈRE et al. 1975; FLEISCHMAJER et al. 1983; MIYAHARA et al. 1984). Interactions with other components such as proteoglycans, polysaccharides, link protein, fibronectin and other types of collagen might also regulate fibrillogenesis (CHANDRASEKHAR et al. 1984; KLEINMAN et al. 1981; LAPIÈRE et al. 1977).

After forming the initial polymer, the collagen monomers in the fibrils undergo progressive packing leading to progressive reduction of extractability. Shortly after synthesis, collagen can be solubilised by isotonic saline at neutral pH and low temperature. With increasing time after synthesis extractability requires the use of increased ionic strength or an acid pH. This situation leads to complete insolubility in the absence of denaturation or proteolysis.

The mechanical resistance of polymeric collagen is ultimately ensured by covalent cross-linking of the monomers. The first step in this process is the oxidative deamination of defined lysyl and hydroxylysyl residues under the action of lysyl oxidase (for review, see SIEGEL 1979). The resultant aldehydes then condense with adjacent lysyl or hydroxylysyl residues to form unstable and later covalent stable bifunctional intra- and intermolecular bridges. Polyfunctional bonds can also form by condensation of unstable cross-links with additional lysyl or histidinyl residues (TANZER 1976). The type of cross-link and the extent of cross-linking vary with the nature of the collagen, the location in the tissue and age.

IV. Degradation

Degradation of polymeric collagen requires a collagenase, a highly specific neutral protease (GROSS and LAPIÈRE 1962), operating a single cleavage across the three polypeptide chains of native collagen. The fragments (a large three-quarter amino terminal and a short one-quarter carboxy terminal) denature at body temperature and the resulting gelatin is further degraded to small peptides and free amino acids by a variety of peptidases. Various types of collagenase have been detected that display specificity for different types of collagen. One collagenase produced by fibroblast in culture is active on the various types of fibrillar collagen but at different rates (I > III > II). One specific collagenase isolated from tumour cells can degrade type IV collagen (LIOTTA et al. 1979), while type V collagen requires still another collagenase for degradation (MAINARDI et al. 1980). A collagenase c-DNA clone has recently been isolated (GROSS et al. 1984).

D. Elastin

I. The Elastic Fibre

The elastic fibres are composed of a microfibrillar glycoprotein and elastin, a very hydrophobic protein. Both constitutive macromolecules are synthesised by various types of cell: fibroblasts, chondroblasts, smooth muscle cells and endothelial cells. Details and references regarding elastic fibre and its components can be found in a review by SANDBERG et al. (1981). In skin, the proportion of fibrillar component and amorphous elastin varies from the subepithelial layers to the reticular dermis and the hypodermis. In respect to their staining properties these polymeric structures containing an increasing proportion of elastin have been called oxytalan, elaunin and elastic fibres (COTTA-PEREIRA et al. 1976).

The elastin gene has been isolated and sequenced. It is characterised by a high intron to exon ratio (15:1), small exons from 27 bp to 114 bp coding separately for hydrophobic and cross-link regions of the molecule (ROSENBLOOM 1984). The amino acid sequence of elastin is considerably different from that of collagen but also shows some similarity. Triplets containing glycine every third residue and hydroxyproline are observed. The major difference is the large concentration of hydrophobic amino acid [mainly glycine (25%), alanine (18%) and valine (10%)] and the lower content of polar amino acids.

II. Biosynthesis, Polymerisation and Degradation

The mRNA (FOSTER et al. 1980; ROSENBLOOM et al. 1980) directs the synthesis of tropoelastin, a polypeptide of 79 Kd that will be modified by co-translational enzyme-mediated reactions leading to the formation of hydroxyproline and lysyl aldehydes. The significance of hydroxyproline in elastin is not known. The process of covalent cross-linking through lysyl aldehydes is similar to that occurring in collagen although the polyfunctional condensation products (desmosine and isodesmosine) are different (PARTRIDGE et al. 1963). Measurements of elastin mRNA by recombinant cDNA clones in several systems demonstrate that the rate of elastin synthesis is regulated at the gene level (BURNETT et al. 1980). The very hydrophobic and globular elastin monomers establish a network of molecules stabilised by interaction of the lysyl aldehydes forming the desmosines. After cross-linking the coacervate of elastin acquires elastic properties that are related to the hydrophobic nature of its polypeptide, the entropy of which is increased, upon traction, by exposure of the internal aspect of the molecules to a hydrated surrounding. This excess energy is released with production of heat upon relaxation.

Degradation of polymeric elastin is performed by specific proteases, elastases (ROBERT and ROBERT 1969), present in various strains of cell and in the exocrine secretion of the pancreas.

E. Proteoglycans and Glycosaminoglycans

I. Molecular Structure

The so-called ground substance is an ultrafiltrate of the plasma structured in the connective tissue by interaction with various charged macromolecules, glycoproteins and glycosaminoglycans (GAGs). Although most of the information about polysaccharides (an old term) arises from the study of cartilage, it is most probable that it in part applies to skin. Except for hyaluronate that occurs as a pure GAG and perhaps heparin that might have lost its polypeptide core, all the GAGs are found under the form of proteoglycans, i.e. complex polymers made of a polypeptide core upon which are bound various types of GAG. Skin contains four main types of GAG: hyaluronate, chondroitin 4-sulphate, dermatan sulphate and heparin. The polypeptide core of proteodermatan sulphate exists in two different sizes (MATSUNAGA and SHINKAI 1986). An additional proteoglycan specific to basement membrane, at least by its polypeptide core, has also been described. Its GAG is heparan sulphate (HASSEL et al. 1980). Heparan sulphate-containing proteoglycans have also been shown to be associated with the cell surface (FRANSSON et al. 1985) and substratum adhesion sites of human skin fibroblasts in vitro (KENT et al. 1986). In the dermis of the calf and of the chick the distribution of the various GAGs is not uniform. Hyaluronic acid is associated in larger concentration with the thin collagen bundles of the papillary dermis while dermatan sulphate is evenly distributed in the dermis (TAJIMA and NAGAI 1980). During development in the chick embryo there is a shift in concentration from hyaluronate to dermatan sulphate with increasing collagen deposition (NAKAMURA and NAGAI

1980). The close association of a dermatan sulphate-rich proteoglycan with collagen fibrils at the d gap as described by SCOTT and HAIGH (1985) is probably closely involved in the regulation of fibrillogenesis.

Hyaluronate has a molecular size in the order of several millions (four to eight). According to FUJII and NAGAI (1981) the proteodermatan sulphate of calf skin has a molecular size of 115000 daltons of which the polypeptide core accounts for 56000 and each of the four GAGs for 17000. In general, proteoglycans of non-cartilaginous origin (skin, periodontal ligament, sclera, aorta) display a much higher ratio of polypeptides to GAGs than those of cartilaginous origin. Their polypeptides also differ from tissue to tissue. The existence of hybrid proteoglycans in these tissues is possible (PEARSON and GIBSON 1982).

All GAGs are repeating disaccharide units of uronate and hexosamine for the major part of the molecules. The connection between the GAG and the protein core is a covalent link between the terminal sugar of a connecting piece of specific sugar composition (-Gal-Gal-Xyl-Ser) and a seryl residue.

All GAGs display a high content of charged anionic groups ionised under physiological conditions. Details and references regarding the chemical and physical structure of the GAGs can be found in the reviews by COMBER and LAURENT (1978) and RODEN and HOROWITZ (1978).

II. Biosynthesis, Organisation and Degradation

The biosynthesis of the proteoglycans requires the synthesis of a polypeptide, while the assembly and processing of the GAGs require the activity of three different classes of enzyme system. The process is initiated by the synthesis of the polypeptide core. The assembly of the adequate sugar nucleotides occurs in the cytosol upon the activity of a first set of enzymes. They catalyse multiple reactions, including oxidations, reductions, inversions and decarboxylations or a mixture of these reactions, and activation to sugar nucleotides. These reactions are not specific for a defined type of GAG. The glycosylation and additional reaction mediated by a second set of enzymes occur within the cisternae of the rough endoplasmic reticulum using sugar nucleotides as co-substrate. A set of three transferases is involved in the synthesis of the connecting piece. The first step is mediated by xylosyl transferase (to serine) followed by galactosyltransferase I and II. This is then followed by the alternate addition of the repeating sugars catalysed by the set of the two transferases specific to the type of GAG being synthesised. An additional and optional third class of enzymes intervene after the synthesis of the proteoglycan to catalyse sulphation that occurs on the growing GAG and might determine the arrest of chain elongation. The transferases and the sulphation enzymes are membrane bound. The finished molecules are packed into secretory vesicles that are secreted after fusion with the plasma membrane. For review and references regarding these processes, see PHELPS (1980).

In the extracellular space the GAGs form a gel related to the large domain that they occupy in the hydrated state. An excellent discussion of this problem and ample literature can be found in a review by PEARSON (1981).

Degradation in the extracellular space can be achieved by destruction of the covalent structure of the proteoglycans under the action of glycanases on the

GAGs or by the activity of endopeptidases acting on the polypeptide backbone. In vitro, cathepsin D, B_1, trypsin, papain or hyaluronidase can perform these functions. It is unlikely that the various cathepsins active at acid pH could be effective in vivo while various proteinases from leucocytes are active at neutral pH (for review, see BARRETT 1975). A hyaluronidase exists in some tissues (TOOLE and GROSS 1971), as do various endoglycosidases terminating the intracellular degradation of the polyose.

F. Structural Glycoproteins

Besides GAGs and proteoglycans, all connective tissues also contain structural glycoproteins that are specific tissue components. Some are well defined, such as fibronectin, laminin and entactin (for review, see GRANT et al. 1981) and the glycoprotein of the elastic fibres; others are vague constituents that have been called non-collagen proteins. This fraction contains various proteins filtrated from the plasma and other glycoproteins more resistant to extraction that are probably specific connective tissue components (for details and references, see ANDERSON 1976).

I. Fibronectin, Laminin, Entactin and Others

Fibronectin or cold insoluble globulin is a large glycoprotein found in most body fluids and in extracellular matrices. It is an adhesive glycoprotein involved in various contact processes such as cell attachment spreading and migration, maintenance of normal cell morphology, opsonisation, differentiation and oncogenic transformation. The structural organisation of this glycoprotein containing four to six oligosaccharide units per polypeptide is well known, being composed of domains displaying specific binding sites for collagen, fibrin, heparin, bacteria, DNA and specific receptors of the cell surface (for review, see YAMADA 1983). The isolation of cDNA clones and the determination of the complete primary structure of fibronectins from various sources and species strongly suggest that the diverse forms of fibronectin (cellular, plasmatic or oncofoetal) are generated by the transcription of a single gene into a common precursor mRNA undergoing alternative splicings (KORNBLIHTT et al. 1985).

Laminin is a large ($9–10 \cdot 10^5$ daltons) cross-shaped glycoprotein which is formed of three disulphide-linked subunits of 440 Kd and 220 Kd and is found in basement membranes. It is located in the lamina rara and also contains functional domains displaying specific binding for the major components of basement membrane, type IV collagen, heparan sulphate proteoglycan, nidogen and cell surface through a specific receptor (for review, see VON DER MARK and KÜHL 1985). It contains terminal mannose and binds to concanavalin A. Antibodies have been raised against this glycoprotein and some of its pepsin-resistant fragments. They react with all basement membranes.

Nidogen is a sulphated glycoprotein of ± 160 Kd that has been isolated from basement membranes of endodermal neoplastic cells (BENDER et al. 1981) or from the Engelbreth-Holm-Swarm tumour (TIMPL et al. 1983). It is different from laminin but forms stable complexes with it and might also play a role in cell adhesion.

The microfibrillar component of the elastic fibres are made of disulphide-linked glycoproteins. Recent data suggest that the major antigen of elastin-associated microfibrils is a 31-Kd glycoprotein (GIBSON et al. 1986). Such glycoproteins are supposed to be secreted and polymerised in the extracellular space to provide the framework within which elastin is deposited.

II. Biosynthesis

The biosynthesis of a glycoprotein starts with the assembly of the polypeptide chain using the usual cell machinery. There are two main types of linkage of the glycoconjugate to the protein, *N*-acetyl-*d*-glucosamine-asparagine (as in most serum glycoproteins and the C-terminal precursor extension of type I collagen) and *N*-acetyl-*d*-galactosamine-serine (or threonine), as in epithelial mucus. In most glycoproteins the sugar chains are branched and multiple. They are modulated post-synthetically by enzyme-mediated addition or subtraction of their termini. Besides galactose, fucose and xylose (but not uronic acid) and hexosamine, glycoproteins often contain mannose and terminal sialic acid. The complete sugar moiety that will be N-linked to the side chain of the amino acid is synthesised in the cytosol by the dolichol pathway, i.e. by linking the first sugar to a polyprenol lipid and successive enzyme-mediated addition of the sugars. The polyprenoid-linked saccharide is transferred to the cisternae of the endoplasmic reticulum and enzymatically bound to the polypeptide. Secretion of the completed glycoprotein occurs via the Golgi vesicular system (for review and details, see PHELPHS 1980).

G. Regulation and Diseases

The density of fibroblasts in the connective tissues and their biosynthetic activity are highly regulated functions. These cells display a metabolic activity and a rate of multiplication that are high in a growing organism and low in normal adult tissues. Metabolic activity and cell multiplication can be markedly increased under pathological or experimental conditions.

I. Fibroblasts

In an adult connective tissue such as the dermis, the fibroblasts are regularly spaced and isolated from one another by matrix components. In adult dermis there are few multiplications of the fibroblasts and few signs of biosynthetic activity. These thin elongated cells are often called fibrocytes. The embryonic, foetal and new-born fibroblasts establish contact in the form of tight junctions. These cells are larger, contain more cytoplasmic organelles, multiply and display signs of important biosynthetic activity (a well developed rough endoplasmic reticulum, uptake of radioactive amino acids, etc.).

Even when collected from adult skin, fibroblasts in culture revert to a very active phase of multiplication, but in the absence of viral or chemical transformation they display a limited potential for replication (HAYFLICK 1965). In these experimental conditions the maximum replicative life span of the fibroblast in culture varies from tissue to tissue (it is higher in the papillary than in the reticular dermis) and with the donor's age. Progeria and Werner's syndrome are characterised by a strikingly reduced replicative capacity of the fibroblast (MARTIN et al. 1965).

In culture the fibroblasts divide until they reach saturation density. This contact inhibition is suppressed by replating the cells at a lower density. Lack of contact inhibition is a characteristic of transformed cells. Under such conditions, the synthesis of collagen and fibronectin are strikingly reduced (HATA and PETERKOSKY 1977; HYNES et al. 1979). When seeded in culture within a three-dimensional collagen matrix, fibroblasts attach to and retract their fibrous support and no longer divide, therefore displaying morphological and functional features somewhat similar to those of the fibrocytes of the adult dermis (BELL et al. 1979; NUSGENS et al. 1984). The retractile properties of fibroblasts are impaired in dermatosparaxis (DELVOYE et al. 1983), a defect perhaps linked to the lack of anchorin, a collagen binding protein, at the surface of these cells (MAUCH and VON DER MARK 1986). Fibroblasts from keloids display increased retractile properties (personal observation).

In granulation tissue, during skin wound healing or in inflammatory granuloma and fibrotic diseases (e.g. Dupuytren's contracture) the fibroblasts assume several new features. These cells have been called myofibroblasts (MAJNO et al. 1971) since they display enhanced contractile properties related to a more abundant actomyosin cytoskeleton and modified biosynthetic capacities (increased synthesis of collagen type III). They are responsible for the contraction of the wound. In vitro they behave like smooth muscle cells in respect to pharmacological mediators (GABBIANI et al. 1972).

II. Collagen

Alterations of the genes coding for collagen polypeptides have been demonstrated in an increasing variety of heritable diseases of the connective tissue (for recent review, see UITTO et al. 1986; TSIPOURAS and RAMIREZ 1986; VUORIO 1986). The regulation of collagen synthesis can operate at the gene level and at a pre-translational stage by modification of the level of mRNA or of its stability. Many examples of such regulation processes have been described. Possible mediators of collagen regulation in vivo are the pro-peptides. The addition of amino terminal sequences of pro-collagen type I (and type III) depresses specifically the biosynthesis of collagen by fibroblasts in culture, and not that of other proteins (WIESTNER et al. 1979). The same compound and some of its fragmentation products selectively inhibit the translation of collagen mRNA without affecting this process for other mRNAs (PAGLIA et al. 1979). Similar results have been obtained with the carboxy pro-peptide of type I pro-collagen (WU et al. 1986). The potential significance of this control in pathological conditions is supported by the observation that the increased synthesis observed in cultured fibroblasts from active scleroderma can be reduced to normal by adding amino terminal precursor sequences (KRIEG et al. 1978).

Post-translational enzyme-mediated modifications of collagen polypeptides might also regulate the amount of secreted collagen. In the absence of adequate prolyl hydroxylation, the triple-helical arrangement of the polypeptides is unstable at body temperature and the random coiled polypeptides are degraded intracellularly (ROSENBLOOM et al. 1973). In the absence of vitamin C or by chelation of Fe^{2+} the hydroxylases do not function correctly and collagen synthesis is ineffective. Scurvy is the disease related to the lack of vitamin C. The absence of lysyl hydroxylation by reduction of activity of the specific hydroxylase has been found in Ehlers-Danlos syndrome type VI (PINNELL et al. 1972). The absence of lysyl hydroxylation does not modify collagen biosynthesis while it interferes with further processes, glycosylation and cross-linking. The level of the enzymes mediating post-translational modifications of the polypeptides undergoes variations that parallel collagen biosynthesis (ANTTINNEN et al. 1973, 1977) as a function of age and in some pathological processes.

The intracellular degradation of collagen is modulated by the level of cAMP and depressed by ammonium chloride (BAUM et al. 1980). All drugs or cell–matrix interactions affecting the cytoskeletal proteins inhibit collagen fibre formation (EHRLICH et al. 1974) and increase intracellular degradation (UNEMORI and WERB 1986).

The reduced activity of lysyl oxidase has been proposed to be responsible for Ehlers-Danlos syndrome type V and some forms of cutis laxa (BORNSTEIN and BYERS 1980). In the absence of ionised copper the function of lysyl oxidase is impaired, as might be the case in Menkes' kinky hair syndrome (DANKS et al. 1972). The activity of pro-collagen peptidases can be inhibited by the presence of serum inhibitors such as α_2-macroglobulin and other glycoproteins. The lack of activity of pro-collagen peptidase type I is responsible for dermatosparaxis (LAPIÈRE et al. 1971) in various animal species and in one type of Ehlers-Danlos syndrome type VII.

Extracellular degradation is also highly controlled. Collagenolysis is a process regulated in multiple steps from transcription of mRNA to inhibition of active enzymes in tissues involving complex interactions between activators and inhibitors (for review, see HARRIS et al. 1984). Increased collagenase activity is responsible for the recessive form of dystrophic epidermolysis bullosa (BAUER et al. 1978). Leucocyte and macrophage collagenase might participate in connective tissue destruction in inflammatory diseases (WERB 1978). Cancer cells produce different types of collagenase among which one is specific for basement membranes and required for tumour growth and metastasis (LIOTTA et al. 1979). Degradation of collagen produces small peptides and free amino acids. A small proportion of these products containing hydroxyproline is excreted in urine and their amount is increased in the presence of enhanced collagen turnover (KIVIRIKKO 1970).

III. Elastin

Biosynthesis is active during development and a close parallel between mRNA template activity and elastin production suggests that the control is at the level of the gene (BURNETT et al. 1980; DAVIDSON et al. 1984). Unlike pro-collagen, inhibition of peptidyl proline hydroxylation does not limit the rate of tropoelastin secretion (ROSENBLOOM 1984).

A depression of lysyl oxidase activity is responsible for impaired cross-linking and defective elastic fibre formation, as in nutritional copper deficiency, Menkes' kinky hair syndrome and some forms of cutis laxa (see review by BORNSTEIN and BYERS 1980; ROSENBLOOM 1984).

Elastin biosynthesis in skin mainly occurs during growth. It seems reduced and even non-existent in the adult. Reconstruction of elastic fibres in scar tissue appears possible (SCHWARTZ 1977). Alterations of the elastic fibres in aging dermis are well documented (BOUISSOU et al. 1984).

Degradation of elastic fibres is mediated by neutral proteases and the most active are elastases. Enzymes of this class are present in the granules of polymorphonuclear leucocytes. The activity of elastases is inhibited by various compounds (for review, see BIETH 1980), among them α_2-macroglobulin and α_1-antitrypsin.

IV. Proteoglycans

Little is known about the regulation of each of the enzymes involved in the assembly of the GAGs. Hyaluronate synthetase has been extracted from skin and its activity has been shown to be modulated by oestrogen in mouse skin (USUKA et al. 1981). Synthetase activity in a cell-free system is increased in fibroblasts cultured from skin of a patient presenting Marfan's syndrome (APPEL et al. 1979).

The rate of turnover of the GAGs in skin seems to be higher than that of the fibrous proteins (collagen and elastin) with which they are associated. This rate is not identical for all types of GAG. This is most obvious during limb bud development characterised by a decreasing ratio of hyaluronate to chondroitin sulphate (TOOLE and GROSS 1971) that depends on increased hyaluronidase activity (TOOLE 1972).

Proteoglycans degradation in inflammation is mediated by proteinases and glycanases of defined specificity (for review, see VAES 1985). Lack of one of the various endoglycosidases catalysing the terminal hydrolysis of the GAG leads to a specific form of storage disease (for review, see VAN HOFF 1974).

V. Structural Glycoproteins

The regulation of structural glycoproteins is poorly defined. Many cell types are able to synthesise proteins of this class that are important in establishing contact with their surrounding support. Fibronectin, one of these proteins, is present in high concentration in the blood. Its serum concentration is reduced in septic shock. Abnormal distribution of fibronectin has been described in various pathological conditions (for review see MOSHER and FURCHT 1981). Ehlers-Danlos syndrome type X was shown to be associated with an alteration of the functional properties of fibronectin (ARNESON et al. 1980).

Many proteases, including thrombin, can degrade fibronectin but regulation at this level is ill defined.

The synthesis of laminin is increased in diabetes while the amount of the heparan sulphate proteoglycan is decreased. This could account for the thickened basement membranes in this disease (ROHRBACH et al. 1982).

H. Interaction Between the Macromolecules of the Connective Tissue

The function of the connective tissue in a multicellular organism depends on its mechanical properties. Each of the macromolecules participating in the composition of the supporting framework is synthesised and secreted as soluble monomers that interact in the extracellular space to form polymeric structures. Under their physiological state in the tissues, except for a very small fraction of newly synthesised molecules and degradation products, these components are aggregated and insoluble. This interaction is potentially very complex since it involves a number of substances, including those forming the cell coat and the cell membrane. An increasing number of studies demonstrate the importance of the cell–matrix interactions in development, growth, differentiation and pathological processes (for review, see SLAVKIN and GREULICH 1975; REDDI 1985).

J. Conclusions

All connective tissues of the organism are differentiated and specialised and these characteristics depend on the phenotypic expression of the cells responsible for the synthesis of the macromolecular components. The various types of fibroblast use families of genes to code for the different types of collagen, proteoglycans, glycoproteins and elastin while the basic molecular mechanism responsible for the post-translational modification of the initial products of synthesis are quite similar in all types of cell. Differentiation results from a modulation of the genetic expression regulating the nature, the amounts and the proportions of the various products secreted in the extracellular space and probably also the timing of their secretion. These factors are of prime importance in conditioning the interaction between the macromolecules that organise in the extracellular space to form a network with adequate mechanical properties. Connective tissue cells, like all types of cell, also interact with their supports by the intermediate of specialised secretion products. Such an interaction is most often required for the expression of their function.

References

Anderson JC (1976) Glycoproteins of the connective tissue matrix. In: Hall DA, Jackson DS (eds) International review of connective tissue research, vol 7. Academic, New York, pp 251–322

Anttinen H, Orava S, Ryhanen L, Kivirikko KI (1973) Assay of protocollagen lysyl hydroxylase activity in the skin of human subjects and changes in the activity with age. Clin Chim Acta 47:289–294

Anttinen H, Oikarinen A, Kivirikko KI (1977) Age-related changes in human skin collagen galactosyltransferase and collagen glucosyltransferase activities. Clin Chim Acta 76:95–101

Appel A, Horwitz AL, Dorfman A (1979) Cell free synthesis of hyaluronic acid in Marfan syndrome. J Biol Chem 254:12199–12203

Arneson MA, Hammerschmidt DE, Furcht LT, King RA (1980) A new form of Ehlers-Danlos syndrome. JAMA 244:144–147

Barrett AJ (1975) The enzymic degradation of cartilage matrix. In: Burleigh PMc, Poole AR (eds) Dynamics of connective tissue macromolecules. Elsevier, Amsterdam, pp 189–226
Bauer EA, Gedde-Dahl T, Eisen AZ (1978) The role of human skin collagenase in epidermolysis bullosa. J Invest Dermatol 68:119–124
Baum BJ, Moss J, Breul SD, Berg RA, Crystal RG (1980) Effect of cyclic AMP on the intracellular degradation of newly synthesized collagen. J Biol Chem 255:2843–2847
Bell E, Ivarsson B, Merrill C (1979) Production of a tissue like structure by contraction of collagen lattices by human fibroblasts of different proliferative potential in vitro. Proc Natl Acad Sci USA 76:1274–1278
Bender BL, Jaffe R, Carlin B, Chung AE (1981) Immunolocalization of entactin, a sulfated basement membrane component, in rodent tissues and comparison with GP-2 (laminin). Am J Pathol 103:419–426
Bieth J (1980) Natural and synthetic inhibitors of pancreatic and leucocyte elastases. In: Robert L (ed) Frontiers of matrix biology, vol 8. Karger, Basel, pp 216–227
Bornstein P, Byers PH (1980) Disorders of collagen metabolism. In: Bondy PK, Rosenberg LE (eds) Metabolic control and disease, 8th edn. Saunders, Philadelphia
Bornstein P, Sage H (1980) Structurally distinct collagen types. Annu Rev Biochem 49:957–1003
Bouissou H, Pierragi MT, Julian M (1984) Dermis ageing. Pathol Res Pract 178:515–517
Burnett W, Eichner R, Rosenbloom J (1980) Correlation of functional elastin messenger ribonucleic acid levels and rate of elastin synthesis in the developing chick aorta. Biochemistry 19:1106–1111
Chandrasekhar S, Kleinman HK, Hassell JR, Martin GR, Termine JD, Trelstad RL (1984) Regulation of type I collagen fibril assembly by link protein and proteoglycans. Collagen Relat Res 4:323–337
Comper WD, Laurent TC (1978) Physiological function of connective tissue polysaccharides. Physiol Rev 58:255
Cotta-Pereira G, Rodrigo FG, Bittencourt-Sampaio S (1976) Oxytalan, elaunin and elastic fibers in the human skin. J Invest Dermatol 66:143
Danks DM, Stevens BJ, Campbell PE, Gillespie JM, Walker-Smith J, Blomfield J, Turner B (1972) Menkes' kinky hair syndrome: an inherited defect in intestinal copper absorption with widespread consequences. Lancet I:1100–1103
Davidson JM, Shibahara S, Boyd C, Mason ML, Tolstoshev P, Crystal R (1984) Elastin mRNA levels during foetal development of sheep nuchae ligament and lung. Biochem J 220:653–661
de Crombrugghe B, Schmidt A, Liau G, Setoyama C, Mudryj M, Yamada Y, McKeon C (1985) Structural and functional analysis of the genes for α_2 (I) and α_1 (III) collagens. Ann NY Acad Sci 460:154–162
Delvoye P, Nusgens BV, Lapière ChM (1983) The capacity of retracting a collagen matrix is lost by dermatosparactic skin fibroblasts. J Invest Dermatol 81:267–270
Ehrlich HP, Ross R, Bornstein P (1974) Effects of antimicrotubular agents on the secretion of collagen. A biochemical and morphological study. J Cell Biol 62:390–405
Fleischmajer R, Olsen BR, Timpl R, Perlish JS, Lovelace O (1983) Collagen fibril formation during embryogenesis. Proc Natl Acad Sci USA 80:3354–3358
Foster JA, Rich CB, Fletcher S, Karr SR, Przybyla A (1980) Translation of chick aortic elastin messenger ribonucleic acid. Comparison to elastin synthesis in chick aorta organ culture. Biochemistry 19:857–864
Fransson LA, Coster L, Carlstedt I, Malmstrom A (1985) Domain structure of proteoheparan sulphate from confluent cultures of human embryonic skin fibroblasts. Biochem J 231:683–687
Fujii N, Nagai Y (1981) Isolation and characterization of a proteodermatan sulfate from calf skin. J Biochem 90:1249–1258
Gabbiani G, Hirschel BJ, Ryan GB, Statkov PR, Majo G (1972) Granulation tissue as a contractile organ. A study of structure and function. J Exp Med 135:719–725
Gay S, Viljanto J, Raikallio J, Penttinen R (1978) Collagen types in early phases of wound healing in children. Acta Chir Scand 144:205–211

Gibson MA, Hughes JL, Fanning JC, Cleary EG (1986) The major antigen of elastin associated microfibrils is a 31 Kd glycoprotein. J Biol Chem 261:11429–11434
Grant ME, Heathcote JG, Orkin RW (1981) Current concepts of basement membrane structure and function. Biosci Rep 1:819–842
Gross J, Lapière ChM (1962) Collagenolytic activity in amphibian tissues: a tissue culture assay. Proc Natl Acad Sci USA 48:1014–1022
Gross RH, Sheldon LA, Fletcher CF, Brinckerhoff CE (1984) Isolation of a collagenase cDNA clone and measurement of changing collagenase mRNA levels during induction in rabbit synovial fibroblasts. Proc Natl Acad Sci USA 81:1981–1985
Harper RA, Grove G (1979) Human skin fibroblasts derived from papillary and reticular dermis: differences in growth potential in vitro. Science 204:526–527
Harris ED, Welgus HG, Krane SM (1984) Regulation of the mammalian collagenases. Review. Collagen Relat Res 4:493–504
Hassell JR, Robey PG, Barrach HJ, Wilczek J, Rennard SI, Martin GR (1980) Isolation of a heparan sulfate containing proteoglycan from basement membrane. Proc Natl Acad Sci USA 77:4494–4498
Hata RI, Peterkofsky B (1977) Specific changes in collagen phenotype of Balb 3T3 cells as a result of transformation by sarcoma viruses or a chemical carcinogen. Proc Natl Acad Sci USA 74:2933–2937
Hatamochi A, Paterson B, de Crombrugghe B (1986) Differential binding of a CCAAT DNA binding factor to the promoters of the mouse α_2 (I) and α_1 (III) collagen genes. J Biol Chem 261:11310–11314
Hayflick L (1965) The limited in vitro lifetime of human diploid cell strains. Exp Cell Res 37:614–636
Helseth DL, Lechner JH, Veis A (1979) Role of the aminoterminal extrahelical region of type I collagen in directing the 4D overlap in fibrillogenesis. Biopolymers 18:3005–3014
Hodge AH, Petruska JA (1963) Recent studies with the electronmicroscope on ordered aggregates of the tropocollagen macromolecules. In: Ramachandran GN (ed) Aspect of proteins structure. Academic, New York, pp 289–311
Hulmes DJS, Miller A (1979) Quasi hexagonal molecular packing in collagen fibrils. Nature 282:878–880
Hynes RO, Destree AT, Perkins ME, Wagner DD (1979) Cell surface fibronectin and oncogenic transformation. J Supramol Struct 11:95–102
Kent WM, Funderburg FM, Culp LA (1986) Proteoglycans in the substratum adhesion sites of human papillary or reticular dermal fibroblasts. Aging in vivo or in vitro. Mech Ageing Dev 33:115–137
Kivirikko KI (1970) Urinary excretion of hydroxyproline in health and disease. In: Hall DA, Jackson DS (eds) International review of connective tissue research, vol 8. Academic, New York, pp 23–72
Kivirikko KI, Myllyla R (1984) Biosynthesis of collagens. In: Piez KA, Reddi AH (eds) Extracellular matrix biochemistry. Elsevier, New York, pp 83–118
Kleinman HK, Wilkes CM, Martin GR (1981) Interaction of fibronectin with collagen fibrils. Biochemistry 20:2325–2330
Kornblihtt AR, Umezawa K, Vibe-Pedersen K, Baralle FE (1985) Primary structure of human fibronectin: differential splicing may generate at least 10 polypeptides from a single gene. EMBO J 4:1755–1759
Krieg T, Horlein D, Wiestner M, Muller PK (1978) Aminoterminal extension peptides from type I procollagen normalize excessive collagen synthesis of scleroderma fibroblasts. Arch Dermatol Res 263:171–180
Lapière ChM, Lenaers A, Kohn L (1971) Procollagen peptidase: an enzyme excising the coordination peptides of procollagen. Proc Natl Acad Sci USA 68:3054–3058
Lapière ChM, Nusgens B, Pierard GE, Hermanns JF (1975) The involvement of procollagen in spatially orientated fibrogenesis. In: Burleigh M, Poole R (eds) Dynamics of connective tissue macromolecules. North Holland, Amsterdam, pp 33–50
Lapière ChM, Nusgens B, Pierard GE (1977) Interaction between collagen type I and type III in conditioning bundles organization. Connect Tissue Res 5:21–29

Liotta LA, Abe S, Robey PG, Martin GR (1979) Preferential digestion of basement membrane collagen by an enzyme derived from a metastatic murine tumor. Proc Natl Acad Sci USA 76:2268–2272
Mainardi CL, Dixit SN, Kang AH (1980) Degradation of type V (basement membrane) collagen by a proteinase isolated from human polymorphonuclear leukocyte granules. J Biol Chem 255:5435–5441
Majno G, Gabbiani G, Hirschel BJ, Ryan GB, Statkov PR (1971) Contraction of granulation tissue in vitro: similarity to smooth muscle. Sciences 173:548–550
Martin GM, Gartler SM, Epstein CJ, Motulsky AG (1965) Diminished lifespan of cultured cells in Werner's syndrome. Fed Proc 24:678
Matsunaga E, Shinkai H (1986) Two species of dermatan sulfate proteoglycans with different molecular size from newborn calf skin. J Invest Dermatol 87:221–226
Mauch C, Von der Mark K (1986) Dermatosparaxis, a molecular defect in collagen binding proteins? 10th Meeting of the federation of European connective tissue societies, July 28th–August 1st, Manchester
Miller EJ (1985) Recent information on the chemistry of collagens. In: Butler WT (ed) The chemistry and biology of mineralized tissues. Ebasco Media, Birmingham, Ala, pp 80–93
Miyahara M, Yayashi K, Berger J, Tanzawa K, Njieha F, Trelstad RL, Prockop DJ (1984) Formation of collagen fibrils by enzymatic cleavage of precursors of type I collagen in vitro. J Biol Chem 259:9891–9898
Mosher DF, Furcht LT (1981) Fibronectin: review of its structure and possible functions. J Invest Dermatol 77:175–180
Nakamura T, Nagai Y (1980) Developmental changes in the synthesis of glycosaminoglycans and collagen in embryonic chick skin. J Biochem (Tokyo) 87:629–637
Nusgens BV, Goebels Y, Shinkai H, Lapière ChM (1980) Procollagen type III N-terminal endopeptidase in fibroblast culture. Biochem J 191:699–706
Nusgens B, Merrill C, Lapière ChM, Bell E (1984) Collagen biosynthesis by cells in a tissue equivalent matrix in vitro. Collagen Rel Res 4:351–364
Paglia L, Wilczek J, Diaz de Leon L, Martin GR, Horlein D, Muller P (1979) Inhibition of procollagen cell free synthesis by aminoterminal extension peptides. Biochemistry 18:5030–5034
Partridge SM, Elsden DF, Thomas J (1963) Constitution of the cross-linkages of elastin. Nature 197:1297
Pearson CH (1981) The ground substance of the periodontal ligament. In: Berkovitz BK, Moxham B, Newman HN (eds) The periodontal ligament in health and disease. Pergamon, Oxford, pp 125–143
Pearson CH, Gibson GJ (1982) Proteoglycans of bovine periodontal ligament and skin. Occurrence of different hybrid sulphated galactosaminoglycans in distinct proteoglycans. Biochem J 201:27–37
Peltonen L, Halila R, Ryhanen L (1985) Enzymes converting procollagens to collagens. J Cell Biochem 28:15–21
Phelps CF (1980) Glycosylation. In: Freedman RB, Hawkins HC (eds) The enzymology of post-translational modification of proteins, vol 1. Academic, New York, pp 105–155
Piez KA, Trus BL (1978) Sequence regularities and packing of collagen molecules. J Mol Biol 122:419–432
Pinnell SR, Krane SM, Kenzora JE, Glimcher MJ (1972) Heritable disorder of connective tissue. Hydroxylysine deficient collagen disease. N Engl J Med 286:1013–1020
Ramirez F, Bernard M, Chu M, Dickson L, Sangiorgi F, Weil D, de Wet W, Junien C, Sobel M (1985) Isolation and characterization of the human fibrillar collagen genes. Ann NY Acad 460:117–129
Reddi AH (1985) Extracellular matrix: structure and function. UCLA Symp Mol Cell Biol. New Series 25:436
Reddi AH, Gay R, Gay S, Miller EJ (1977) Transitions in collagen types during matrix induced cartilage, bone, and bone marrow formation (immunofluorescence connective tissue). Proc Natl Acad Sci USA 74:5589–5592

Robert M, Robert L (1969) Determination of elastolytic activity with 125 I and 131 I labelled elastin. Eur J Biochem 11:62–67

Roden L, Horowitz MI (1978) Structure and biosynthesis of connective tissue proteoglycans. Mammalian glycoproteins, glycolipids and proteoglycans. In: Horowitz MI, Pigman W (eds) The glycoconjugates, vol 2. Academic, New York, pp 3–71

Rohrbach DH, Hassell JR, Kleinman H, Martin GR (1982) Alterations in the basement membrane (heparan sulfate) proteoglycan in diabetic mice. Diabetes 31:185–188

Rosenbloom J (1984) Elastin: relation of protein and gene structure to disease. Lab Invest 51:605–623

Rosenbloom J, Harsch M, Jimenez S (1973) Hydroxyproline content determines the denaturation temperature of chick tendon collagen. Arch Biochem Biophys 158:478

Rosenbloom J, Harsch M, Cywinski A (1980) Evidence that tropoelastin is the primary precursor in elastin biosynthesis. J Biol Chem 255:100–106

Sandberg LB, Soskel NT, Leslie JG (1981) Elastin structure, biosynthesis and relation to disease states. N Engl J Med 304:566–579

Schmidt A, Rossi P, de Crombrugghe B (1986) Transcriptional control of the mouse α_2 (I) collagen gene: functional deletion analysis of the promoter and evidence for cell-specific expression. Mol Cell Biol 6:347–354

Schwartz D (1977) The proliferation of elastic fibres after skin incisions in albino mice and rats: a light and electron microscopic study. J Anat 124:401–412

Scott JE, Haigh M (1985) Proteoglycan type I collagen fibril interactions in bone and noncalcifying connective tissues. Biosci Rep 5:71–81

Sear CHJ, Grant ME, Jackson DS (1981) The nature of the microfibrillar glycoproteins of elastin fibres. A biosynthetic study. Biochem J 194:587–598

Sengel P (1976) Morphogenesis of skin. In: Abercrombie M, Newth DR, Torrey JG (eds) Development and cell biology series. Cambridge University Press, Cambridge

Siegel RC (1979) Lysyl oxidase. In: Hall DA, Jackson DS (eds) International review of connective tissue research, vol 8. Academic, New York, pp 73–118

Slavkin H, Greulich R (1975) Extracellular matrix influences on gene expression. Academic, New York, p 833

Tajima S, Nagai Y (1980) Distribution of macromolecular components in calf dermal connective tissue. Connect Tissue Res 7:65–71

Tanzer ML (1976) Cross-linking. In: Ramachandran GN, Reddi AH (eds) Biochemistry of collagen. Plenum, New York, pp 137–162

Timpl R, Dziadek M, Fujiwara S, Nowack H, Wick G (1983) Nidogen: a new self aggregating basement membrane protein. Eur J Biochem 137:455–465

Toole BP (1972) Hyaluronate turnover during chondrogenesis in the developing chick limb and axial skeleton. Dev Biol 29:321–329

Toole BP, Gross J (1971) The extracellular matrix of the regenerating new limb: synthesis and removal of hyaluronate prior to differentiation. Dev Biol 25:57–77

Trelstad RL (1982) Multistep assembly of type I collagen fibrils. Cell 28:197–198

Tsipouras P, Ramirez F (1986) Genetic disorders of collagen. J Med Genet 24:2–8

Uitto J, Murray L, Blumberg B, Shamban A (1986) Biochemistry of collagen in diseases. Ann Intern Med 105:740–756

Unemori EN, Werb Z (1986) Reorganization of polymerized actin: a possible trigger for induction of procollagenase in fibroblast cultured in and on collagen gels. J Cell Biol 103:1021–1031

Usuka M, Nakajima K, Ohta S, Mori Y (1981) Induction of hyaluronic acid synthetase by estrogen in the mouse skin. Biochem Biophys Acta 673:387–393

Vaes G (1985) Macrophage secretory products and connective tissue remodeling: role of macrophage enzymes and matrix regulatory monokines. In: Dean RT, Stahl P (eds) Developments in cell biology, vol 1. Secretory processes. Butterworths, London, pp 99–117

Van Hoof F (1974) Mucopolysaccharidoses and mucolipidoses. J Clin Pathol [Suppl] 8:1–11

Vogeli G, Avvedimento EV, Sullivan M, Maizel JV, Lozano G, Adams SL, Pastan I, de Crombrugghe B (1980) Isolation and characterization of genomic DNA coding for alpha 2 type I collagen. Nucl Acids Res 8:1823–1837

Von der Mark K, Kühl U (1985) Laminin and its receptor. Biochim Biophys Acta 823:147–160
Vuorio E (1986) Connective tissue diseases – mutations of collagen genes. Ann Clin Res 18:234–241
Werb Z (1978) Pathways for the modulation of macrophage collagenase activity. In: Horton JE, Tarpley TM, Davis WF (eds) Mechanisms of localized bone loss. National Institutes of Health, Washington, pp 213–228
Wiestner P, Krieg T, Horlein D, Glanville RW, Fietzek P, Muller PK (1979) Inhibiting effect of procollagen peptides on collagen biosynthesis in fibroblast cultures. J Biol Chem 254:7016–7023
Williams BR, Gelman RA, Poppke DC, Piez KA (1978) Collagen fibril formation. Optimal in vitro conditions and preliminary kinetic results. J Biol Chem 253:6578–6585
Wu CH, Donovan CB, Wu GY (1986) Evidence for pretranslational regulation of collagen synthesis by procollagen propeptides. J Biol Chem 261:10482–10484
Yamada KM (1983) Cell surface interactions with extracellular materials. Annu Rev Biochem 52:761–799

CHAPTER 6

Dermal Blood Vessels and Lymphatics

D. I. ABRAMSON

This chapter deals with the gross and microscopic anatomy, physiology and pharmacology of the blood vessels and lymphatics of the skin.

A. Dermal Blood Circulation

I. Functions of Dermal Vascular Bed

The cutaneous blood circulation in the extremities has several unrelated but important functions, foremost among which is to supply different structures with oxygen and nutritive substances and to remove metabolites. However, the rate of blood flow through this vascular bed is far in excess of that necessary to satisfy the constant and low metabolic requirements of skin. The explanation is that in the case of the limbs, particularly the distal portions, the cutaneous circulation also plays a role in the preservation of a steady body temperature in the face of marked variations in body heat production and a wide range of ambient temperatures (see Sect. A.III.1).

Besides the above-mentioned functions, the many vascular plexuses, the capillary and venular beds, and other types of vessels in the skin act as a reservoir when there is a need for shunting blood to inactive tissues. At the same time, a proper relationship is maintained between fluid and circulating blood volumes. Dermal blood vessels also contribute to the response to inflammation, the capillaries and the subpapillary venous plexuses participating in skin disorders characterised by erythema.

II. Anatomy of Dermal Blood Vessels

The arrangement of the cutaneous vasculature is complex, influenced by the region of the body, the thickness of the panniculus adiposus and the relation of the skin to the underlying bone or fascia. In the finger, for example, the vascularity is developed beyond the degree necessary for the metabolic needs of the comparatively thin epidermis and the small number of adnexal structures found in this site (WILKINS et al. 1938). In the hand proper, the volume of the volar dermal vascular bed is approximately twice that of the dorsal dermal vascular bed. The presence of systems of interconnections at all levels of the skin suggests that blood flows both vertically towards the surface and horizontally in the different vascular plexuses.

Fig. 1. Candelabrum-like arteries of the skin, in posterior section of a finger-nail, sagittal view, obtained from a 60-year-old man (FLEISCHHAUER and HORSTMANN 1955). Benzidin stain, ×45

1. Distributing Arteries

The cutaneous arteries arise from the subcutaneous tissue and then enter the dermis to form an extensive anastomosing system, the deep plexus (cutaneous arterial plexus). This network lies between the deep reticular portion of the dermis and the subcutaneous tissue (HORSTMANN 1957). Some of its branches extend to the subcutaneous layer, whereas others ascend through the reticular portion of the dermis, branching throughout to form candelabrum-type arrangements (FLEISCHHAUER and HORSTMANN 1955). Each of the latter anastomoses with similar neighbouring structures (Fig. 1). A second more superficial network (subpapillary arteriolar plexus), also having its origin from branches of the deep plexus, gives off terminal arterial twigs that no longer anastomose but instead end in capillary loops running in the connective tissue of the papillae. Arterial arcades are formed by a regular pattern of intercommunicating links, from which arise most of the precapillaries (arterioles). Another series of small arteries joins veins through the intermediary of arteriovenous shunts (see Sect. A.II.4).

2. Arterioles and Metarterioles

Arterioles

Arterioles are widely distributed in the dermis. Two types have been described in the fingertips, based on morphological differences. Those in the deep subcutaneous tissue are similar to comparable vessels in other organs. In the other group, the chief differences are a thick endothelial lining; a large number of endothelial nuclei with respect to vessel size; the presence of bundles of filaments, rod-shaped granules and vesicular bodies in the endothelial cytoplasm; and the absence of an elastic lamina (HIBBS et al. 1958). Many such arterioles are associated with sweat glands.

Metarterioles

Metarterioles consist of an intimal layer of endothelium, a discontinuous layer of smooth muscle cells in the media, and an adventitia formed by a scant amount of connective tissue. No elastic laminae are noted.

3. Capillary Bed

Cutaneous capillaries, originating from a succession of arterioles, do not anastomose freely. Instead they present as separate loops supplying the dermis and the

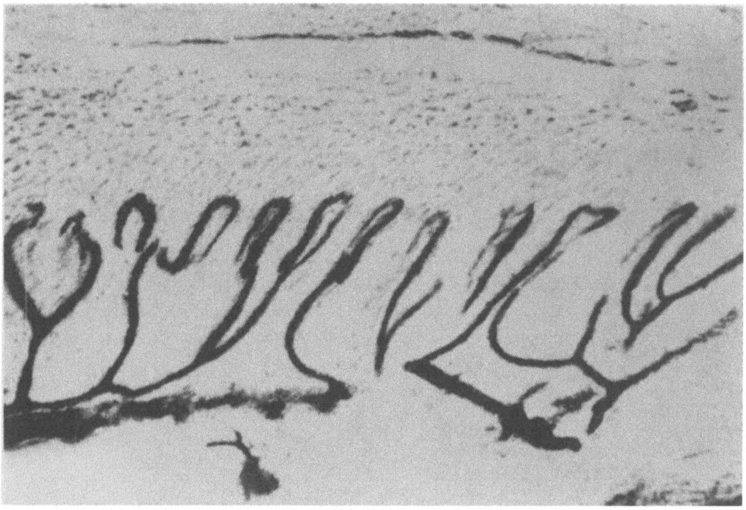

Fig. 2. Regularly spaced attenuated loops forming the vascular bed for the mid-portion of the finger-nail (ELLIS 1961)

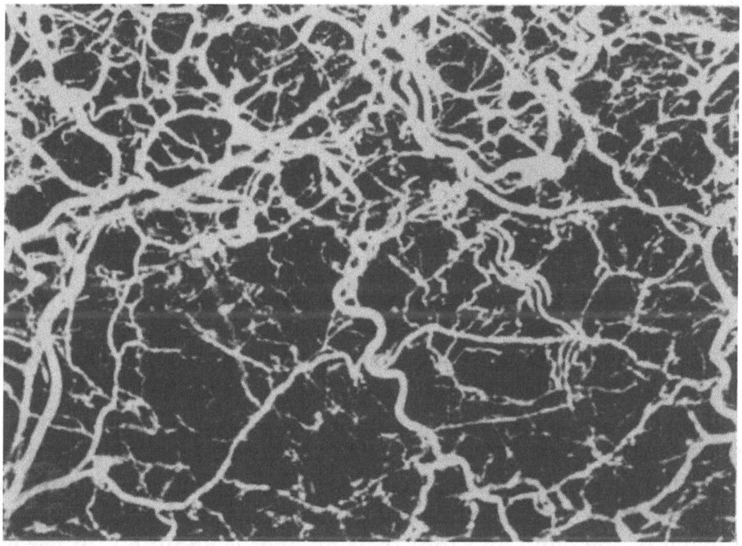

Fig. 3. Microangiogram showing arterioles, venules and the capillary network of the skin of the forearm of a foetus (SAUNDERS 1961). × 36

basal layer of the epidermis (Fig. 2) (ELLIS 1961). Each forms a fine connecting link between the arteriole and the venule (Figs. 3, 4). The loops are oriented in a plane perpendicular to the surface of the skin and consist of an ascending narrow arteriolar limb with an average diameter of 0.010 mm and a more dilated descending venular portion with an average diameter of 0.015 mm (DAVIS and LAWLER

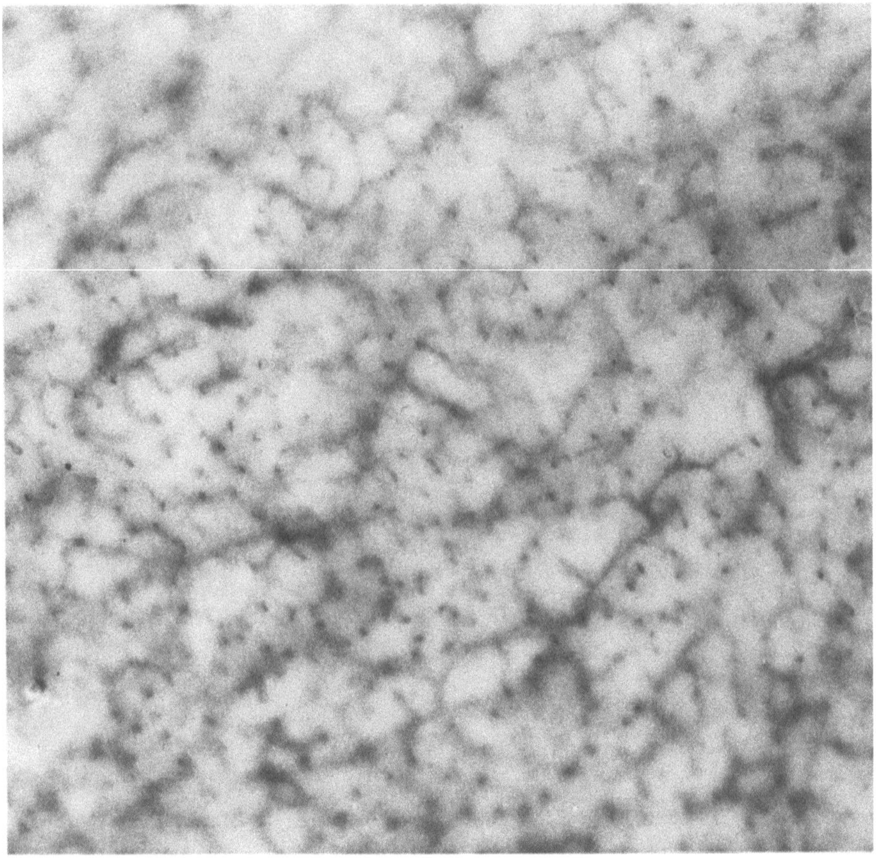

Fig. 4. Normal capillary bed of the dorsum of the hand; also visualised is the subpapillary venous plexus (reproduced by courtesy of Dr. ALLAN L LORINCZ)

1958; DAVIS et al. 1960). Because the apical portion of each loop has the same diameter as the venular limb, it is considered to be part of the latter.

Dermal capillaries are relatively scanty [maximum of 65–70 per sq mm of cross-section (WETZEL and ZOTTERMAN 1926) as compared with 2000 per mm^2 in underlying voluntary muscle (KROGH 1929) and they demonstrate differences in number and morphology depending upon the vascular bed studied. For example, on the dorsum of the hand and foot their number is about 60–70 per mm^2 and their length varies from 0.15 to 0.20 mm. In contrast, in the nail fold, only 20 capillaries are found in an equivalent-sized area and their length is approximately 0.30 mm. In the connective tissue of the dermis there are no capillaries (HORSTMANN 1957), this tissue receiving its blood supply from the microcirculation that accompanies the larger arteries, veins and nerves. Each of the papillae of the skin may have one of several capillaries.

The diameter of the capillary lumen varies in different portions, from 5–7.5 μm in the descending extrapapillary part (BRAVERMAN and YEN 1977) to only 3.5–6.0 μm in the intrapapillary segment.

Microscopic and Electron Microscopic Structure

The capillary consists essentially of an endothelial membrane, one cell thick, which may encircle the entire circumference of the vessel, and a small amount of delicate connective tissue containing pericytes and other cells. In larger capillaries, there may be a single layer of two to three flat endothelial cells curving around the lumen of the vessel, with their serrated edges interdigitating with each other.

On electron microscopic examination, endothelial cells are found to be very complex, acting as the interfacing structure between the blood vessel lumen and the blood (MAJNO 1965). They are found to be linear in appearance, with a raised central area, most likely representing the nucleus (CLARK and GLAGOV 1976). They contain a unique organelle, the rod-shaped tubular body or Weibel-Palade body, approximately 0.1 µm in diameter and up to 3 µm in length, giving it an oblique shape. The origin of this highly organised structure is not clear, but it may be derived from the Golgi apparatus (SENGEL and STOEBNER 1970). The luminal surface of the endothelial cells in dermal capillaries often demonstrates microvilli.

4. Arteriovenous Anastomoses (Shunts)

Arteriovenous anastomoses arise from small arteries or arterioles and end in small veins (Figs. 5,6) (DANIEL and PRITCHARD 1956; HALE and BURCH 1960; PRITCHARD and DANIEL 1954). The afferent artery or arteriole, the connecting loop (arteriovenous anastomosis), the neuroreticular and vascular structures around the vessels, and the efferent vein are collectively termed the glomus.

Arteriovenous shunts are found in the stratum reticulare of the skin of the hands and feet, the greatest number being present in the nail bed of the fingers and toes (as many as 500 per cm^2). The fingertips, finger pad, palmar aspects of the fingers, sole of the foot, and thenar and hypothenar eminences of the hand contain slightly fewer of these structures. They appear histologically as short channels with a lumen approximately 220 µm in diameter. The wall of the vessels is about 20–40 µm thick, which is about three times as great as that of an artery of similar-sized lumen. The wall consists of a layer of endothelial cells, a subendothelial reticular layer and a media containing an outer loosely arranged collagenous reticulum and smooth muscle. No internal elastic membrane is present. In the contractile layer are found the characteristic glomus cells, which are epithelioid in appearance (PRITCHARD and DANIEL 1954). Separating the glomus from the other structures of the cutis is coarse, lamellated collagenous tissue. (For sympathetic innervation of the arteriovenous anastomosis, see Sect. A.IV.2).

5. Venules

The cutaneous venules form four separate networks that run parallel to the surface of the skin but at different levels within the dermis. The most superficial plexus, which lies just below the papillae, is developed from small venules which drain the papillary loops. The second plexus is located a little deeper than the first

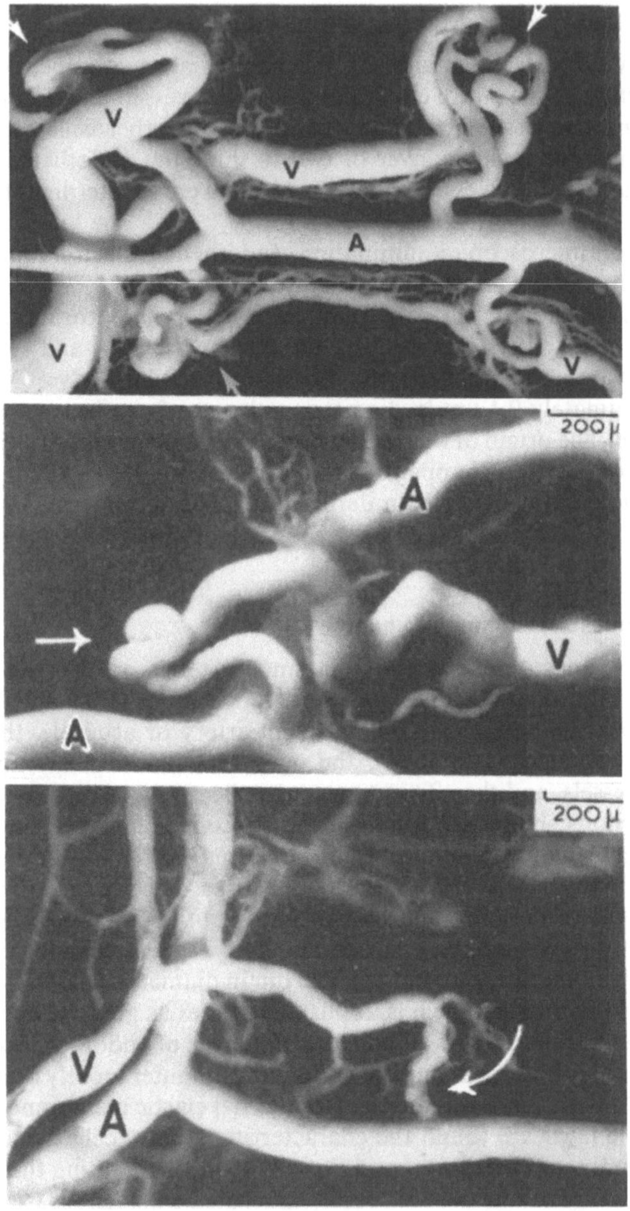

a b c

Fig. 5 a–c. Arteriovenous anastomoses in the external ear of different species of animals visualised by the use of a neoprene latex cast. **a** Vessels of the sheep. Four arteriovenous shunts are present, demonstrating the tortuous nature of these vessels and their large calibre as compared with capillaries. The arteriovenous shunt in the right upper quadrant resembles the complexity of the architecture of the glomus found in the human finger or toe. **b** Vessels of the pig. The arteriovenous anastomosis is much less complex than in the case of the sheep. Its size is much greater than that of the capillaries. **c** Vessels of the dog. Demonstrated is an arteriovenous anastomosis of relatively simple form, commonly found in this species. *A*, artery; *V*, vein. *Arrows* point to arteriovenous anastomoses. (DANIEL and PRITCHARD 1956)

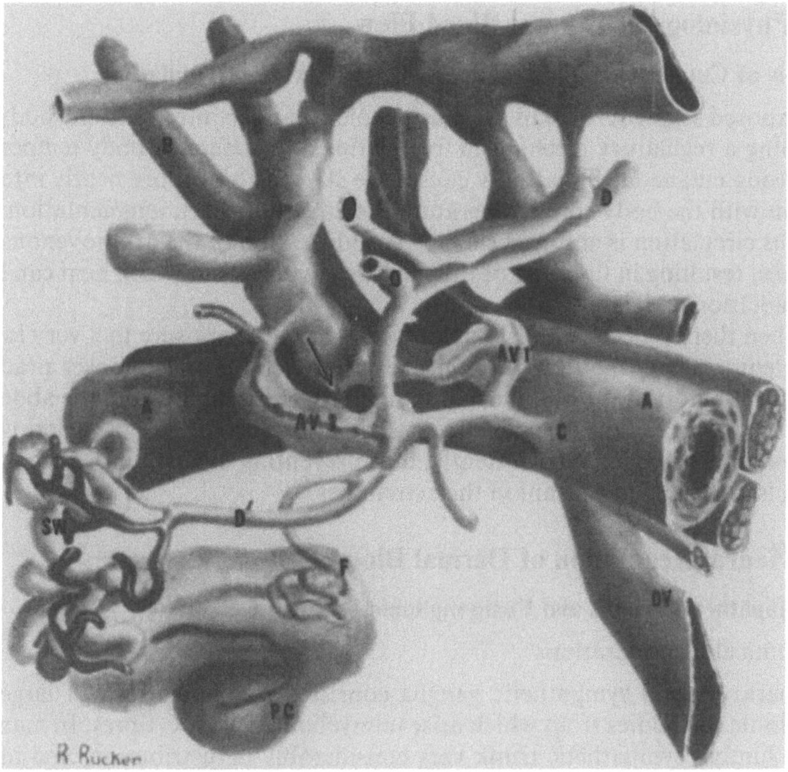

Fig. 6. Steriogram illustrating the reconstructed spatial relations of arteriovenous anastomoses obtained from the volar dermis of the finger of a 2½-month-old infant. Length of arched artery (*A*), running parallel to the skin, is 400 mm. Perforating arteries (*B*) and (*D*) arising from main artery (*A*) run perpendicular to skin surface. *AV1* and *AV2*, arteriovenous anastomoses. Perforating branch (*B*), from which arises the arteriovenous shunt, can be traced to the capillary rete of the skin and surrounding sweat gland (*SW*) and to a pacinian corpuscle (*PC*). (HALE and BURCH 1960)

and is in close association with the subpapillary arteriolar network. The blood in the two plexuses drains into a third one which is located approximately in the middle of the dermis. The fourth or deepest system of venules is found between the dermis and subcutaneous tissue, in the vicinity of the cutaneous arterial plexus. Into it enter many of the branches draining the sweat glands and the adipose tissue.

Venules are made up of single layers of endothelial cells adherent to thin zones of collagen fibres, with pericytes lying just outside the endothelium. Smooth muscle and elastic fibres are generally only found in the larger vessels. Electron microscopy has revealed the presence of thick filaments in the cytoplasm of the cells, similar to those seen in the second type of arteriole (see Sect. A.II.2) (HIBBS et al. 1958). The blood collected in the fourth venular plexus empties into large subcutaneous veins and the deep venous system which accompanies the arteries.

III. Physiology of Dermal Blood Flow

1. Role of Cutaneous Circulation in Body Temperature Regulation

The exposed segments of skin are much cooler than the interior of the body, thus affording a regulatory mechanism for the fine adjustment of body temperature. Increasing cutaneous blood flow causes the skin to come more nearly into equilibrium with the body core temperature. For the most part, augmentation of cutaneous circulation is accomplished by vasodilatation of the arteriovenous anastomoses, resulting in flooding of the superficial veins from which heat can be lost by irradiation and conduction.

When there is need for heat conservation, as with exposure to a very low ambient temperature, the flow through the arteriovenous anastomoses practically ceases, with most of the blood being shunted to deeper tissues and the abdominal viscera. However, if the vasoconstricting stimulus is prolonged, the arteriovenous anastomoses will open intermittently, thus preventing overcooling. Such a mechanism is especially important in the extremities.

IV. Neural Regulation of Dermal Blood Flow

1. Sympathetic Trunks and Postganglionic Pathways

Anatomical Considerations

The paravertebral sympathetic ganglia consist of accumulations of large postganglionic cell bodies from which arise unmyelinated class C fibres. In man those in the lumbar sympathetic trunk vary considerably in distribution, and rarely is one ganglion consistently associated with each vertebra. In some instances several cell bodies may condense into a single large ganglion. Whether there are cross-communications between the two lumbar trunks has not been proved (Webber 1958).

The unmyelinated postganglionic sympathetic axons may travel up or down the sympathetic trunk for several segments and then leave through the grey communicating ramus (sympathetic root) to re-enter the spinal nerve near the origin of the white communicating ramus (Fig. 7). Generally, however, they exit one or two segments below the point of entrance of the preganglionic fibre with which they have made synapse. On reaching the spinal nerve and then the peripheral mixed nerve, the postganglionic fibres may follow varied pathways by accompanying somatic sensory nerves to reach the periphery (Fig. 7). Their distribution among the cutaneous vessels is usually not strictly in accordance with the sensory dermatomes.

2. Adrenergic Sympathetic Control of Dermal Vessels

Anatomical Considerations

The muscular arteries and arterioles (precapillary resistance vessels) of the skin are richly supplied with adrenergic nerves (Ehinger et al. 1967; Norberg and Hamberger 1964). In contrast, the large elastic arteries have a sparse sympathetic innervation and, as a result, the consequence of neurogenic smooth muscle activation of this type of vessel is primarily a reduction in wall compliance.

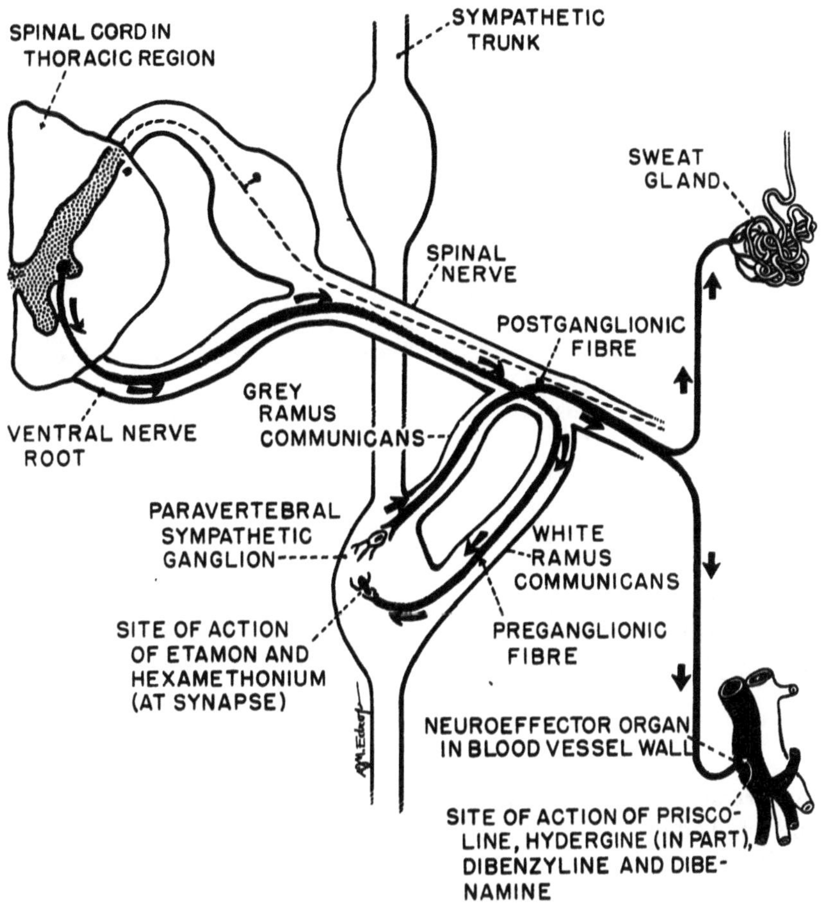

Fig. 7. Schematic representation of peripheral sympathetic pathways. (ABRAMSON 1957)

Arteriovenous anastomoses are also profusely innervated by unmyelinated and myelinated sympathetic fibrils located in the loose, finely fibrillar collagenous tissue which surrounds the wall of the anastomotic canal (Fig. 8) (GRANT and THOMPSON 1963). These nerve structures enter the adventitia and smooth muscle cells forming the outer layer of the arteriovenous anastomosis, but they do not appear to penetrate deeply into the centrally situated mass of epithelioid cells.

The cutaneous veins are less well innervated by adrenergic nerves than the arteries. The only exceptions are the vessels in the lower limbs, exposed to high hydrostatic pressure loads, with the result that in them the smooth muscle media is comparatively thick and the adrenergic sympathetic innervation is very profuse.

Physiological Considerations

Compared with most other regions of the body, the cutaneous vascular bed, particularly that in the fingers and toes, is under a much greater control of the sym-

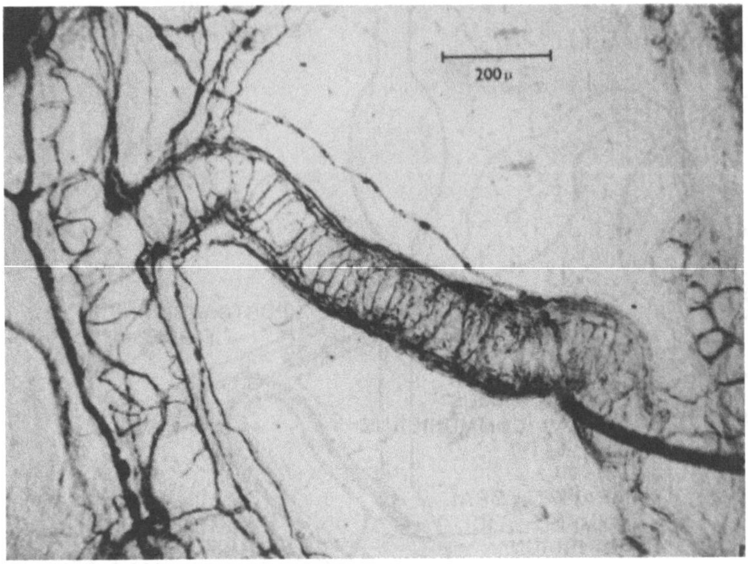

Fig. 8. Perichondrium of normal ear of a rabbit, formalin fixed under pressure and incubated 20 h in acetylthiocholine. Nerve plexuses are visualised surrounding an arteriovenous anastomosis arising on the left from an artery and passing into a vein on the right side. (GRANT and THOMPSON 1963)

pathetic nervous system. As a consequence, a mechanism is readily available for initiating rapid adjustments of tone to the cutaneous vessels, a response achieved through the release of noradrenaline at the adrenergic nerve terminals.

Although the anatomical finding of a rich supply of sympathetic fibres to arteriovenous anastomoses would suggest adrenergic innervation of these structures (WARIS et al. 1980), many of the nerve fibres have also been found to contain cholinesterase (HURLEY and MESCON 1956), suggesting that arteriovenous anastomoses may be innervated by cholinergic, as well as adrenergic, sympathetic nerves. This is in contrast to arterioles whose nervous control is solely adrenergic. However, opposed to such a possibility is the finding that in the hand the cutaneous vasodilator effect of body heating is not abolished by atropine (RODDIE et al. 1957b).

Despite the fact that sympathetic innervation in the venous tree is minimal, the diameter of venules and small veins can still be actively regulated through the adrenergic nervous system, so that these capacitance vessels can function as an adjustable reservoir.

3. Adrenergic Neuroeffector End Organs

The α-receptor endings (ALLQUIST 1948), the adrenergic terminal perivascular nerves, are found in large numbers in cutaneous vessels, being identified by histochemical and electron microscopic techniques.

Anatomical Considerations

The terminal nerve junction is usually located in the adventitiomedial border of the vessel wall, the adrenergic nerve endings entering the smooth muscle cells in the form of simple neurovascular contacts, lacking membrane specialisation. The synaptic cleft, the tissue space between the terminal and the cell on which the neurotransmitter acts, is wide. Only the outermost layer of smooth muscle cells is in direct contact with the nerve terminals.

Physiological Considerations

Neuronal activation includes both diffusion of the transmitter substance and myogenic conduction through low resistance nexus formations. The latter also couple the smooth muscle cells mechanically (LJUNG 1976).

When an adrenergic nerve potential reaches the nerve terminal, it releases noradrenaline from the granules (varicosities) stored in the sympathetic effector organ (NORBERG and HAMBERGER 1964). The hormone, visualised by means of a highly specific fluorescence method (NORBERG and HAMBERGER 1964; LINDVALL and BJÖRKLUND 1972) and by electron microscopic techniques (GOODMAN 1972), then enters the synaptic cleft where it binds to specific smooth muscle membrane receptors in the blood vessel wall. The interaction between the receptor and the transmitter molecule can be considered the triggering event in the response of the cells. By such a means, a swift vasoconstrictor action is possible because of the rapid liberation of noradrenaline and its rapid elimination at the receptor site through a local re-uptake mechanism. Only negligible amounts of the hormone escape into the bloodstream by the slower process of diffusion.

4. Cutaneous Vasodilator Sympathetic Nerves

Under most physiological states, an increase in local cutaneous blood flow is usually elicited either by inhibition of normal or abnormal vasoconstrictor tone or by some locally induced autoregulatory mechanism. There is also some evidence to indicate that such a response may likewise follow stimulation of sympathetic cholinergic vasodilator nerves (BRODIE and SHAFFER 1970; EDHOLM et al. 1957; RODDIE et al. 1957a). According to such a view, impulses passing over these structures activate γ-receptors through the production of acetylcholine.

a) Forearm Innervation

Vasodilator nerves have been shown to innervate the cutaneous vessels in the forearm. For example, it has been demonstrated that the usual increase in blood flow that occurs in this vascular bed with body heating is eliminated if the peripheral mixed nerves supplying the skin of the forearm are previously blocked by a local anaesthetic (EDHOLM et al. 1957). Moreover, performing such a step after vasodilatation has been elicited by body heating causes a reduction in blood flow. Further evidence has been obtained by administering adrenaline after an α-adrenergic blocking agent (phenoxybenzamine) has been given (Fig. 9b) (ALLWOOD and GINSBURG 1961).

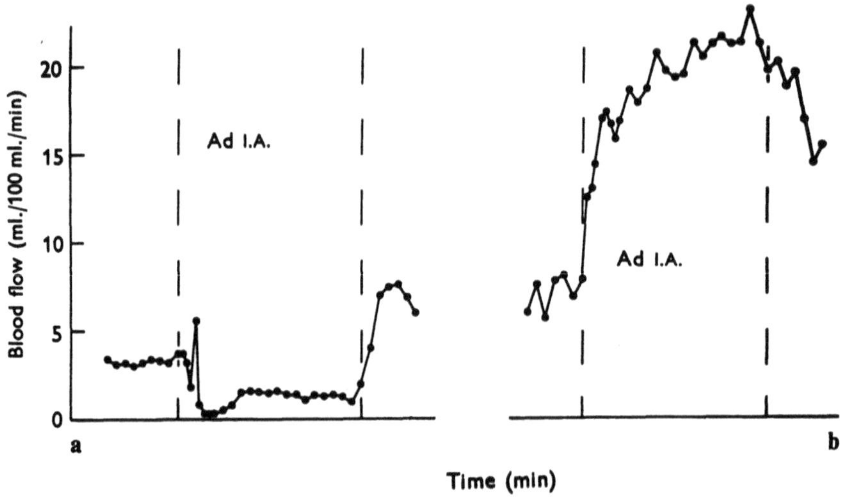

Fig. 9 a, b. Changes in blood flow in the forearm produced by adrenaline. Graphs represent averaged results from six subjects. **a** Effect of the drug injected intra-arterially (2 μg/min over a 10-min period). Initial transient vasodilatation followed immediately by vasoconstriction which persisted during the remainder of period of administration. Following this there was vasodilatation which reached a maximum about 2 min after end of infusion. **b** Preceding the injection of adrenaline, phenoxybenzamine was given. Now adrenaline produced an immediate and marked increase in blood flow, rising from a resting level of 8.1 ml/min/100 ml limb vol to 21.4 ml. (Modified from ALLWOOD and GINSBURG 1961)

On the basis of the available evidence, it has been concluded that the cutaneous circulation in the forearm is controlled mainly by a neural vasodilator mechanism, with inhibition of vasoconstrictor (adrenergic) nerve activity evidently playing only a minor role.

b) Hand and Foot Innervation

There is no convincing evidence that a vasodilator mechanism influences cutaneous circulation of the hand and foot (GASKELL 1956; SARNOFF and SIMEONE 1947; UVNÄS 1960). It plays no part in the increased blood flow that occurs in the distal segments of limb with body heating. Moreover, adrenaline, which acts on both vasodilator and vasoconstrictor effector endings of the sympathetic nervous system, produces no increase in circulation in the cutaneous vessels of the hand if preceded by the administration of an α-receptor (vasoconstrictor) blocking agent (DUFF and GINSBURG 1956). It can be concluded, then, that if vasodilator nerves are present in the hand and foot, they have a very limited physiological function under normal conditions (ARNOTT and MACFIE 1948; SARNOFF and SIMEONE 1947).

V. Other Mechanisms for Regulation of Dermal Blood Flow

1. Hormonal Control

There are a number of vasoactive hormone systems which function as local humoral vascular control mechanisms, as well as having a systemic effect. These are discussed below.

a) Adrenaline

Although under normal conditions vasomotor regulation is dominated by sympathetic vasomotor influence, when there is need for extensive and massive vasoconstrictor activation, the hormonal link in the sympatho-adrenal system comes into play to reinforce the direct effect of neural innervation (CELANDER 1954). Involved in this process is the liberation of adrenaline from the adrenal medulla, its distribution throughout the body via the bloodstream, and its local stimulative action on α-receptors in the blood vessel wall. (For the pharmacological responses to adrenaline, see Sect. A.VI.2).

b) Noradrenaline

Noradrenaline plays a much less important role than adrenaline in supporting the direct effect of neural innervation. Its concentration in the blood plasma is dependent largely upon the quantity of overflow from the synaptic cleft following adrenergic sympathetic activity (see Sect. A.IV.3). Nevertheless, when an active phaeochromocytoma exists, the blood-borne hormone can become important in producing an extensive end organ response. As in the case of adrenaline, the action of circulating noradrenaline is through stimulation of α-receptors in the blood vessel wall. (For the pharmacological reactions to noradrenaline see Sect. A.VI.2).

c) Renin–Angiotensin System

Renin, a proteolytic glycoprotein enzyme, interacts with an α-globulin substrate to form decapeptide angiotensin I. The latter is hydrolysed by a converting enzyme to the octapeptide, angiotensin II (angiotonin), this being the main effector hormone of the system. Angiotensin II is a powerful pressor agent, causing widespread constriction of resistance vessels by a direct effect upon vascular smooth muscle. It probably also acts upon the central nervous system, giving rise to increased sympathetic outflow (PEACH 1977). Recent studies support the view that renin has a local vasoconstrictor action on cutaneous vessels (SWALES 1980).

Arterial infusion of angiotonin causes skin pallor and a significant fall in blood flow in the hand, reducing the circulation by an average of 45% (DALY and DUFF 1960). The readings remain low for several minutes and then gradually return to the control level in about 20 min.

The renin–angiotensin system appears to perform an important role in maintaining blood pressure when the circulation is threatened by sodium and fluid depletion or by orthostatic hypotension. Many tissues, including the arterial wall, contain enzymes that will split the substrate molecule in vitro to yield angiotensin I (GANTEN et al. 1976; HACKENTHAL et al. 1978). Whether such a reaction occurs in vivo has not been established.

d) Hormones Causing Peripheral Inhibition of Sympathetic Activity

Histamine and acetylcholine, vasoactive amines, have been found to produce vasodilatation through peripheral inhibition of perivascular adrenergic sympathetic nerves. Serotonin also appears to inhibit adrenergic neurotransmission (as produced by electrical stimulation of sympathetic nerve endings) through a decrease of released adrenergic transmitter (MCGRATH 1977). It has been suggested that this hormone, which normally circulates in association with blood platelets and is released during their aggregation, may have a local role in the maintenance of vascular tone (JARROTT et al. 1975). Like prostaglandins and angiotensin, it may be an important mediator for human cutaneous resistance vessels. (For the pharmacological responses to serotonin, see Sect. A.VI.2).

2. Local Control

That removal of vasomotor tone to cutaneous vessels does not elicit maximal augmentation in local blood flow is supported by the vascular response to topical heat. Even in the recently sympathectomised hand or foot, this stimulus invariably produces a further relaxation of vessels already passively dilated by removal of the influence of adrenergic sympathetic impulses.

Axon Reflex Vasodilatation

The local erythema produced in skin following the application of a noxious stimulus (axon reflex vasodilatation) results from antidromic stimulation of intact cutaneous sensory nerves. However, the mechanism responsible for the vascular phenomenon has not as yet been elucidated.

A possibility that has been proposed is that the cutaneous vasodilatation is mediated through the release of vasoactive humoral substances originating in sensory nerves. Among the agents suggested as possible intermediaries are histamine, bradykinin, prostaglandins (BERGSTRÖM et al. 1965), dopamine (BELL et al. 1975), angiotensin, adenine nucleotides, nucleosides (HADDY and SCOTT 1968) and substance P (SP). The latter agent (SP) is the undecapeptide putative neurotransmitter released from the terminals of primary sensory neurones in the spinal cord (BURNSTOCK 1977).

JANSCO et al. (1967) reported that topically applied capsaicin (*trans*-8-methyl-*N*-vanillyl-6-nonenamide), found in the seeds and membranes of some species of plants of the nightshade family, causes the skin of man to become insensitive to various types of chemical pain agents, a response similar to that following denervation. These investigators also noted that antidromic stimulation of sensory nerves could be inhibited by capsaicin. Such evidence has supported the view that this agent interferes with either the production, the storage or the release of one or more vasoactive "neurohumors" released by chemical or antidromic stimulation of chemosensitive primary sensory neurons (BERNSTEIN and LORINCZ 1984). Also proposed is the possibility that substance P is a transmitter of primary sensory neurones and that it might be the vasoactive "neurohumor" responsible for axon reflex vasodilatation. Further evidence for this theory is the demonstration by JESSELL et al. (1979) that parenterally administered capsaicin results in a loss of substance P from the spinal cord terminals of primary sensory neurones. Moreover, it has been found that substance P is capable of releasing histamine from mast cells and that it is itself a very potent vasodilator agent (JOHNSON and ERDOS 1973; BURCHER et al. 1977). Furthermore, it has been noted that topical application of capsaicin to human skin prevents the vasodilator effect of such substances as histamine, bradykinin, prostaglandin E and papain (SWIFT et al. 1979; BERNSTEIN et al. 1981 a, b). The other responses, such as itching and the wheal, still are present, as are pinprick, touch and temperature sensation. However, in the development of axon reflex vasodilatation, it has not been possible to distinguish a direct effect by substance P and one mediated through histamine released from dermal mast cells (JORIZZO et al. 1981).

VI. Therapeutic Modulation of Dermal Blood Flow by Drugs

1. Vasodilator Agents

Drugs may act to augment dermal blood flow through various mechanisms: by depressing higher centres, by interrupting the pathway through which adrenergic (constrictor) impulses travel to reach the peripheral blood vessels, and by causing direct relaxation of the vascular musculature.

a) Drugs Acting on Higher Centres

The barbiturates and veratrum alkaloids inhibit the formation or block the transmission of vasoconstricting impulses in higher centres (above the level of the vasomotor centre), thus permitting the peripheral arterial tree to dilate passively, resulting in increase in local circulation.

Depression of the vasomotor centre is caused by ethyl alcohol, intravenous procaine and chlorpromazine. Alcohol approximately doubles the circulation in the hand by causing vasodilatation of cutaneous arterioles (ABRAMSON et al. 1941).

b) Ganglionic Blocking Agents

Ganglionic blocking agents block synaptic transmission in paravertebral sympathetic ganglia by competing with acetylcholine, the ganglionic neurotransmitter. As a result, sympathetic vasoconstrictor impulses are prevented from passing from preganglionic to postganglionic fibres, thus causing widespread passive dilatation and increased local cutaneous blood flow, particularly in the digits. Associated with the peripheral vascular changes are inhibition of reflex cardiac mechanisms and a marked degree of orthostatic hypotension due to abolition of reflex vasoconstriction normally elicited by standing. Unfortunately, because of the unwanted side-effects, such ganglionic blocking agents as tetraethylammonium chloride (Etaman), pentolinium, hexamethonium bromide, mecamylamine hydrochloride (Inversine) and chlorisondamine chloride (Ecolid) have little if any clinical application.

c) Adrenergic Blocking Agents

Adrenergic blocking agents cause cutaneous vasodilatation by competitive inhibition at the α-receptor endings in the blood vessel wall. Among such agents are imidazoline compounds, including tolazoline (Priscoline) and phentolamine (Rogitine), derivatives of ergot and the β-haloalkylamines, such as phenoxybenzamine hydrochloride (Dibenzyline) (Fig. 9b) (ALLWOOD and GINSBURG 1961).

d) Drugs Antagonising Catecholamines

Among the agents which counteract the action of circulating vasopressor substances or noradrenaline are phentolamine and benzodioxine. Such drugs as guanethidine sulphate (Ismelin), bethanidine hydriodide (Esbatal) and bretylium tosylate (Bretylol) selectively accumulate in postganglionic adrenergic neurones, where they inhibit noradrenaline release at the adrenergic synaptic cleft by depressing adrenergic nerve terminal excitability.

Reserpine (Serpasil) is believed to deplete the sympathetic nerve endings in the vessel wall of noradrenaline although the mechanism whereby this is accomplished is disputed (LINDMAR and MUSCHOLL 1964; WHELAN and SKINNER 1963).

e) Myovascular Relaxants

Several drugs have a direct inhibitory or paralysing action on the vascular smooth muscle, including nicotinic acid, xanthine derivatives (caffeine, theophylline and theobromine), kinins, the benzyllisoquinoline group of alkaloids (papaverine), nitrites, histamine and cyclandelate.

f) Rubefacients

For many years rubefacients have played a role as therapy for muscle and joint pains. Those commonly used are various nicotinic acid esters, oil of cloves and mustard oils. When rubbed into the skin, they produce erythema, hyperemia and a local increase in cutaneous temperature (GROSS and MERZ 1948; LANGE and WEINER 1949). The predominant vascular response to these substances is dilatation of arterioles and muscular venules, with no change in calibre of capillaries or small venules (FULTON et al. 1959). The resulting response in muscular vessels is prompt, extending over a larger portion of vascular channels than that exposed to direct contact with the drug.

Since all segments of vessels appear to dilate simultaneously, the possibility has been raised that the response is mediated by a conduction system with nerve-like properties. That the vascular alterations in human skin to rubefacients do not depend upon the integrity of peripheral nerves is supported by the finding that a benzyl ester of nicotinic acid applied to the forearm still produces normal vasodilatation even after the cutaneous nerves in the area had been blocked by a local anaesthetic (CRISMAN et al. 1959).

2. Vasoconstrictor Agents

a) Adrenaline

Adrenaline has a highly specific action on the effector endings of the entire sympathetic nervous system. When injected intracutaneously, it produces strong constriction of metarterioles and precapillary sphincters. Continuous intravenous infusion of the medication causes an increase in frequency of the normal intermittent constriction of these vessels, as well as of terminal capillary loops (vasomotion), as determined by direct observation of the minute vascular bed in the nail fold (GREISMAN 1952). Whether the changes in the capillaries result from a local effect of the amine on these vessels or follow constriction of the parent metarteriole has not been definitely determined.

Intravenous administration of adrenaline produces a decrease in circulation in the hand and foot (DE LARGY et al. 1950; KUNKEL et al. 1939; WILKINS et al. 1938), sites composed primarily of skin (ABRAMSON and FERRIS 1940). Injection into the brachial artery results in a reduction in circulation in the hand to about 40% of the control level (DALY and DUFF 1960), with a similar change in the foot (BELL and STEAD 1952). In the forearm, the vasoconstrictor response is much less marked (Fig. 9 a). There is no evidence that vasodilatation occurs in the hand as a result of stimulation of cholinergic vasodilator fibres, a response which is noted in the forearm (Fig. 9 b) (ALLWOOD and GINSBURG 1961).

Adrenaline acts primarily on blood vessels which contain smooth muscle, to produce constriction. The previous administration of an α-receptor blocking agent reveals a vasodilator response in the forearm. Reductions in true capillary lumen size observed with adrenaline are probably a passive response due to changes occurring proximally.

The finding that adrenaline causes facial pallor supports the belief that it produces constriction of the subpapillary venous plexus. For example, it has been found that healthy venules of the skin react to the point of complete closure, with a resulting stasis in the capillary bed (HEIMBERGER 1925). On the other hand, if these vessels become more or less permanently dilated through exposure to noxious agents, they become unresponsive to adrenaline (LEWIS 1927).

Adrenaline also acts upon the large cutaneous veins to raise venous tone. This may occur even in the face of an increase in systemic blood pressure and in local arterial circulation to the underlying muscles (DE LARGY et al. 1950).

b) Noradrenaline

Noradrenaline is secreted locally, acting as the mediator in sympathetic nerves (see Sect. A.IV.3 above) (VON EULER 1956). In general the vascular responses to it are not as great as those produced by adrenaline (DE LARGY et al. 1950).

The continuous intravenous infusion of noradrenaline produces acceleration of vasomotion (normal intermittent constriction of metarterioles and precapillary segments of the vascular bed), resulting in increasing frequency of periods of interruption of flow through the arteriolar portions of the capillary loops in the nail fold (GREISMAN 1954). With more rapid administration of the drug, the duration of cessation of flow becomes progressively longer until the entire loop disappears from view for periods of 2 s to several hours. However, as in the case of adrenaline (see above), there is no conclusive evidence that noradrenaline acts directly on the capillaries, since the observed changes could have resulted passively as a consequence of alterations in calibre of supplying vessels proximal to the capillary bed.

Infusion of noradrenaline into the brachial artery consistently causes a marked reduction in blood flow in the hand (DALY and DUFF 1960; DUFF and GINSBURG 1956). Intravenous administration of the drug has a similar effect (DE LARGY et al. 1950), and at the same time there is a decrease in finger circulation (JONES and BASHOUR 1965). In contrast to the action of adrenaline, the α-receptor blocking agent phenoxybenzamine does not uncover a vasodilator response in the forearm to intra-arterial injection of noradrenaline, although the vasoconstrictor effect of the drug is reduced; with intravenous administration it is abolished (ALLWOOD and GINSBURG 1961).

It would appear, therefore, that the direct action of noradrenaline on the cutaneous arterial circulation is vasoconstrictor, with a resultant reduction in local blood flow and a rise in peripheral resistance. The drug has its effect mainly on vessels containing circular smooth muscle.

c) Serotonin (5-Hydroxytryptamine)

Serotonin has a twofold effect: constriction of cutaneous resistance vessels and dilatation of the cutaneous microcirculation. For example, following intra-arterial injection in human subjects, there is a decrease in total blood flow, associated with marked erythema of the skin (RODDIE et al. 1955). The drug also has a marked direct vasoconstricting action on large subcutaneous veins (DEMIS et al. 1960; LORINCZ and PEARSON 1959; REID 1952) but not on smaller ones or on the venules of the subpapillary plexus (DEMIS et al. 1960). Intradermal administration in the human forearm produces local erythema, a prominent protracted flare and, less often, cyanosis (DEMIS et al. 1960). Despite the ability of serotonin to increase capillary permeability in rats, this effect is not noted to any extent in human skin (DEMIS et al. 1960). Nor is the rate of cutaneous capillary blood flow affected by the drug, at a time when there is a slowing of local circulation. (For the physiological response to serotonin, see Sect. A.V.1)

d) Corticosteroids

Besides their anti-inflammatory effect, corticosteroids, topically applied, also elicit a vasoconstrictor response in the cutaneous microcirculation (STOUGHTON 1972). This action may contribute to the therapeutic effect of such medications.

B. Dermal Lymphatic Circulation

The cutaneous lymphatic system does not demonstrate an obvious pattern of distribution, thus giving the appearance of a vast unstructured but elaborate mesh or network throughout the connective tissue (MONTAGNA and PARAKKAL 1974). It parallels the major vascular plexuses but is independent of them.

Since the dermal lymphatics are extremely frail, filled with transparent lymph, and generally in a collapsed state in histological sections, they are not easily detected, except for the fact that even the smaller ones have numerous valves (see Sect. B.I.2). In the subcutaneous tissue layer, the collecting channels are more obvious, being similar in structure to blood vessels except that they have thinner walls relative to diameter. In the dermis, a single lymphatic is generally found in each papilla. Channels are also noted in the walls of arteries, running up to the media but not entering it; in veins, intramural lymphatics appear to be a normal occurrence (JOHNSON 1969).

I. Anatomical Considerations

The introduction of electron microscopy has shed considerable light on the structure of normal lymphatics, particularly the capillary bed (CASLEY-SMITH and FLOREY 1961). As a preliminary step, the vessels of experimental animals are identified by injecting them with a mixture of equal parts of 25% thorium dioxide (Thorotrast) and 5% Pontamine blue in physiological saline, using a micro-injection technique. Small blocks of tissue are then cut, placed in a fixative, embedded, sectioned and finally mounted (CASLEY-SMITH and FLOREY 1961).

1. Dermal Lymph Capillaries (Initial Lymphatics)

a) General Characteristics

Lymph capillaries are found in large numbers in the skin and subcutaneous tissue, forming a peripheral meshwork within the connective tissue. They appear as blind end tubes (LEAK 1972) or bulbous saccules (CLIFF and NICOLL 1970), having no continuity with the interstitial spaces and running along with and in proximity to the blood capillaries. Characteristically the lymph capillaries are irregular in shape and possess a wide lumen (15–25 µm). Their walls are made up of a single-layered endothelium which resembles that of blood capillaries.

b) Intercellular Junctions

Electron microscopic studies have revealed that there are spaces between the endothelial cells (intercellular junctions) which are larger than those found in cutaneous blood capillaries. Moreover, they are less firmly closed and frequently lack the distinct adhesion devices common to blood capillaries (CASLEY-SMITH 1968). They vary considerably in form, some being perpendicular to the basement membrane and as short as 500 Å, whereas others are almost parallel to the membrane and up to 5 µm in length. A number are sinuous, with several deep curves (CASLEY-SMITH and FLOREY 1961). The intercellular spaces usually have a density about the same as that of the general interstitial ground substance. Injected ferritin has been found in them (WISSIG 1958).

c) Vesicles and Caveolae

Among other electron microscopic findings in lymphatic capillaries are many isolated vesicles and caveolae intercellularis in the cytoplasm of the endothelial cells. Some of the caveolae are so deep that they almost penetrate the cell from side to side (CASLEY-SMITH and FLOREY 1961). Both types of structure are usually 200–

500 Å diameter. As in the case of the intercellular junctions, injected ferritin (Wissig 1958) and other foreign bodies (Leak 1971) have been found in vesecles, lying free in the cytoplasm.

d) Basement Membrane

Although a basement membrane may be clearly visible beneath the endothelial cells in some sites, in most locations it appears to be irregularly broken up or not even identifiable. When present, the basement membrane is 150–250 Å thick and is usually separated from the plasma membrane by a gap of 100–200 Å, this being increased in some locations to as much as 1000 Å. Such a finding is in contrast to neighbouring blood capillaries, which have a much smaller gap between endothelium and basement membrane. The basement membrane in lymphatic capillaries has an electron density which is about the same as that of the general interstitial ground substance.

e) Supporting Structure

Loosely arranged fine bundles or networks of scattered collagen fibres are attached to the outer surface of lymphatic capillaries (Mislin 1961); these pass into surrounding connective tissues and tie the vessels to them (Leak and Burke 1968a). Elastic tissue is occasionally found among the collagen fibres, whereas pericytes are rarely noted. The fenestrae present in blood capillaries are not observed in lymphatic capillaries.

2. Dermal Collecting Lymph Channels and Trunks

a) General Characteristics

Lymph in the cutaneous lymph capillaries drains into a plexus of superficial post-capillary fine-calibred lymph channels (100–200 μm in diameter) which end in a deep dermal lymphatic plexus. The latter, in turn, terminates in lymphatic trunks.

In contrast to lymphatic capillaries, the large lymphatic channels tend to demonstrate fewer and fewer open junctions until the largest ones and the great lymphatic trunks have none. The deep lymphatic plexus and the vessels into which it empties possess a well-developed basement membrane which becomes more and more prominent as the channels enlarge, finally giving way to an elastic lamina in the largest vessels. Connective tissue elements and smooth muscle cells are readily identified in the walls of the lymph channels. Amyelinated nerves can be traced to the muscle fibres (Casley-Smith 1969).

b) Valves

Valves are found in the deep dermal lymphatic plexus and in the large vessels. They have been described as tricuspid, although some studies using stereomicroscopy and scanning electron microscopy indicate that funnel-like valvular structures dominate, at least in the pulmonary lymphatic bed (Lauweryns 1971). Besides differences in shape, the valves also demonstrate variations in size. Of importance in distinguishing middle-sized lymphatic vessels from veins is that the former contain valves, while veins of comparable size do not.

3. Dermal Lymphatic System in the Limbs

General Characteristics

In the limbs there are fine superficial cutaneous lymph channels, an intermediate plexus and a deep network (FORBES 1938). The deep system of lymphatics is found at the junction with the subcutaneous vessels and is furnished with valves. The intermediate plexus is located between the superficial and the deep networks. The subcutaneous lymph channels enter the main lymphatic trunks lying close to the deep fascia. Communications exist between the dermal lymphatics located superficial to the deep fascia and the muscular lymphatics running in the muscle sheaths. These structures are generally found in the vicinity of the antecubital, axillary and regional lymph nodes, although there are also a few others elsewhere (MÁLEK et al. 1964). It has been proposed that direct connections likewise exist between the lymph nodes and the adjacent veins (PRESSMAN and SIMON 1961). Under physiological conditions, however, these structures probably play a minor role (KINMONTH 1965).

II. Physiological Considerations

The lymphatic system of the skin and subcutaneous tissue is an important determinant of tissue volume in these structures (TAYLOR et al. 1973; TAYLOR and GIBSON 1975; GUYTON et al. 1975; GRANGER 1979).

1. Methods for Study of Lymph Flow

In man, the function of the dermal lymphatics has been studied mainly by the injection of vital dyes into the subcutaneous tissue. Among the materials used are ^{131}I-labeled albumin, ^{24}NaCl and Na^{131}I, in each instance the rate of disappearance of the substance being studied. The assumption on which the method is based is that the vital dyes are removed mainly by lymphatic vessels. This is true at least for ^{131}I-labeled albumin, its disappearance rate being considerably slower than that of ^{24}NaCl or of Na^{131}I (HOLLANDER et al. 1961).

2. Mechanisms Involved in Passage of Fluids and Particulate Matter into Lymphatic Capillaries

a) Role of Ultracirculatory System

There is good evidence to believe that fine connective tissue fibres found in the interstitium contribute to the filling of lymphatic capillaries with lymph (CASLEY-SMITH 1979a; HAUCK 1972). This extensive system of tissue channels (fine or ultracirculatory system) ultimately is the path by which all materials pass from the blood and tissue cells into the lymphatics. It is the final link in the chain connecting the most remote cells with each other and with the exterior (CASLEY-SMITH 1979a).

b) Role of Lymphatic Capillaries

In a sense, lymphatic capillaries are simply an extension of the ultracirculatory system, except that they are provided with endothelium. On the basis of studies using vital dyes, they are found to be more permeable to large molecules than are blood capillaries (YOFFEY and COURTICE 1970).

Electron microscopic study of the lymphatic capillaries in both superficial and deep areas of guinea-pig dermal tissue has revealed that two morphologically definable pathways exist for the passage of fluids and various particulate substances from connective tissue areas into lymphatic lumina (LEAK and BURKE 1968b). One consists of intercellular clefts of patent junctions (intercellular junctions) and the other is across lymphatic endothelium within membrane-bounded vessels (endothelial gaps). The finding of cytolysosomes (autophagic vacuoles) within lymphatic endothelial cells suggests that the cells may also be involved in the intracellular digestion of foreign materials removed from adjoining tissues (LEAK and BURKE 1968b). In this regard, it has been shown that lymphatic endothelium, like blood capillary endothelium, can engulf particles by means of pinocytotic vesicles (CASLEY-SMITH and FLOREY 1961; FRENCH et al. 1960). These latter structures may also be involved in the transport of material across the cell.

c) Movement of White Blood Cells

Besides the transport of fluid and particulate matter into the lymphatics, there is a large scale migration of blood cells from the blood vascular system into the lymph. Lymphocytes in particular continually leave the blood circulation through the postcapillary venules located in the lymph nodes and enter the lymphatics (SOKOLOWSKI et al. 1978). They also move from the blood vessels in the subcutaneous tissue to lymphatic channels and then into lymph nodes.

3. Factors Involved in Transport of Lymph

Fluid in the lymphatic capillaries is propelled forward into collecting and larger lymphatics by a number of extrinsic and intrinsic mechanisms. Among the former are respiratory movements and contraction of neighbouring skeletal muscle, as during physical work. Also, elevated metabolic activity frequently causes swelling of interstitial tissues, resulting in dilatation of lymphatic capillaries through the development of greater tension in the fibrils joining these vessels to the surrounding structures (CASLEY-SMITH 1979a).

Among the intrinsic factors is rhythmic contraction of the smooth muscle in the walls of the major collecting lymphatic channels and in lymph nodes, pumping fluid from the interstitium through the vessels and ultimately delivering it to the bloodstream (CAMPBELL and HEATH 1973; HUTH and BERNHARDT 1977; KINMONTH and TAYLER 1956; NICOLL and TAYLOR 1977; SILBERBERG 1979; SMITH 1949). This intrinsic contractility is myogenic in origin and is the major factor responsible for lymph propulsion (CAMPBELL and HEATH 1973). The relationship between contractility and flow enables the rate of lymph movement to keep pace with the rate of lymph production. Another important intrinsic mechanism is the action of the numerous valves in the walls of the lymphatics which permits unidirectional motion from the periphery toward the thoracic duct. Further refinement of lymph flow is supplied by the contractibility that exists in individual segments of vessel located between two valves, which is mediated by autonomous neural stimulation. Such a means permits functional adaptation to the quantity of lymph in the lymph channel by varying the number of contractions (HUTH and BERNHARDT 1977). Although the flow of lymph in subcutaneous tissue is very low, the average intralymphatic pressure is above tissue pressure (HALL et al. 1965).

4. Alterations in Concentration of Lymph in Its Passage Through the Lymphatics

The lymphatic capillaries, filled with tissue fluid, initially concentrate the plasma protein within their lumina, thus developing an osmotic force across the lymphatic wall that pulls in water. Despite such a response, the lymph leaving the lymph capillaries is still concentrated above that in the tissues. However, as it courses through the larger lymphatics, fluid is withdrawn from the surrounding tissues, with the result that the lymph is now diluted to equal the concentration of fluids found in tissue spaces. The fact that the large lymphatics have closed junctions makes them poorly permeable to large molecules (CASLEY-SMITH 1969), and as a consequence, the latter are moved through the vessels with little loss. Such a situation is essential for the proper functioning of the lymphatics as a transport system.

5. Response of Lymphatics to Inflammation

Typical changes occur in the lymphatic capillaries in the presence of inflammation (CASLEY-SMITH 1973), associated with an increase in fluid content in intercellular spaces. The latter causes the tissues to be placed under tension, which, in turn, exerts pull on the filaments attaching the lymphatic capillaries to the surrounding structures. Hence, these vessels become dilated, causing more endothelial junctions to open than is normally the case (CASLEY-SMITH 1965). This type of response readily allows much of the protein-laden interstitial fluid to enter the vessels. Although a considerable amount is thus pumped away, eventually the increasing load placed on the lymphatic capillaries becomes too great and so oedema develops.

III. Pharmacological Considerations

Very little is known regarding the responses of lymphatics to various types of drugs. In the case of lymph capillaries, potassium cyanide has been found to open many of their intercellular junctions. Heat and histamine act to cause contraction of endothelial cells, which, in turn, also results in enlargement of intercellular junctions. Lymph flow in skin and subcutaneous tissue rises with histamine infusion, burns and endotoxins, all of which increase permeability in the vascular capillary bed (TAYLOR and GRANGER 1984).

With regard to collecting lymphatics, anaesthetics depress the smooth muscle of their walls, whereas noradrenaline increases the amplitude and frequency of their contraction (Casley-Smith 1979b). Papaverine and ergotamine decrease both of the latter factors, whereas histamine and caffeine increase frequency but lower the amplitude of the contractions.

Benzopyrones augment the flow of lymph by improving the efficiency of contraction of the walls of collecting lymphatic channels, but only when some abnormality of the lymphatic system exists. For example, in the presence of B-complex avitaminosis or P avitaminosis, there is a tendency for the lymphatic channels to leak protein, a response which can be prevented by these drugs (Casley-Smith 1979b). However, in the case of normal lymphatic channels already working maximally, no beneficial effect can be expected from the administration of the benzopyrones.

References

Abramson DI (1957) Peripheral vascular disorders. In: Zimmerman LM, Levine R (eds) Physiologic principles of surgery, chap 16. Saunders, Philadelphia, pp 419–426

Abramson DI, Ferris EB Jr (1940) Responses of blood vessels in the resting hand and forearm to various stimuli. Am Heart J 19:541

Abramson DI, Zazeela H, Schkloven N (1941) The vasodilating action of various therapeutic procedures which are used in the treatment of peripheral vascular disease: a plethysmographic study. Am Heart J 21:756

Allquist RP (1948) A study of the adrenotropic receptors. Am J Physiol 153:586

Allwood MJ, Ginsburg J (1961) The effect of phenoxybenzamine (Dibenzyline) on the vascular response to sympathomimetic amines in the forearm. J Physiol (Lond) 158:219

Arnott WM, Macfie JM (1948) Effect of ulnar nerve block on blood flow in the reflexly vasodilated digit. J Physiol (Lond) 107:233

Bell C, Conway EL, Lang WJ, Padany R (1975) Vascular dopamine receptors in the canine hindlimb. Br J Pharmacol 55:167

Bell DM, Stead EA Jr (1952) Effects of epinephrine on the vessels of the calf: observations on the period of initial vasodilatation. J Appl Physiol 5:228

Bergström S, Carlson LA, Ekelund I, Oro L (1965) Cardiovascular and metabolic response to infusions of prostaglandin E to simultaneous infusions of noradrenaline and prostaglandin E in man. Acta Physiol Scand 64:332

Bernstein JE, Lorincz AL (1984) Pharmacology of cutaneous blood circulation. In: Abramson DI, Dobrin PV (eds) Blood vessels and lymphatics in organ systems, chap 18. Academic Press, New York, pp 608–613

Bernstein JE, Swift RM, Saltani K, Lorincz AL (1981a) Inhibition of axon-reflex vasodilatation with capsaicin. J Invest Dermatol 76:394

Bernstein JE, Hamill JR, Soltani K (1981b) Bradykinin, substance P, prostaglandin E_2, and papain induced flare suppressed by capsaicin. Clin Res 29(4):787A

Braverman IM, Yen A (1977) Ultrastructure of the human dermal microcirculation. II. The capillary loops of the dermal papillae. J Dermatol 68:44

Brodie MJ, Shaffer RA (1970) Distribution of vasodilator nerves in the canine hindlimb. Am J Physiol 218:470

Burcher E, Atterhog JH, Pernow B, Rosell S (1977) Cardiovascular effects of substance P. In: Von Euler US (ed) Substance P. Effects on the heart and regional blood flow in the dog. Raven, New York, pp 261–268

Burnstock G (1977) Autonomic neuroeffector junction reflex vasodilatation of the skin. J Invest Dermatol 69:47

Campbell T, Heath T (1973) Intrinsic contractility of lymphatics in sheep and in dogs. Q J Exp Physiol 58:207

Casley-Smith JR (1965) Endothelial permeability. II. The passage of particles through the lymphatic endothelium of normal and injured ears. Br J Exp Pathol 46:35

Casley-Smith JR (1968) How the lymphatic system works. Lymphology 1:77
Casley-Smith JR (1969) The structure of normal large lymphatics. How this determines their permeabilities and their ability to transport lymph. Lymphology I:15
Casley-Smith JR (1973) The lymphatic system in inflammation. In: Zweifach BW, Grant L, McCluskey RT (eds) The inflammatory process, 2nd edn, vol 2. Academic, New York, pp 161–204
Casley-Smith JR (1979a) The fine structure of the microvasculature in inflammation. Bibl Anat 17:36
Casley-Smith JR (1979b) Pharmacology of lymphatics and tissue proteolysis. Bibl Anat 18:119
Casley-Smith JR, Florey HW (1961) The structure of normal small lymphatics. Q J Exp Physiol 46:101
Celander O (1954) The range of control exercised by the "sympathicoadrenal system". A quantitative study on blood vessels and other smooth-muscle effectors in the cat. Acta Physiol Scand [Suppl 116] 32:1
Clark JM, Glagov S (1976) Evaluation and publication of scanning electron micrographs. Science 192:1360
Cliff WJ, Nicoll PA (1970) Structure and function of the bat's wing. Q J Exp Physiol Cogn Med Sci 55:112
Crisman JM, Fox RH, Goldsmith R et al. (1959) Forearm blood flow after inunction of rubefacient substances. J Physiol (Lond) 145:47P
Daly JJ, Duff RS (1960) Direct effects of angiotonin on peripheral vessels of subjects with normal and raised blood pressures. Clin Sci 19:457
Daniel PM, Pritchard ML (1956) Arterio-venous anastomoses in the external ear. Q J Exp Physiol 41:107
Davis MJ, Lawler JC (1958) The capillary circulation of the skin. AMA Arch Dermatol 77:690
Davis MJ, Demis DJ, Lawler JC (1960) The microcirculation of the skin. J Invest Dermatol 34:31
De Largy C, Greenfield ADM, McCorry RL et al. (1950) The effects of intravenous infusion of mixtures of *l*-adrenaline and *l*-noradrenaline on the human subject. Clin Sci 9:71
Demis DJ, Davis MJ, Lawler JC (1960) A study of the cutaneous effects of serotonin. J Invest Dermatol 34:43
Duff RS, Ginsburg J (1956) Antagonism between chlorpromazine and noradrenaline in blood vessels of the hands. Br J Pharmacol 11:318
Edholm OG, Fox RH, Macpherson RK (1957) Vasomotor control of the cutaneous blood vessels in the human forearm. J Physiol (Lond) 139:455
Ehinger B, Falck B, Sporrong B (1967) Adrenergic fibres in the heart and to peripheral vessels. Bibl Anat 8:35
Ellis RA (1961) Vascular patterns of the skin. In: Montagna W, Ellis RA (eds) Advances in biology of skin, vol 2, chap 2. Pergamon, New York, p 20
Fleischhauer K, Horstmann E (1955) Der Papillarkörper und die Kapillaren des Perionychium. Z Zellforsch 42:213
Forbes G (1938) Lymphatics of the skin, with a note on lymphatic watershed areas. J Anat 72:399
French JE, Florey HW, Morris B (1960) The absorption of particles by the lymphatics of the diaphragm. Q J Exp Physiol 45:88
Fulton GP, Farber EM, Moreci AP (1959) The mechanism of action of rubefacients. J Invest Dermatol 33:317
Ganten D, Schelling P, Vecsei P et al. (1976) Iso-renin of extrarenal origin: the tissue angiotensinogenase systems. Am J Med 60:760
Gaskell P (1956) Are there sympathetic vasodilator nerves to the vessels of the hands? J Physiol (Lond) 131:647
Goodman TF (1972) Fine structure of the cells of the Suquet-Hoyer canal. J Invest Dermatol 59:363

Granger HJ (1979) Role of the interstitial matrix and lymphatic pump in regulation of transcapillary fluid balance. Microvasc Res 18:209
Grant RT, Thompson RHS (1963) Cholinesterase and the nerve supply to blood vessels in the rabbit's external ear. J Anat 97:7
Greisman SE (1952) The reactivity of the capillary bed of the nailfold to circulating epinephrine and norepinephrine in patients with normal blood pressure and with essential hypertension. J Clin Invest 31:782
Greisman SE (1954) The reaction of the capillary bed of the nailfold to the continuous intravenous infusion of levo-nor-epinephrine in patients with normal blood pressure and with essential hypertension. J Clin Invest 33:975
Gross F, Merz E (1948) Pharmakologische Eigenschaften des Trafuril eines neuen Nikotinsäureesters mit hyperämisierender Wirkung. Schweiz Med Wochenschr 78 (pt 2):1151
Guyton AC, Taylor AE, Granger HJ (1975) Circulatory physiology, vol 2. Saunders, Philadelphia, pp 18–22
Hackenthal E, Hackenthal R, Hilgenfeldt U (1978) Isorenin, pseudorenin, cathepsin D and renin: a comparative enzymatic study of angiotensin-forming enzymes. Biochim Biophys Acta 522:574
Haddy FJ, Scott JB (1968) Metabolically-linked vasoactive chemicals in local regulation of blood flow. Physiol Rev 48:688
Hale AR, Burch GE (1960) The arteriovenous anastomoses and blood vessels of the human finger: morphological and functional aspects. Medicine (Baltimore) 39:191
Hall JG, Morris B, Wooley G (1965) Intrinsic rhythmic propulsion of lymph in the unanaesthetized sheep. J Physiol (Lond) 180:336
Hauck G (1972) Pathways between capillaries and lymphatics. Pflugers Arch 336 (Suppl):55–57
Heimberger H (1925) Über die Contractilität der kleinsten Venen. Z Ges Exp Med 48:179
Hibbs RG, Burch GE, Phillips JH (1958) The fine structure of the small blood vessels of normal human dermis and subcutis. Am Heart J 56:662
Hollander W, Reilly P, Burrows BA (1961) Lymphatic flow in human subjects as indicated in the disappearance of I^{131} labelled albumin from the subcutaneous tissue. J Clin Invest 40:223
Horstmann E (1957) Blutgefäße der Haut. In: Von Mollendorff W (ed) Handbuch der Mikroskopischen Anatomie des Menschen, vol 3, part 3. Springer, Berlin Göttingen Heidelberg, p 198
Hurley HJ Jr, Mescon H (1956) Cholinergic innervation of the digital arteriovenous anastomoses of human skin: a histochemical localization of cholinesterase. J Appl Physiol 9:82
Huth F, Bernhardt D (1977) The anatomy of lymph vessels in relation to function. Lymphology 10:54
Jansco N, Jansco-Gabor A, Szolesanyi J (1967) Direct evidence for neurogenic inflammation and its prevention by denervation and by pretreatment with capsaicin. Br J Pharmacol 31:138
Jarrott B, McQueen A, Graf L et al. (1975) Serotonin levels in vascular tissue and the effect of serotonin synthesis inhibitor on blood pressure in hypertensive rats. Clin Exp Pharmacol Physiol (Suppl) 2:201
Jessell TM, Tsunoo A, Kanazawa L, Otsuka M (1979) Substance P: depletion in the dorsal horn of rat spinal cord after section of the peripheral processes of primary sensory neurons. Brain Res 168:247
Johnson AR, Erdos EG (1973) Release of histamine from mast cells by vasoactive peptides. Proc Soc Exp Biol Med 142:1252
Johnson RA (1969) Lymphatics of blood vessels. Lymphology 2:44
Jones RE, Bashour FA (1965) Digital blood flow. II. The effects of cigarette smoking and vasopressor agents using electrical impedance method. Diseases of Chest 47:470
Jorizzo JL, Coutts AA, Greaves MW, Burnstock G (1981) A comparison of substance P and adenosine triphosphate as mediators of cutaneous inflammation. J Invest Dermatol 76:315 (Abstract)
Kinmonth JB (1965) Primary lymphoedema of the lower limb. Proc R Soc Med 58:1021

Kinmonth JB, Taylor GW (1956) Spontaneous rhythmic contractility in human lymphatics. J Physiol (Lond) 133:3P
Krogh A (1929) The anatomy and physiology of capillaries. Revised edn. Yale University Press, New Haven
Kunkel P, Stead EA Jr, Weiss S (1939) Blood flow and vasomotor reactions in the hand, forearm, foot, and calf in response to physical and chemical stimuli. J Clin Invest 18:225
Lange K, Weiner D (1949) The effect of certain hyperkinemics on the blood flow through the skin (evaluation of counterirritants). J Invest Dermatol 12:263
Lauweryns JM (1971) Stereomicroscopic funnel-like architecture of pulmonary lymphatic valves. Lymphology 4:125
Leak LV (1971) Studies on the permeability of lymphatic capillaries. J Cell Biol 50:300
Leak LV (1972) The fine structure and function of the lymphatic vascular system. In: Meessen H (ed) Handbuch der allgemeinen Pathologie. Springer, Berlin Heidelberg New York, pp 149–196
Leak LV, Burke JF (1968 a) Ultrastructural studies on the lymphatic anchoring filaments. J Cell Biol 36:129
Leak LV, Burke JF (1968 b) Electron microscopic study of lymphatic capillaries in the removal of connective tissue fluids and particulate substances. Lymphology 1:39
Lewis T (1927) The blood vessels of the human skin and their responses. Shaw, London
Lindmar R, Muscholl E (1964) Die Wirkung von Pharmaka auf die Elimination von Noradrenalin aus der Perfusionsflüssigkeit und die Noradrenalinaufnahme in das isolierte Herz. Naunyn Schmiedebergs Arch Exp Pathol 247:469
Lindvall O, Björklund A (1972) The glyoxylic acid fluorescence histochemical method. A detailed account of the methodology for the visualization of central catecholamine neurons. Histochemistry 39:97
Ljung B (1976) Physiological patterns of neuroeffector control mechanisms. In: Bevan JA, Burnstock G, Johansson B, Maxwell RA, Nedergaard OA (eds) Vascular neuroeffector mechanisms. Karger, Basel, p 143
Lorincz AL, Pearson RW (1959) Studies on axon reflex vasodilatation and cholinergic urticaria. AMA Arch Dermatol 32:429
Majno G (1965) Ultrastructure of the vascular membrane. In: Hamilton W, Dow P (eds) Handbook of physiology, sect 2, vol 3. American Physiological Society, Bethesda, pp 2293–2362
Málek P, Belán A, Kocandrle VE (1964) The superficial and deep lymphatic system of the lower extremities and their mutual relationship under physiological and pathological conditions. J Cardiovasc Surg Torino 5:686
McGrath MA (1977) 5-Hydroxytryptamine and neurotransmitter release in canine blood vessels: inhibition by low and augmentation by high concentrations. Circ Res 41:428
Mislin H (1961) Experimenteller Nachweis der autochthonen Automatie der Lymphgefäße. Experientia 17:29
Montagna W, Parakkal PF (1974) The structure and function of skin. Academic Press, New York
Nicoll PA, Taylor AE (1977) Lymph formation and flow. Annu Rev Physiol 39:73
Norberg K-A, Hamberger B (1964) The sympathetic adrenergic neuron: some characteristics revealed by histochemical studies on the intraneuronal distributor of the transmitter. Acta Physiol Scand [Suppl] 238:1
Peach MJ (1977) Renin-angiotensin system: biochemistry and mechanisms of action. Physiol Rev 57:313
Pressman JJ, Simon MB (1961) Experimental evidence of direct communications between lymph nodes and veins. Surg Gynecol Obstet 113:537
Pritchard MML, Daniel PM (1954) Arterio-venous anastomoses in the tongue of the sheep and the goat. Am J Anat 95:203
Reid G (1952) Circulatory effects of 5-hydroxytryptamine. J Physiol (Lond) 118:435
Roddie IC, Shepherd JT, Whelan RF (1955) The action of 5-hydroxytryptamine on the blood vessels of the human hand and forearm. Br J Pharmacol 10:445

Roddie IC, Shepherd JT, Whelan RF (1957a) The vasomotor nerve supply to the skin and muscle of the human forearm. Clin Sci 16:67
Roddie IC, Shepherd JT, Whelan RF (1957b) The contribution of constrictor and dilator nerves to the skin vasodilatation during body heating. J Physiol (Lond) 136:489
Sarnoff SJ, Simeone FA (1947) Vasodilator fibers in the human skin. J Clin Invest 26:453
Saunders RL de CH (1961) X-ray projection microscopy of the skin. In: Montagna W, Ellis RA (eds) Advances in biology of skin, vol 2, chap 3. Pergamon, New York, p 38
Sengel A, Stoebner P (1970) Golgi origin of tubular inclusions in endothelial cells. J Cell Biol 44:223
Silberberg A (1979) Microcirculation and the extravascular space. Bibl Anat 17:54
Smith RO (1949) Lymphatic contractility. A possible intrinsic mechanism of lymphatic vessels for the transport of lymph. J Exp Med 90:497
Sokolowski J, Jakobsen E, Johannessen JV (1978) Cells in peripheral leg lymph of normal men. Lymphology 11:202
Stoughton RB (1972) Bioassay system for formulations of topically applied glucocorticoids. Arch Dermatol 106:825
Swales JD (1980) Local vascular renin activity as a factor in circulatory control. Cardiovasc Rev Rep I:309
Swift RM, Bernstein JE, Soltani K, Lorincz AL (1979) Inhibition of axon reflex vasodilatation in human skin by topically applied capsaicin. Clin Res 27:245A
Taylor AE, Gibson H (1975) Concentrating ability of lymphatic vessels. Lymphology 8:43
Taylor AE, Granger DN (1984) Exchange of macromolecules across the circulation. In: Renkin EM, Michel CC (eds) Handbook of physiology, chap 11. American Physiological Society, Bethesda, pp 467–520
Taylor AE, Gibson WH, Granger HJ, Guyton AC (1973) The interaction between intracapillary and tissue forces in the overall regulation of interstitial fluid volume. Lymphology 6:192
Uvnäs B (1960) Sympathetic vasodilator system and blood flow. Physiol Rev [Suppl 4] 40:69
Von Euler US (1956) Noradrenaline: chemistry, physiology, pharmacology and clinical aspects. Thomas, Springfield/Ill
Waris T, Kyösola K, Partanen S (1980) The adrenergic innervation of arteriovenous anastomoses in the subcutaneous fascia of rat skin. Scand J Plast Reconstr Surg 14:215
Webber RH (1958) A contribution on the sympathetic nerves in the lumbar region. Anat Rec 130:581
Wetzel NC, Zotterman Y (1926) On difference in the vascular colouration of various regions of the normal human skin. Heart 13:357
Whelan RF, Skinner SL (1963) Autonomic transmitter mechanisms. Br Med Bull 19:120
Wilkins RW, Doupe J, Newman HW (1938) The rate of blood flow in normal fingers. Clin Sci 3:403
Wissig SL (1958) An electron microscope study of the permeability of capillaries in muscle. Anat Rec 130:467 (Abstract)
Yoffey JM, Courtice FC (1970) Lymphatics, lymph, and the lymphomyeloid complex. Academic Press, London

CHAPTER 7

Blood Flow – Including Microcirculation

M. J. FORREST and T. J. WILLIAMS

Changes in blood flow occur continually in health and disease and measurement can provide valuable information on endogenous and exogenous substances involved in regulating flow, on the progress of a disease process and on the efficacy of therapeutic agents. This chapter is concerned with some of the many techniques available for measuring blood flow in skin.

A. Visual Assessment

Dermal inflammatory reactions to locally applied antigens have long been employed in the determination of hypersensitivity. Area of erythema provides a quick, reliable and very sensitive index of the severity of the vasodilator response to putative mediators. The major practical difficulty is in defining the area of the reaction, which may be quite irregular. One cause of irregularity is the non-homogeneous nature of the response. LEWIS (1927) observed that in man the intradermal injection of histamine caused a central wheal due to increased microvascular permeability, local reddening believed to be a direct action of histamine on arterioles and a surrounding axon reflex flare. The flare reaction, which can be mimicked by a number of mediators, generally has the most irregular shape. The intradermal injection of prostaglandins produces an approximately circular area of reddening with pseudopodia (JUHLIN and MICHAELSSON 1969) or streaks (CRUNKHORN and WILLIS 1971) which may result from diffusion along lymphatics. As a consequence of these irregularly shaped responses, different authors have used different measurements to estimate the magnitude of the response. These have included (a) measuring the longest axis (COOK and SHUSTER 1981); (b) assuming the shape of the response is a regular ellipse, and then calculating its area from the measurements of the shortest and longest diameters (JUHLIN and MICHAELSSON 1969); (c) calculating the area of a circle using the mean value of the longest and shortest diameters (FOREMAN et al. 1982); and (d) tracing the response onto graph paper and counting the number of squares (ANAND et al. 1983).

Reddening indicates an increase in the blood capacity of a tissue, not an increase in blood flow per se, and may be caused by several distinct factors. Changes in tissue blood flow may be due to dilatation of existing vessels or the opening of new channels. A diversion of flow to superficial vessels will result in reddening without necessarily increasing total flow. In many skin diseases a change in both number and pattern of vessels supplying a tissue is responsible for wide variations in the colour and appearance. Reddening also occurs when tissue blood capacity is high and flow is low, e.g. stasis, where there is a complex relationship between reddening and flow.

B. Thermal Measurements

I. Thermometry and Thermography

Cohnheim (1889) first demonstrated that heat given off from an inflamed tissue was caused by an increase in local blood flow and Lewis (1927) used thermocouples for assessing blood flow. Thermocouples using a copper–constantan junction are less sensitive than those employing a copper–tellurite junction (Hatfield 1950), and a sensitive skin thermometer was developed by Holti (1955) using the latter. Thermocouples have now been largely superseded by thermistors, ceramic semi-conductors whose resistance decreases with increases in temperature. Thermistors can be constructed in the form of small beads or discs of low thermal capacitance, which makes it possible to measure rapid temperature changes. The advantages of thermometry are that it is non-invasive, the equipment is inexpensive and, using small probes, readings can be taken rapidly. The disadvantages are that readings can only be taken from a limited number of sites at one time and variations in physical contact between a probe and a subject affect readings. More importantly, there is no simple relationship between temperature and flow, and at high rates of flow an increase in flow produces only a small increase in temperature.

An alternative to measuring temperature by direct skin contact is to measure the infra-red emission from the skin, which acts as an almost perfect black body emitter. Infra-red detectors with a constant acceptance angle can be used to measure the temperature of a fixed area of skin (Williams et al. 1960; Ring and Cosh 1968). Alternatively, a scanning infra-red sensitive camera may be used to measure contours of different skin temperature, the output being displayed in either black and white or colour on a cathode ray tube. The principle and use of such a thermographic camera are described by Macey and Oliver (1972). The advantages of measuring infra-red emission rather than temperature directly are that skin contact is avoided and infra-red sensitive detectors have a far more rapid response than do thermistors or thermocouples. However, infra-red thermography suffers from the same drawbacks as contact temperature detectors in that at high rates of flow, little change in infra-red emission occurs. A change in the skin's surface temperature may not necessarily reflect an increase in skin blood flow but may result from heat conducted to the surface from an underlying structure, e.g. breast carcinoma or joint inflammation.

II. Thermal Clearance (Conductance)

Skin temperature is related to total skin blood flow. Because total flow is high in comparison with the flow supplying the upper dermis, a change in so-called nutrient flow to the upper dermis may not be reflected by a change in skin temperature. The technique of thermal clearance principally measures flow through the papillary dermis. Measurements are made by comparing the temperature of a heated copper disc with that of a surrounding ring of unheated copper, both in contact with the skin surface (CHALLONER 1975; HOLTI and MITCHELL 1978). Heat is transferred from the central disc to the outer ring by conduction through the epidermis and through the body of the probe and by clearance from blood flowing through superficial dermal vessels. At high rates of flow, more heat is cleared and the temperature difference between disc and outer ring is reduced, whilst at low flow rates the temperature difference increases. This technique has been used in assessing patients with Raynaud's phenomenon (HOLTI 1978). The disadvantage of this method is the physical contact between the probe and the subject: the resultant transfer of heat from the disc to the subject may influence the result.

All these techniques based on skin temperature are influenced by ambient temperature, so that equilibration of the subject in a temperature-controlled environment is desirable.

C. Radioisotopic Techniques

Both the uptake of tracers from the circulation and the clearance of locally injected tracers are used in the determination of blood flow.

I. Isotope Extraction

SMITH and MORALES (1944 a) derived theoretical equations for the uptake of inert gases by tissues from the blood and showed how these equations may be applied in measuring peripheral blood flow non-invasively (SMITH and MORALES 1944 b); mathematical analysis is complicated, however, because different constants have to be used to calculate flow in different tissues. In the techniques described by SAPIRSTEIN (1956, 1958) a radioisotope is injected intravenously, the animal is killed after a defined time period (up to 2 min) and the distribution of the isotope within different tissues is determined. SAPIRSTEIN used the monovalent cations ^{42}K and ^{86}Rb, which he assumed were taken up equally by different tissues for a given blood flow. Provided the venous drainage of the isotope is negligible, it is proposed that the fractional distribution of the isotope among the tissues corresponds to the fractional distribution of the cardiac output. MCDONAGH et al. (1978) suggest that the assumption of equal uptake is not valid and therefore isotope distribution is not dependent only on flow. Further, they showed that the concentrations of isotope in the venous drainage of each organ are not negligible and that the levels vary from organ to organ.

An alternative to the use of labelled cations is the use of microspheres with a typical diameter of about 15 μm which can be labelled with a variety of metallic radionuclides. The microspheres, which are too large to pass through capillaries, lodge within a tissue's microcirculation following a single pass of blood and because there is no venous drainage of the isotope, the proportion of microspheres within individual tissues is proportional to the fractional distribution of the cardiac output. The principle and validity of the technique has been discussed by many workers (NEUTZE et al. 1968; BUCKBERG et al. 1971; HEYMANN et al. 1977; GROSS et al. 1981; MOORE et al. 1981; OFJORD et al. 1981). The following is a summary of the major precautions which need to be taken when using microspheres to determine blood flow:

1. There must be rapid, complete mixing of the spheres following their injection. This can be achieved by placing a catheter into one of the chambers of the heart or into the aortic arch. Poor mixing will result in an uneven distribution of spheres within the arterial circulation.
2. The distribution of microspheres to the tissues must be directly proportional to the blood flow to these tissues, i.e. there must be no preferential uptake of microspheres or selective barriers to microspheres within the tissues.
3. Microspheres must be completely trapped within the microcirculation following a single pass of blood such that no radioactivity appears in the venous drainage. In addition, each microsphere should block flow completely in the vessel where it lodges.
4. An adequate number of microspheres must be present in tissue samples to minimise the random deviation in sphere distribution. This factor is more important than an adequate amount of radioactivity, since low radioactive samples can always be counted for a longer period.
5. The number of spheres must not be such that the microcirculation is significantly impaired.

Having satisfied these criteria there are two methods of calculating tissue blood flow (BUCKBERG et al. 1971). If the cardiac output is determined, then the fraction of microspheres within a tissue to the number of microspheres injected can be multiplied by the cardiac output to give the tissue blood flow; thus:

$$\text{tissue blood flow} = \frac{\text{number of spheres in tissue}}{\text{number of spheres injected}} \times \text{cardiac output (ml/min)}\,.$$

An alternative method is the reference sample method. A sample of blood can be withdrawn from an artery, at a known constant rate. Then, using this known rate, the amount of radioactivity in the reference sample and the amount of radioactivity in a particular tissue, the tissue's flow rate can be calculated from:

$$\text{tissue blood flow} = \frac{\text{counts tissue}}{\text{counts reference}} \times \text{flow reference (ml/min)}\,.$$

The use of radioactive microspheres has been extensively validated for use in measuring flow to whole organs. However, its validation for measuring changes in microcirculatory function is not substantial (HALES and CLIFF 1977). In particular, the direct visualisation of microspheres within the microcirculation has revealed the presence of chains of microspheres stacked one behind the other, indicating plasma leakage past the spheres (HARELL et al. 1977). The behaviour of microspheres within the microcirculation must therefore be taken into consideration if microspheres are to be used for such studies.

II. Clearance of Locally Injected Radiolabels

KETY (1949) developed the principle of isotope clearance as an alternative to existing isotope uptake methods described by SMITH and MORALES (1944a, b). Mathematically, the clearance of locally injected isotope was found to be far easier to analyse. Further, the total amount of radioactivity the body receives is low, although locally there is a higher dosage. In practice, the use of the principle can be accomplished by one of two methods. Following a single local injection of isotope, the initial amount of radioactivity present and the amount present at timed intervals can be monitored using an external gamma-probe collimated to the site of injection (SEJRSEN 1969; CHALLONER 1972; YOUNG and HOPEWELL 1980). Clearance constants are then calculated from the washout curve. This method allows only one estimate of flow per probe at any one time. Thus, comparisons between different areas of skin become less reliable and very time consuming. The advantage of this technique is that many readings over a period of days or weeks at one site may be made. This has obvious benefits in the assessment of perfusion in skin grafts.

An alternative method for use in animals is the multi-site clearance technique of WILLIAMS (1979). This method has some advantages over many techniques used in the experimental determination of skin blood flow. The technique involves the intradermal injection of test agents mixed with ^{133}Xe into a random pattern of injection sites on the dorsal clipped skin of rabbits. After a defined time interval, usually 15–20 min, the animal is killed, the dorsal skin removed and injection sites excised using a steel punch. The initial amount of radioactivity injected and the amounts remaining in injection sites are counted in an automatic gamma-counter. Changes in blood flow relative to control sites can be calculated assuming a single exponential clearance of the isotope. This method allows up to 50 different injection sites to be made in one animal. It is therefore possible to compare blood flow at sites of experimental inflammatory reactions, or the effect of vasodilators and vasoconstrictors. Because a number of determinations are made simultaneously in one experiment, a suitable statistical analysis of the data is made possible. Further, the comparisons between treatments become more reliable and there is a considerable saving in time.

Changes in blood flow within a tissue or relative differences in blood flow between tissues can be expressed in terms of clearance constants. This is adequate for many purposes. Absolute flow can be calculated from clearance constants which can be employed where necessary (SEJRSEN 1969).

D. Red Blood Cell Velocity Measurements

The basis of red blood cell (RBC) velocity determinations as an estimate of blood flow velocity in single microvessels dates back to 1674 when LEEUWENHOEK introduced small grains of sand into the circulation and measured the distance a single grain travelled in the time it took to pronounce a four syllable word. The utilisation of this principle requires that the vessels in which measurements are to be made either are accessible, e.g. human retinal vessels and human nail-fold capillaries, or can be made accessible, e.g. cat omentum and rabbit mesentery. The tissue is illuminated and viewed with a microscope to which is attached either cinematographic or television equipment. The advantage of television is that the recorded image can be directly retrieved for electronic analysis.

Velocity of RBCs may be measured by the time an RBC takes to move a fixed distance or the distance an RBC moves in a fixed period. The first method has been used extensively in microcirculatory RBC velocity studies (WAYLAND and JOHNSON 1967; INTAGLIETTA et al. 1975; FAGRELL et al. 1977). Measurements are made by concurrently viewing in two windows an upstream and downstream section of capillary separated by a known distance. The period between a sequence of RBCs and plasma spaces appearing in one window and reappearing in the second window is determined by counting the number of video frames which elapse between these two events. Measurements can only be made if the same sequence of RBCs is seen in both windows. Thus, not only is a great deal of video monitoring information discarded, but also measurements can only be made if the flow is sufficiently slow that a sequence of RBCs is not already beyond the downstream window by the time of the next video frame. It may therefore be a matter of seconds between two velocity determinations being obtained, thus preventing the measurement of rapid changes in flow.

An alternative method (ANLIKER et al. 1977; TYML and SHEREBRIN 1980) is to use a single window through which the position of an RBC is monitored by integrating its image intensity along each television line. This can be compared in consecutive frames to measure the distance a cell has moved within a single frame period, e.g. 20 ms with a video speed of 50 frames per second. The advantages of this method are that both greater velocities and more rapid changes in velocity may be determined.

A problem with RBC velocity determinations is that the illumination light source may heat the tissue despite filters. This can be overcome by lasers whose monochromatic light can be focussed onto individual capillaries without heating. The velocity of RBC in individual microvessels has been computed by determining the Doppler shift of back-scattered light (TANAKA et al. 1974; EINAV et al. 1975; LE-CONG and ZWEIFACH 1979).

E. Doppler Shift Techniques

The principle of the Doppler technique is the increase in electromagnetic wavelength when reflected off a moving surface.

I. Ultrasound Doppler

The ultrasound Doppler technique is used in the estimation of flow velocity in large vessels. An estimate of volume flow is only possible in exposed or superficial vessels. Early methods (FRANKLIN et al. 1961) measured blood flow velocity in the descending aorta of the anaesthetised dog by placing piezoelectric crystals on either side of the exposed vessel. The Doppler shift measured as the difference between the continuously transmitted and reflected ultrasound frequencies is related to the velocity of the RBCs. Significant improvements on this early design followed. For example, RUSHMER et al. (1966) produced an instrument capable of the transcutaneous estimation of velocity flow, while MCCLEOD (1967) and STRANDNESS et al. (1969) incorporated design features in their instruments which would differentiate between periods of forward and reverse flow. A limitation of ultrasound is that it estimates velocity flow. To estimate volume flow it is necessary to measure the diameter of the vessel and the angle between the acoustic axis of the probe and the blood flow axis (RUSHMER et al. 1966). Vessel diameter has been measured with a pulsed Doppler signal (PERONNEAU et al. 1969) but the measurement of the angle between acoustic and blood flow axes has proved more difficult (BAKER et al. 1974; ANGELSEN and BRUBAKK 1976). In peripheral arteries where the vessels lie parallel to the skin surface, it is possible to use a dual transducer system to estimate this angle (LEVENSON et al. 1981).

II. Laser Doppler

In 1975 STERN suggested the use of light from a helium/neon laser and the principle has been developed extensively (HOLLOWAY and WATKINS 1977; POWERS and FRAYER 1978; WATKINS and HOLLOWAY 1978; OBERG et al. 1979; NILSSON et al. 1980 a, b). The advantages of the laser Doppler system over other techniques for measuring skin blood flow are that the system is non-invasive and flow can be measured continuously and repeatedly. Although the measurements are reproducible it is less certain that they are a true representation of flow. Laser Doppler flowmeters produce an average velocity of RBCs within the microcirculation, but as vessel diameter and the angle between the laser beam axis and the flow axis are infinitely variable in the microcirculation, direct calculation of volume flow is not possible. An indirect assessment can be made by calibrating the output against volume flow measured by a different technique. The calibration will not be valid if the different technique measures a different aspect of the skin blood flow. An example is the use of the clearance of intradermal or epicutaneously applied ^{133}Xe as a standard by which to compare laser Doppler flowmeters (STERN et al. 1977; ENGELHART and KRISTENSEN 1983). There is evidence (ENGELHART and KRISTENSEN 1983) that the clearance of ^{133}Xe is primarily a measure of nutritive blood flow, whereas the laser Doppler flowmeter is sensitive to both nutritive (via capillaries) and non-nutritive (via anastomoses) components of skin blood flow. The extent to which laser Doppler flowmeters detect non-nutritive blood flow is related to the depth to which the flow probe can detect. This depth is reported to be only 1–2 mm but is likely to vary according to pigmentation, epithelial thickness and optical properties of skin, all of which may vary in health and disease.

F. Plethysmography

The principle of plethysmography is to measure the initial rate of change in volume of an organ or part of the body during rapid venous occlusion. This technique gives a measure of total blood flow.

The intensity of light back-scattered from, or transmitted through, a vascular bed is related to the blood volume within the tissue (HERTZMAN 1938), and reflectance plethysmography can be used to measure differences in skin blood volume (HERTZMAN and RANDALL 1948). Skin colour has an effect on reflectance and transmittance but the effect can be minimised by selection of a suitable wavelength (ROLFE 1979).

G. Electromagnetic Flowmeters

Blood flow measurements can be made based on Faraday's law of electromagnetic induction and Lenz's law of direction of an induced electromotive force. Combined, these state that a conductor (in this case, blood) moving at a given velocity through a magnetic field will generate an electromotive force in a direction perpendicular to both the direction of motion of the conductor and the magnetic field. There are three operative types of electromagnetic flow probes: probes positioned in vessels, probes applied to the outside of exposed vessels and non-invasive cuffs placed around the whole limb. Catheter-type probes (MILLS and SHILLINGFORD 1967) have the advantage that they can be accurately positioned in vessels. However, they record flow velocity and not flow volume and the presence of the catheter within the vessel is likely to cause disturbances to flow. Cuff probes positioned around surgically exposed vessels do not interfere with flow; they have been used extensively in clinical surgery (CAPPELEN and HALL 1967) and experimentally, and are often considered as a standard by which to judge other methods (VATNER et al. 1970). A principal advantage is that continuous measurements of flow in known vessels can be made. Early designs produced calibration difficulties and inaccuracies when used on small vessels (CAPPELEN and HALL 1967) but there are probes which can be used on vessels with lumen diameters of less than 1 mm (NIGRA et al. 1981). The invasive nature of the technique limits its use. A non-invasive electromagnetic flowmeter has been described (LEE et al. 1980). It is unable to localise particular vessels but provides a value for total arterial blood flow.

H. Conclusion

Different techniques measure different aspects of skin blood flow, e.g. total volume flow, nutritive flow, blood velocity. Further, different methods measure flow at different depths in skin so that knowledge of changes in the architecture of skin microvasculature is important. However, for some problems even a simple technique can give a satisfactory result and, in selecting a technique, matching sophistication to need is probably the most important consideration.

References

Anand P, Bloom SR, McGregor GP (1983) Topical capsaicin pretreatment inhibits axon reflex vasodilation caused by somatostatin and vasoactive intestinal polypeptide in human skin. Br J Pharmacol 78:665–669

Angelsen BAJ, Brubakk AO (1976) Transcutaneous measurement of blood flow velocity in the human aorta. Cardiovasc Res 10:368–379

Anliker M, Casty M, Friedli P, Kubli R, Kreller H (1977) Non-invasive measurement of blood flow. In: Hwang NHC, Norman NA (eds) Cardiovascular flow dynamics and measurements. University Park Press, Baltimore, pp 43–88

Baker DW, Johnson SL, Strandness DE (1974) Prospects for quantitation of transcutaneous pulsed Doppler techniques in cardiology and peripheral vascular diseases. In: Reneman RS (ed) Cardiovascular applications of ultrasound. Elsevier-North Holland, Amsterdam, p 108

Buckberg GD, Luck JC, Payne DB, Joffman JIE, Archie JP, Fixler DE (1971) Some sources of error in measuring regional blood flow with radioactive microspheres. J Appl Physiol 31:598–604

Cappelen C, Hall KV (1967) Electromagnetic blood flowmetry in clinical surgery. Acta Chir Scand [Suppl] 368:3–27

Challoner AVJ (1972) Measurement of cutaneous blood flow by isotope clearance and thermal conductance methods. In: Ryan TJ, Jolles B, Holti G (eds) Methods in microcirculation studies. Lewis, London, p 43

Challoner AVJ (1975) Accurate measurement of skin blood flow by a thermal conductance method. Med Biol Eng 13:196–201

Cohnheim J (1889) Lectures on general pathology. The New Sydenham Society, London

Cook LJ, Shuster S (1981) The effect of placebo and "aggro" on the weal and flare response to histamine. Br J Dermatol 104:27–29

Crunkhorn P, Willis AL (1971) Cutaneous reaction to intradermal prostaglandins. Br J Pharmacol 41:49–56

Einav S, Berman HS, Fuhro RC, DiGiovanni PR, Fine S, Fridman JD (1975) Measurement of velocity profiles of red blood cells in the microcirculation by laser Doppler anemometry. Biorheology 12:203–205

Engelhart M, Kristensen JK (1983) Evaluation of cutaneous blood flow responses by ^{133}xenon washout and a laser-Doppler flowmeter. J Invest Dermatol 80:12–15

Fagrell B, Fronek A, Intaglietta M (1977) A microscope-television system for studying flow velocity in human skin capillaries. Am J Physiol 283:H318–H321

Foreman JC, Jordan CC, Piotrowski W (1982) Interaction of neurotensin with the substance P receptor mediating histamine release from rat mast cells and the flare in human skin. Br J Pharmacol 77:531–539

Franklin DL, Schlegel W, Rushmer RF (1961) Blood flow measured by Doppler frequency shift of back-scattered ultrasound. Science 134:564–565

Gross PM, Marcus ML, Heistad DD (1981) Measurement of blood flow to bone and marrow in experimental animals by means of the microsphere technique. J Bone Joint Surg [Am] 63:1028–1031

Hales JRS, Cliff WF (1977) Direct observations of the behaviour of microspheres in microvasculature. Bibl Anat 15:87–91

Harell GS, Dickhoner WA, Breiman RS (1977) The simultaneous visualization of microspheres and blood flow in the microvascular bed of the hamster cheek pouch. Microvasc Res 13:203–210

Hatfield HS (1950) A heat flow meter. J Physiol (Lond) 111:10P

Hertzman AB (1938) The blood supply of various skin areas as estimated by the photoelectric plethysmograph. Am J Physiol 124:328–340

Hertzman AB, Randall WC (1948) Regional differences in the basal and maximal rates of blood flow in the skin. J Appl Physiol 1:234–241

Heymann MA, Payne DB, Hoffman JIE, Rudolph AM (1977) Blood flow measurements with radionuclide labelled particles. Prog Cardiovasc Dis 20:55–79

Holloway GA, Watkins DW (1977) Laser Doppler measurement of cutaneous blood flow. J Invest Dermatol 69:306–309

Holti G (1955) The copper–tellurite–copper thermocouple adapted as a skin thermometer. Clin Sci 14:137–141

Holti G (1978) Experimentally controlled evaluation of vasoactive drugs in digital ischemia. Angiology 29:89–94

Holti G, Mitchell KW (1978) Estimation of the nutrient skin blood flow using a segmented thermal clearance probe. Clin Exp Dermatol 3:189–198

Intaglietta M, Silverman NR, Tompkins WR (1975) Capillary flow measurements in vivo and in situ by television methods. Microvasc Res 10:165–179
Juhlin L, Michaelsson G (1969) Cutaneous vascular reactions to prostaglandins in healthy subjects and in patients with urticaria and atopic dermatitis. Acta Derm Venereol (Stockh) 49:251–261
Kety SS (1949) Measurement of regional circulation by the local clearance of radioactive sodium. Am Heart J 38:321–328
Le-Cong P, Zweifach BW (1979) In vivo and in vitro velocity measurements in microvasculature with a laser. Microvasc Res 17:131–141
Lee BY, Trainor FS, Thoden WR, Kauner D, Madden JL (1980) Use of non-invasive electromagnetic flowmetry in the assessment of peripheral arterial disease. Surg Gynecol Obstet 150:342–346
Leeuwenhoek A van (1674) Select works containing his microscopical discoveries in many of the works of nature, vol 1 (translated by Samuel Hoole and W. Nichol, 1798–1807). London, pp 89–112
Levenson JA, Peronneau PA, Simon A, Safar ME (1981) Pulsed Doppler: determination of diameter, blood flow velocity and volumic flow of brachial artery in man. Cardiovasc Res 15:164–170
Lewis T (1927) The blood vessels of the human skin and their responses. Shaw, London
Macey DJ, Oliver R (1972) Infra-red thermography. In: Ryan TJ, Jolles B, Holti G (eds) Methods in microcirculation studies. Lewis, London, p 57
McCleod FD (1967) Directional Doppler demodulation. Proc Annu Conf Eng Med Biol 9:27.1
McDonagh PF, Salel AF, Krohn KA, Rhode EA, Mason DT (1978) Analysis of myocardial and total body integral extraction ratios of rubidium-86. Am J Physiol 235:H794–H802
Mills CJ, Shillingford JP (1967) A catheter tip electromagnetic velocity probe and its evaluation. Cardiovasc Res 1:263–273
Moore CD, Gewertz BL, Wheeler HT, Fry WJ (1981) An additional source of error in the microsphere measurement of regional blood flow. Microvasc Res 21:377–383
Neutze JM, Wyler F, Rudolph AM (1968) Use of radioactive microspheres to assess distribution of cardiac output in rabbits. Am J Physiol 215:486–495
Nigra CAL, Andrews DF, McKee NH (1981) Quantitative assessment of low blood flow using an electromagnetic flowmeter. J Surg Res 31:201–209
Nilsson GE, Tenland T, Oberg PA (1980a) Evaluation of a laser Doppler flow meter for measurement of tissue blood flow. IEEE Trans Biomed Eng 27:597–604
Nilsson GE, Tenland T, Oberg PA (1980b) A new instrument for continuous measurement of tissue blood flow by light beating spectroscopy. IEEE Trans Biomed Eng 27:12–19
Oberg PA, Nilsson GE, Tenland T, Holmstrom A, Lewis DH (1979) Use of a new laser Doppler flowmeter for measurement of capillary blood flow in skeletal muscle after bullet wounding. Acta Chir Scand [Suppl] 489–145–150
Ofjord ES, Clausen G, Aukland K (1981) Skimming of microspheres in vitro: implications for measurement of intrarenal blood flow. Am J Physiol 241:H342–H347
Peronneau PP, Deloche A, Hung B-M, Hinglais J (1967) Débitmetrie ultrasonore – Développementes et applications experimentales. Eur Surg Res 1:147–156
Powers EW, Frayer WW (1978) Laser Doppler measurement of blood flow in the microcirculation. Plast Reconstr Surg 61:250–255
Ring EFJ, Josh JA (1968) Skin temperature measurement by radiometry. Br Med J 4:448
Rolfe P (1979) Theoretic aspects of skin blood flow estimation using thermal, optical and electrical impedance techniques. Birth Defects 15:135–147
Rushmer RF, Baker DW, Stegall HF (1966) Transcutaneous Doppler flow detection as a nondestructive technique. J Appl Physiol 21:554–566
Sapirstein LA (1956) Fractionation of the cardiac output of rats with isotopic potassium. Circ Res 4:689–692
Sapirstein LA (1958) Regional blood flow by fractional distribution of indicators. Am J Physiol 193:161–168

Sejrsen P (1969) Blood flow in cutaneous tissue in man studied by washout of radioactive xenon. Circ Res 25:215–229
Smith RE, Morales (1944a) On the theory of blood-tissue exchanges. I. Fundamental equations. Bull Math Biophys 6:125–131
Smith RE, Morales MF (1944b) On the theory of blood-tissue exchanges. II. Applications. Bull Math Biophys 6:133–139
Stern MD (1975) In vivo evaluation of microcirculation by coherent light scattering. Nature 254:56–58
Stern MD, Lappe DL, Bowen PD, Chimosky JE, Holloway GA, Keiser HR, Bowman RL (1977) Continuous measurement of tissue blood flow by laser-Doppler spectroscopy. Am J Physiol 232:H441–H448
Strandness DE, Kennedy JW, Judge TP, McLeod FD (1969) Transcutaneous directional flow detection: a preliminary report. Am Heart J 78:65–74
Tanaka T, Riva C, Ben-Sira I (1974) Blood velocity measurements in human retinal vessels. Science 186:830–831
Tyml K, Sherebrin MH (1980) A method for on-line measurements of red cell velocity in microvessels using computerized frame by frame analysis of television images. Microvasc Res 20:1–8
Vatner SF, Franklin D, Van Citters RL (1970) Simultaneous comparison and calibration of the Doppler and electromagnetic flowmeters. J Appl Physiol 29:907–910
Watkins D, Holloway GA (1978) An instrument to measure cutaneous blood flow using the Doppler shift of laser light. IEEE Trans Biomed Eng 25:28–33
Wayland H, Johnson PC (1967) Erythrocyte velocity measurement in microvessels by a two-slit photometric method. J Appl Physiol 22:333–337
Williams KL, Williams FL, Handley RS (1960) Infra-red radiation thermometry in clinical practice. Lancet II:958–959
Williams TJ (1979) Prostaglandin E_2, prostaglandin I_2 and the vascular changes of inflammation. Br J Pharmacol 65:517–524
Young CMA, Hopewell JW (1980) The evaluation of an isotope clearance technique in the dermis of pig skin: a correlation of functional and morphological parameters. Microvasc Res 20:182–194

CHAPTER 8

Immunopharmacology of Mast Cells

M.K. CHURCH, R.C. BENYON, L.S. CLEGG, and S.T. HOLGATE

The mast cell was first described by Paul EHRLICH in 1876, as a tissue fixed cell containing many granules which exhibited metachromasia when exposed to basic dyes such as toluidine blue. This histochemical characteristic indicates the presence of the highly acidic proteoglycan, heparin, one of the many preformed chemical mast cell mediators which are secreted in response to cell activation. At the turn of the century, the structural elucidation of histamine (WINDAUS and VOGT 1907), its association with the mast cell (BEST et al. 1927) and its release following anaphylactic reactions in animal models (DALE 1910) established a role for the mast cell in mediating the type 1 or immediate hypersensitivity response associated with allergic reactions. This type of immunological reaction has been implicated in the pathogenesis of skin diseases such as eczema and urticaria. Indeed, the skin has been used as the primary site at which to undertake allergen testing, in the form of intradermal or prick tests, and into which allergens may be introduced in hyposensitisation treatment. Apart from the immediate hypersensitivity reaction involving the reaginic antibody IgE, mast cells play a contributory role in the defence against neoplasia (GOTO et al. 1984), in regulating fibroblast growth and maturation (GUPTA 1970; KAWANAMI et al. 1985) and in the elimination of nematode parasites (WELLS 1977). The recognition of a wider role for the mast cell in the pathogenesis of human disease has stimulated renewed interest in this cell. As in the gastrointestinal and respiratory tracts, the potential importance of mast cells in human skin is reflected by the large numbers that are present in this tissue. Because of the relative inaccessibility of tissue mast cells, much of the knowledge that has been gained about mast cell structure and function has been derived from studies of rat peritoneal mast cells, which can be recovered in large numbers and purified to homogeneity. Recently, however, the use of a variety of enzyme digestion techniques has enabled mast cells from other sources to be dispersed. These techniques have provided overwhelming evidence that mast cells from different species, and even from different sites within the same species, exhibit heterogeneity with respect to both structure and function (CHURCH et al. 1982; BENYON et al. 1987; LOWMAN et al. 1987). Conclusions drawn from studies of mast cells of one particular animal or body site may not, therefore, be applicable to mast cells of another species or site.

In this chapter, we discuss the morphology, distribution, structure and possible functions of human mast cells, particularly those of human skin, and compare them with those of the more widely studied rodent mast cells.

A. Mast Cell Content of Human Skin

The number of mast cells in human skin has been the subject of conflicting reports. MIKHAIL and MILLER-MILINSKA (1964), using toluidine blue to stain mast cells in human skin, found approximately 7×10^3 mast cells/mm^3 with no significant variation in relation to age, sex, race or body region. This is in marked contrast to earlier studies by HELLSTROM and HOLMGREN (1950), who showed 7×10^3 mast cells/mm^3 in the corium of newborn babies but only 1×10^3 mast cells/mm^3 in the same sites of 70- to 80-year-old subjects. Furthermore, BINAZZI and RAMPICHINI (1959) showed considerable variation in the mast cell populations from area to area in human skin, ranging from 46 mast cells/mm^3 in skin from the leg to 177 mast cells/mm^3 in the scrotal skin.

Under normal conditions, mast cells are not found in the epidermis of healthy skin, though they move into this site during various disease states. Mast cells in the dermis are not randomly distributed but are grouped around blood vessels, nerves and appendages (EADY et al. 1979). Using a careful mapping technique to define more precisely the distribution of mast cells within human skin, COWEN et al. (1979) showed that in skin biopsy specimens from the forearm and upper arm, the greatest density of mast cells occurred in the superficial dermal zone just below the dermo-epidermal junction. A close correlation has been found between mast cell numbers and histamine content of human skin (EADY et al. 1979), though extreme methods have proven necessary to ensure that all the histamine was extracted from the skin. Using repeated boiling to extract histamine from the tissue, SONDERGAARD and ZACHARIAE (1968) found significant amounts of this mediator in the epidermis of human skin, a site which is histologically devoid of mast cells. With the recognition that both rodent and human tissues may contain atypical mast cells which require special fixation procedures for their visualisation (ENERBACK 1966a; STROBEL et al. 1981), it can be concluded that the discrepancies in histamine content and mast cell numbers at various sites in human skin may not necessarily indicate a non-mast cell source of histamine; rather there may be mast cells or mast cell precursors in human skin which cannot be visualised with metachromatic stains. Alternatively, the presence of epidermal histamine may represent the uptake of mast cell granules or histamine by phagocytic cells in this skin layer.

B. Mast Cell Structure

The metachromasia of mast cell granules exhibited on binding of basic dyes has meant that the light microscopic appearance of the cells has been well described. Mast cells located in the corium are about 5–15 μm in length, though occasionally cells as large as 30 μm have been reported. Their shape has been variously described as polyhedral, fusiform, ovoid, rectangular and rectangular, with a cytoplasm containing a variable number of lysosomal granules 0.1–0.5 μm in diameter. When free from the constraints of the surrounding tissue elements, dispersed skin mast cells are rounded; this allows more accurate size estimates to be made, and according to BENYON et al. (1987) the cells are 4–18 μm in diameter, with 79%

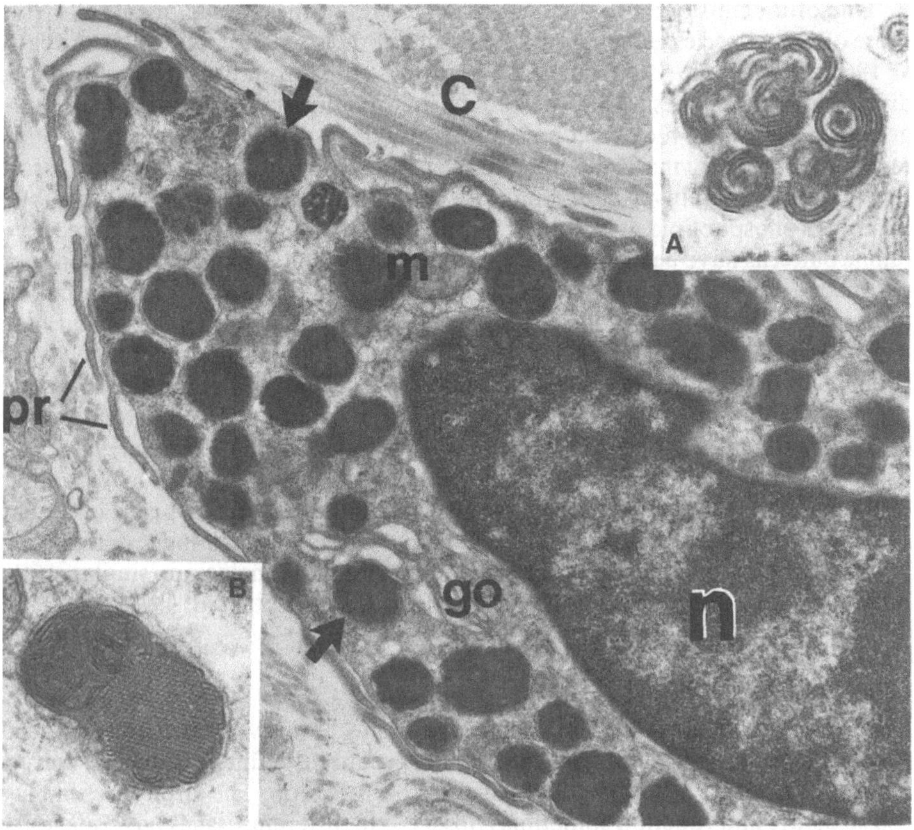

Fig. 1. Electron micrograph illustrating the morphological features of the human skin mast cell. The cytoplasm contains numerous rounded electron-dense granules (*arrowed*). Mitochondria (*m*), Golgi apparatus (*go*), cell nucleus (*n*) and cytoplasmic processes are illustrated. The cell is surrounded by collagen (*c*) (×16300). *Insert A* shows a granule with scroll-like inclusions (×82000). *Insert B* illustrates another granule with scroll-like and lattice-like crystalline inclusions (×62800). (Reproduced by kind permission of Dr. R. A. J. EADY)

of them measuring between 9 and 13 μm. Under the electron microscope, mast cells in the skin, in common with mast cells from other sites, contain many membrane-bound lysosomal granules which are specifically modified in this cell type for secretion. The lysosomal granules may vary from 0.2 to 0.5 μm in diameter, with each cell containing 50–500 (Fig. 1). Mast cell granules originate from the Golgi apparatus, which is responsible for the synthesis and organisation of the performed mediators contained therein. Other normal cytoplasmic constituents include a single nucleus which is round or ovoid, with distinct chromatin margination and one, or sometimes two, small nucleoli. There are relatively few mitochondria within mast cells, though cytoplasmic microtubules and perinuclear intermediate filaments are a striking feature in this cell type (CAULFIELD et al. 1980).

One of the most striking features which distinguishes human mast cells from those of other species is the crystalline nature of their secretory granules. As early as 1960, HIBBS et al. (1960) suggested that mast cells from human tissues can be divided into three types with respect to the internal fine structure of their secretory granules. Following the development of methods to disperse mast cells from human lung tissue enzymatically, the ultrastructural morphology of mast cells from this tissue has been investigated in more detail. Each granule is surrounded by a bilaminar phospholipid membrane. Within the phospholipid envelope three crystalline patterns have been described, scrolls, lattices and gratings, with a common periodicity of 75 and 150 Å (CAULFIELD et al. 1980). Since all three granule patterns may be found within the same granule, it is suggested that the various patterns do not represent granule heterogeneity but rather variations in the way the subunits of the granule matrix are assembled. Since mast cell granules contain the preformed inflammatory mediators which are secreted upon cell activation, the crystalline matrix granules of resting mast cells indicates that these mediators are organised in a highly ordered array possibly to aid their storage. The whorl and scroll patterns of lung mast cells have also been observed in mast cells of human skin (BOWYER 1968; LAGUNOFF 1972) (Fig. 1). ORR (1977) has suggested that the granules of skin mast cells contain frequent crystalline inclusions similar to the lattice structure described in lung mast cells by CAULFIELD et al. (1980).

C. The Ontogeny of Mast Cells

Unlike the blood basophil, which is recognised to be derived from a granulocyte precursor, the progenitor of the mast cell has yet to be defined. However, considerable advances have been made recently in this direction. The differences in mast cell structure and function sometimes observed in different tissues have suggested that there may be more than one stem cell giving rise to mast cells. In 1966, ENERBACK (1966a, b) described two distinct mast cell phenotypes within the gastrointestinal tract of the rat which differ in their sensitivity to formaldehyde fixation and the staining of their granules with basic dyes. Mast cells located predominantly in connective tissue stain well with astra blue, alcian blue and toluidine blue, irrespective of the tissue fixative used. In contrast, those mast cells restricted mostly to the gastrointestinal mucosa fail to stain if formalin fixation is used, but stain well when fixed with Carnoy's fixative or glutaraldehyde. Using a sequential staining technique it has also been demonstrated that safranin O is able to displace alcian blue from the granules of rodent connective tissue mast cells but not mucosal mast cells, causing them to stain red (ENERBACK 1966a, b). The differences in the staining characteristics of mast cell granules in the various anatomical sites possibly reflects differences in granule proteoglycan structure (SPICER 1963; ENERBACK 1986). Adoption of these staining techniques with human skin has shown the presence of both formalin-resistant and formalin-sensitive cells, the former predominating. However, as the structures of human mast cell heparin and proteases are different from those of the rat, the validity of the extrapolation of rodent methods to human cells has not been proven.

Over the last few years the concept of rodent mast cell heterogeneity has been extended to biochemical and functional studies. Mucosal and connective tissue mast cells are known to have differences in their content of preformed mediators. Mast cells of skeletal muscle, like those isolated from the peritoneal cavity of rats, contain large quantities of a neutral protease which exhibits chymotryptic enzyme activity and is, therefore, termed chymase (rat mast cell protease, RMCPI) (WOODBURY et al. 1978). This 27000-dalton enzyme, in being strongly electropositive, binds tightly to the glycosaminoglycan side chains of heparin and stays associated with heparin complex when released from the granules following degranulation (SCHWARTZ et al. 1981 a). Mast cells isolated from the gastrointestinal mucosa of the rat contain an antigenically distinct chymotryptic protease (RMCPII) which differs from RMCPI in being only weakly associated with the granule matrix (WOODBURY and MILLER 1982). Recent studies have shown that RMCPII may be released both into the gut lumen and into the circulation following gastrointestinal challenge of rats previously sensitised to *Nippostrongylus brasiliensis* (MILLER et al. 1983).

Like mast cells of the rat, human mast cells from different tissue sites contain different neutral proteases, suggesting heterogeneity at the level of mediator content. Human lung mast cells have as their major neutral protease a tryptic enzyme (human lung tryptase) (SCHWARTZ et al. 1981 b), whereas large quantities of a chymotryptic enzyme which has been identified as a mast cell product (SCHECHTER et al. 1986) are present in human skin (SCHECHTER et al. 1983). Using antibodies raised against human mast cell tryptase and chymase, IRANI et al. (1986) have demonstrated that the majority of mast cells of human lung and intestinal mucosa contain only tryptase, whereas mast cells containing both tryptase and chymase predominate in human skin and intestinal submucosa. On the basis of different protease content, it is clear then that there is heterogeneity within the human mast cell population. However, studies of the possible functional differences between these cytochemically different mast cells are still at an early stage.

Recently, there has been great interest in further defining the factors that influence mast cell growth and maturation. A number of workers have demonstrated that mast cells may be cultured from the bone marrow of mice (NAGAO et al. 1981) and rats (HAIG et al. 1983), provided that the cells are exposed to growth factors derived from stimulated T-lymphocytes (NABEL et al. 1981). This important mast cell growth factor has been isolated from stimulated mouse T-lymphocytes and purified to homogeneity (IHLE et al. 1982). It consists of a glycoprotein of 28000 daltons and, because it is a lymphokine derived from T-lymphocytes, it has been called interleukin-3 (IL-3) (IHLE et al. 1983). IL-3 possesses growth-promoting properties that have advanced our understanding of mast cell ontogeny. Cells in tissue culture that differentiate from mouse bone marrow in the presence of IL-3 have many properties ascribed to mast cells. Their cell surface contains up to 30000 specific receptors for the C4-domain of IgE and their granules contain histamine, neutral proteases, exoglycosidases and an acidic proteoglycan (SREDNI et al. 1983). However, since the proteoglycan of bone marrow-derived mast cells has an entirely different structure in containing less acidic glycosaminoglycan side chains (chondroitin sulphate E) rather than heparin, it is likely that these cells relate more to the mucosal-type mast cells described by ENERBACK (1966 a, b, 1986) than to those of connective tissue. Further evidence for this may be inferred from the studies of MILLER (1980), who has shown that the mucosal mast cells of the parasitised rat gastrointestinal tract require a T-lymphocyte factor in order to differentiate. If rats are rendered deficient in T-lymphocytes or unable to generate IL-3 following pre-treatment with corticosteroids, then mucosal mast cell differentiation is greatly impaired. While histochemically the mucosal-type mast cells of the rat and mouse gastrointestinal tracts appear similar, glycosaminoglycan chains of the rat proteoglycan consist of chondroitin sulphate-di B rather than chondroitin sulphate E (STEVENS et al. 1986). Similarities and differences between mucosal-type mast cells of the mouse and rat and connective tissue mast cells of the same species are shown in Table 1.

At present, there is relatively little information on the ontogeny of connective tissue mast cells. At one time it was suggested that these cells originate and differentiate at their final tissue sites, though more recently this view has been modified by the demonstration that heparin-containing mast cells may be isolated from progenitor cells of rodent lymph nodes when grown on fibroblast monolayers (GINSBURG et al. 1978; LEVI-SCHAFFER et al. 1985). In patients with systemic mastocytosis, the finding of large numbers of mast cells in the bone marrow, spleen and liver might suggest that the reticulo-endothelial system may also contain large quantities of progenitor connective tissue mast cells. The demonstration by HORTON and O'BRIEN (1983) that colonies of mast cells may be grown from the bone marrow of patients with mastocytosis in the presence of certain growth factors strongly implicates the bone marrow as the origin of the precursor cell for connective tissue mast cells. Histamine-containing basophiloid-like cells have been cultured from adult (CZARNETZKI et al. 1983) and neonatal (ISHIZAKA et al. 1985) monocytes in the presence of T-lymphocyte derived growth factors, but these cells are more like basophils than mast cells (SELDIN et al. 1985). Hyperplasia of connective tissue mast cells can be induced by agents which promote tu-

Table 1. Comparison of rodent mucosal gastrointestinal and connective tissue serosal mast cells

	Mucosal	Connective tissue
1) Staining	Sensitive to fixation Orthochromasia Alcian blue > safranin O	Insensitive to fixation Metachromasia Safranin O > alcian blue
2) Size	Small	Large
3) Mediators	Histamine 1–2 pg/cell No 5-HT RMCPII Chondroitin sulphate-di B	Histamine 8–12 pg/cell 5-HT RMCPI (chymase) Heparin
4) Growth	T-lymphocyte dependent	T-lymphocyte independent
5) Secretagogues	Antigen-IgE, ionophore A23187, ? hormones	Antigen-IgE, ionophore A23187, Co 48/80, basic polyamines
6) Inhibitory drugs	Nedocromil, steroids	Sodium cromoglycate β_2-Adrenoceptor agonists Methylxanthines

morigenesis (GOTO et al. 1984) and fibrosis (GUPTA 1970). For example, bleomycin treatment of rat lungs has been shown to increase the mast cell population up to tenfold preceding the onset of fibrosis. In human fibrosing lung diseases such as extrinsic allergic and cryptogenic fibrosing alveolitis, mast cell hyperplasia has also been reported (KAWANAMI et al. 1985). It seems likely, therefore, that an intimate relationship exists between tissue fibroblasts and mast cells but the factor(s) governing their regulation remain to be defined.

The distinct possibility that all mast cells may originate from a single progenitor cell in the bone marrow and that its phenotype depends on the effects of secondary maturation principles peculiar to its local environment is an attractive hypothesis (NAKANO et al. 1985; KITAMURA et al. 1986a, b). This would not only unify the mechanisms of development of both typical (connective tissue) and atypical (mucosal) mast cells but would also explain the more subtle differences between apparently similar mast cells from different tissues (PEARCE 1986; LOWMAN et al. 1987).

D. Preformed Granule-Associated Mediators

Although it has long been known that the secretory granules of human skin mast cells contain both histamine and heparin, there is surprisingly little information on other preformed inflammatory mediators. One of the reasons for this is that methods to disperse and purify mast cells from human skin for their more detailed biochemical study have only recently been developed (BENYON et al. 1986a, 1987). Most of the information available on the chemical content of human mast cell

Table 2. Preformed mediators of rat serosal, human lung and human skin mast cells

	Rat serosal	Human lung	Human skin
Amines	5-HT	–	–
	Histamine	Histamine	Histamine
Exoglycosidases	Arylsulphatase A	Arylsulphatase B	?
	β-Glucuronidase	β-Glucuronidase	?
	β-Hexosaminidase	β-Hexosaminidase	?
	β-Galactosidase	?	?
Neutral protease	Chymase	Tryptase	Tryptase
		Kininogenase	Kininogenase
	Carboxypeptidase A	Carboxypeptidase B	Carboxypeptidase B
	Dipeptidase	Cathepsin G	? Cathepsin
		Elastase	Chymotryptase
Chemotactic factors	ECF-A oligopeptides	ECF-A oligopeptides	ECF-A oligopeptides
	HMW-NCF	HMW-NCF	HMW-NCF
Proteoglycans	Heparin	Heparin	Heparin
Miscellaneous	Superoxide dismutase	Superoxide dismutase	?
	Peroxidase	Peroxidase	?
	Prostaglandin releasing factor	Prostaglandin releasing factor	?

granules has derived from studies on mast cells isolated from human lung tissue by mechanical and enzymatic means. Using a combination of affinity chromatography (SCHULMAN et al. 1982), density gradient centrifugation (ISHIZAKA et al. 1983) and countercurrent elutriation (SCHULMAN et al. 1982) it has been possible to obtain almost homogeneous mast cell preparations.

Table 2 illustrates the array of preformed granule-associated mediators which have been found in mast cells of the rat peritoneal cavity, human lung and human skin. In the following sections, granule-derived mediators will be discussed under the headings of biogenic amines, neutral proteases, acid hydrolases, other mast cell enzymes, chemotactic factors and proteoglycans.

I. Biogenic Amines

On a molar basis, the biogenic amines, which include histamine and 5-hydroxytryptamine (serotonin), form the major components of mast cell secretory granules.

1. Histamine

Rat serosal mast cells contain 10–20 μg (HUMPHREY et al. 1963) and human lung and skin mast cells 1–5 μg (CHURCH et al. 1982; BENYON et al. 1987) histamine per 10^6 cells. Histamine is synthesised in mast cells from histidine and is stored within the mast cell secretory granule by forming a complex with the glycosaminoglycan side chains of heparin. The association between these two mediators, however, is loose, and upon mast cell degranulation histamine rapidly leaves the granule by cation exchange with extracellular sodium ions (UVNAS 1967).

Histamine, on reaching the extracellular environment, produces its physiological effects by stimulating specific cell surface receptors designated H_1 and H_2. A wide variety of pharmacological effects are produced which include increased vascular permeability, vasodilatation and contraction of airways and visceral smooth muscle. All of these effects are mediated through stimulation of H_1-receptors. Stimulation of H_2-receptors results in increased gastric acid secretion by parietal cells, inhibition of mediator release from basophils, neutrophils and lymphocytes, and augmentation of T cell suppressor activity. Following the intradermal injection of histamine, a characteristic triple response occurs as originally described by LEWIS (1927). This response consists of local erythema, caused by a direct vasodilator effect of histamine, a wider area of erythema due to axon reflex induced vasodilatation, and finally the characteristic oedematous weal, resulting from the opening of tight junctions between capillary endothelial cells, thereby increasing vascular permeability. There is now convincing evidence that the axon reflex involves antidromic nervous conduction along afferent sensory nerves in the skin with release of substance P from peptidergic nerve endings (HAGERMARK et al. 1978). The role of mast cells in the weal and flare reaction is discussed later.

Histamine-mediated vasodilatation and increases in vascular permeability have been shown to synergise with chemotactic agents such as leukotriene B_4 in causing neutrophil infiltration similar to that observed in antigen-induced late phase responses (SOLLEY et al. 1976). In addition, the sulphidopeptide leukotriene, LTC_4 (which increases vascular permeability without causing vasodilatation), is able to synergise physiologically with intradermal PGD_2 in causing the weal response (LEWIS et al. 1981). Enhancement of vascular permeability facilitates the interaction of high molecular weight serum components, such as kininogen, angiotensinogen and C3, with mast cell proteases to generate a variety of secondary mediators, such as bradykinin, angiotensin and C3a (PROUD et al. 1985; WINTROUB et al. 1986).

Once released, histamine is rapidly metabolised by one of two pathways: by N-methylation to methyl histamine under the influence of *N*-methyltransferase and subsequently by monoamine oxidation to methylimidazole acetic acid or by diamine oxidase catalysed oxidation, which degrades histamine to imidazole acetic acid. Both of these degradative pathways are efficient at removing histamine so that it only has a short half-life of 1–2 min within the circulation.

2. 5-Hydroxytryptamine

5-Hydroxytryptamine (serotonin, 5-HT) is stored in the granules of rat mast cells, where it is found in a concentration of 1 μg per 10^6 cells. 5-Hydroxytryptamine added to rat mast cells is selectively taken up and concentrated within the cell (FRISK-HOLMBERG and UVNAS 1972). It is a unique granule mediator in that it may be released from the granule by a transcytoplasmic transport mechanism without degranulation per se (THEOHARIDES et al. 1982). Unlike human platelets, mast cells of human lung and skin do not contain any 5-hydroxytryptamine (Table 2).

II. Neutral Proteases

Rat connective tissue mast cells contain predominantly chymase (RMCPI), a chymotryptic enzyme, which in being highly electropositive has a close association with the electronegative glycosaminoglycan side chains of heparin (YURT and AUSTEN 1977). As discussed earlier, mast cells of rat mucosal surfaces contain predominantly RMCPII of molecular weight 25200 daltons, which, in being less charged than RMCPI, is found loosely associated with the granular proteoglycan from which it can be easily solubilised. Although both rat mast cell proteases exhibit chymotryptic enzyme activity, they have slightly different substrate profiles (MILLER 1980).

Human skin contains a number of neutral proteases, most of which are localised in the dermis. Small amounts of protease isolated from the epidermis consist of plasminogen activator which is thought to derive from epidermal cells rather than mast cells (BROWN and CHAMBERS 1984). Human skin contains two neutral proteases which are extractable in high ionic salt solutions, one with chymotrypsin-like and the other with trypsin-like enzymatic profiles (FRAKI and HOPSU-HAVU 1975). The chymotryptic protease of human skin, which has a molecular weight of 30000 daltons, has been reported to be distinct from cathepsin G, a chymotryptic enzyme of similar molecular weight, isolated from human neutrophil granules (SCHECHTER et al. 1983, 1986). Levels of chymotryptic protease are markedly increased in mastocytoma tissue, suggesting, but not proving, a mast cell origin for this enzyme (KATAYAMA and ENDE 1965; SCHECHTER et al. 1983). More convincing evidence that the chymotryptic enzyme is mast cell associated derives from the observations of SCHECHTER et al. (1986) that enzyme-linked antibodies raised against the enzyme stained only those cells which stained with the metachromatic dye toluidine blue in human skin sections.

The trypsin-like protease of human skin has been difficult to study since it appears to be labile when highly purified. When extracted from human skin in high salt buffers it has a molecular weight of 120000 daltons on gel filtration and is composed of several subunits (FRAKI and HOPSU-HAVU 1975). Like the chymase, human skin tryptase levels increase during mastocytosis. The physicochemical characteristics of this protease and its inability to be inhibited by α_1-trypsin inhibitor or α_2-macroglobulin make it almost identical to the trypsin-like enzyme called tryptase, which is the major neutral protease isolated from the granules of human lung mast cells (SCHWARTZ et al. 1981 b). Tryptase is a serine esterase of molecular weight 144000 daltons, composed of two subunits of molecular weight 37000 daltons and two subunits of molecular weight 35000 daltons. Both pairs of subunits bind ^{3}H-DFP, indicating the presence of an active esteratic site. Recently, several other neutral proteases have been isolated from human lung mast cells. These include elastase, cathepsin G (MEIER et al. 1985) and a chymotrypsin-like enzyme with angiotensin-converting enzyme (WINTROUB et al. 1986). Each of these enzymes is released from mast cells activated by an immunological stimulus and the latter enzyme may be identical to the human skin chymase.

Tryptase and chymase enzymes appear to be common constituents of human mast cells. Immunohistochemical techniques employing anti-human lung tryptase and anti-human skin chymase antibodies have demonstrated, however, that mast cells of different human tissues differ in their content of these two enzymes (IRANI et al. 1986). Using a sequential staining technique, IRANI et al. demonstrated that tryptase was ubiquitously distributed in mast cells of human lung, intestinal mucosa, skin and intestinal submucosa. However, chymase was identified in $<10\%$ of dispersed lung mast cells and intestinal mucosal mast cells and, in contrast, in approximately 90% of skin and submucosal mast cells. On the basis of protease content, it therefore appears that two distinct types of mast cell exist within human tissues. Interestingly, human skin mast cells, which are predominantly chymase positive, are activated for histamine secretion by a variety of non-immunological stimuli to which mast cells of the other three tissues, including the predominantly chymase-positive submucosal cells, are unresponsive (BENYON et al. 1987; LOWMAN et al. 1987; REES et al. 1987). This argues against the hypothesis formulated from experiences with rodent mast cells that different protease content reflects different secretory and possibly functional characteristics. As regards the possible biological role of mast cell neutral proteases, human lung mast cell tryptase has some interesting enzyme activities. It is able to degrade C3 in vitro to C3b, generating the anaphylatoxin C3a (SCHWARTZ et al. 1982), which itself activates basophil leucocytes for histamine secretion (GRANT et al. 1986). It is also able to inactivate human high molecular weight kininogen, thereby inhibiting its capacity to release bradykinin in response to kallikrein (MAIER et al. 1983; PROUD and LICHTENSTEIN 1984) and degrade fibrin and fibrinogen (SCHWARTZ 1984).

Elastase and cathepsin G are effective at degrading connective tissue (collagen, elastin, proteoglycans) whilst chymase might act to generate the vasoactive peptide angiotensin II in perivascular sites in close proximity to vascular smooth muscle targets (WINTROUB et al. 1986). Human skin chymase may function in the separation of dermis and epidermis during the formation of blisters (BRIGGAMAN et al. 1984).

III. Acid Hydrolases

Rodent and human mast cell granules contain a variety of exo- and endoglycosidases which have the capacity to degrade glycosaminoglycans, glycoproteins and complex glycolipids in an acid environment (SCHWARTZ and AUSTEN 1981). β-Hexosaminidase occurs as two isomers, A and B, each composed of four subunits. The A isomer comprises two alpha and two beta subunits and the B isomer four beta subunits (SCHWARTZ and AUSTEN 1981; SCHWARTZ et al. 1981c). The alpha and beta subunits differ structurally and arise from different genes. Only the B isomer of the enzyme has been found in human mast cells, whereas rat mast cells contain predominantly the A-type isoenzyme. The ratio of β-hexosaminidase to histamine in human lung mast cell granules is approximately 40 times greater than in those of rat mast cells.

β-Glucuronidase occurs in two isomeric forms, a microsomal and a lysosomal type, each being composed of tetramers of molecular weight 75000 daltons. The amount of β-glucuronidase in human lung mast cells is about one-fifth that in rat serosal mast cells, with about 65% of the total enzyme activity being localised to the secretory granules (SCHWARTZ and AUSTEN 1981).

Arylsulphatases catalyse the hydrolysis of aromatic sulphate esters. Two lysosomal forms of the enzyme, A and B, have been identified with differing molecular weights and isoelectric points. Rat serosal mast cells contain approximately equal amounts of the two isomers, but only the A subtype can be immunologically released, suggesting that it is this type that is localised to the secretory granule (SCHWARTZ and AUSTEN 1981). Human pulmonary mast cells also contain arylsulphatase but the isoenzyme form has not been characterised. β-Galactasidase has been identified in rat mast cells (SCHWARTZ and AUSTEN 1981), but whether it occurs in human mast cells is at present not known.

IV. Other Mast Cell Enzymes

A small quantity of superoxide dismutase, which accelerates the conversion of superoxide anion to hydrogen peroxide, may be immunologically released from rat cells, but it is not known whether this enzyme is present in the granules of human mast cells (HENDERSON and KALINER 1978). Similarly, peroxidase, which in the presence of hydrogen peroxide causes chemical oxidation of suitable substrates, has also been reported to be released from immunologically activated rat mast cells in small amounts (HENDERSON and KALINER 1979). The enzyme is tightly bound to the mast cell heparin, where it exhibits greater enzyme activity than when freely soluble. Peroxidases have the capacity to inactivate leukotrienes and in concert with dipeptides may have an important role in the bioregulation of tissue concentrations of these newly formed mast cell mediators.

V. Chemotactic Factors

A chemotactic factor is one which is able to stimulate the directed migration of mobile inflammatory cells along a concentration gradient, while an agent which promotes chemokinesis increases the cells' random movement. Continual exposure of a susceptible migratory cell to a chemotactic agent results in tachyphylaxis, often referred to as deactivation. Accumulations of neutrophils, monocytes and eosinophils have been observed in human skin sites when injected with antigen (SOLLEY et al. 1976) and after cold challenge of the skin of patients with cold urticaria (CENTER et al. 1979) and necrotising veneolitis (SOTER et al. 1978). Chemotactic activity for eosinophils has been demonstrated in human blister fluid following challenge of exposed dermal sites with ragweed antigen (TING et al. 1981). While a number of chemotactic peptides and proteins have been described to be more or less specific for eosinophils and neutrophils, the acidic eosinophil chemotactic tetrapeptides, val-gly-ser-glu and leu-gly-ser-glu, are now no longer considered to be major constituents of human or rat mast cell granules, and probably represent cleavage products from high molecular weight eosinophil chemotactic factors have been described with differing molecular weights, but it is the high molecular weight NCF that has attracted most attention (KAY and LEE 1982). This material is released following exercise challenge in cholinergic urticaria and following cold challenge of a limb in cold urticaria (KAY and LEE 1982). The fact that NCF is released in parallel to the release of histamine has led to the suggestion that this mediator is mast cell derived (KAY and LEE 1982). O'DRISCOLL et al. (1983) have demonstrated the release from antigen-challenged human lung fragments of small amounts of high molecular weight NCF with an apparent molecular weight by gel filtration of 750000 daltons and an isoelectric point approaching neutrality. However, the recent demonstration of release of a similar material from IgE-dependent activation of blood monocytes and lymphocytes suggests that this mediator or group of mediators is not specific to mast cells (CUNDELL and DAVIES 1985). A 1400-dalton peptide has been isolated from rat mast cell granules which, when injected into rat skin, elicits a biphasic influx of neutrophils at 2–8 h and mononuclear cells at 24 h (KALINER and LEMANSKE 1984a) that could be related to the late phase reaction. Whether human mast cells possess a similar peptide requires further elucidation.

VI. Proteoglycans

The mast cell is the sole source of heparin proteoglycan. Its characteristic histochemical property of interacting with basic dyes in the form of metachromasia forms one of the major histochemical criteria for identifying these cells. Heparin is a proteoglycan glycosaminoglycan (GAG) composed of uronic acid and either glucuronic or iduronic acid which are linked in disaccharide units in a 1–4 configuration. These disaccharide units are in turn linked to glycosamine in a beta 1–4 configuration (HELTING and LINDAHL 1972). The GAG side chains are attached to a core peptide which, in the case of rat mast cell heparin, consists of alternating serine and glycine residues. The GAG side chains are highly sulphated, both on the hydroxyl groups of glycosamine and iduronic acid, and on the nitrogen atoms of glycosamine. GAG side chains are synthesised within the Golgi apparatus of mast cells by the sequential polymerisation of uridine diphosphatidyl glucuronic acid and uridine diphosphatidyl-*N*-acetylglycosamine. The GAGs are attached to priming sites on the peptide core by a specific carbohydrate linkage of ser-xyl-gal-gal-glucuronic acid. The expanding GAGs are then deacylated followed by O and N sulphation (SILBERT 1967; HOOK et al. 1975; JACOBSON et al. 1979; JACOBSON and LINDAHL 1980).

Rat serosal mast cells contain approximately 28 pg of heparin per cell while human lung mast cells contain about one-seventh of this (METCALFE et al. 1979). The molecular weight of rat mast cell heparin is in excess of 750 000 daltons, while that of human pulmonary mast cells is 60 000 daltons (YURT et al. 1977). The difference in molecular weight between the proteoglycans of the two species may be largely accounted for by the length of the GAG side chains, which are of 50 000–100 000 daltons in rat mast cell heparin (METCALFE et al. 1980) and around 20 000 daltons in human mast cell heparin (ROBINSON et al. 1978).

Heparin forms the major supporting matrix of mast cell granules, the highly charged side chains offering an economical method of packaging other preformed mediators in such a way that they are rendered biologically inactive (SCHWARTZ and AUSTEN 1984). At physiological salt concentrations, human mast cell tryptase and carboxypeptidase B bind strongly to heparin on account of their relatively high isoelectric points. It is thought that this association may well give rise to the characteristic crystalline patterns of the human mast cell granules (CAULFIELD et al. 1980).

The finding that mast cells of the rat mucosa contain predominantly chondroitin sulphate di-B rather than heparin (STEVENS et al. 1986), while cultured bone marrow derived mast cells contain a unique proteoglycan whose side chains are composed of chondroitin sulphate E (RAZIN et al. 1982), has been linked to the difficulty in staining these cells after formalin fixation (ENERBACK 1986). As it has been shown that in the presence of *p*-nitrophenyl-*β*-D-xyloside rat serosal mast cells can be switched from heparin synthesis to that of chondroitin sulphate (STEVENS and AUSTEN 1982), it remains possible that various subdivisions of rodent mast cells represent various stages in mast cell maturation. Indeed, in a mast cell deficient mouse strain, cultured T cell dependent "mucosal mast cells", when injected into the skin or peritoneal cavity, have been shown to take on all the histochemical characteristics of classical rat connective tissue mast cells (NAKANO et al. 1985; KITAMURA et al. 1986a). The local tissue microenvironment is therefore likely to be of major importance in the final differentiation of mast cells and their subsequent morphological and functional heterogeneity.

Heparin of human skin appears to be very similar to that of lung mast cells. The heparin content of human skin is greatly increased in lesional areas in mastocytosis and exhibits the same physicochemical characteristics as that isolated from normal mast cells (METCALFE et al. 1981).

Heparin has long been known as a potent anticoagulant, mediating its effects on the clotting system by potentiating the inhibitory effects of anti-thrombin 3 on a number of coagulation components, namely thrombin, fragments of factor 12 and factors 9a, 10a, and 11a, kallikrein and plasmin. Anti-thrombin 3 binds to the sulphated radicals of glycosamine, an interaction which may be inhibited in the presence of cationic proteins such as protamine, the major basic protein of eosinophil granules and platelet factor 4. Heparin also inhibits C1q binding to immune complexes, the activation of C1 esterase and the amplification of the alternative pathway C3b convertase (SCHWARTZ and AUSTEN 1984). Outside the coagulation and complement system, heparin has been shown to potentiate the elastase activity of granulocytes (LONKY et al. 1978), lipoprotein lipase (ENHOLM et al. 1975) and the binding of fibronectin to collagen (JILEK and HORMANN 1979). Within the mast cell granule, the GAG side chains of heparin inhibit the biological activity of other preformed mediators and, upon degranulation of the cell, the control rate of their release into the extracellular environment. Soluble mediators with isoelectric points approaching neutrality, such as histamine, the chemotactic peptides and the exoglycosidases, leave the granule matrix rapidly, while tryptase, carboxypeptidase B and peroxidase stay granule associated. Recent evidence accrued from studies in rat skin indicates that the released mast cell heparin–protease matrix (which may contain other tightly bound mediators) may be an important stimulus for the late phase infiltration of skin with neutrophils and eosinophils (KALINER and LEMANSKE 1984b). Further work has shown that fibroblasts are capable of phagocytosing these complexes and in doing so become activated (PILLARISETTI et al. 1983). Since mast cells are found in greatly increased numbers in association with sites of skin healing (PESSINGER et al. 1983) and during the induction of fibrosing lung disease (GOTO et al. 1984), an important role for the mast cell in the regulation of fibroblast function must be considered. An effect of heparin pertinent to wound healing is the stimulation of capillary endothelial cell migration (MARKS et al. 1986), which also might be relevant to the vascularisation of neoplasms, where increased mast cell numbers are also found.

E. Newly Generated Inflammatory Mediators

In addition to releasing granule-derived mediators it has become recognised that activated mast cells have a capacity to generate a wide variety of other highly potent chemical mediators, most of which are lipid derived. The most widely studied and probably the most abundant newly generated mast cell-derived mediators originate from the oxidation of arachidonic acid, which in turn is liberated from the phospholipid plasma membrane following activation of membrane-located phospholipases (KENNERLY et al. 1979) (Fig. 2). Arachidonic acid may be metabolised along one of two routes, the cyclo-oxygenase and the lipoxygenase pathways. Both metabolic routes as have been described in human mast cells will be considered separately.

I. Cyclo-oxygenase Products of Arachidonic Acid

Arachidonic acid is a 20-carbon fatty acid which may undergo oxidation at the C-11 and C-15 positions, forming a cyclic unstable intermediate, PGG (HAMBERG

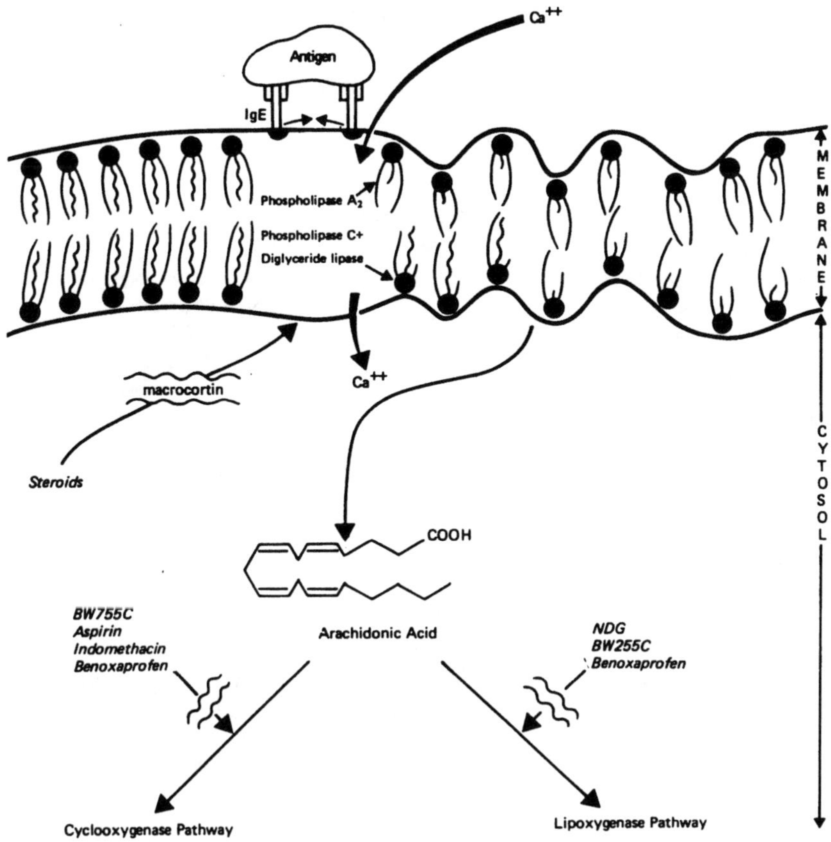

Fig. 2. Release of arachidonic acid from mast cell membrane phospholipids

et al. 1974). In vitro microsomal cyclo-oxygenase requires haemoglobin, myoglobin or catalase to serve as a proton acceptor, while in the presence of tryptophan and other aromatic compounds, the 15-hydroperoxy group of PGG undergoes reduction to yield PGH (OHKI et al. 1979), another unstable endoperoxide which plays a key role in the further metabolism towards the prostanoids as shown in Fig. 3. Depending upon the cell type, a wide variety of prostanoids may be generated from PGH_2. These include PGE_2, $PGF_{2\alpha}$, PGD_2, PGI_2 (prostacyclin) and thromboxane A_2. Immunological activation of enzymatically dispersed human lung cells containing 1%–5% mast cells generates PGD_2 and TXA_2 as the major cyclo-oxygenase products of arachidonic acid, with smaller amounts of $PGF_{2\alpha}$, PGE_2 and the 6-keto $PGF_{1\alpha}$ metabolite of PGI_2 (HOLGATE et al. 1984). Cell fractionation studies of human lung cells have demonstrated that mast cells generate almost entirely PGD_2 in amounts up to 50 ng per 10^6 mast cells (HOLGATE et al. 1984). Most of the TXA_2 originates from activated macrophages and monocytes, while the other prostanoids are derived from other secondary cell types within human lung.

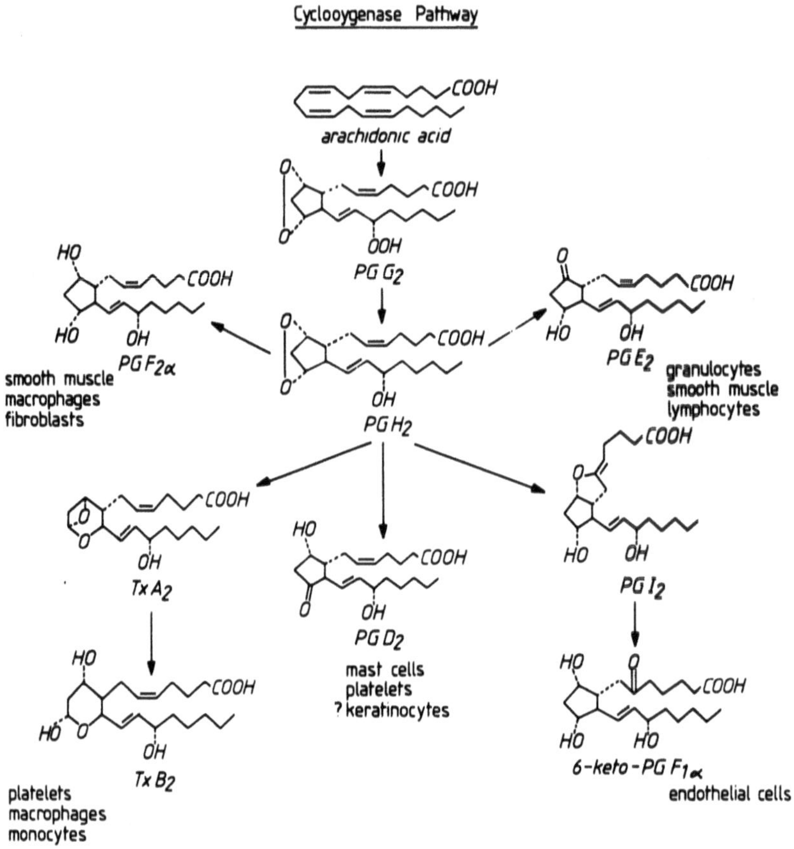

Fig. 3. Cyclo-oxygenase metabolism of arachidonic acid to prostanoids

KINGSTON and GREAVES (1976) have localised the PG synthetase activity of homogenised rat skin to the microsomal fraction, with PGE_2 being generated as the major product followed by PGD_2 and $PGF_{2\alpha}$. It should be recognised, however, that PGD_2 synthetase is a cytosolic enzyme whose activity would be greatly underestimated if only the microsomal fraction were investigated. In skin from patients with mastocytosis dermographic wealing occurs despite total H_1-receptor blockade with astemizole (KRAUSE and SHUSTER 1985), and enhanced PGD_2 release has been implied by the excessive urinary secretion of PGD_2 and PGD_2 metabolites in the urine (ROBERTS et al. 1980). Further evidence for the mast cell origin of PGD_2 is provided by the finding that large amounts of the major PGD_2 metabolite α-hydroxy,11,15-di-oxo-2,3,4,5-tetranor-prostain-1,20-dioic acid are present in the urine of patients with systemic mastocytosis when compared with levels in urine from normal subjects (ROBERTS et al. 1980). At least eight other PGD_2 and ten $PGF_{2\alpha}$ metabolites have also been detected in increased amounts in the urine of mastocytosis patients (ROBERTS and SWEETMAN 1985). The large amounts of PGD_2 generated by the unstable mast cells in mastocytosis may lead to a variety of systemic features of the disease which are not amenable to H_1 and

H_2 antihistamine blockade. In such patients who experience hypotensive episodes with syncopy, blockade of the cyclo-oxygenase pathway with large doses of non-steroidal anti-inflammatory drugs has been shown to be highly effective (ROBERTS et al. 1980).

Release of PGD_2 into the circulation has been demonstrated following ultraviolet B radiation (BLACK et al. 1981), antigen provocation (KORO et al. 1986a), heat urticaria (KORO et al. 1986b) and cold urticaria (HEAVEY et al. 1986). Its association with histamine in these reactions again suggests the mast cell to be its source of origin. Using dispersed skin mast cells enriched to >80% we have demonstrated highly significant ($P<0.001$) correlations between PGD_2 production, histamine release and mast cell numbers following activation of mast cells with calcium ionophore A23187. The production of PGD_2 under these conditions was 40–60 ng/10^6 mast cells and the histamine–PGD_2 ratio was 19:1 (unpublished observations). These results are similar to those obtained in lung mast cells (HOLGATE et al. 1984). A single report by RUZICKA and PRINTZ (1982) has suggested that PGD_2 is the major cyclo-oxygenase product of epidermal cells which are essentially devoid of mast cells. To our knowledge this has not been confirmed.

Although PGD_2 is the only prostaglandin to be generated by mast cells in appreciable quantities, additional cyclo-oxygenase products may be generated by other cells in vivo. Examples of this are the additional generation of PGE_2, PGF_2 and 6-oxo-$PGF_{1\alpha}$ following ultraviolet B radiation (BLACK et al. 1981) and of these products plus thromboxane following antigen challenge of human skin in vivo (OIKARINEN et al. 1981).

II. The Lipoxygenase Pathway

It has long been recognised that anaphylactic stimulation of animal and human tissues in vitro generates in highly potent contractile substance named "slow reacting substance of anaphylaxis" (SRS-A). The structural elucidation of this contractile activity as a mixture of 6-sulphidopeptide leukotrienes of arachidonic acid has opened up an intensive area of research into the definition of new pro-inflammatory mediators. The 5-lipoxygenase pathway has attracted most attention since it is this route of arachidonic acid oxidation that leads to the formation of the sulphidopeptide leukotrienes (reviewed by PIPER 1983). The metabolic pathway for leukotriene biosynthesis is illustrated in Fig. 4. Arachidonic acid is first oxidised by a calcium-dependent 5-lipoxygenase enzyme to the unstable intermediate 5-hydroperoxyeicosatetraenoic acid (5-HPETE). 5-HPETE may be reduced to 5-hydroxyeicosatetraenoic acid, a moderately active chemotactic factor (GOETZL et al. 1980), or undergo epoxide formation leading to the unstable second intermediate, leukotriene (LT) A_4 (5,6-oxidoeicosatetraenoic acid). Leukotriene A_4 may undergo one of two further enzymatic reactions. Enzymatic hydrolysis of LTA_4 leads to formation of LTB_4, 5S,12R-dihydroxyeicosatetraenoic acid, with potent neutrophil and eosinophil chemotactic properties (FORD-HUTCHINSON et al. 1980). Alternatively, conjugation of the epoxide with the tripeptide, glutathione, catalysed by glutathione S transferase yields the first of a series of six sulphidopeptide leukotrienes called LTC_4 (HAMMARSTROM 1983).

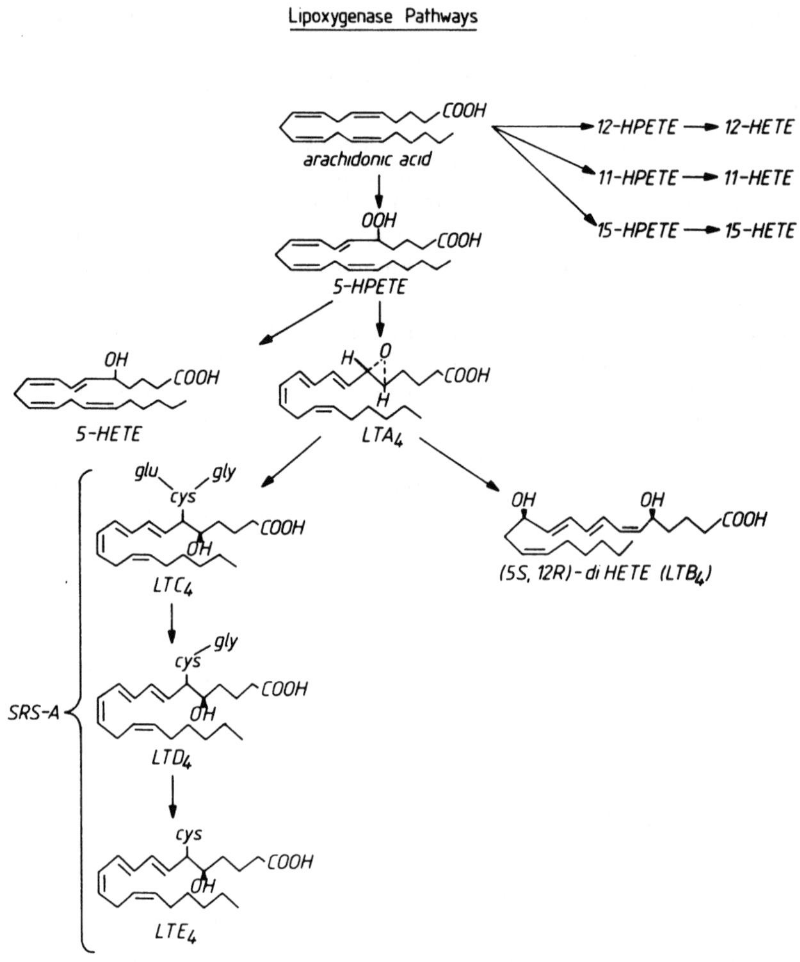

Fig. 4. Lipoxygenase metabolism of arachidonic acid to hydroperoxyeicosatetraenoic acids (*HPETE*), hydroxyeicosatetraenoic acids (*HETE*) and leukotrienes (*LT*)

Cleavage of the terminal glutamyl residue of LTC_4 by γ-glutamyl transpeptidase results in the formation of LTD_4, while leukotriene E_4 may subsequently arise by the removal of the glycyl residue by a dipeptidase. The recognition that purified human lung mast cells have the capacity to generate large quantities of LTC_4 (MACGLASHAN et al. 1982) provides direct evidence that these highly potent mediators may play an important role in the pathogenesis of allergic reactions (LEITCH and DRAZEN 1984).

The first evidence of a functional 5-lipoxygenase in skin was presented by ZIBOH et al. (1983), who showed that calcium ionophore A23187-stimulated mouse keratinocytes release products with chromatographic behaviour of 5-HETE and LTD_4. BRAIN et al. (1982) have further shown that cultured human keratinocytes can generate appreciable amounts of LTB_4. Our own preliminary studies with partially purified human skin mast cells have indicated that they do not generate large numbers of leukotrienes. Of those produced following calcium ionophore activation, LTC_4 is the major product (unpublished observations). The paucity of information on the capacity of human skin and in particular skin mast cells to generate lipoxygenase products of arachidonic acid is in marked contrast to the large amount of data that has been generated concerning their pharmacological actions, which will be considered later in this chapter.

III. Platelet Activating Factor

Platelet activating factor, which was first recognised in 1966 from activated rabbit leucocytes (BARBARO and ZVAIGLER 1966), has now been identified as 1-O-alkyl-2-acetyl-sn-glyceryl-3-phosphorylcholine (AGEPC or PAF-acether) (DEMOPOULIS et al. 1979). AGEPC is synthesised by human neutrophils, monocytes, platelets and mast cells following their specific activation. However, in the majority of cells, including mast cells, most of the AGEPC synthesised by the cells is not released into the external environment, thus suggesting it to be an intracellular mediator (LICHTENSTEIN et al. 1984; SNYDER 1985). In biological fluids it is extremely unstable and undergoes rapid hydrolysis of the acyl side chain to the lyso-derivative (FARR et al. 1982). Intracutaneous administration of AGEPC in humans causes a weal and flare reaction associated with pain, pruritis and a late inflammatory response of neutrophil infiltration similar to that described for LTB_4. However, while it is tempting to invoke the release of AGEPC in allergic reactions involving human skin, as with the sulphidopeptide leukotrienes, its definitive release by resident cells of this tissue has yet to be demonstrated.

F. Mechanisms of Mast Cell Activation

The discovery by PRAUSNITZ and KUSNER (1921) that the immediate skin test to a specific allergen could be passively transferred into human skin represented a major advance in the understanding of the immediate hypersensitivity reaction. The active principle responsible for the transfer of hypersensitivity, called reagin, was later demonstrated to be a unique immunoglobulin IgE by ISHIZAKA and ISHIZAKA (1967). The discovery of the human E-myeloma protein established a physicochemical property of this protein in being a gamma-1 glycoprotein with a sedimentation coefficient of 8S and a molecular weight of 190000. In both human and monkey skin E-myeloma protein was able to block the PCA reaction though this was lost if the IgE antibody was heated to 56 °C for 2–4 h. These observations therefore formed the basis of our understanding of IgE-dependent mediator release from mast cells.

I. Mechanisms of IgE-Dependent Mediator Secretion from Mast Cells

Human lung mast cells contain >130000 specific receptors which bind to the C4 domain of the Fc region of the heavy chains of IgE (ISHIZAKA and ISHIZAKA 1984). Over the last 10 years Ishizaka and co-workers have carried out some elegant studies employing an antibody to the IgE receptor of rat basophil leukaemia cells and have unequivocally demonstrated that activation of mast cells and basophils for mediator secretion occurs through the cross-bridging of IgE Fc receptors. Whereas anti-receptor antibody and its Fab'2 fragments induce both skin reactions in normal rats and histamine release from mast cells, monomer fragments of the antibody are unable to do so. The mechanisms whereby cross-linking of cell-bound IgE by specific allergens and cross-linking of their corresponding Fc receptors initiates mediator secretion have been intensively investigated over the last 5 years. The first step in the reaction sequence appears to be the activation of a serine esterase which can be inhibited by DFP (AUSTEN and BROCKLEHURST 1960). The serine esterase activity can also be inhibited by a variety of chymotryptic and tryptic enzyme inhibitors and substrates (ISHIZAKA and ISHIZAKA 1984). The observation that chymotrypsin and chymase are able to activate mediator secretion from rat peritoneal mast cells might represent a similar initiating mechanism for mediator secretion (LAGUNOFF et al. 1975). The serine esterase may represent a component of the IgE receptor complex, but not that portion that binds specifically to IgE. Figure 5 illustrates the current model of the IgE receptor complex, with its three components, α_1 and α_2 which bind to the Fc fragment of IgE, and two other components beta and gamma which are associated with but not covariantly linked to the alpha component. The whole receptor spans the cell membrane and presumably, therefore, the beta and gamma components are responsible for the transmembrane linkage of IgE receptor activation to cell secretion (reviewed by FROESE 1984). Whether either the beta or the gamma component represents a serine esterase activity is not yet known.

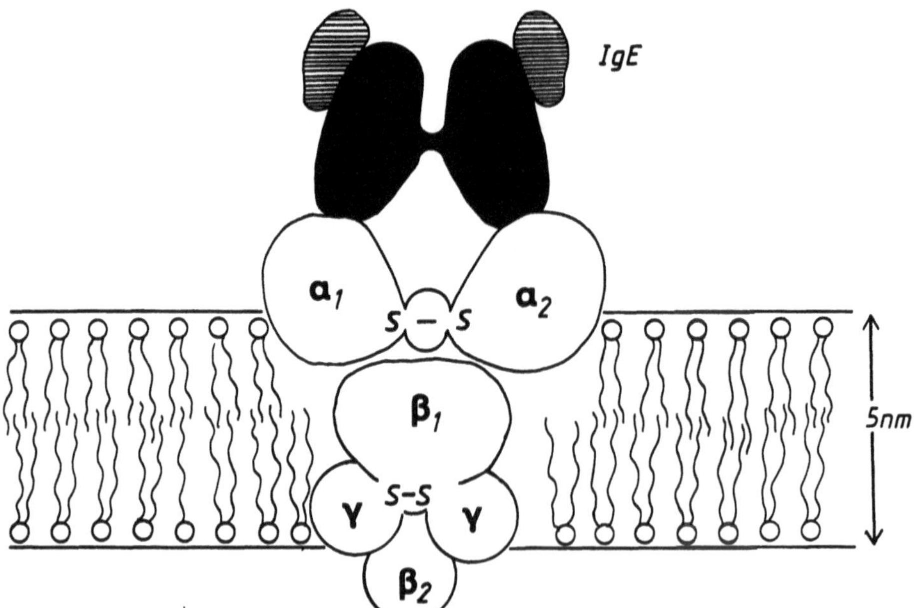

Fig. 5. Proposed structure of the IgE Fc receptor of the rat leukaemic basophil

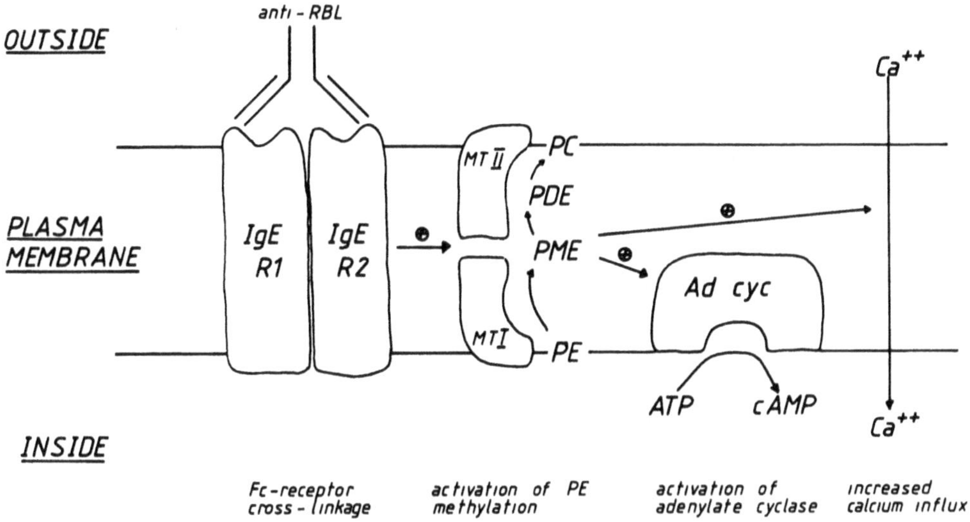

Fig. 6. Phospholipid methylation in IgE-dependent mast cell activation. *R1* and *R2*, IgE Fc receptors; *MTI* and *MTII*, methyltransferases I and II; *PE*, phosphatidylethanolamine; *PME*, monomethylphosphatidylethanolamine; *PDE*, dimethylphosphatidylethanolamine; *PC*, phosphatidylcholine; *Ad cyc*, adenylate cyclase

It has long been known that mediator secretion from rodent and primate mast cells requires the presence of extracellular calcium ions (Mongar and Schild 1958) and it has therefore reasonably been suggested that cross-linkage of cell-bound IgE with receptor activation initiates a series of biochemical events in the cell membrane leading to the opening up of calcium channels (Ishizaka et al. 1979). The observation that calcium ionophores such as A23187, which bind with and then transport calcium ions directly into the mast cell, initiate mediator secretion adds support to the concept that the secretory mechanism is always intact and requires for its initiation only an increase in cellular calcium (Pearce 1982). However, it is the physiological mechanism of formation of the calcium channels that has aroused so much recent interest. Ishizaka and co-workers have suggested that the next step following serine esterase activation is a progressive methylation of membrane phospholipids and in particular the conversion of phosphatidylethanolamine (PE) to phosphatidylcholine (PC) (Fig. 6). It is suggested that this occurs in two steps, the first being the donation of a single methyl group to PE forming monomethylphosphatidylethanolamine (MPE), catalysed by a methyltransferase located in the inner leaflet of the cell membrane. The second step, donation of a further two methyl groups, follows translocation of the now slightly less polar MPE to another methyltransferase orientated to the outer leaflet of the cell membrane. The subsequent increase in the membrane concentration of methylated phospholipids is presumed to be coupled with an increase in membrane fluidity and the creation of calcium channels. In addition, the newly formed PC may provide substrate for phospholipase A_2 with the subsequent release of arachidonic acid and its subsequent oxidative metabolism to newly generated medi-

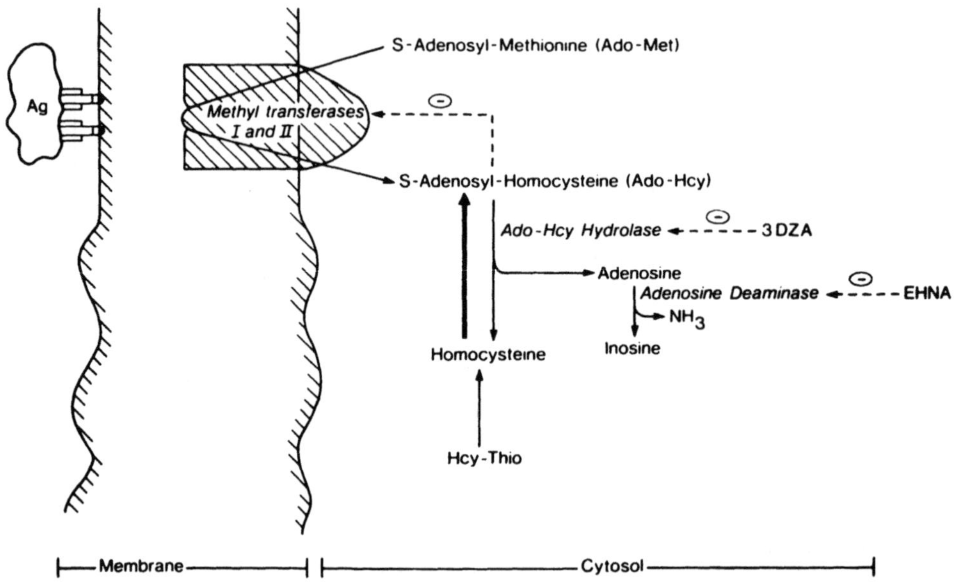

Fig. 7. Proposed effect of 3-deazaadenosine (*3-DZA*) and erythro-9-(2-hydroxy-3-nonyl)-adenine (*EHNA*) on phospholipid methylation

ators. The further demonstration by Ishizaka and co-workers (ISHIZAKA et al. 1980; ISHIZAKA and ISHIZAKA 1984) that the methylation inhibitor 3-deaza-adenosine (Fig. 7) inhibited both IgE-dependent phospholipid methylation and histamine secretion provided evidence that this biochemical pathway was essential for IgE-dependent mediator secretion from both rodent and human mast cells.

While the biochemical mechanism depicted in Fig. 6 is attractive in explaining the initial sequence of mast cell activation–secretion coupling, there have been major problems associated with it. Firstly, the amount of phospholipid that undergoes this biochemical transformation is considered to be small and far less than that required for the subsequent generation of arachidonic acid (VANCE and DE KRUIJFF 1980). Secondly, and most importantly, a number of laboratories have been unable to demonstrate enhanced phospholipid methylation consequent upon IgE receptor cross-linkage, despite being able to initiate significant histamine secretion (BOAM et al. 1984; MOORE et al. 1984; BENYON et al. 1986b). Thirdly, an alternative mechanism has been found for the actions of 3-deazaadenosine. This compound has the ability to elevate concentrations of cyclic AMP in rat and human mast cells (BENYON et al. 1985), probably by inhibiting cyclic AMP phosphodiesterase (ZIMMERMAN et al. 1980). As increased levels of this cyclic nucleotide generally inhibit histamine secretion, the actions of 3-deazaadenosine need not be interpreted on the basis of its effects on phospholipid methylation pathway putatively required for secretion. Thus, while phospholipid methylation as a step in the linking of cell surface activation to the secretion of mast cell mediators may occur, its obligatory nature is doubtful.

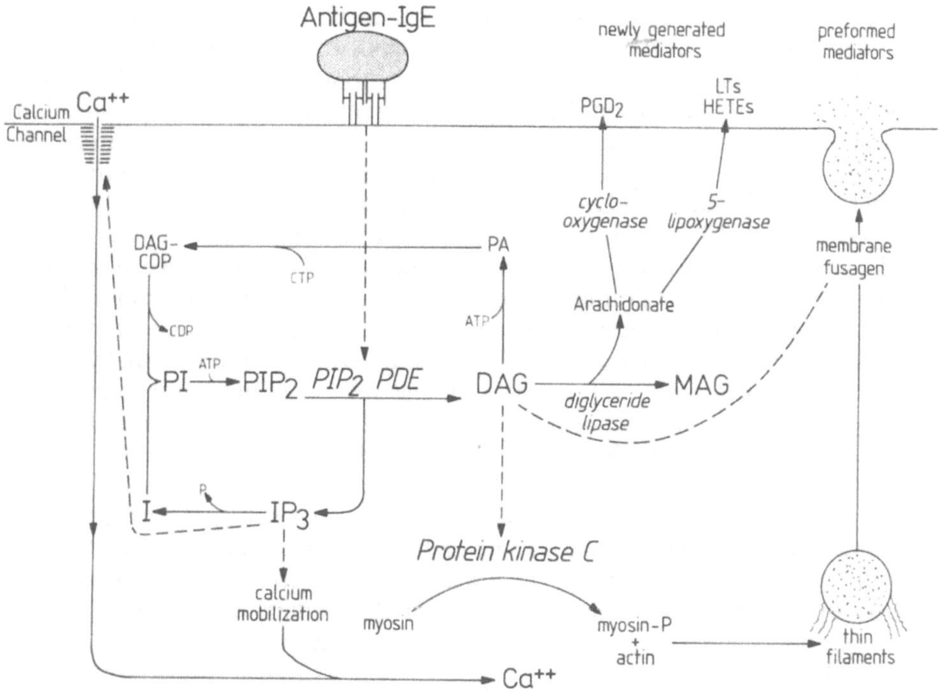

Fig. 8. Proposed phosphatidylinositol turnover in the mast cell. *PI*, phosphatidylinositol; *PIP*$_2$, polyinositol bisphosphate; *PIP*$_2$ *PDE*, polyinositol bisphosphate phosphodiesterase; *DAG*, diacylglycerol; *MAG*, monoacylglycerol; *IP*$_3$, inositol trisphosphate; *I*, inositol; *PA*, phosphatidic acid; *CTP*, cytidine trisphosphate; *CDP*, cytidine diphosphate

An alternative mechanism that has been suggested in promoting the formation of calcium channels in a wide variety of cell types, including the mast cell, is the stimulation of the phosphatidic acid–phosphatidylinositol cycle (PA-PI cycle) (SULLIVAN 1981). In this reaction, illustrated in Fig. 8, cross-linkage of cell surface IgE receptors results in the stimulation of a phospholipase C-like polyphosphoinositide phosphodiesterase which catalyses the conversion of phosphatidylinositol bisphosphate to diacylglycerol (DAG) and inositol trisphosphate (IP_3) (BERRIDGE 1984). DAG may then undergo one of two biochemical transformations, conversion to monoacylglycerol releasing arachidonic acid for subsequent oxidative metabolism, or condensation with ATP to form PA. IP_3 has been implicated in the mobilisation of intracellular calcium and possibly the formation of physiological membrane calcium channels (MICHELL 1982). In addition, further DAG may be generated by cleavage of PI by phospholipase C. This reaction, which is unlikely to be a primary event in activation–secretion coupling, serves to provide additional DAG but not IP_3. The rapid turnover of the PA–PI cycle following IgE receptor bridging has been clearly demonstrated in rat peritoneal mast cells (KENNERLY et al. 1979) and more recently in cultured human basophils (ISHIZAKA and WHITE 1986). DAG, in addition to becoming substrate for the enzyme diglyceride lipase forming monoacylglycerol and arachidonic acid, may also activate the cytoplasmic protein kinase C, which is known to phosphorylate a number of mast cell proteins (KATAKAMI et al. 1984). Translocation of the kinase to the plasma and perigranular membranes is thought to be associated with the phosphorylation of the light chain of myosin which comprises a thin contractile filament. Phosphorylation of myosin light chains is obligatory for their interaction in the presence of ATP and calcium with actin and contraction of the filament. In addition DAG and monoacylglycerol are membrane fusagens (KENNERLY et al. 1979) and thus their increased concentration in close relation to the plasma membrane would facilitate the fusion of the perigranular and plasma membrane as the granules approach each other and the cell surface. Thus, in the presence of increased intracellular calcium following IgE-receptor activation, contraction of thin filaments could be thought of as the mechanism by which mast cell granules move within the cell, the release of mediators occurring through the process of chemiosmosis as the fused granules approach each other and the cell surface, where the two membranes undergo the well-described fusion. These biochemical processes require the consumption of energy provided by the glycolytic cycle and also the influx of anions and water across the perigranular membrane which helps solubilise some of the preformed mediators within the granules prior to degranulation. The arachidonic acid provided as a cleavage of DAG, together with the possible activation of phospholipase A_2, forms the substrate for the cyclo-oxygenase and 5-lipoxygenase enzymes leading to the generation of PGD_2 and LTC_4.

An additional biochemical event that occurs following perturbation of the IgE Fc receptor is a transient stimulation of mast cell adenylate cyclase and an increase in the levels of cellular cyclic AMP (ISHIZAKA et al. 1983; SULLIVAN et al. 1976; HUGHES et al. 1983). The function of this intracellular cyclic AMP has yet to be clearly defined, but a number of studies in the early 1980s suggested that it may be involved in the facilitation of mast cell mediator secretion through the activation of cyclic AMP-dependent protein kinases (HOLGATE et al. 1980). An alternative hypothesis is that the early rise in cyclic AMP and also possibly cyclic GMP activates protein kinases which terminate the mast cell secretory response, as a physiological regulatory mechanism. This bidirectional control of mast cell mediator secretion, involving calcium ions in one direction and cyclic nucleotides in the other, may, however, be too simplistic. Recent evidence suggests the importance of guanine nucleotide binding proteins in the regulation of mast cell mediator secretion (GOMPERTS 1983; COCKCROFT and GOMPERTS 1985) and indeed these proteins are also involved in the regulation of adenylate and guanylate cyclases. It remains possible that the early changes in cellular levels of cyclic nucleotides represent alterations in the biochemical expression of these guanine nucleotide binding proteins, which perform a number of biochemical actions within the mast cell and in mediator secretion. The recent demonstration that the primary action of mast cell "stabilisers" such as sodium cromoglycate and nedocromil is by inhibition of protein kinase in the melanophore (LUCAS and SHUSTER 1987) suggests a similar mechanism in the mast cell.

II. Human Mast Cell Activation by IgE-Dependent and IgE-Independent Stimuli

Rat connective tissue mast cells may be activated for histamine release by a plethora of chemical agents (LAGUNOFF et al. 1983) (Fig. 9). Some of these, such as specific antigen, anti-IgE and concanavalin A, activate these cells by cross-linkage of cell-surface IgE, and mimic the mechanism whereby mast cells respond to allergens during the acute allergic reaction. However, other secretagogues act independently of IgE receptors. Included in this group are a variety of polybasic compounds, such as oligomers of basic amino acids (polylysine and polyarginine) as well as compound 48/80 (a condensation product of methoxyphenylethylamine and formaldehyde). In addition, a variety of neuropeptides are effective histamine liberators, including substance P, vasoactive intestinal polypeptide, dynorphin and calcitonin gene-related peptide (GOETZL et al. 1985). The mast cell receptor(s) for these non-IgE-dependent cell activators is only poorly characterised, but in many cases there are interactions between the stimuli, possibly suggesting a receptor of low specificity (GROSMAN 1981). There are great differences in the secretory response to IgE-dependent and -independent stimuli. The latter cause a much more rapid release of histamine and are relatively independent of extracellular calcium (PEARCE 1982).

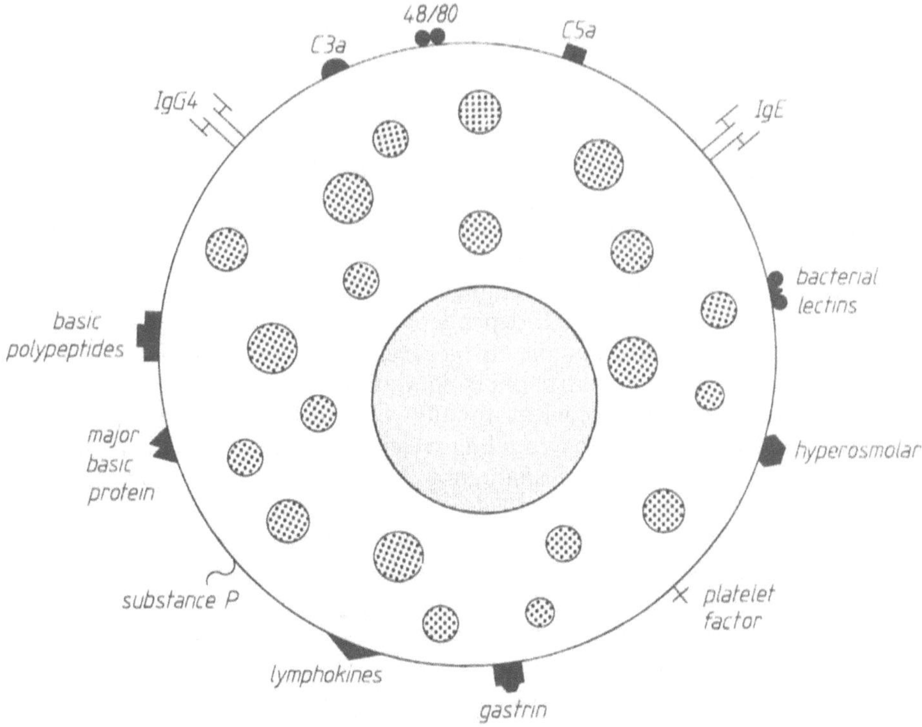

Fig. 9. Agents capable of activating mast cells for mediator secretion

It has become increasingly apparent that human mast cells are quite distinct from those of the rat with regard to their responsiveness to secretagogues. Due to the availability of enzymatic techniques for their dispersion, the most widely studied human mast cells have been those of human lung. These secrete histamine only in response to IgE-dependent stimuli and calcium ionophore A23187 (CHURCH et al. 1982). Activation by the former stimulus demonstrates that lung mast cells can participate in immediate hypersensitivity or type I allergic reactions. IgE-dependent histamine release from these cells is complete within 3–5 min, is dependent on intact pathways of glycolysis and oxidative phosphorylation, and is entirely dependent on the presence of extracellular calcium (CHURCH et al. 1982). These characteristics are shared by the ionophore stimulus.

Until recently, studies of human mast cells from other tissue sites have been hampered by the absence of techniques for their dispersion from solid tissue. However, digestion of tissue under controlled conditions by proteolytic enzymes, particularly collagenase, has allowed dispersion of tissue into discrete cell suspensions, ideally suited for the study of mast cell biochemistry (CHURCH et al. 1983; BENYON et al. 1987; LOWMAN et al. 1987; REES et al. 1987). Studies by these workers using dispersed mast cells of human tonsil, adenoid, intestinal mucosa, intestinal submucosa and skin have shown that they are all responsive to anti-IgE and calcium ionophore A23187. However, skin mast cells alone are unique amongst these cells in being, like rat mast cells, responsive to a variety of non-immunological stimuli. Responsiveness to substance P, compound 48/80, polylysine and morphine (CLEGG et al. 1985; BENYON et al. 1987) clearly distinguishes skin mast cells from those of human lung, tonsil, adenoid, intestinal mucosa as shown in Fig. 10 (LOWMAN et al. 1987) and intestinal submucosa (REES et al. 1987).

As in rat peritoneal mast cells, there are major differences between the characteristics of IgE-dependent and non-immunological histamine secretion from skin mast cells (BENYON et al. 1987). IgE-dependent release is complete within 5–7 min and is completely dependent on the presence of extracellular calcium ions. In contrast, secretion activated by substance P, compound 48/80, polylysine and morphine is far more rapid, being complete in <20 s, and is largely independent of extracellular calcium. Both IgE-dependent and non-immunological secretion are energy dependent, being completely blocked by pretreatment of cells with inhibitors of glycolysis and oxidative phosphorylation. The overall similarity of the response to each non-immunological stimulus suggests that they share a common pathway of activation–secretion coupling which is distinct from that of IgE-dependent stimulation. That these stimuli may share a common receptor site is suggested by the observation that [D-Pro^{4},D-Trp^{7,9,10}]-substance P fragment 4-11 peptide, which is believed to antagonise substance P binding to skin mast cells (PIOTROWSKI et al. 1984), reduces, to non-significant levels, histamine release by all four non-immunological stimuli (unpublished observations).

It is important to consider the possible physiological stimulus which activates skin cells via the non-immunological mechanism. Intradermal injection of substance P into human skin induces a weal and flare reaction (HAGERMARK et al. 1978; FOREMAN et al. 1982). That the flare is inhibited by histamine H_1 antagonists (HAGERMARK et al. 1978) suggests that the observed vasodilatation is due to

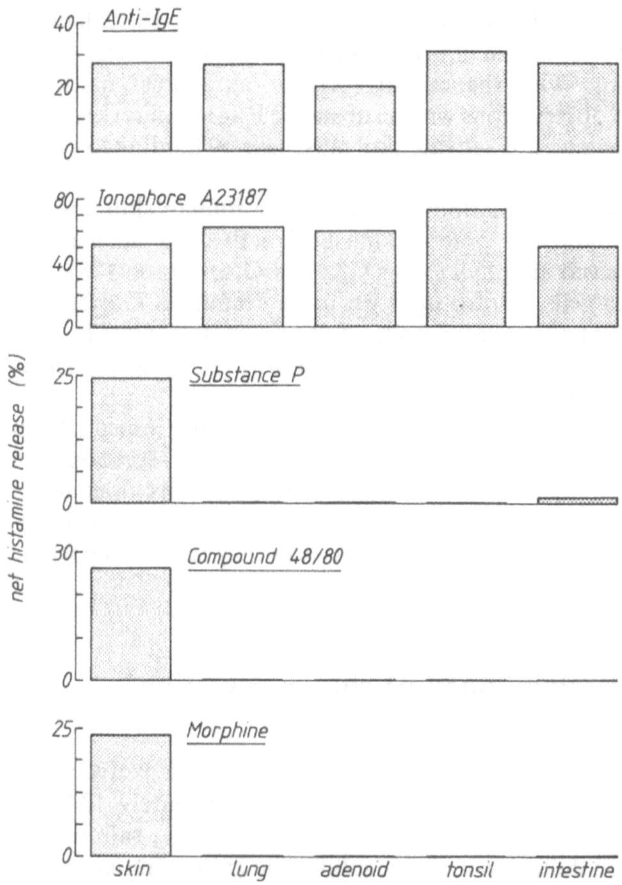

Fig. 10. Responsiveness of dispersed human mast cells to IgE-dependent and non-immunological stimuli

mast cell histamine release. In combination with observations that substance P directly activates human skin mast cell histamine secretion, these results suggest that this neuropeptide effects local changes in skin blood flow via mast cell histamine secretion. Skin mast cells are ideally situated to perform a role, being closely associated with nerves and blood vessels (EADY et al. 1979; WIESNER-MENZEL et al. 1981). The possible involvement of other mast cell newly generated mediators in increasing skin blood vessel size and permeability has been discussed above. However, although mast cells purified from human skin digests (BENYON et al. 1987) have the capacity to generate 40–60 ng PGD_2 per 10^6 cells (unpublished observations) following challenge with calcium ionophore A23187, their ability to do so following non-immunological stimulation has not yet been assessed. Whether substance P represents the only physiological stimulus of skin mast cells for non-immunological histamine secretion is unknown. The demonstration of opioid peptides in human dermis (GIBBINS et al. 1985), their ability to induce mast cell histamine release in vitro (BENYON et al. 1987) and in vivo (CASALE et al. 1984) and the observation that urticaria patients show enhanced responsiveness to intradermal codeine (COHEN and ROSENSTREICH 1986) suggest a role for this axis in skin disease. However, this is probably due to the enhanced responsiveness of the skin to histamine in urticaria (KRAUSE and SHUSTER 1985). Likewise, preliminary results with the opioid receptor antagonist naloxone (BENYON et al. 1987) and the observations with a substance P antagonist described above suggest that morphine may not activate skin mast cells by binding to a classical non-opioid receptor.

Several other agents cause vasodilation and oedema in human skin and are therefore, either directly or indirectly, potential mast cell activators. Such agents include the complement-derived anaphylatoxins C3a and C5a (HUGLI 1981), although their effects on mast cells in vitro have yet to be examined. Gastrin has been reported to activate histamine release from adult human skin slices in vitro (THARP et al. 1984), although we have been unable to confirm this effect in either skin slices or dispersed mast cells (BENYON et al. 1987).

Human skin mast cells, therefore, differ in several important respects from those of other human tissue sites in being activated for histamine secretion by a variety of non-immunological stimuli. The ability of skin mast cells alone to secrete histamine in response to neuropeptides and their close anatomical relationships with nerves and blood vessels suggest that these cells may fulfil homeostatic functions in addition to IgE-mediated defence.

G. Pharmacological Modulation of Mediator Secretion from Skin Mast Cells

It has long been known that mast cells of human lung are very sensitive to the inhibitory actions of β_2-adrenoceptor agonists (IC_{50} approximately 10^{-7} *M*) (CHURCH and YOUNG 1983). In vivo the β_2-adrenoceptor agonist, salbutamol, when injected locally into the skin, is also able to inhibit the antigen-induced weal and flare reaction, but when this drug is taken systemically in a dose of 8 mg, no effect on the antigen-induced dose–response curve can be demonstrated (AKAM and HOWARTH 1984). Similar results have been reported by TING et al. (1983a) using terbutaline. The weal and flare reaction is a complex physiological response to allergen which involves not only the release of mast cell mediators but also the effect of these mediators on the target cells, particularly the vascular endothelium. It remains possible that by the oral route insufficient concentrations of β-agonist gain access to mast cells in human skin or that high concentrations are required to inhibit their mediator release and to diminish the opening of gap junctions in the endothelium as a consequence of mediator release. In chopped human skin, β_2-agonists are at least 100 times less efficacious in inhibiting IgE-dependent histamine secretion compared with their effects on dispersed lung mast cells (CLEGG et al. 1985), supporting the findings in vivo. Further differences may also be found concerning the action of the putative mast cell stabilising agent, sodium cromoglycate. This chromone has a potent activity in inhibiting mediator secretion from isolated rat peritoneal mast cells (ORR 1975), though it is less active on histamine secretion from either chopped human lung fragments (CHURCH and YOUNG 1983) of dispersed human lung mast cells (CHURCH and HIROI 1987). In human skin, sodium cromoglycate has no effect in blocking the allergen-induced immediate weal and flare reaction, histamine release into skin chambers or ultrastructural changes in mast cells evoked by challenge (TING et al. 1983b), and no effect has been demonstrated following IgE-dependent activation of mast cells in skin slices in vitro (PEARCE et al. 1974; CLEGG et al. 1985). Furthermore the more recently introduced sodium nedecromil, which inhibits release from lung mast cells, causes histamine release in skin (HUMPHREYS and SHUSTER 1987). Thus, human skin mast cells exhibit a number of pharmacological differences from mast cells of human lung, which provides support for the concept of heterogeneity proposed for the action of mast cell secretogogues.

The weal and flare reaction is a complex physiological response requiring the participation of a number of inflammatory mediators. The recent introduction of potent H_1-receptor antagonists which are free of anticholinergic and sedative side-effects has allowed the role of histamine to be further evaluated in the skin weal and flare reaction. Treatment of patients with astemizole 10 mg daily for 5 days prior to skin weal testing with histamine or allergen led to complete inhibition of the immediate reaction to 10 μg/ml histamine, but only 50% inhibition of the allergen-induced response (HOWARTH et al. 1984). Following astemizole treatment, HUMPHREYS et al. (1987) found a 47% inhibition of wealing with house dust mite and 68% inhibition of compound 48/80 wealing while histamine wealing was inhibited by 83%; these observations might imply that while histamine has a role in mediating the weal and flare reaction, other mediators are also involved, as suggested by KRAUSE and SHUSTER (1984). Of particular relevance may be generation of prostaglandin D and leukotriene C_4, both of which have been associated with degranulation of skin mast cells (HEAVEY et al. 1986; DOLLERY et al. 1985; our unpublished observations), which can interact to promote oedema formation. Indomethacin has been said to attenuate the late skin response with allergen challenge, implying that cyclo-oxygenase products of arachidonic acid, possibly prostaglandin D_2, are important in this response also (GRONNENBERG and ZETTERSTROM 1985). However, HUMPHREYS et al. (1987) showed that topical indomethacin did not affect the rate of formation, the magnitude, or the rate of disappearance of weals induced by histamine, compound 48/80, or house dust mite extract.

It has long been known that corticosteroids, when administered prior to skin prick testing with specific allergen, have little effect on the immediate response but attenuate the late response. The fact that they have the capacity to inhibit the later reaction implies that lipid-deprived mediators are involved, and that inhibition of the late reaction occurs through the suppressive effect of steroids through macrocortin on phospholipase A_2 and the generation of oxidative products of arachidonic acid and AGEPC (OERTEL and KALINER 1981). Through a similar mechanism, the inhibitory effect of coricosteroids on the late reaction may involve the recruitment of secondary effector cells, namely neutrophils and eosinophils, and their subsequent activation by mast cell-dependent mediators.

H. Conclusions

Mast cells are abundant in human skin, particularly in the upper layers of the dermis, where they are in close association with blood vessels and nerves. Their content of inflammatory mediators and proteases coupled with their ability to generate newly formed mediators such as PGD_2 and LTC_4 makes them strong candidates for participation in skin diseases, particularly urticaria. The detection of histamine and PGD_2 release in association with provocation of urticarial reactions provides strong circumstantial evidence for mast cell involvement in the pathogenesis of urticaria. The ability of human skin mast cells to be activated for secretion by substance P, opioids and compound 48/80, effects possibly mediated through a common receptor site on the cell surface, clearly distinguishes these cells from mast cells of other human tissues or organs. Two recent studies have suggested that enhanced mast cell reactivity to these non-immunological stimuli may be a crucial factor in urticaria. COHEN and ROSENSTREICH (1986) demonstrated a 100-fold increase in the sensitivity of urticaria patients to wealing provoked by intradermal codeine, but not antigen, when compared with normal, rhinitis and asthma patients. In a further study, BEDARD and colleagues (1986) showed that histamine release into skin chamber fluid stimulated by compound 48/80 was markedly enhanced in urticaria patients. Although these studies provide strong suggestive evidence of a mast cell abnormality in urticaria, further definitive evidence is necessary before we can be certain.

References

Akam RM, Howarth PH (1984) Differential effects of oral and inhaled salbutamol on mast cell mediated bronchoconstriction in asthma (abstr). Respiration 46 (Suppl 1):2

Austen KF, Brocklehurst WE (1960) Anaphylaxis in chopped guinea pig lung. 1. Effect of peptidase substrates and inhibitors. J Exp Med 113:521–539

Barbaro JF, Zvaifler NJ (1966) Antigen induced histamine release from platelets of rabbits producing homologous PCA antibody. Proc Soc Exp Biol Med 122:1245–1247

Bedard PM, Brunet C, Pelletier G, Hebert J (1986) Increased compound 48/80 induced local histamine release from nonlesional skin from patients with chronic urticaria. J Allergy Clin Immunol 78:1121–1125

Benyon RC, Church MK, Holgate ST, Hughes PJ (1985) Methyltransferase inhibitors may inhibit histamine released by elevating cyclic AMP in mast cells and basophils (abstr). Br J Pharmacol 86 (Proc Suppl):407P

Benyon RC, Church MK, Clegg LS, Holgate ST (1986a) Dispersion and characterization of mast cells from human skin. Int Arch Allergy Appl Immunol 79:332–334

Benyon RC, Church MK, Holgate ST (1986b) IgE-dependent activation of mast cells is not associated with enhanced phospholipid methylation. Biochem Pharmacol 35:2535–2544

Benyon RC, Lowman MA, Church MK (1987) Human skin mast cells: their dispersion, purification and secretory characteristics. J Immunol 138:861–867

Berridge MJ (1984) Inositol trisphosphate and diacylglycerol as second messengers. Biochem J 220:345–360

Best CH, Dale HH, Dudley HW, Thorpe WV (1927) The nature of vasodilator constituents of certain tissue extracts. J Physiol (Lond) 62:397–417

Binazzi M, Rampichini L (1959) Investigations on the regional distribution of mast cells in human skin. Ital Gen Rev Dermatol 1:17–21

Black AK, Fincham N, Greaves MW, Hensby CN (1981) Changes in levels of arachidonic acid and prostaglandins D_2, E_2, F_2, and 6-oxo-$PGF_{1\alpha}$ in human skin within 48 hours following ultraviolet B radiation. Br J Dermatol 105:353–354

Boam DSW, Stanworth DR, Spanner SG, Ansell GB (1984) Is the stepwise methylation of phosphatidyl ethanolamine relevant to the release of histamine from the mast cell. Biochem Soc Trans 12:782–783
Bowyer A (1968) Observations on the granularity of mast cells in human skin. Acta Derm Venereol (Stockh) 48:574–577
Brain SD, Camp RDR, Leigh IM, Ford-Hutchinson AW (1982) The synthesis of leukotriene B4-like material by cultured human keratinocytes (abstr). J Invest Dermatol 78:328
Briggaman RA, Schechter NM, Fraki JE, Lazarus GS (1984) Degradation of the epidermal-dermal junction by a proteolytic enzyme from human skin and human polymorphonuclear leukocytes. J Exp Med 160:1027–1042
Brown JM, Chambers DA (1984) Effects of biological response modifiers on plasminogen activator activity in epidermal cell culture. Br J Dermatol 111(Suppl):252–256
Casale TB, Bowman S, Kaliner M (1984) Induction of human cutaneous cell degranulation by opiates and endogenous opioid peptides: evidence for opiate and nonopiate receptor participation. J Allergy Clin Immunol 73:775–781
Caulfield JP, Lewis RA, Hein A, Austen KF (1980) Secretion of dissociated human pulmonary mast cells: evidence for solubilization of granule contents before discharge. J Cell Biol 85:299–311
Center DM, Soter NA, Wasserman SI, Austen KF (1979) Inhibition of neutrophil chemotaxis in association with experimental angioedema in patients with cold urticaria: a model of chemotactic deactivation in vivo. Clin Exp Immunol 35:112–118
Church MK, Hiroi J (1987) Inhibition of IgE-dependent histamine release from human lung mast cells by anti-allergic drugs and salbutamol. Br J Pharmacol 90:421–429
Church MK, Young KD (1983) The characteristics of inhibition of histamine release from human lung fragments by sodium cromoglycate, salbutamol and chlorpromazine. Br J Pharmacol 78:671–679
Church MK, Pao GJ-K, Holgate ST (1982) Characterization of histamine secretion from dispersed human lung mast cells: effects of anti-IgE, calcium ionophore A23187, compound 48/80 and basic polypeptides. J Immunol 129:2116–2121
Church MK, Mageed RAK, Holgate ST (1983) Human tonsillar mast cells: characterization of histamine secretion and methods of dispersion. Int Arch Allergy Appl Immunol 72:188–190
Clegg LS, Church MK, Holgate ST (1985) Histamine secretion from human skin slices induced by anti-IgE and artificial secretagogues and the effects of sodium cromoglycate and salbutamol. Clin Allergy 15:321–328
Cockcroft S, Gomperts BD (1985) Role of guanine nucleotide binding protein in the activation of polyphosphoinositide phosphodiesterase. Nature 314:534–536
Cohen RW, Rosenstreich DL (1986) Discrimination between urticaria-prone and other allergic patients by intradermal skin testing with codeine. J Allergy Clin Immunol 77:802–807
Cowen T, Trigg P, Eady RAJ (1979) Distribution of mast cells in human dermis: development of a mapping technique. Br J Dermatol 100:635–640
Cundell DR, Davies RJ (1985) NCA release from human blood lymphocytes: effects of salbutamol and sodium cromoglycate on this release (abstr). J Allergy Clin Immunol 75:109
Czarnetzki BM, Kruger G, Sterry W (1983) In vitro generation of mast cell-like cells from human peripheral mononuclear phagocytes. Int Arch Allergy Appl Immunol 71:161–167
Dale HH (1910) Croonian lectures on some chemical factors in the control of the circulation. Lancet I:1179, 1233, 1285
Demopoulos CA, Pinckard RN, Hanahan DJ (1979) Platelet activating factor: evidence for 1-*O*-alkyl-2-acetyl-sn-glyceryl-3-phosphorylcholine as the active component (a new class of lipid chemical mediators). J Biol Chem 254:9355
Dollery CT, Heavey DJ, Richmond R, Taylor GW, Vial J (1985) Histamine and peptidoleukotriene release from immunologically challenged guinea-pig skin (abstr). Br J Pharmacol 85 (Proc Suppl):277P

Eady RAJ, Cowen T, Marshall TF, Plummer V, Greaves MW (1979) Mast cell population density, blood vessel density and histamine content of normal human skin. Br J Dermatol 100:623–633
Ehrlich P (1876) Beiträge zur Kenntnis der Anilinfärbungen und ihrer Verwendung in der mikroskopischen Technik. Arch Mikr Anat 13:263–277
Enerback L (1966a) Mast cells in the gastrointestinal mucosa. I. Effects of fixation. Acta Pathol Microbiol Scand 66:289–302
Enerback L (1966b) Mast cells in the gastrointestinal mucosa. II. Dye binding and metachromatic properties. Acta Path Microbiol Scand 66:303–312
Enerback L (1986) Mast cell heterogeneity: the evolution of the concept of a specific mucosal mast cell. In: Befus AD, Bienenstock J, Denburg JA (eds) Mast cell differentiation and heterogeneity. Raven, New York, pp 1–26
Enholm C, Shaw W, Greten H, Brown WV (1975) Purification from human plasma of a heparin-released lipase with activity against triglyceride and phospholipid. J Biol Chem 250:6756–6761
Farr RS, Wardlow ML, Cox CP, Meng KE, Greene DE (1982) Human serum acid labile factor (ALF) is an acyl hydrolase that inactivates platelet activating factor (PAF) (abstr). Fed Proc 41:733
Ford-Hutchinson AW, Bray MA, Doig MV, Shipley ME, Smith MJH (1980) Leukotriene B_4, a potent chemotactic and aggregating substance released from polymorphonuclear leucocytes. Nature 286:264–265
Foreman JC, Jordan CC, Piotrowski W (1982) Interaction of neurotensin with the substance P receptor mediating histamine release from rat mast cells and the flare in human skin. Br J Pharmacol 77:531–539
Fraki JE, Hopsu-Havu VK (1975) Human skin proteases. Separation and characterization of two alkaline proteases, one splitting trypsin and the other chymotrypsin substrates. Arch Dermatol Res 253:261–276
Frisk-Holmberg M, Uvnas B (1972) The influence of chlorpromazine on the uptake of biogenic amines by rat mast cells in vitro. Acta Physiol Scand 86:1–11
Froese A (1984) Receptors for IgE on mast cells and basophils. Prog Allergy 34:142–187
Gibbins IL, Furness JB, Costa M, MacIntyre I, Hillyard C, Girgis S (1985) Coexistence of calcitonin gene-related peptide, dynorphin and cholecystokinin in substance P-containing dorsal root ganglion neurones of the guinea-pig. Neurosci Lett 19 (Suppl):S65
Ginsburg H, Nir I, Hammel I, Eren R, Weissman B, Naot Y (1978) Differentiation and activity of mast cells following immunization in cultures of lymph node cells. Immunology 35:485–502
Goetzl EJ, Austen KF (1975) Purification and synthesis of eosinophilotactic tetrapeptides of human lung tissue. Identification as eosinophil chemotactic factor of anaphylaxis. Proc Natl Acad Sci USA 72:4123–4127
Goetzl EJ, Brash AR, Tauber AI, Oates JA, Hubbard WC (1980) Modulation of human neutrophil function by monohydroxyeicosatetraenoic acids. Immunology 39:491–501
Goetzl EJ, Chernov T, Renold F, Payan DG (1985) Neuropeptide regulation of the expression of immediate hypersensitivity. J Immunol 135:802s–805s
Gomperts BD (1983) Involvement of guanine nucleotide binding protein in the gating of calcium by receptors. Nature 306:64–66
Goto T, Befus D, Low R, Bienenstock J (1984) Mast cell heterogeneity and hyperplasia in bleomycin-induced pulmonary fibrosis of rats. Am Rev Respir Dis 130:797–802
Grant JA, Lett-Brown MA, Warner JA, Plaut M, Lichtenstein LM, Haak-Frendscho M, Kaplan AP (1986) Activation of basophils. Fed Proc 45:2653–2658
Gronnenberg R, Zetterstrom O (1985) Inhibition of the late phase response to anti-IgE in humans by indomethacin. Allergy 40:36–41
Grosman N (1981) Histamine from isolated rat mast cells: effect of morphine and related drugs and their interaction with compound 48/80. Agents Actions 11:196–203
Gupta RK (1970) Mast cell variations in prostate and urinary bladder. Arch Pathol 89:302–305
Hagermark O, Hokfelt T, Pernow B (1978) Flare and itch induced by substance P in human skin. J Invest Dermatol 71:233–235

Haig DM, McMenamin C, Gunneberg C, Woodbury R, Jarrett EEE (1983) Stimulation of mucosal mast cell growth in normal and nude rat bone marrow cultures. Proc Natl Acad Sci USA 80:4499–4503

Hamberg M, Svensson J, Wakabayashi T, Samuelsson B (1974) Isolation and structure of two prostaglandin endoperoxides that cause platelet aggregation. Proc Natl Acad Sci USA 71:345–349

Hammarstrom S (1983) Leukotrienes. Ann Rev Biochem 52:355–377

Heavey DJ, Kobza-Black A, Barrow SE, Chappell CG, Greaves MW, Dollery CT (1986) Prostaglandin D_2 and histamine release in cold urticaria. J Allergy Clin Immunol 78:458–461

Hellstrom B, Holmgren H (1950) Numerical distribution of mast cells in the human skin and heart. Acta Anat (Basel) 10:81–107

Helting T, Lindahl U (1972) Biosynthesis of heparin. Transfer of *N*-acetylglucosamine and glucuronic acid to low molecular weight heparin fragments. Acta Chem Scand 26:3515–3523

Henderson WR, Kaliner M (1978) Immunologic and non-immunologic generation of superoxide from mast cells and basophils. J Clin Invest 61:187–196

Henderson WR, Kaliner M (1979) Mast cell granules peroxidase; location, secretion, and SRS-A inactivation. J Immunol 122:1322–1328

Hibbs RG, Burch GE, Phillips JM (1960) Electronicmicroscopic observations on human mast cells. Am Heart J 60:121–127

Holgate ST, Lewis RA, Austen KF (1980) 3′,5′-cyclic adenosine monophosphate dependent protein kinase of the rat serosal mast cell and its immunologic activation. J Immunol 124:2093–2099

Holgate ST, Burns GB, Robinson C, Church MK (1984) Anaphylactic and calcium dependent generation of prostaglandin D_2 (PGD_2), thromboxane B_2 and other cyclooxygenase products of arachidonic acid by dispersed human lung cells and relationship to histamine release. J Immunol 133:2138–2144

Hook M, Lindahl U, Hallen A, Backstrom G (1975) Biosynthesis of heparin: studies on the microsomal sulfation process. J Biol Chem 250:6065–6071

Horton MA, O'Brien HAW (1983) Characterization of human mast cells on long term culture. Blood 62:1251–1260

Howarth PH, Emanuel MB, Holgate ST (1984) Astemizole, a potent histamine H_1-receptor antagonist: effects in allergic rhino-conjunctivitis, on antigen and histamine induced skin weal responses and relationship to serum levels. Br J Clin Pharmacol 18:1–8

Hughes PJ, Holgate ST, Roath S, Church MK (1983) The relationship between cyclic AMP changes and histamine release from basophil-enriched human leucocytes. Biochem Pharmacol 32:2557–2563

Hugli TE (1981) The structural basis for anaphylatoxin in chemotactic functions of C3a and C5a. CRC Crit Rev Immunol 1:321–366

Humphreys F, Shuster S (1987) The effect of nedecromil on weal reactions in human skin. Br J Clin Pharmacol 24:405–408

Humphreys F, Krawe LB, Shuster S (1987) The effect of astemizole and indomethacin on weal and flare reactions to histamine, 48/80 and house dust mite antigen. Br J Dermatol 116:435

Humphreys SH, Austen KF, Rapp HJ (1963) In vitro studies of reversed anaphylaxis with rat cells. Immunology 6:225–245

Ihle JN, Keller J, Henderson L, Klein F, Palaszynski EW (1982) Procedures for the purification of interleukin 3 to homogeneity. J Immunol 129:2431–2436

Ihle JN, Keller J, Oroszlan S, Henderson LE, Copeland TD, Fitch F, Prytsowsky MB, Goldwasser E, Schrader JW, Palaszynski E, Dy M, Lebel B (1983) Biological properties of homogenous interleukin 3: demonstration of WEHI-3 growth factor activity, mast cell growth factor activity, P-cell stimulating factor activity, colony-stimulating activity and histamine producing cell stimulating factor activity. J Immunol 131:282–287

Irani AA, Schechter NM, Craig S, DeBlois G, Schwartz LB (1986) Two types of human mast cells that have distinct neutral protease compositions. Proc Natl Acad Sci USA 83:4464–4468

Ishizaka K, Ishizaka T (1967) Identification of gamma-E antibodies as a carrier of reaginic activity. J Immunol 99:1187–1198

Ishizaka T, Ishizaka K (1984) Activation of mast cells for mediator release through IgE receptors. Prog Allergy 34:188–235

Ishizaka T, White JR (1986) Triggering mechanisms of mast cells and basophils. In: Reed CE (ed) Proceedings of the XII international congress of allergy and clinical immunology. J Allergy Clin Immunol 159–163

Ishizaka T, Foreman JC, Sterk AR, Ishizaka K (1979) Induction of calcium flux across the rat mast cell membrane by bridging IgE receptors. Proc Natl Acad Sci USA 76:5858–5862

Ishizaka T, Hirata F, Ishizaka K, Axelrod J (1980) Stimulation of phospholipid methylation, Ca^{2+} influx, and histamine release by bridging of IgE receptors on rat mast cells. Proc Natl Acad Sci USA 77:1903–1906

Ishizaka T, Conrad DH, Schulman ES, Sterk AR, Ishizaka K (1983) Biochemical analysis of initial triggering events of IgE-mediated histamine release from human lung mast cells. J Immunol 130:2357–2362

Ishizaka T, Dvorak AM, Conrad DH, Niebyl JR, Marquette JP, Ishizaka K (1985) Morphologic and immunologic characterization of human basophils developed in cultures of cord blood mononuclear cells. J Immunol 134:532–540

Jacobson I, Lindahl U (1980) Biosynthesis of heparin: concerted action of late polymer-modification reactions. J Biol Chem 255:5094–5100

Jacobson I, Hook M, Petterson I, Lindahl U, Larm O, Wiren E, Von Figura K (1979) Identification of *N*-sulphated disaccharide units in heparin-like polysaccharides. Biochem J 179:77–87

Jilek F, Hormann H (1979) Fibronectin (cold insoluble globulin). VI. Influence of heparin and hyaluronic acid on the binding of native collagen. Hoppe Seyler's Z Physiol Chem 360:597–603

Kaliner M, Lemanske R (1984a) Inflammatory response to mast cell granules. Fed Proc 43:2846–2851

Kaliner M, Lemanske R (1984b) Mast cell-derived inflammatory factors and late-phase allergic reactions. In: Kay AB, Austen KF, Lichtenstein LM (eds) Asthma: physiology, immunopharmacology and treatment. Third international symposium. Academic, London, pp 229–244

Katakami Y, Kaibuchi K, Sawamura M, Takai Y, Nishizuka Y (1984) Synergistic action of protein kinase C and calcium for histamine release from rat peritoneal mast cells. Biochem Biophys Res Commun 121:573–578

Katayama Y, Ende N (1965) Esterase studies on dog mast cell tumors. Nature 205:190–191

Kawanami O, Ferrans VJ, Fulmer JD, Crystal RG (1985) Ultrastructure of pulmonary mast cells in patients with fibrotic lung disorders. Lab Invest 40:717–734

Kay AB, Lee TH (1982) Neutrophil chemotactic factor of anaphylaxis. J Allergy Clin Immunol 70:317–320

Kennerly DA, Sullivan TJ, Parker CW (1979) Activation of phospholipid metabolism during mediator release from stimulated rat mast cells. J Immunol 122:152–159

Kingston WP, Greaves MW (1976) Factors affecting prostaglandin synthesis by rat microsomes. Prostaglandins 12:51–69

Kitamura Y et al. (1986a) Probable transdifferentiation between connective tissue and mucosal mast cells. In: Befus AD, Bienenstock J, Denburg JA (eds) Mast cell differentiation and heterogeneity. Raven, New York, pp 135–140

Kitamura Y, Nakano T, Kanakura Y, Matsuda H (1986b) Factors influencing mast cell differentiation. In: Reed CE (ed) Proceedings of the XII international congress of allergy and clinical immunology. J Allergy Clin Immunol 154–158

Koro O, Francis RM, Barr AK, Black AK, Numata T, Greaves MW (1986a) Antigen-induced release of prostaglandin D_2 from human skin in vivo (abstr). J Invest Dermatol 87:151

Koro O, Dover JS, Francis DM, Kobza Black A, Kelly RW, Barr RM, Greaves MW (1986b) Release of prostaglandin D_2 and histamine in a case of localized heat urticaria and effects of treatments. Br J Dermatol 115:721–728

Krause L, Shuster S (1984) H_1 receptor active histamine not sole cause of chronic idiopathic urticaria. Lancet II:929
Krause LB, Shuster S (1985) Minimal effect of complete H_1 receptor blockade on urticaria pigmentosa. Acta Derm Venereol (Stockh) 65:338–340
Lagunoff D (1972) Contributions of electron microscopy to the study of mast cells. J Invest Dermatol 58:296–311
Lagunoff D, Chi EY, Wan H (1975) Effects of chymotrypsin and trypsin on rat peritoneal mast cells. Biochem Pharmacol 24:1573–1578
Lagunoff D, Martin TW, Read G (1983) Agents that release histamine from mast cells. Annu Rev Pharmacol Toxicol 23:331–351
Leitch AG, Drazen JM (1984) Pulmonary mechanical response to leukotriene administration in vivo. In: Kay AB, Austen KF, Lichenstein LM (eds) Asthma: physiology, immunopharmacology and treatment. Third international symposium. Academic, London, pp 85–99
Levi-Schaffer F, Austen KF, Caulfield JP, Hein A, Bloes WF, Stevens RL (1985) Fibroblasts maintain the phenotype and viability of the rat heparin-containing mast cell in vitro. J Immunol 135:3454–3462
Lewis RA, Drazen JM, Corey EJ, Austen KF (1981) Structural and functional characteristics of the leukotriene components of slow reacting substance of anaphylaxis. In: Piper PJ (ed) SRS A and leukotrienes. Wiley, Chichester, pp 101–117
Lewis T (1927) The blood vessels of the human skin and their responses. Shaw, London
Lichtenstein LM, Schleimer RP, MacGlashan DW, Peters SP, Schulman ES, Proud D, Creticos PS, Naclerio RM, Kagey-Sobotka A (1984) In vitro and in vivo studies of mediator release from human mast cells. In: Kay AB, Austen KF, Lichtenstein LM (eds) Asthma: physiology, immunopharmacology and treatment. Third international symposium. Academic, London, pp 1–18
Lonky SA, Marsh J, Wohl H (1978) Stimulation of human granulocyte elastase by platelet factor 4 and heparin. Biochem Biophys Res Commun 85:1113–1118
Lowman MA, Rees PH, Benyon RC, Church MK (1987) Human mast cell heterogeneity: histamine release from mast cells dispersed from skin, lung, adenoids, tonsils and intestinal mucosa in response to IgE-dependent and non-immunological stimuli. J Allergy Clin Immunol 81:590–597
Lucas A, Shuster S (1987) Cromolyn inhibition of protein kinase activity. Biochem Pharmacol 36:561–562
MacGlashan DW, Schleimer RP, Peters SP, Schulman ES, Adams GK, Newball HH, Lichtenstein LM (1982) Generation of leukotrienes by purified human lung mast cells. J Clin Invest 70:747–751
Maier M, Spragg J, Schwartz LB (1983) Inactivation of human high molecular weight kininogen by human mast cell tryptase. J Immunol 130:2352–2356
Marks RM, Roche WR, Czerniecki M, Penny R, Nelson DS (1986) Mast cell granules cause proliferation of human microvascular endothelial cells. Lab Invest 55:289–294
Meier HL, Heck LW, Schulman ES, MacGlashan DW (1985) Purified human mast cells and basophils release elastase and cathepsin G by an IgE-mediated mechanism. Int Arch Allergy Appl Immunol 77:179–183
Metcalfe DD, Lewis RA, Silbert JE, Rosenberg RD, Wasserman SI, Austen KF (1979) Isolation and characterisation of heparin from human lung. J Clin Invest 4:1537–1543
Metcalfe DD, Smith JA, Austen KF, Silbert JE (1980) Polydispersity of rat mast cell heparin: implications for proteoglycan assembly. J Biol Chem 255:11753–11758
Metcalfe DD, Kaliner M, Donlon MA (1981) The mast cell. CRC Crit Rev Immunol 1:23–74
Michell RH (1982)Is phosphatidylinositol really out of the calcium gate. Nature 296:492–493
Mikhail GR, Miller-Milinska A (1964) Mast cell population in human skin. J Invest Dermatol 43:249–254
Miller HRP (1980) The structure, origin and function of mucosal mast cells: a brief review. Biol Cell 39:249–254

Miller HRP, Woodbury RG, Huntley JF, Newlands GFJ (1983) Systemic release of mucosal mast cell protease in primed rats challenged with *Nippostrongylus brasiliensis*. Immunology 49:471–479
Mongar JL, Schild HO (1958) The effect of calcium and pH on the anaphylactic reaction. J Physiol (Lond) 140:272–284
Moore JP, Johannsson A, Hesketh TR, Smith GA, Metcalfe JC (1984) Calcium signals and phospholipid methylation in eukaryotic cells. Biochem J 221:675–684
Nabel G, Galli SJ, Dvorak AM, Dvorak HF, Cantor H (1981) Inducer T-lymphocytes synthesise a factor that stimulates proliferation of cloned mast cells. Nature 291:332–334
Nagao K, Yokoro K, Aaronson SA (1981) Continuous lines of basophil/mast cells derived from normal mouse bone marrow. Science 212:333–335
Nakano T, Sonoda T, Hayashi C, Yamatodani A, Kanayama Y, Yamamura T, Asai H, Yonezawa T, Kitamura Y, Galli SJ (1985) Fate of bone marrow-derived cultured mast cells after intracutaneous, intraperitoneal and intravenous transfer into genetically mast cell deficient W/Wv mice: evidence that cultured mast cells can give rise to both connective tissue type and mucosal mast cells. J Exp Med 162:1025–1043
O'Driscoll BRC, Lee TH, Cromwell O, Kay AB (1983) Immunologic release of neutrophil chemotactic activity from human lung tissue. J Allergy Clin Immunol 72:695–701
Oertel HL, Kaliner M (1981) The biologic activity of mast cell granules. III. Purification of inflammatory factors of anaphylaxis (IF-A) responsible for causing late-phase reactions. J Immunol 127:1398–1402
Ohki S, Ogino N, Yamamoto K, Hayaishi O (1979) Prostaglandin hydroperoxidase as an integral part of prostaglandin synthetase from bovine vesicular gland microsomes. J Biol Chem 254:829–836
Oikarinen A, Viinikka L, Rytsala H, Kiistla U, Ylikorkala O (1981) Prostacyclin, thromboxane and prostaglandin $F_{2\alpha}$ in suction blister fluid of human skin: effect of systemic aspirin and glucocorticoid treatment. Life Sci 29:391–396
Orr TSC (1975) Recent developments concerning the mast cell and the mode of action of disodium cromoglycate. Acta Allergol 12 (Suppl):13–29
Orr TSC (1977) Fine structure of the mast cell with special reference to human cells. Scand J Respir Dis 98 (Suppl):1–7
Pearce CA, Greaves MW, Plummer VM, Yamamoto S (1974) Effect of disodium cromoglycate on antigen evoked histamine release in human skin. Clin Exp Immunol 17:437–440
Pearce FL (1982) Calcium and histamine secretion from mast cells. Prog Med Chem 19:59–109
Pearce FL (1986) Functional differences between mast cells from different locations. In: Befus AD, Bienenstock J, Denburg JA (eds) Mast cell differentiation and heterogeneity. Raven, New York, pp 215–222
Pessinger MA, Lepage P, Simard JP, Parker GH (1983) Mast cell numbers in incisional wounds in rat skin as a function of distance, time and treatment. Br J Dermatol 108:179–187
Pillarisetti V, Rao S, Friedman MM, Atkins FM, Metcalfe DD (1983) Phagocytosis of mast cell granules by cultured fibroblasts. J Immunol 130:341–349
Piotrowski W, Devoy MAB, Jordan CC, Foreman JC (1984) The substance P receptor on rat mast cells and in human skin. Agents Actions 14:420–423
Piper PJ (1983) Pharmacology of leukotrienes. Br Med Bull 39:255–259
Prausnitz C, Kusner H (1921) Studien über die Überempfindlichkeit. Zentralbl Bakteriol 86:160–169
Proud D, Lichtenstein LM (1984) Human lung mast cell kininogenase: apparent identity to tryptase (abstr). Fed Proc 43:1807
Proud D, MacGlashan DW, Newball HH, Schulman ES, Lichtenstein LM (1985) IgE-mediated release of a kininogenase from purified human lung mast cells. Am Rev Respir Dis 132:405–408
Razin E, Stevens RL, Akiyama F, Schmid K, Austen KF (1982) Culture from mouse bone marrow of a subclass of mast cells possessing a distinct chondroitin sulfate proteoglycan with glycosaminoglycans rich in *N*-acetylgalactosaminase-4,6-disulphate. J Biol Chem 257:7229–7236

Rees PH, Hillier K, Church MK (1987) The secretory characteristics of mast cells isolated from human intestinal mucosa and submucosa (to be published)
Roberts LJ, Sweetman BJ (1985) Metabolic fate of endogenously synthesised prostaglandin D_2 in a human female with mastocytosis. Prostaglandins 30:383–401
Roberts LJ, Sweetman BJ, Lewis RA, Austen KF, Oates JA (1980) Increased production of prostaglandin D_2 in patients with systemic mastocytosis. N Engl J Med 303:1400–1404
Robinson HC, Horner AA, Hook M, Ogren S, Lindahl U (1978) A proteoglycan form of heparin and its degradation to single chain molecules. J Biol Chem 253:6687–6693
Ruzicka T, Printz MP (1982) Arachidonic acid metabolism in guinea-pig skin. Biochim Biophys Acta 711:391–397
Schechter NM, Fraki JE, Geesin JC, Lazarus GS (1983) Human skin chymotryptic proteinase. Isolation and relation to cathepsin G and rat mast cell protease. J Biol Chem 258:2973–2978
Schechter NM, Choi JK, Slavin DA, Deresienski DT, Sayama S, Dong G, Lavaker RM, Proud D, Lazarus GS (1986) Identification of a chymotrypsin-like proteinase from human mast cells. J Immunol 137:962–970
Schulman ES, MacGlashan DW, Peters SP, Schleimer RP, Newball HH, Lichtenstein LM (1982) Human lung mast cells: purification and characterization. J Immunol 129:2662–2667
Schwartz LB (1984) Tryptase from human pulmonary mast cells. In: Kay AB, Austen KF, Lichtenstein LM (eds) Asthma: physiology, immunopharmacology and treatment. Third international symposium. Academic, London, pp 19–37
Schwartz LB, Austen KF (1981) Acid hydrolases and other enzymes of rat and human mast cell secretory granules. In: Becker EL, Simon AS, Austen KF (eds) Biochemistry of the acute allergic reactions. Liss, New York, pp 103–121
Schwartz LB, Austen KF (1984) Structure and function of the chemical mediators of mast cells. Prog Allergy 34:271–321
Schwartz LB, Riedel C, Caulfield JP, Wasserman SI, Austen KF (1981 a) Cell association of complexes of chymase, heparin proteoglycan and protein after degranulation by rat mast cells. J Immunol 126:2071–2078
Schwartz LB, Lewis RA, Austen KF (1981 b) Tryptase from human pulmonary mast cells: purification and characterization. J Biol Chem 256:11939–11943
Schwartz LB, Lewis RA, Seldin D, Austen KF (1981 c) Acid hydrolases and tryptase from secretory granules of dispersed lung mast cells. J Immunol 126:1290–1294
Schwartz LB, Schratz JJ, Vik D, Fearon DT, Austen KF (1982) Cleavage of human C3 by human mast cell tryptase (abstr). Fed Proc 41:487
Seldin DC, Adelman S, Austen KF, Stevens RL, Hein A, Caulfield JP, Woodbury RG (1985) Homology of the rat basophilia cell and the rat mucosal mast cell. Proc Natl Acad Sci USA 82:3871–3875
Seppa HEJ, Jarvinen M (1978) Rat skin main neutral protease: purification and properties. J Invest Dermatol 70:84–89
Silbert JE (1967) Biosynthesis of heparin. J Biol Chem 242:5146–5152
Snyder F (1985) Chemical and biochemical aspects of platelet activating factor: a novel class of acetylated ether-linked choline-phospholipids. Med Res Rev 5:107–140
Solley GO, Gleich GJ, Jordan RE, Schroeter AL (1976) The late phase of the immediate wheal and flare skin reaction: its dependence on IgE antibodies. J Clin Invest 58:408–420
Sondergaard J, Zachariae H (1968) Epidermal histamine. Arch Klin Exp Dermatol 233:323–328
Soter NA, Mihm MC, Dvorak HF, Austen KF (1978) Cutaneous necrotising veneolitis: a sequential analysis of the morphological alterations occurring after mast cell degranulation in a patient with a unique syndrome. Clin Exp Immunol 32:46–58
Spicer SS (1963) Histochemical properties of mucopolysaccharide and basic protein in mast cells. Ann NY Acad Sci USA 103:322–332
Sredni B, Friedman MM, Bland CE, Metcalfe DD (1983) Ultrastructural, biochemical and functional characteristics of histamine-containing cells cloned from mouse bone marrow: tentative identification as mucosal mast cells. J Immunol 131:915–922

Stevens RL, Austen KF (1982) Effect of *p*-nitrophenyl-*β*-D-xyloside on proteoglycan and glycosaminoglycan biosynthesis in rat serosal mast cell cultures. J Biol Chem 257:253–259

Stevens RL, Katz HR, Seldin DC, Austen KF (1986) Biochemical characteristics distinguish subclasses of mammalian mast cells. In: Befus AD, Bienenstock J, Denburg JA (eds) Mast cell differentiation and heterogeneity. Raven, New York, pp 183–203

Strobel S, Miller HRP, Ferguson A (1981) Human intestinal mucosal mast cells: evaluation of fixation and staining techniques. J Clin Pathol 34:851–858

Sullivan TJ (1981) Diacylglycerol metabolism and the release of mediators from mast cells. In: Becker EL, Simon AS, Austen KF (eds) Biochemistry of the acute allergic reactions. Liss, New York, pp 229–238

Sullivan TJ, Parker KL, Kulczycki A, Parker CW (1976) Modulation of cyclic AMP in purified rat mast cells. III. Studies on the effects of concanavalin A and anti-IgE during histamine release. J Immunol 117:713–716

Tharp MD, Thirlby R, Sullivan TJ (1984) Gastrin induces histamine release from human cutaneous mast cells. J Allergy Clin Immunol 74:159–165

Theoharides TC, Bondy PK, Tsakalos ND, Askenase PW (1982) Differential release of serotonin and histamine from mast cells. Nature 297:229–231

Ting S, Zweiman B, Lavker RM, Dunsky EH (1981) In vivo release of eosinophil chemoattractant activity in human allergic skin reactions. J Immunol 127:557–560

Ting S, Zweiman B, Lavker RM (1983a) Terbutaline modulation of human allergic skin reactions. J Allergy Clin Immunol 71:437–441

Ting S, Zweiman B, Lavker RM (1983b) Cromolyn does not modulate human allergic skin reactions in vivo. J Allergy Clin Immunol 71:12–17

Uvnas B (1967) Mode of binding and release of histamine in mast cell granules of the rat. Fed Proc 26:219–221

Vance DE, de Kruijff B (1980) The possible functional significance of phosphatidylethanolamine methylation. Nature 288:277–278

Wells PD (1977) *Nippostrongylus braziliensis:* lung mast cell populations in repeatedly inoculated rats. Exp Parasitol 43:326–335

Wiesner-Menzel L, Schulz B, Vakilzadeh F, Czarnetzki BM (1981) Electron microscopical evidence for a direct contact between nerve fibres and mast cells. Acta Derm Venereol (Stockh) 61:465–469

Windaus A, Vogt W (1907) Synthese des Imidazolylathylamins. Ber Dtsch Chem Ges 3:3691–3695

Wintroub BU, Kaempfer CE, Schechter NM, Proud D (1986) Human lung mast cell chymotrypsin-like enzyme: identification and partial characterization. J Clin Invest 77:196–201

Woodbury RG, Miller HRP (1982) Quantitative analysis of mucosal mast cell protease in the intestine of *Nippostrongylus*-injected rats. Immunology 46:487–495

Woodbury RG, Gruzenski GM, Lagunoff D (1978) Immunofluorescent localization of a serine protease in rat small intestine. Proc Natl Acad Sci USA 75:2785–2789

Yurt RW, Austen KF (1977) Preparative purification of rat mast cell chymase. Characterization and interaction with granule components. J Exp Med 146:1405–1419

Yurt RW, Leid RW, Austen KF, Silbert JE (1977) Native heparin from rat peritoneal mast cells. J Biol Chem 252:518–521

Zibho VA, Casebolt T, Marcelo CL, Voorhees JJ (1983) Enhancement of 5′-lipoxygenase activity in soluble preparations of human psoriatic plaque preparation. J Invest Dermatol 80:359

Zimmerman TP, Schmitges CJ, Wolberg G, Deeprose RD, Duncan GS, Cuatrecasas P, Elion GB (1980) Modulation of cyclic AMP metabolism by *S*-adenosylhomocysteine and *S*-3-deazaadenosylhomocysteine in mouse lymphocytes. Proc Natl Acad Sci USA 77:5639–5643

CHAPTER 9

Lymphocytes

J. Morley

A. Introduction

Classically, the involvement of lymphocytes in inflammatory skin lesions has been associated with immunological responses, whereby the host expresses, or develops, enhanced reactivity to an inciting allergen. Yet it is also known that lymphocytes can contribute to inflammatory responses induced by non-allergic stimuli. These observations may be reconciled by considering the allergic response to be an enhanced version of a non-selective inflammatory response involving lymphocyte activation.

Over the last 20 years it has proved possible to classify lymphocytes into subsets, whose activation determines different aspects of the immunological response. Immunocompetent cells originating in the thymus are termed T cells; they participate as effector cells in reactions of cellular immunity, but additionally they provide regulatory suppressor or helper mechanisms which determine the nature and intensity of the immunological response. Immunocompetent cells originating in the bursa of Fabricius of birds, or the bone marrow of mammals, are termed B cells; they serve as a source of the antibody-secreting plasma cells. T cells and B cells share a common progenitor stem cell, whose differentiation into competent T cells or B cells is determined largely by events in the thymus. The hormonal signal directing development to T cells is provided successively by thymopoietin, formed by thymic epithelial cells, and by a secretory product of thymic macrophages. The hormonal signal directing development to B cells is less certainly established, but a colony stimulatory factor, generated by thymocytes, has been proposed.

Distinction between B cells and T cells has a well accepted functional basis, since B cells effect immunocompetence by secretion of specific antibody, a property lacking in T cells. Consistent with this distinction is evidence of surface and other markers distinguishing B cells from T cells. In mice, these include Thy-1 antigen (T cells) and surface markers that precede specific immunoglobulins of IgM and IgD (B cells), as well as the Ly series (Ly 1,2,3,4) and the enzyme terminal deoxyribonucleotidyl transferase, whose activity during early stages of precursor cell differentiation corresponds to T cell commitment (high) or B cell commitment (low). Such clear distinction, coupled with the gross difference between antibody-mediated and cell-mediated reactions, has caused the functional attributes of T cells and B cells to be viewed as separate entities; yet more recent evidence has clearly established T cell regulation of B cell activation.

B. Lymphokines

B cell responses to specific antigen are mediated by antibodies which interact with antigen in a highly selective manner. The biological response consequent to this interaction is effected by secondary amplification mechanisms involving release of preformed mediators (e.g. mast cell degranulation, lysosomal enzyme secretion) or initiation of mediator formation (e.g. complement activation, generation of prostaglandins and leukotrienes). In T cell responses, several manifestations of cellular immunity might readily be ascribed to cytotoxicity with non-specific inflammatory sequelae. Only more recently has it become apparent that T-cell responses to specific antigens may depend upon the elaboration of mediators.

The in vitro test that most closely parallels delayed hypersensitivity (cellular immunity) is that of macrophage migration inhibition (GEORGE and VAUGHAN 1962). In 1964, DAVID et al. showed that the inhibitory effect of antigen on macrophage migration was wholly dependent upon the presence of sensitised lymphocytes, leading in 1966 to the demonstration that the inhibitory effect of antigen on macrophage migration could be attributed to the generation of a secretory product of lymphocytes, macrophage migration inhibition factor (MIF) (DAVID 1966; BLOOM and BENNETT 1966). The migration of macrophages from animals exhibiting antibody-mediated hypersensitivity is not influenced by antigen (GEORGE and VAUGHAN 1962), nor is the presence of antigen necessary for MIF to exhibit its biological activity, so making it highly unlikely that this product of lymphocyte activation can be related to classical antibody, or to complement components. Hence, products of T cell activation (such as MIF) can be viewed as putative mediators of cellular immunity that are more comparable to the long established mediators of acute inflammation (e.g. histamine, kinins, prostaglandins) than to antibodies, with their dependence upon selective recognition of antigen. Subsequently, it became apparent that lymphocyte activation by antigen or by polyclonal mitogen could result in secretion of material affecting other target cells, including lymphocytes, neutrophils, eosinophils, osteoclasts and vascular endothelium. With such a plethora of factors, it became appropriate to introduce a generic term (lymphokine) to identify the material exerting such activities, especially as it had proved particularly difficult to separate these biological activities by contemporary techniques of protein fractionation. More recently, substantial purification has been achieved for a number of lymphokine activities. This topic has been reviewed at length elsewhere (HANSON et al. 1982).

C. Lymphokines as Mediators of Cellular Immunity

Reactions of cellular immunity are of major importance, since they exhibit both protective manifestations (as in destruction of tumour cells or in suppression of viral, bacterial, protozoan and metazoan parasites) and deleterious manifestations (as in auto-immune disease, graft rejection and contact sensitivity). Thus, there is practical importance in resolving the mechanism underlying such reactions. Lymphokines exhibit a range of biological effects appropriate to a mediator of cellular immunity and are especially noteworthy for their capacity to stimulate

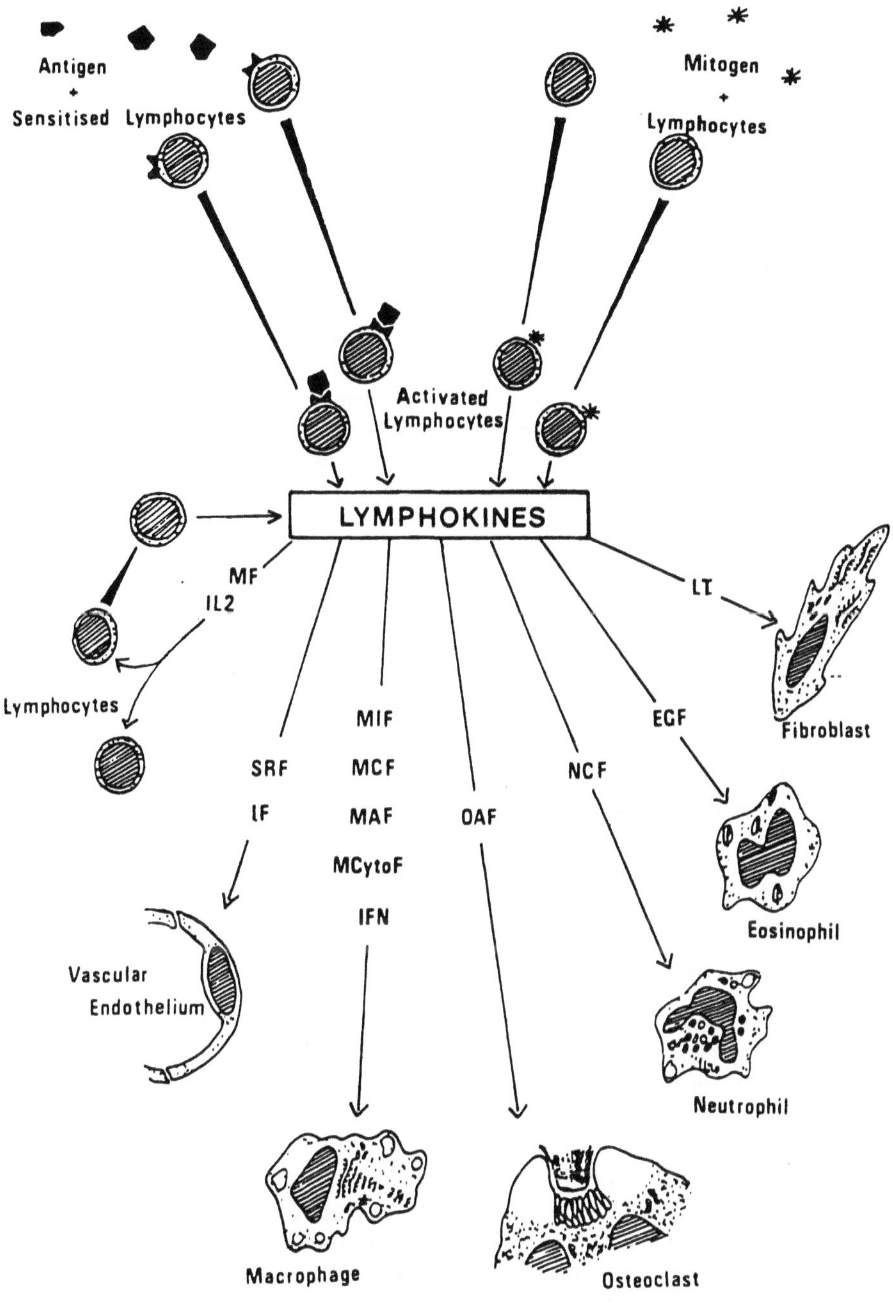

Fig. 1. Schematic representation of lymphokine generation. *LT*, lymphotoxin; *ECF*, eosinophil chemotactic factor; *NCF*, neutrophil chemotactic factor; *OAF*, osteoclast activating factor; *MIF*, macrophage migration inhibition factor; *MCF*, macrophage chemotactic factor; *MAF*, macrophage activating factor; *MCytoF*, macrophage cytotoxic factor; *IFN*, interferon; *SRF*, skin reactive factor; *IF*, inflammatory factor; *MF*, mitogenic factor; *IL2*, interleukin 2 (Hanson et al. 1982)

mononuclear cells. It has long been appreciated that only a minor proportion of activated lymphocytes in a lesion of cellular immunity are selectively sensitive to the eliciting antigen. Lymphokines, by virtue of their capacity to stimulate other lymphocytes in a non-selective manner, afford a potential mechanism for such amplification, and their ability to effect chemotaxis for lymphocytes in vitro is consistent with a role both in recruitment and in activation of lymphocytes. Similarly, lymphokines are chemotactic for monocytes in vitro and serve to activate macrophages, thereby enhancing the capacity of these cells to kill intracellular parasites and effect tissue damage. The inhibition of macrophage migration is a manifestation of this process of macrophage activation and it seems likely that certain lymphokines affecting macrophage function share a common identity. These in vitro observations provide strong circumstantial evidence that lymphokines are mediators of reactions of cellular immunity, a viewpoint that is reinforced by the capacity of antisera raised against MIF-rich preparations to inhibit skin reactions of cellular immunity in the guinea-pig.

Although histological evidence and cell transfer experiments establish a primary role for mononuclear cells, particularly lymphocytes, in reactions of cellular immunity, it is apparent that other cell types are recruited into these lesions and activated therein. The properties of lymphokines are appropriate to this subsidiary role in reactions of cellular immunity; thus, lymphokines are chemotactic to neutrophils, eosinophils and basophils. Additionally, by acting upon vascular endothelium they effect sustained increased vascular permeability, a phenomenon which may be pertinent to cell accumulation in vivo. Whilst there is no rigorous proof that these various properties of lymphokines underly cellular immunity, such a spectrum of effects provides a convenient mechanism to account for many aspects of the sequence of events evident in cellular immunity (Fig. 1).

D. Regulation of Lymphocyte Activation

I. Role of Arachidonic Acid

Activation of lymphocytes with consequent lymphokine generation provides an amplification mechanism to allow the relatively sparse population of specifically sensitive cells to recruit and activate other lymphocytes. Implicit in such a powerful positive feedback mechanism is the need for control of lymphocyte activation. PGE_2 is a potent inhibitor of lymphocyte activation in vitro, so that the substantial capacity of macrophages to generate PGE_2 led to the suggestion that this process provided a mechanism for regulating lymphocyte activation and hence limiting the extent and duration of cellular immune responses (MORLEY 1974). PGE_2 will inhibit the mitogenic response of lymphocytes as well as the generation of lymphokines; however, the latter process is much more sensitive to suppression by PGE_2, so that any controlling influence of PGE_2 during lymphocyte activation seems more likely to operate at the level of lymphokine secretion (Fig. 2).

It has been observed that arachidonic acid serves as a stimulant of lymphocyte activation and that the arachidonic acid analogue, eicosatetraenoic acid (ETEA), can wholly inhibit the activation process, implying a central role for arachidonic

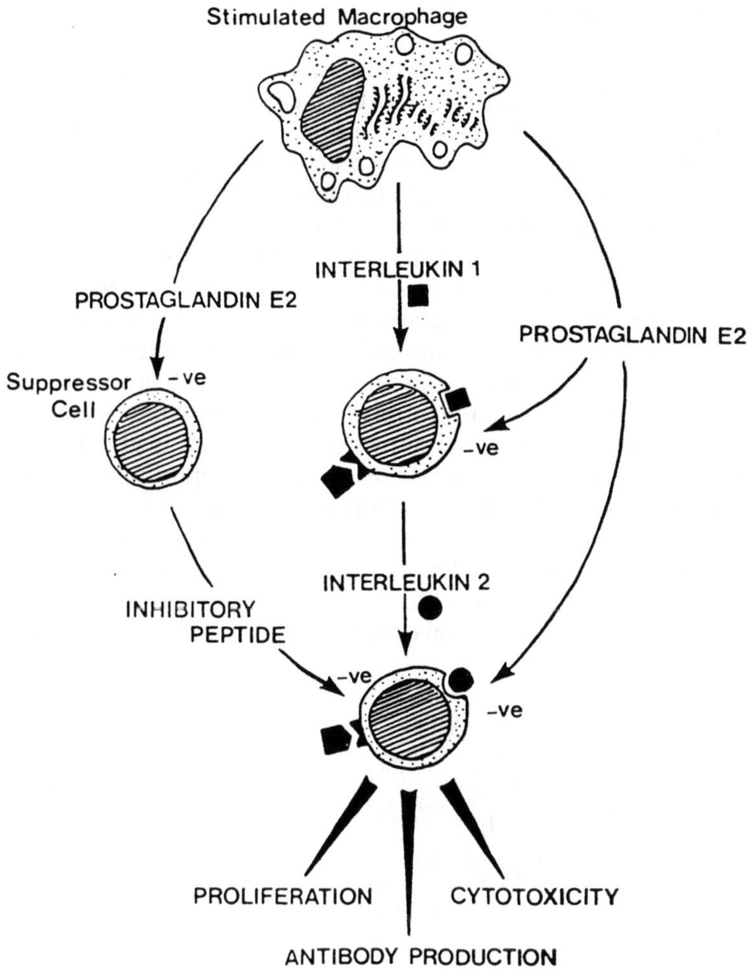

Fig. 2. Regulatory effects of PGE_2 (HANSON et al. 1982)

acid metabolism during lymphocyte activation. This effect of ETEA might be taken to imply a fundamental role for products of arachidonic acid metabolism, via lipoxygenase or cyclo-oxygenase pathways. However, non-steroidal anti-inflammatory drugs (NSAIDs) and related drugs fail to produce effects consistent with this action of ETEA (ALI et al., unpublished observations). Hence, it is likely that ETEA inhibits activation by affecting some other route of arachidonic acid metabolism. Activation of lysolecithin acetyl CoA:acyl transferase may provide the primary transducer mechanism of mitogen (and hence antigen) activation of lymphocytes, for following activation of this enzyme there is insertion of arachidonic acid into membrane phospholipids which activates certain membrane-bound enzymes in a manner closely paralleling that observed in intact lymphocytes (SZAMEL and RESCH 1981).

II. Role of Interleukins

Macrophages are known to produce immunoregulatory molecules other than arachidonic acid and its metabolites. Indeed, the participation of macrophages in responses of cellular immunity has been appreciated since 1972, when Gery and Waksman demonstrated that lymphocyte activation was facilitated by a non-dialysable macrophage product which they termed lymphocyte activating factor (LAF). For some time, it was considered that LAF augmented lymphocyte activation (see Oppenheim et al. 1980), but more recently it has become evident that LAF acts rather as an *obligatory* co-factor in lymphocyte activation, so as to permit the secretion of mitogenic lymphokine. Currently, this macrophage product is termed interleukin 1 and the mitogenic lymphokine is termed interleukin 2 (Fig. 3). The stimulatory effects of the mitogenic lymphokine (interleukin 2), coupled with the inflammatory potential of other lymphokines acting via macrophages, neutrophils and vascular endothelium, imply that inadvertent activation of small numbers of lymphocytes could result in a disproportionate inflammatory response. It seems likely that the obligatory nature of interleukin involvement in the activation process provides a powerful safeguard against inadvertent lymphocyte activation, since successful stimulation of effector lymphocytes requires additional and coincident activation both of helper lymphocytes and of macrophages. This system provides a regulatory mechanism for limiting lymphocyte ac-

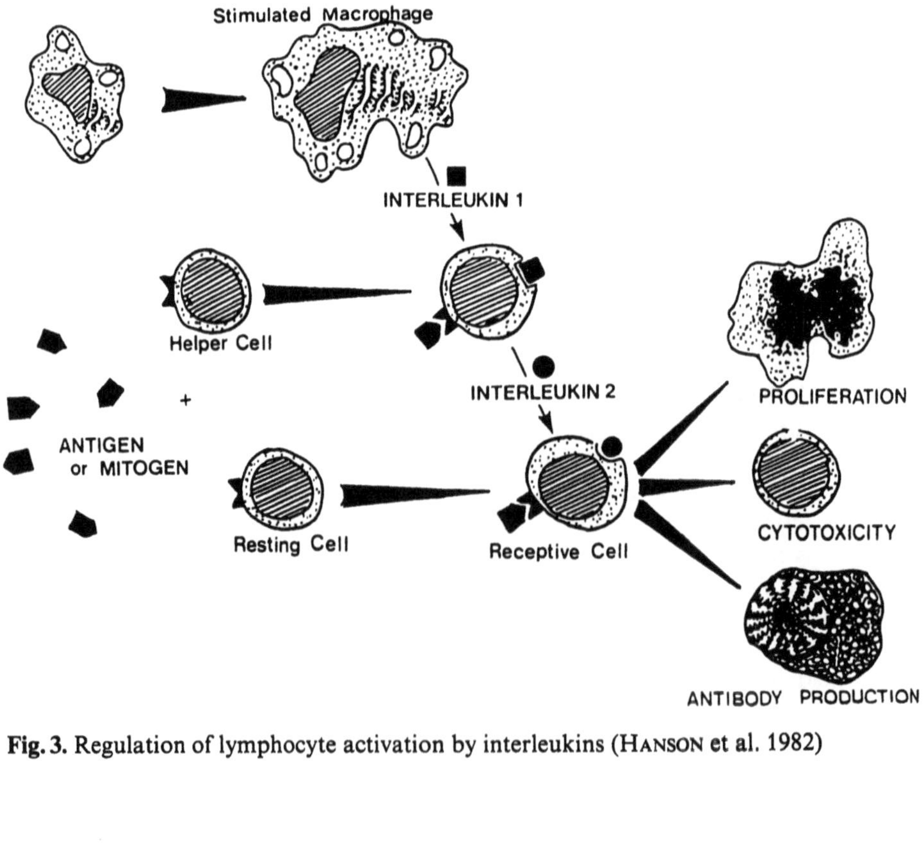

Fig. 3. Regulation of lymphocyte activation by interleukins (Hanson et al. 1982)

tivation by reduction or suppression of interleukin 2 production. Such a reduction may be achieved by impaired production of interleukin 1 or by generation of materials which inhibit the activity of the helper cell. PGE_2 falls into this latter category and it would seem reasonable to consider its role in lymphocyte activation as being a modulator (inhibitor) of interleukin 2 formation.

E. Conclusion

The regulation of lymphocyte activation is an area of intense research activity. It is now apparent that lymphokines have properties appropriate to mediators of cellular immunity, but that certain of these materials (interleukins) also serve a further function of controlling lymphocyte activation. This clarification of the regulatory role of these substances, together with the availability of highly purified lymphokines, should greatly assist the development of drugs with highly selective actions on lymphocyte activation.

References

Bloom BR, Bennett B (1966) Mechanisms of a reaction in vitro associated with delayed type hypersensitivity. Science 153:80–82

David JR (1966) Delayed hypersensitivity in vitro: its mediation by cell-free substances formed by lymphoid cell antigen interaction. Proc Natl Acad Sci USA 56:72–77

David JR, Lawrence HS, Thomas L (1964) Delayed hypersensitivity in vitro. II. Effect of sensitive cells on normal cells in the presence of antigen. J Immunol 93:274–278

George M, Vaughan JH (1962) In vitro cell migration as a model for delayed hypersensitivity. Proc Soc Exp Biol Med 111:514–521

Hanson JM, Rumjanek VM, Morley J (1982) Mediators of cellular immune reactions. Pharmacology and therapeutics 17:165–198

Morley J (1974) Prostaglandins and lymphokines in arthritis. Prostaglandins 8:315–326

Oppenheim JJ, Mizel SB, Meltzer MS (1979) Biological effects of lymphocyte and macrophage-derived mitogenic "amplication" factors. In: Cohen S, Pick E, Oppenheim JJ (eds) Biology of the lymphokines. Academic, New York, pp 291–323

Szamel M, Resch K (1981) Modulation of enzyme activities in isolated lymphocyte plasma membranes by enzymatic modification of phospholipid fatty acids. J Biol Chem 256:11618–11623

CHAPTER 10

Structure, Function and Control: Afferent Nerve Endings in the Skin

B. LYNN

A. Introduction

The skin is an important sensory structure concerned with the physical nature of our immediate environment (e.g. its temperature or roughness) and its own internal state (e.g. the degree of stretch or the presence of inflammatory chemical agents). In Sects. C–E the structure and function of the specialised nerve endings concerned with innocuous mechanical stimuli (Sect. C), with innocuous temperature changes (Sect. D) and with noxious, potentially or actually damaging, stimuli (Sect. E) will be considered in turn. Most of our knowledge of the function of cutaneous nerve endings has been deduced from recordings from afferent nerve fibres some distance from the receptor endings themselves. The fibre composition of cutaneous nerves will thus be considered first in Sect. B. Finally, Sect. G will consider what is currently known about the responses of cutaneous receptors to drugs.

Cutaneous nerves may also contain efferent autonomic fibres running to vascular smooth muscle and, in some areas, to sweat glands and arrector pili muscles. Information about these nerve endings will be found in sections of this book dealing with the target structures.

B. Fibre Composition of Cutaneous Nerves

Cutaneous nerve fibres may be myelinated or unmyelinated. Physiological measurements of conduction velocities indicate that the unmyelinated axons conduct at 0.4–2 m/s and the myelinated axons at 4–70 m/s. Several terminologies are in use to describe sensory axons of different diameter. The usual system designates myelinated axons A and unmyelinated C. In this chapter the terminology Aβ (=faster conducting myelinated fibre group), Aδ (=slower conducting myelinated fibre group) and C (=unmyelinated) will be used.

C. Mechanoreceptors

I. Introduction

Recordings from afferent nerve fibres have revealed the existence of several distinctive classes of mechanoreceptor unit from the skin. Rapidly adapting units that show no maintained response after the stimulus has been applied to the skin for more than 2–3 s will be considered first. Slowly adapting units that continue to discharge for many seconds, or even for several minutes, when the skin undergoes steady displacement will be considered second. Cutaneous mechanoreceptors of all types signal predominantly about changing, rather than static, events.

II. Rapidly Adapting Mechanoreceptors

1. Pacinian Corpuscles

The pacinian corpuscle is a large encapsulated nerve ending supplied by a single $A\beta$ axon and is most commonly found in the hands and feet of primates and carnivores. Pacinian corpuscles are readily excited by rapid skin movements. They fire well to fast transient stimuli and to high frequency (50–500 Hz) sinusoidal vibration. They are insensitive to slow movements because the capsule around the nerve ending acts as a mechanical "high pass" filter (HUBBARD 1958). Despite their location deep in the skin or in subcutaneous tissues, they can still respond to movements of the skin surface of only 1 μm or less (LYNN 1969). Because of their deep location and high transient sensitivity, Pacinian corpuscles have extensive receptive fields (LYNN 1971; JOHANSSON 1978).

2. RA and Field Receptors

Receptor units with a high sensitivity for skin movement but restricted receptive fields have been described in hairy and glabrous skin. In glabrous skin such units have been designated RA or rapidly adapting (JANIG et al. 1968; TALBOT et al. 1968; VALLBO and JOHANSSON 1978). In the human hand these units are common and their distribution bears some similarities to the distribution of Meissner's corpuscles.

In hairy skin similar rapidly adapting mechanoreceptor units have been designated Field units (BURGESS et al. 1968). From a physiological point of view, RA and Field units appear to be very similar.

3. Hair Follicle Receptors

All hair follicles appear to be innervated, usually by several myelinated fibres (SINCLAIR 1981) but not by unmyelinated fibres. The responses evoked by hair movement are predominantly rapidly adapting. Hair follicle units can be subdivided into several subclasses in terms of their sensitivity, receptive field size and the type of hair innervated (BROWN and IGGO 1967; BURGESS and PERL 1973). The morphology of nerve endings in hair follicles has been reviewed by ANDRES and VON DUERING (1973) and MUNGER (1982). Recent ultrastructural investigations have revealed a unique association between nerve terminals and pairs of Schwann cell processes in some hair follicles (ANDRES and VON DUERING 1973; CAUNA 1976).

The specialised sinus hairs or vibrissae in the facial region of many mammals show a more complex and extensive innervation than is found with the general body hairs (ANDRES and VON DUERING 1973), and 150 fibres may run to a single rat vibrissa (VINCENT 1913). Physiological investigations have revealed both slowly adapting and rapidly adapting responses from sinus hairs (ZUCKER and WALKER 1969; GOTTSCHALDT et al. 1973; DYKES 1975). The rapidly adapting units are relatively more common in most samples (DYKES 1975) and can be divided into two main subgroups, one with high and one with low velocity threshold. All these units have large myelinated axons and a single axon only innervates a single vibrissa.

III. Slowly Adapting Mechanoreceptors

1. Type SA I

In the hairy skin of several mammals, including man, small elevated areas are found, often in association with large Tylotrich hairs. These structures contain the terminals of a single large myelinated axon located in close contact with Merkel cells in the deepest part of the epidermis. Various names (e.g. Touch domes, haarscheibe) have been used for these structures, but the term Pinkus corpuscle is perhaps most satisfactory (SINCLAIR 1981). A single afferent axon may branch to supply up to seven corpuscles, but usually the number is one to three. Slight pressure applied to the corpuscle elicits a slowly adapting discharge in the afferent axon whilst strong pressure of adjacent skin or lateral stretch is ineffective.

In glabrous skin no Pinkus corpuscles are present but groups of Merkel cells and the associated terminals of large myelinated nerve fibres are found in the deep epidermis adjacent to the dermal papillae. The designation slowly adapting type I (SA I), first used for the Pinkus corpuscle units in hairy skin (IGGO 1966), is now used to describe all slowly adapting units with the properties of (a) small, often multi-point, receptive fields, (b) irregular adapted discharge, (c) marked velocity dependence of transient firing and (d) little response to lateral stretch of the skin (e.g. see IGGO 1977; KNIBESTOL 1975). Despite the lack of response to stretch, SA I units do respond when textured stimuli, e.g. Braille dot patterns, are moved tangentially across their receptive fields (JOHNSON and LAMB 1981).

The relation between Merkel cells and afferent nerve terminals has attracted considerable attention since it is a unique association in the somatosensory system. One possibility is that the Merkel cell plays the role of transducer of the mechanical stimulus into an electrical signal and in turn excites the nerve terminal via a chemical synaptic link.

2. Type SA II

A second distinct class of slowly adapting mechanoreceptor unit with Aβ axons, the type SA II, has been described in both hairy and glabrous skin (IGGO 1966, 1977; KNIBESTOL 1975). These units are sensitive to lateral skin stretch, give a regular discharge during steady skin displacements, often show regular spontaneous firing, have single-centred receptive fields with diffuse borders and show a less marked velocity dependence of their firing to transients than do SA I units. In cat hairy skin SA II units appear to be derived from encapsulated Ruffini endings in the dermis (CHAMBERS et al. 1972). A further group of slowly adapting units with many of the attributes of SA II units respond to forces applied to the nails in man (JOHANSSON 1978). Some SA II units show marked preferences for skin stretch in a particular direction (CHAMBERS et al. 1972; JOHANSSON 1978).

3. Slowly Adapting Hair Follicle Units

The large mystacial and carpal sinus hairs are innervated by slowly adapting mechanoreceptor units (NILLSON 1969; ZUCKER and WALKER 1969; GOTTSCHALDT et al. 1973). Slowly adapting responses from non-sinus hairs on the hands and face of monkeys have also been described (MERZENICH and HARRINGTON 1969; BIEMESDERFER et al. 1978).

4. C-Mechanoreceptors

In proximal skin of densely furred mammals (e.g. cat) large numbers of very sensitive, slowly adapting mechanoreceptors with unmyelinated axons are present (IGGO 1960; BESSOU et al. 1971). These C-mechanoreceptor units have small receptive fields, fire best to slow stroking of the skin, often with a pronounced afterdischarge, and fatigue readily with repeated stimulation. C-mechanoreceptor units are uncommon in less densely furred species (e.g. rat, LYNN and CARPENTER 1982; monkey, KUMAZAWA and PERL 1977) and have not been found in glabrous skin (BESSOU et al. 1971) or in man (TOREBJORK and HALLIN 1974). These receptors would be excellent "bug detectors" and so may be important for grooming in densely furred areas.

IV. Summary

Table 1 summarises the situation for both glabrous and hairy skin areas. In most skin areas four classes of mechanoreceptor unit, all with Aβ axons, are found. These are *Pacinian corpuscle* units, *RA/Field* units, *SA I* units associated with

Table 1. Summary of main types of afferent unit from mammalian skin

Type	Axon calibre	Adaptation rate	Frequency of occurrence	
			Glabrous skin	Hairy skin
Mechanoreceptor units				
PC	Aβ	Very rapid	++	+
RA/Field	Aβ	Rapid	++	++
Hair follicle	Aβ, Aδ	Rapid	o	+++
SA I	Aβ	Slow	++	++
SA II	Aβ	Slow	++	++, +
C-mechanoreceptor	C	Slow	o	o, +, ++[a]
Thermoreceptor units				
Cold	Aδ, C	Slow	+, ++	+
Warm	C	Slow	+	+
Nociceptor units				
HTM	Aδ	Slow	++	++
Polymodal (PMN)	C[b]	Slow	+++	+++

Key: o, absent; +, uncommon; ++, common; +++, very common.
[a] Varies markedly between species; o, man; +, rat, monkey; ++, cat, rabbit.
[b] In man some PMN units have Aδ axons.

Merkel cells and *SA II* units. RA/Field units are less common in hairy skin and instead large numbers of rapidly adapting *hair follicle* receptor units are found. Some hair follicle units have rather slowly conducting (Aδ) axons. Sensitive *C-mechanoreceptor* units are also present in hairy skin of most mammals except man.

D. Thermoreceptors

I. Cold Units

Afferent units selectively excited by small falls in skin temperature have been isolated from many cutaneous nerves (HENSEL 1981; SPRAY 1986). These units have small receptive fields comprising a single point or zone. Although sensitivity is greatest for falling temperatures, such units also fire tonically at temperatures between about 15° and 40 °C, with best firing at 20°–30 °C. Many cold units also fire when the skin temperature is raised to around 45 °C, the so-called paradoxical firing.

Cold units are present in most skin areas but are more common on the distal limb and the face than on the proximal limb. Some cold units have C fibres, others Aδ fibres, but the properties of both types are similar (IGGO 1969). In general, units with Aδ axons are most common in glabrous skin and in primates.

II. Warm Units

A second distinctive group of skin thermoreceptor units has a high sensitivity to small temperature increases, with significant firing occurring for changes of as little as 0.1 °C (HENSEL 1981; JOHNSON et al. 1979). Like cold units, warm units have small receptive fields and show static as well as dynamic firing. The static firing typically covers the range 32°–48 °C, with a best temperature of 40°–46 °C. At temperatures of around 35 °C both cold and warm units are often active. Warm units are common in the skin of the face and scrotum, and the glabrous skin of the distal limb; they are rather uncommon in hairy skin of the limb. The majority of warm units have unmyelinated (C) axons.

E. Nociceptors

Two different types of specialised nociceptive unit occur in hairy and glabrous skin of several species (BURGESS and PERL 1973). These are high threshold mechanoreceptor (HTM) units with Aδ axons (BURGESS and PERL 1967) and polymodal nociceptor (PMN) units with C axons (BESSOU and PERL 1969). The properties of these two common types of nociceptive unit are summarised in Table 2.

I. High Threshold Mechanoreceptor Units

High threshold mechanoreceptor units have two very distinctive properties: they only respond to strong pressure and this must be applied to discrete points (up

Table 2. Summary of the properties of two major classes of nociceptive unit (LYNN 1983)

	High threshold mechanoreceptor	Polymodal nociceptor
Axon diameter	Aδ	C[a]
Response to strong pressure	+	+
Response to heat	o	+
Response to irritant chemicals	o	+
Receptive field	Multiple points	Single small zone[b]

Key: +, responds well; o, no response.
[a] In man some polymodal-type units have Aδ axons.
[b] In primates, some PMN units with fields over 1 cm across and comprising more than one sensitive zone have been found.

to 20) spread over an area of skin that can exceed 1 cm^2. Responses to pressure are slowly adapting with little or no dynamic firing during stimulus onset. Most HTM units have Aδ axons, whilst a few have faster-conducting axons in the Aβ range.

In undamaged skin HTM units only rarely respond to skin heating. However, after heating the skin to noxious levels, these units may become responsive to heat stimulation (FITZGERALD and LYNN 1977; CAMPBELL et al. 1979; MEYER and CAMPBELL 1981). HTM units do not respond to irritant chemicals or to strong cooling (BURGESS and PERL 1967). A new type of intra-epidermal nerve ending has been described as the receptor for HTM units (KRUGER et al. 1981).

II. Polymodal Nociceptor Units

Polymodal nociceptor units are characterised by a slowly adapting response to firm pressure, a short latency response to noxious heat and excitation by a range of irritant chemicals (see Fig. 1). Some PMN units also respond if the skin is made very cold (GEORGOPOULOS 1976; CHERY-CROZE 1981). Receptive fields are usually single spots or small zones (<2 mm^2) in non-primates (BESSOU and PERL 1969; c.f. TOREBJORK 1974). PMN units account for about half of all sensory C units in nerves to cat hairy skin and over 80% of C units in rat and primate. They are the most numerous single class of afferent unit from the skin, with a density of about 10/mm^2 on the dorsal surface of the rat's foot (LYNN 1984).

Responses to heating are graded and firing accelerates rapidly with increasing skin temperature (LYNN 1979). The sensitivity to heating can be strongly influenced by the past history of stimulation. For example, in cat, monkey and rabbit hairy skin, responses are enhanced for long periods after a single strong heating of the receptive field (as shown in Fig. 1) (BESSOU and PERL 1969; KUMAZAWA and PERL 1977; LAMOTTE 1979; LYNN 1979). However, sensitisation by mild heat injury does not occur with any regularity in PMN units from rat hairy skin or from

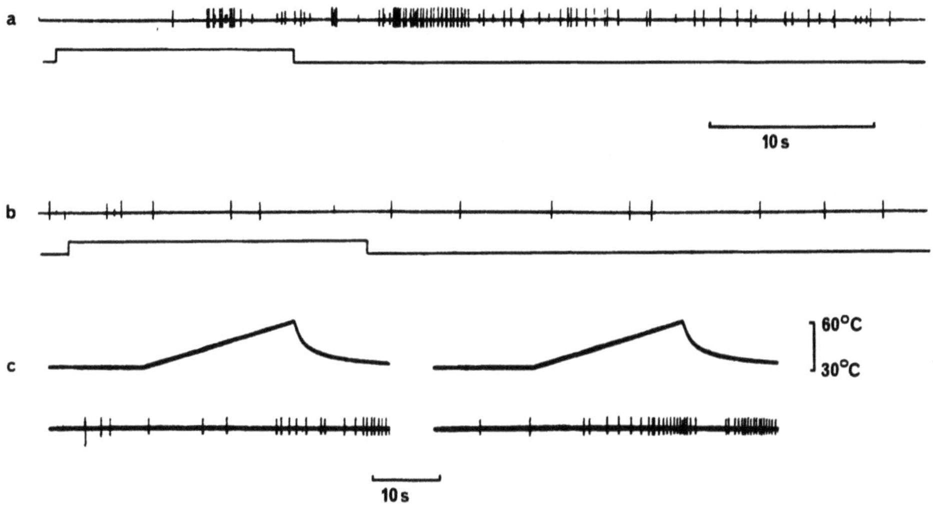

Fig. 1 a–c. Effects of histamine H_1 antagonist on C-fibre responses. **a** Multi-fibre preparation from rabbit sural nerve. Response to close-arterial injection of 30 µg histamine (indicated by lower, marker, trace). Several C units are excited. **b** Same recording. Second histamine injection 4 min after an intravenous injection of 4 mg/kg mepyramine maleate. The C units no longer respond. **c** Single fibre preparation from the same nerve 50 min after mepyramine injection. Response of PMN unit (conduction velocity 0.75 m/s) to repeat skin heatings. *Left traces,* first stimulus; *right traces,* next stimulus, 4 min later. Second response shows typical sensitisation. Subsequent tests with histamine showed that responses were still blocked. (M. FITZGERALD and B. LYNN, unpublished experiments)

primate glabrous skin (LYNN and CARPENTER 1982; MEYER and CAMPBELL 1981). In rabbit, PMN heat sensitivity can also be increased (a) if nearby skin, up to 1 cm away, is injured and (b) by antidromic stimulation of the other C-fibres in the same nerve (FITZGERALD 1979). PMN sensitisation parallels the hyperalgesia seen in human skin after mild heat injury and is presumably a major factor in producing this state (LYNN 1977). However, the sensitisation of HTM units after strong skin heating may also play a part, particularly where PMN units do not readily sensitise, such as in primate glabrous skin (MEYER and CAMPBELL 1981).

F. Overview: Types of Cutaneous Mechanoreceptor, Thermoreceptor and Nociceptor

Ten major types of afferent unit have been found in mammalian skin: six types of mechanoreceptor, two classes of thermoreceptor and two classes of nociceptor. The types of axon and the relative abundance of these groups of afferent units are indicated in Table 1. Units with fast-conducting $A\beta$ axons are all sensitive mechanoreceptors of various types. Units with $A\delta$ axons are predominantly either from cold receptors or from HTM nociceptors in glabrous skin. In hairy skin, however, many sensitive mechanoreceptors from hair follicles also have $A\delta$ axons. Amongst C-fibres, a small proportion are thermoreceptors, either warm or cold units, but the great majority are usually PMN units.

G. Modulation of Sensitivity of Cutaneous Receptors by Drugs

I. Introduction

A practical problem for studies in vivo is the route of administration. Agonists are usually given by injection into the skin or into the artery feeding the relevant skin area ("close arterially"). However, neither of these routes allows accurate estimation of the drug concentration at the receptor endings. This, in turn, can make it difficult to compare doses used in different preparations. Only a few studies using perfused or in vitro preparations have so far been undertaken (e.g. Perl et al. 1976; Reeh 1986); detailed pharmacological investigations will probably require the more extensive use of such preparations.

II. Catecholamines

The actions of adrenaline and noradrenaline have been examined for several types of cutaneous receptor and have been reviewed by Paintal (1964) and Akoev (1980).

1. Mechanoreceptors

Multi-unit recordings from frog skin nerves show an increase in mechanoreceptor activity when the inner surface of the skin is bathed in 10^{-8}–10^{-6} adrenaline or noradrenaline (Loewenstein 1956). Single-unit studies in the frog have shown that following sympathetic stimulation, slowly adapting units become more sensitive, although rapidly adapting mechanoreceptors become less sensitive (Calof et al. 1981). In mammals, Nilsson (1972) showed that vibrissae units became less excitable following injections of 1.5–5 µg/kg noradrenaline. Responses of rapidly adapting D- and G-hair units are depressed by sympathetic stimulation, although some C-mechanoreceptors, SA I and SA II units are excited (Pierce and Roberts 1981; Barasi and Lynn 1986; Roberts et al. 1985).

The dense-cored granules in Merkel cells are similar to amine-containing synaptic vesicles. However, these granules are not affected by reserpine treatment (Iggo and Muir 1969).

2. Thermoreceptors

Cold units in frog skin increase in excitability following application of 10^{-8}–10^{-4} noradrenaline, adrenaline or ephedrine to the inner surface of the skin (Spray 1974). Parallel effects are produced by sympathetic stimulation and these are blocked by both α and β blocking drugs (10^{-9} phentolamine or propranolol) (Spray 1974). In the rat trigeminal region, clonidine (an α_2-agonist) produced mixed excitatory and suppressive effects on cold receptor inputs to the trigeminal nucleus (Davies 1985).

3. Nociceptors

There do not appear to have been any direct studies of the effects of catecholamines on nociceptor unit activity. Local sympathetic stimulation has only small effects on the sensitivity of PMN units in the rabbit pinna (BARASI and LYNN 1986), does not affect reflex blood pressure changes following noxious skin heating (ROBERTS and LINDSAY 1981), but can excite some sensitised Aδ-HTM units (ROBERTS and ELARDO 1985).

4. Nerve Endings in Neuromas

If the regeneration of a sectioned peripheral nerve is impeded, a neuroma forms. The nerve endings in sciatic nerve neuromas in the rat have been found to be very sensitive to catecholamines (WALL and GUTNICK 1974). An α-adrenergic mechanism is implicated since these actions, and similar effects from stimulating the sympathetic fibres in the nerve, can be blocked by phentolamine, but not by propranolol (DEVOR and JANIG 1981; WALL and GUTNICK 1974). In addition, the β-agonist isoprenaline does not excite the nerve endings in rat neuromas. This sensitivity of injured nerve trunks to catecholamines may explain why some peripheral nerve lesions are painful and why pain from such lesions is often relieved by sympathectomy (e.g. see LOH and NATHAN 1978).

III. Inflammatory Mediators

Since inflammatory mediators produce pain or hyperalgesia in man (KEELE and ARMSTRONG 1964; FERREIRA 1972), most studies have examined their effects on nociceptive units. There have also been a limited number of studies of the effects of antagonists to these agents on nociceptive unit responses to mechanical or thermal stimuli, including responses in skin subjected to mild heat injury.

1. Histamine

Close-arterial injections of 2–50 μg excite PMN units in cat (BELCHER 1979; FJALLBRANDT and IGGO 1961) and rabbit (FITZGERALD 1978) (see Fig. 1). Human PMN units are also excited by intracutaneous histamine (TOREBJORK and HALLIN 1974) as well as by nettle stings, one of whose main ingredients is histamine (VAN HEES and GYBELS 1972; TOREBJORK 1974). PMN units also respond weakly to intracutaneous injections of the histamine releaser 48/80 (BESSOU and PERL 1969). Close-arterial histamine also excites C-mechanoreceptor units and Aβ SA mechanoreceptor units in cat (FJALLBRANDT and IGGO 1961) and rabbit (FITZGERALD 1978), including units of the SA I type (IGGO and MUIR 1969).

The excitatory effects of histamine on PMN units can be blocked by an intravenous injection of mepyramine maleate, indicating that the responses are mediated via H_1-receptors (see Fig. 1). Similar doses of mepyramine do not affect mechanical or thermal responses of PMN units nor do they diminish the sensitisation that follows mild heat injury, as shown in Fig. 1. Addition of anti-histamines to the perfusion fluid also failed to alter PMN heat sensitisation in the isolated rabbit ear (PERL et al. 1976).

2. Serotonin

Serotonin (5-hydroxytryptamine) has very similar effects to histamine. Close-arterial injections of 4–40 μg into the cat hind limb excite PMN, C-mechanoreceptor and SA I units, but not hair follicle and other rapidly adapting mechanoreceptor units (FJALLBRANDT and IGGO 1961; BECK and HANDWERKER 1974). Unlike histamine, some HTM units were excited by 5-HT (BECK and HANDWERKER 1974). In a preliminary report the anti-serotonin agent methysergide was found to depress heat-induced sensitisation of PMN units in the perfused rabbit ear (KING et al. 1976).

3. Bradykinin

Close-arterial injections of 0.2–15 μg bradykinin produce several minutes of low or moderate frequency firing in PMN units (BECK and HANDWERKER 1974; BELCHER 1979; SZOLCSANYI 1980). Bradykinin is thus more potent than histamine or serotonin. At low doses bradykinin is also rather more selective for PMN units than histamine and serotonin (SZOLCSANYI 1980), and this is confirmed by the selective activation of dorsal horn neurones with nociceptive input by peripheral injections of bradykinin (BESSON et al. 1972; BELCHER 1979). However, like the other mediators considered above, bradykinin can also excite other types of unit, including C-mechanoreceptors, $A\beta$ SA mechanoreceptors and $A\delta$ HTM units (BECK and HANDWERKER 1974). Repeated injections of bradykinin produce a reduction of responses to subsequent injections (tachyphylaxis). After producing tachyphylaxis, responses to 5-HT and to heating are unaffected (BECK and HANDWERKER 1974).

4. Prostaglandins

Responses of C-PMN units in the cat's paw to bradykinin and to heat are increased by PGE_1 or PGE_2 (HANDWERKER 1976). Responses of cat dorsal horn cells to bradykinin, histamine and skin heating were also found to be increased by close-arterial injections of PGE_1 into the limb circulation (BELCHER 1979). Responses of some pressure-sensitive mechanoreceptors may be elevated by the inflammatory changes produced by injection of PGE_2 into nearby skin (PATEROMICHELAKIS and ROOD 1981).

Heat responses of PMN units in the cat's paw were unaffected by an intravenous dose of 180 mg lysine acetylsalicylate (HANDWERKER 1976). Heat sensitisation of PMN units in perfused rabbit ears was also not reduced by agents that block prostaglandin synthesis (aspirin, 5.10^{-4} g/ml; indomethacin, 10^{-5} g/ml) (PERL et al. 1976; KING et al. 1976).

5. Mediators and Sensitisation

Many of the experiments with inflammatory mediators and their antagonists have aimed to determine whether these agents are involved in the heat-induced sensitisation of nociceptor responses. Negative results with anti-histamines and with prostaglandin synthesis inhibitors indicate that neither histamine nor prostaglandins play a major role in generating sensitisation. Stopping the blood flow to the skin does not enhance heat sensitisation (LYNN 1979), so a stable, diffusible mediator such as serotonin that would normally be removed by the blood supply does not appear to be involved. However, methysergide, a serotonin antagonist, has been reported to reduce heat sensitisation (KING et al. 1976). No convincing evidence is available for or against a role for bradykinin.

Overall, therefore, no clear picture has emerged concerning the role of mediators in sensitisation. Kinins could play a large role, or several agents may each make a partial contribution. It is also possible that sensitisation is due to a local change at the nerve endings themselves and does not involve mediators at all.

IV. Other Agents that Excite Cutaneous Receptors

1. Capsaicin

Capsaicin, the active principle in hot peppers, excites PMN units, with as little as 0.5 μg (close-arterial) being effective (SZOLCSANYI 1985; KENINS 1981; LYNN and CARPENTER 1982). The action is highly selective for PMN units, and other types of C-fibre unit (mechanoreceptor and thermoreceptor) and all types of A-fibre unit (except a few Aδ mechano-heat units) are not excited (FOSTER and RAMAGE 1976; SZOLCSANYI 1985; KENINS 1981). The capsaicin derivative *n*-nonanoyl-vanillylamide (VAN) has similar actions (FOSTER and RAMAGE 1976). PMN heat responses are also enhanced following capsaicin (SZOLCSANYI 1980; KENINS 1981). These results are consistent with capsaicin's ability to produce hyperalgesia and burning pain in human skin (SZOLCSANYI 1977).

As well as its immediate excitatory action, capsaicin can, in both the neonatal and adult rat, produce a long-lasting loss of function and even degeneration of some afferent C-fibres, particularly those of the PMN type (JANSCO et al. 1977, 1980, 1985; LYNN et al. 1984, 1987).

2. Substance P

Close-arterial injections of the undecapeptide substance P cause weak excitation of a range of cutaneous receptors, including PMN units, C-mechanoreceptors and SA mechanoreceptors (FITZGERALD and LYNN 1979). Heat responses of PMN units are, however, not significantly enhanced following substance P injections. The strong pain produced by substance P extracts is now thought to have been due to K^+ and kinin contamination (LEMBECK et al. 1977); the synthetic peptide does not produce pain in human skin (STEWART et al. 1976).

There is good evidence that substance P release from cutaneous nerve terminals is a major factor in the generation of the flare component of the triple response (HAGERMARK et al. 1978; GAMSE et al. 1980; CARPENTER and LYNN 1981). However, in view of its failure to cause sensitisation, substance P is a poor candidate for involvement in the hyperalgesia that occurs in the flare zone.

3. Acetylcholine

Small (20 μg) amounts of acetylcholine (ACh) injected close-arterially excite afferent fibres of all sizes, both myelinated and unmyelinated, from cat hairy skin

(DOUGLAS and RITCHIE 1960). These effects involve nicotinic receptors since they are also produced by nicotine and are blocked by hexamethonium and tubocurarine, but not by atropine (DOUGLAS and GRAY 1953; DOUGLAS and RITCHIE 1960). Similar effects occur with rapidly adapting mechanoreceptors in frog skin (JARRETT 1956). Single unit studies have shown that SA I (FJALLBRANDT and IGGO 1961; SMITH and CREECH 1967), C-mechanoreceptor (FJALLBRANDT and IGGO 1961) and C-polymodal (BURGESS and PERL 1973) units are amongst those involved. In the tongue, warm and cold thermoreceptor units are facilitated by local injection of ACh or acetyl-β-methylcholine (DODT et al. 1953), while in frog skin low concentrations of ACh or nicotine (10^{-7}–10^{-8}) enhanced cold responses whereas larger doses depressed them (SPRAY and GALANSKY 1975). For a review of the actions of acetylcholine on all types of mechanoreceptor, see AKOEV (1980) and PAINTAL (1964).

Regenerating nerve fibres develop a sensitivity to ACh comparable to that found at cutaneous nerve endings, and, as with receptors, this sensitivity is due to a nicotinic mechanism (DIAMOND 1959). Although responses to ACh are readily blocked by nicotinic antagonists, these agents do not affect the sensitivity of cutaneous receptors to mechanical stimulation (DOUGLAS and GRAY 1953; DOUGLAS and RITCHIE 1960). However, complex effects of cholinergic antagonists on the responses to cooling recorded in frog skin nerves have been reported (SPRAY and GALANSKY 1975).

Histochemical studies have revealed both specific acetylcholinesterase (AChE) and non-specific butyrylcholinesterase (BuChE) activity in organised receptor endings in skin, most of which are probably mechanoreceptors (e.g. see MONTAGNA 1960; KOELLE 1963; WINKELMANN 1968). However, inhibition of cutaneous cholinesterases by application of diisopropyl fluorophosphate has no obvious effect on tactile and pinprick sensitivity (HURLEY and KOELLE 1958) or on itch (SHELLEY and ARTHUR 1957). In addition, the experiments described above on cholinergic blockers indicate that AChE plays no role in normal mechanoreceptor function. What part, if any, might be played by cholinesterases in mechanoreceptor function is therefore still not known. In view of present interest in the physiological roles of peptides in the skin, the demonstration of peptidase activity by pituitary BuChE suggests new possibilities for this enzyme (DAS et al. 1982).

4. Other Irritants

A large number of substances produce pain or itch when applied to the skin, and several extensive reviews listing these agents are available (e.g. KEELE and ARMSTRONG 1964; CHAHL and KIRK 1975). The effect of a rather limited number of irritants has been examined on cutaneous afferent nerve preparations. One of particular interest is itch powder.

Itch powder [Cowhage (*Mucuna pruriens*) spicules] excites some, but not all, C-PMN units from cat and human hairy skin (TOREBJORK 1974; TUCKETT 1980). Interestingly, application of itch powder to cat skin is reported not to cause scratching (SHELLEY and ARTHUR 1957). Microstimulation of human afferent C-fibres of the PMN type produces sensations, referred to the appropriate cutaneous field, of *either* itch *or* pain, but not both, even when a range of stimulus frequencies is tested (TOREBJORK and OCHOA 1981). There is thus now some direct evidence for a subclass of PMN units that is especially concerned with itch.

V. Other Agents that Modulate the Responses of Cutaneous Nerve Endings

Several agents that can modify the sensitivity of cutaneous receptors without exciting them have already been considered (e.g. adrenaline, prostaglandins). Other agents known to have such effects will now be described.

Local anaesthetics are well known to depress or abolish the firing of sensory nerve terminals. Their actions at terminals are essentially similar to those on nerve fibres in general and have been extensively reviewed elsewhere (e.g. RITCHIE and GREENE 1980).

One general anaesthetic, halothane, has been found to increase the heat sensitivity of nociceptive afferents in the monkey whilst barbiturates had no such effect (CAMPBELL et al. 1984).

Opiate agonists, both peptide and non-peptide, can reduce neurogenic inflammatory responses that depend on activity in afferent C-fibre terminals (BARTHO and SZOLCSANYI 1981) and can reduce inflammatory pain by a peripheral mechanism (FERREIRA 1980). Enkephalins have also been shown to be present in Merkel cells (HARTSCHUCH et al. 1979).

Subcutaneous administration of *vincristine,* an inhibitor of mitosis and of axonal transport, produces a large reduction in the sensitivity of SA I units in the rat 1–34 days later (LEON and MCCOMAS 1980). This action is consistent with the known effects of denervation on the integrity of some Merkel cells, the accessory cells associated with SA I fibre terminals.

VI. Overview: Drugs and Cutaneous Receptors

Table 3 summarises the qualitative effects of the more extensively studied agents on various classes of cutaneous receptor unit. A range of excitatory and facilitatory actions have been discovered and some agents with useful selectivity for PMN units (e.g. bradykinin, capsaicin) have been described. Only a few agents produce depression of responses, notably catecholamines on some RA mechanoreceptor units.

From a physiological point of view, the interesting question is whether any of the many endogenous agents examined actually plays a role in generating or modulating normal receptor sensitivity. Here the results are disappointing. No clear evidence is available for a role of any endogenous chemical in normal receptor function. For thermoreceptors and for most mechanoreceptors, normal function appears to be solely determined by the properties of the nerve terminals themselves and these are well insulated from chemicals in the surrounding tissues by encapsulation and by Schwann cells. The Merkel cell–nerve ending complex of the SA I mechanoreceptor, with its synapse-like organisation, looks promising. In practice, however, apart from a general tendency to respond more easily to injected chemicals than other mechanoreceptors, no distinctive pharmacology has been found that might relate to normal excitation.

Table 3. Summary of the effects of widely studied agents on the sensitivity of different classes of cutaneous afferent unit

	A-fibre units			C-fibre units	
	Hair follicle	SA	HTM	Mechano-receptor	PMN
Catecholamines	–	o, +	o	o	o
Acetylcholine		+		+	+
Histamine	o	+	o	+	+
Serotonin (5-HT)	o	+	o	+	+
Bradykinin	o	o, +	o, +	o	+
PG E_1 or PG E_2					+
Capsaicin	o	o	o	o	+

Key: +, excited or facilitated; o, no effect; –, depressed; blank, not known.

The influence of inflammatory mediators on nociceptors has beeen reviewed in Sect. G.III.5. For heat-induced sensitisation the evidence for a role for the agents studied so far (histamine, serotonin, bradykinin and prostaglandins) is meagre or negative, despite considerable efforts by several laboratories.

Finally, mention should be made of the special sensitivity demonstrated by regenerating cutaneous nerve fibres to ACh and catecholamines. These changes, especially the increased sensitivity to noradrenaline, may be of importance in generating the chronic pain that can occur after peripheral nerve injuries.

References

Akoev GN (1980) Catecholamines, acetylcholine and the excitability of mechanoreceptors. Prog Neurobiol 15:269–294

Andres KH, Von Duering M (1973) Morphology of cutaneous receptors. In: Iggo A (ed) Handbook of sensory physiology, vol 2. Springer, Berlin Heidelberg New York, pp 3–28

Barasi S, Lynn B (1986) Effects of sympathetic stimulation on mechanoreceptive and nociceptive afferent units from the rabbit pinna. Brain Res 378:21–27

Bartho L, Szolcsanyi J (1981) Opiate agonists inhibit neurogenic plasma extravasation in the rat. Eur J Pharmacol 73:101–104

Beck PW, Handwerker HO (1974) Bradykinin and 5-HT effects on various types of cutaneous nerve fibres. Pflügers Arch 347:209–222

Belcher G (1979) The effects of intra-arterial bradykinin, histamine, acetylcholine and prostaglandin E_1 on nociceptive and non-nociceptive dorsal horn neurones of the cat. Eur J Pharmacol 56:385–395

Besson JM, Conseiller C, Hamann K-F, Maillard M-C (1972) Modifications of dorsal horn cell activities in the spinal cord, after intra-arterial injection of bradykinin. J Physiol (Lond) 221:189–205

Bessou P, Perl ER (1969) Response of cutaneous sensory units with unmyelinated fibres to noxious stimuli. J Neurophysiol 32:1025–1043

Bessou P, Burgess PR, Perl ER, Taylor CB (1971) Dynamic properties of mechanoreceptors with unmyelinated (C) fibres. J Neurophysiol 34:116–131

Biemesderfer D, Munger GL, Binck J, Dubner R (1978) The pilo-Ruffini complex: a non-sinus hair and associated slowly adapting mechanoreceptor in primate facial skin. Brain Res 142:197–222
Brown AG, Iggo A (1967) A quantitive study of cutaneous receptors and afferent fibres in the cat and rabbit. J Physiol (Lond) 193:707–733
Burgess PR, Perl ER (1967) Myelinated afferent fibres responding specifically to noxious stimulation of the skin. J Physiol (Lond) 190:541–562
Burgess PR, Perl ER (1973) Cutaneous mechanoreceptors and nociceptors. In: Iggo A (ed) Handbook of sensory physiology, vol 2. Springer, Berlin Heidelberg New York, pp 29–78
Burgess PR, Petit D, Warren RM (1968) Receptor types in cat hairy skin supplied by myelinated fibres. J Neurophysiol 31:833–848
Calof AL, Jones RB, Roberts WJ (1981) Sympathetic modulation of mechanoreceptor sensitivity in frog skin. J Physiol (Lond) 310:481–499
Campbell JN, Meyer RA, LaMotte RH (1979) Sensitization of myelinated nociceptive afferents that innervate monkey hand. J Neurophysiol 42:1669–1679
Campbell JN, Raja SN, Meyer RA (1984) Halothane sensitises cutaneous nociceptors in monkeys. J Neurophysiol 52:762–770
Carpenter SE, Lynn B (1981) Vascular and sensory responses of human skin to mild injury after topical treatment with capsaicin. Br J Pharmacol 73:755–758
Cauna N (1976) Morphological basis of sensation in hairy skin. Prog Brain Res 43:35–45
Chahl LA, Kirk EJ (1975) Toxins which produce pain. Pain 1:3–49
Chambers MR, Andres KH, Von Duering M, Iggo A (1972) The structure and function of the slowly-adapting type II mechanoreceptor in hairy skin. Q J Exp Physiol 57:417–445
Chery-Croze S (1981) La douleur thermique cutanee. Caracteristiques psychophysiques et mecanismes nerveux peripheriques. Ph D thesis, Claude Bernard University, Lyon
Das S, Edwardson JA, Hughes D, McDermott JR (1982) Butyrylcholinesterase (BuChE)-positive glial cells in the pituitary intermediate lobe: evidence for their role in the formation of pituitary colloid and that BuChE is an endopeptidase. J Physiol (Lond) 327:45P
Davies SN (1985) Sympathetic stimulation of cold-receptive neurones in the trigeminal system of the rat. J Physiol (Lond) 366:315–329
Devor M, Janig W (1981) Activation of myelinated afferents ending in a neuroma by stimulation of the sympathetic supply in the rat. Neurosci Lett 24:43–47
Diamond J (1959) The effect of injecting acetylcholine into normal and regenerating nerves. J Physiol (Lond) 145:611–629
Dodt E, Skouby AP, Zotterman Y (1953) The effect of cholinergic substances on the discharges from thermal receptors. Acta Physiol Scand 28:101–114
Douglas WW, Gray JAB (1953) The excitant action of acetylcholine and other substances on cutaneous sensory pathways and its prevention by hexamethonium and D-tubocurarine. J Physiol (Lond) 119:118–128
Douglas WW, Ritchie JM (1960) The excitatory action of acetylcholine on cutaneous non-myelinated fibres. J Physiol (Lond) 150:510–514
Dykes RW (1975) Afferent fibers from mystacial vibrissae of cats and seals. J Neurophysiol 38:650–662
Ferreira SH (1972) Prostaglandins, aspirin-like drugs and analgesia. Nature [New Biol] 240:200–203
Ferreira SH (1980) Peripheral analgesia: mechanism of the analgesic action of aspirin-like drugs and opiate antagonists. Br J Clin Pharmacol 10:237S–245S
Fitzgerald M (1978) The sensitization of cutaneous nociceptors. PhD Thesis, University of London
Fitzgerald M (1979) The spread of sensitization of polymodal nociceptors in the rabbit from nearby injury and by antidromic nerve stimulation. J Physiol (Lond) 297:207–216
Fitzgerald M, Lynn B (1977) The sensitization of high threshold mechanoreceptors with myelinated axons by repeated heating. J Physiol (Lond) 265:549–563

Fitzgerald M, Lynn B (1979) The weak excitation of some cutaneous receptors in cats and rabbits by synthetic substance P. J Physiol (Lond) 293:66–67P
Fjallbrandt N, Iggo A (1961) The effect of histamine, 5-hydroxytryptamine and acetylcholine on cutaneous afferent fibres. J Physiol (Lond) 156:578–590
Foster RW, Ramage AG (1976) Evidence for specific somatosensory receptor in the cat skin that responds to irritant chemicals. Br J Pharmacol 57:436–437P
Gamse R, Holzer P, Lembeck F (1980) Decrease of substance P in primary afferent neurones and impairment of neurogenic plasma extravasation by capsaicin. Br J Pharmacol 68:207–213
Georgopoulos AP (1976) Functional properties of primary afferent units probably related to pain mechanisms in primate glabrous skin. J Neurophysiol 39:71–83
Gottschaldt K-M, Iggo A, Young DW (1973) Functional characteristics of mechanoreceptors in sinus hair follicles of the cat. J Physiol (Lond) 235:287–315
Hagermark O, Hokfelt T, Pernow B (1978) Flare and itch induced by substance P in human skin. J Invest Dermatol 71:233–235
Handwerker HO (1976) Influences of algogenic substances and prostaglandins on the discharges of unmyelinated cutaneous nerve fibers identified as nociceptors. Adv Pain Res Ther 1:41–45
Hartschuh W, Weihe E, Buchler M, Helmstaedter V, Feurle GE, Forssmann WG (1979) Met-enkephalin-like immuno-reactivity in Merkel cells. Cell Tissue Res 201:343–348
Hensel H (1981) Thermoreception and temperature regulation. Academic, London
Hubbard SJ (1958) A study of rapid mechanical events in a mechanoreceptor. J Physiol (Lond) 141:198–218
Hurley HJ, Koelle GB (1958) The effect of inhibition of non-specific cholinesterase on perception of tactile sensation in human volar skin. J Invest Dermatol 31:243–245
Iggo A (1960) Cutaneous mechanoreceptors with afferent C fibres. J Physiol (Lond) 152:337–353
Iggo A (1966) Cutaneous receptors with a high sensitivity to mechanical displacement. In: de Reuck AVS, Knight J (eds) Touch, heat and pain. CIBA Found Symp, pp 237–255
Iggo A (1969) Cutaneous thermoreceptors in primates and sub-primates. J Physiol (Lond) 200:403–430
Iggo A (1977) Cutaneous and subcutaneous sense organs. Br Med Bull 33:97–102
Iggo A, Muir AR (1969) The structure and function of a slowly adapting touch corpuscle in hairy skin. J Physiol (Lond) 200:763–796
Jancso G, Kiraly E, Jancso-Gabor A (1977) Pharmacologically-induced selective degeneration of chemosensitive primary sensory neurones. Nature 270:741–743
Jancso G, Kiraly E, Jancso-Gabor A (1980) Chemosensitive pain fibres and inflammation. Int J Tissue React 11:57–66
Jancso G, Ferencsik M, Such G, Kiraly E, Nagy A, Bujdoso M (1985) Morphological effects of capsaicin and its analogues in newborn and adult mammals. In: Hakanson R, Sundler F (eds) Tachykinin antagonists. Elsevier, Amsterdam, pp 35–44
Janig W, Schmidt RF, Zimmerman M (1968) Single unit responses and the total afferent inflow from the cat's foot pad upon mechanical stimulation. Exp Brain Res 6:100–115
Jarrett AS (1956) The effect of acetylcholine on touch receptors in frog's skin. J Physiol (Lond) 133:243–254
Johansson RS (1978) Tactile sensibility in the human hand: receptive field characteristic of mechanoreceptive units in the glabrous skin area. J Physiol (Lond) 281:101–123
Johnson KO, Lamb GD (1981) Neural mechanisms of spatial tactile discrimination: neural patterns evoked by Braille-like dot patterns in the monkey. J Physiol (Lond) 310:117–144
Johnson KO, Darian-Smith L, LaMotte C, Johnson B, Oldfield S (1979) Coding of incremental changes in skin temperature by a population of warm fibres in the monkey: correlation with intensity discrimination in man. J Neurophysiol 42:1322–1353
Keele CA, Armstrong D (1964) Substances producing itch and pain. Arnold, London
Kenins P (1981) Responses of single sensory nerve fibres to capsaicin applied to the skin. Proc Aust Physiol Pharmacol Soc 12:212P

King JS, Gallant P, Myerson V, Perl ER (1976) The effects of anti-inflammatory agents on the responses and the sensitization of unmyelinated (C) fiber polymodal nociceptors. In: Zotterman Y (ed) Sensory functions of the skin in primates. Pergamon, Oxford, pp 441–454

Knibestol M (1975) Stimulus-response functions of slowly adapting mechanoreceptors in the human glabrous skin area. J Physiol (Lond) 245:63–80

Koelle GB (1963) Cytological distribution and physiological function of cholinesterase. In: Koelle GB (ed) Cholinesterases and anti-cholinesterase agents. Springer, Berlin Heidelberg New York, pp 187–298 (Handbook of experimental pharmacology, vol 15)

Kruger L, Perl ER, Sedivic MJ (1981) Fine structure of myelinated mechanical nociceptor endings in cat hairy skin. J Comp Neurol 198:137–154

Kumazawa T, Perl ER (1977) Primate cutaneous units with unmyelinated (C) afferent fibers. J Neurophysiol 40:1325–1338

LaMotte RH (1979) Intensive and temporal determinants of thermal pain. In: Kenshalo DR (ed) Sensory functions of the skin in humans. Plenum, New York, pp 327–359

Lembeck F, Gamse R, Juan H (1977) Substance P and sensory nerve endings. In: von Euler US, Pernow B (eds) Substance P. Raven, New York, pp 169–181

Leon J, McComas AJ (1980) Effect of vincristine on touch dome function. J Physiol (Lond) 298:34P

Loewenstein WR (1956) Modulation of cutaneous mechanoreceptors by sympathetic stimulation. J Physiol (Lond) 132:40–60

Loh L, Nathan PW (1978) Painful peripheral states and sympathetic blocks. J Neurol Neurosurg Psychiatry 41:664–671

Lynn B (1969) The nature and location of certain phasic mechanoreceptors in the cat's foot. J Physiol (Lond) 201:765–773

Lynn B (1971) The form and distribution of the receptive fields of Pacinian corpuscles found in and around the cat's large foot pad. J Physiol (Lond) 217:755–771

Lynn B (1977) Cutaneous hyperalgesia. Br Med Bull 33:103–108

Lynn B (1979) The heat sensitization of polymodal nociceptors in the rabbit and its independence of the local blood flow. J Physiol (Lond) 287:493–507

Lynn B (1983) Cutaneous sensation. In: Goldsmith LA (ed) Biochemistry and physiology of the skin. Oxford University Press, New York, pp 654–684

Lynn B (1984) The detection of injury and tissue damage. In: Wall PD, Melzack R (eds) Textbook of pain. Churchill Livingstone, Edinburgh, pp 19–33

Lynn B, Carpenter SE (1982) Primary afferent units from the hairy skin of the rat hind limb. Brain Res 238:29–43

Lynn B, Carpenter SE, Pini A (1984) Capsaicin and cutaneous afferents. In: Chahl LA, Szolcsanyi J, Lembeck F (eds) Antidromic vasodilatation and neurogenic inflammation. Akademiai Kaido, Budapest, pp 83–92

Lynn B, Pini A, Baranowski R (1987) Injury of somatosensory afferents by capsaicin: selectivity and failure to regenerate. In: Pubols LM, Sessle BJ (eds) Effects of injury on spinal and trigeminal somatosensory systems. Liss, New York, pp 115–124

Merzenich MM, Harrington T (1969) The sense of flutter-vibration evoked by stimulation of the hairy skin of primates: comparison of human sensory capacity with the responses of mechanoreceptive afferents innervating hairy skin of monkeys. Exp Brain Res 9:236–260

Meyer RA, Campbell JN (1981) Myelinated nociceptive afferents account for the hyperalgesia that follows burn to the hand. Science 213:1527–1529

Montagna W (1960) Cholinesterases in the cutaneous nerves of man. Adv Biol Skin 1:74–87

Munger BL (1982) Multiple afferent innervation of primate facial hairs – Henry Head and Max von Frey revisited. Brain Res Rev 4:1–43

Nilsson BY (1969) Structure and function of the tactile hair receptors on the cat's foreleg. Acta Physiol Scand 77:396–416

Nilsson BY (1972) Effects of sympathetic stimulation on mechanoreceptors of cat vibrissae. Acta Physiol Scand 85:390–397

Paintal AS (1964) Effects of drugs on vertebrate mechanoreceptors. Pharmacol Rev 16:341–380

Pateromichelakis S, Rood JP (1981) Prostaglandin E_2 increases mechanically evoked potentials in the peripheral nerve. Experentia 37:282–283
Perl ER, Kumazawa T, Lynn B, Kenins P (1976) Sensitization of high threshold receptors with unmyelinated (C) afferent fibers. Prog Brain Res 43:263–276
Pierce JP, Roberts WJ (1981) Sympathetically induced changes in the responses of guard hair and type II receptors in the cat. J Physiol (Lond) 314:411–428
Reeh PW (1986) Sensory receptors in mammalian skin in an in vitro preparation. Neurosci Lett 66:141–146
Ritchie JM, Greene NM (1980) Local anaesthetics. In: Gilman AG, Goodman LS, Gilman A (eds) The pharmacological basis of therapeutics, 6th edn. Macmillan, New York, pp 300–320
Roberts WJ, Elardo SM (1985) Sympathetic activation of A-delta nociceptors. Somatosens Res 3:33–44
Roberts WJ, Lindsay AD (1981) Sympathetic activity shown to have no short-term effect on polymodal nociceptors. Soc Neurosci Abstr 7:227
Roberts WJ, Elardo SM, King KA (1985) Sympathetically induced changes in the responses of slowly-adapting type I receptors in cat skin. Somatosens Res 2:223–236
Shelley WB, Arthur RP (1957) The neurohistology of the itch sensation in man. Arch Dermatol 76:296–323
Sinclair D (1981) Mechanisms of cutaneous sensation. Oxford University Press, Oxford
Smith KR, Creech BJ (1967) Effects of pharmacological agents on the physiological responses of hair discs. Exp Neurol 19:477–482
Spray DC (1974) Characteristics, specificity and afferent control of frog cutaneous cold receptors. J Physiol (Lond) 237:15–38
Spray DC (1986) Cutaneous temperature receptors. Annu Rev Physiol 48:625–638
Spray DC, Galansky SH (1975) Effects of cholinergic agonists and antagonists on frog cutaneous cold receptors. Comp Biochem Physiol (C) 50:97–103
Stewart JM, Getto CJ, Neldner K, Reeve EB (1976) Substance P and analgesia. Nature 262:784–785
Szolcsanyi J (1977) A pharmacological approach to elucidation of the role of different nerve fibres and receptor endings in mediating pain. J Physiol (Paris) 73:251–259
Szolcsanyi J (1980) Effect of pain-producing chemical agents on the activity of slowly conducting afferent fibres. Acta Physiol Acad Hung 56:86
Szolcsanyi J (1985) Sensory receptors and the antinociceptive effects of capsaicin. In: Hakanson R, Sundler F (eds) Tachykinin antagonists. Elsevier, Amsterdam, pp 45–54
Talbot WH, Darian-Smith I, Kornhuber HH, Mountcastle VB (1968) The sense of flutter vibration: comparison of the human capacity with the response patterns of mechanoreceptive afferents from the monkey hand. J Neurophysiol 31:301–334
Torebjork HE (1974) Afferent C units responding to mechanical, thermal and chemical stimuli in human non-glabrous skin. Acta Physiol Scand 92:374–390
Torebjork HE, Hallin RG (1974) Identification of afferent C units in intact human skin nerves. Brain Res 67:387–403
Torebjork HE, Ochoa JL (1981) Pain and itch from C-fibre stimulation. Soc Neurosci Abstr 7:228
Tuckett RP (1980) Responses of cutaneous receptors to a pruritic stimulus. Soc Neurosci Abstr 6:428
Vallbo AB, Johansson RS (1978) The tactile sensory innervation of the glabrous skin of the human hand. In: Gordon G (ed) Active touch. Pergamon, Oxford, pp 29–54
Van Hees J, Gybels JM (1972) Pain related to single afferent C fibres from human skin. Brain Res 48:397–400
Vincent SB (1913) The tactile hair of the white rat. J Comp Neurol 23:1–34
Wall PD, Gutnick M (1974) Ongoing activity in peripheral nerves: the physiology and pharmacology of impulses originating from a neuroma. Exp Neurol 43:580–593
Winkelmann RK (1968) New methods for the study of nerve endings. In: Kenshalo DR (ed) The skin senses. Thomas, Springfield, pp 38–57
Zucker E, Walker WJ (1969) Coding of somatic sensory input by vibrissae neurons in the rat's trigeminal ganglion. Brain Res 12:138–156

CHAPTER 11

Sweat Glands: Eccrine and Apocrine

K. J. COLLINS

RANVIER (1879) distinguished two main classes of gland in mammalian skin, the "holocrine" glands (such as sebaceous glands), in which cellular disintegration provides the secretory material, and the "merocrine" glands, in which the cells do not lose their structural integrity. The terms "eccrine" and "apocrine" [Κρίνειν, to separate (secrete); ἐκ, out of; ἀπο from] were introduced by SCHIEFFERDECKER (1917) to describe two types of simple tubular exocrine gland within the merocrine category. Eccrine glands were considered histologically to be those which discharged a fluid secretion without loss of cytoplasmic material, while the apocrine glands were in fact "semi-holocrine", with the luminal secretory cells forming protuberances which ruptured and discharged into the lumen with some of the cell contents. The "necrobiotic" stage in apocrine gland secretion was thought by SCHIEFFERDECKER to be succeeded by a phase during which a simple fluid secretion was produced while the cells returned to a low cylindrical form and recommenced their growth. Thus apocrine secretion appeared to share features of both holocrine and merocrine glands. This histological distinction between the eccrine and apocrine glands is not universally accepted: some consider the evidence that a necrobiotic phase occurs in the sweat glands of man or any other species to be equivocal (JENKINSON 1967). Further, electron microscopy has indicated that the histological differences are not so precise as was originally thought. An alternative terminology was suggested by BLIGH (1967) which differentiated sweat glands in all species by their association or otherwise with hair follicles. By this definition "epitrichial" (associated with a hair follicle/sebaceous gland unit) and "atrichial" become synonymous with apocrine and eccrine glands respectively.

Since the earlier classical reviews (ROTHMAN 1954; KUNO 1956; WEINER and HELLMANN 1960; MONTAGNA 1962) the major advances in elucidating the structure and function of the eccrine gland have come from electron microscopy (see HASHIMOTO 1978), micropuncture methods (SCHULZ 1969) and micro-induction of sweat secretion from the isolated gland in vitro (SATO 1977, 1987) allied to biochemical and biophysical studies of the secretion process (e.g. SLEGERS 1967). Few investigations have been performed to add to the fundamental studies of HURLEY and SHELLEY (1960) on apocrine secretion. The ultrastructure has been investigated in some detail but little work has been done on the physiology of apocrine sweat, one of the main difficulties being that it is easily contaminated by both sebaceous and eccrine secretions.

A. Anatomical Features

There are a number of morphological characteristics which demonstrate a close resemblance and others which clearly distinguish the two types of gland (Fig. 1). In the human, eccrine glands are found in great numbers (totalling more than 2×10^6) virtually over the entire body surface. Apocrine glands are concentrated in clusters in particular regions such as axillary, pubic, circumanal, ceruminal and circumareolar areas. The secretory segment of the eccrine gland is a tightly spiralled closed coil, while that of the apocrine gland is more loosely coiled, sometimes bearing diverticula. The apocrine secretory lumen is wide and in some animals such as cattle and sheep the apocrine gland may be only a simple sac-like structure. Both types have clearly differentiated ducts, the duct of the eccrine gland being more complicated and composed of a coiled segment in proximity to the secretory coil, a straight segment and a corkscrew terminal channel leading to an aperture on the skin surface. The apocrine gland duct, by contrast, is a short straight tube running up alongside the hair follicle to open into the follicular canal above the sebaceous gland. Apocrine glands are sometimes described as being larger than eccrine glands and this appears to be true for the functional glands in the human skin. Apocrine glands are also generally located deeper in the dermis.

There are two particular features of cell structure which are common to both types of sweat gland. Between the epithelial secretory layer and the basement

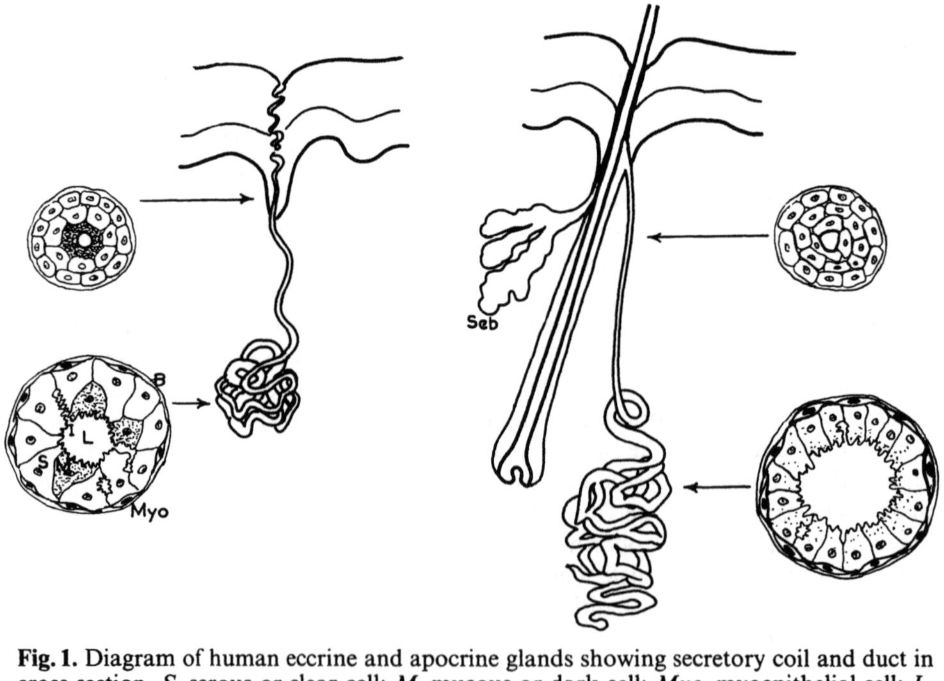

Fig. 1. Diagram of human eccrine and apocrine glands showing secretory coil and duct in cross-section. *S*, serous or clear cell; *M*, mucous or dark cell; *Myo*, myoepithelial cell; *L*, lumen; *I*, intercellular canaliculus; *B*, basement membrane; *Seb*, sebaceous gland. (Modified from Weiner and Hellmann 1960)

membrane in both varieties of gland there are longitudinally disposed smooth muscle fibres, the myoepithelium, whose function is considered below. In addition, the secretory cells of both eccrine and apocrine glands are found to have intercellular channels (canaliculi) communicating with the lumen (HASHIMOTO 1978).

Embryologically the eccrine glands develop as a cord of epithelial cells growing downward from the epidermal ridge early in embryonic life (12–16 weeks). Their formation is completed in late foetal life and physiological function commences at birth. HEY and KATZ (1969) found that the eccrine sweating mechanism was poorly developed in infants of less than 215 days post-conceptual age and that the thermoregulatory sweat response of the full-term infant increased markedly during the first 2 weeks after birth. Apocrine glands develop from the epithelium of hair germs and grow into the dermis in the 15th–20th week of embryonic life. The apocrine glands remain rudimentary and in most skin regions disappear after birth. In the "apocrine areas", however, they remain intact but do not develop further or become active until puberty.

B. Fine Structure

I. Eccrine Glands

The secretory coil contains three distinct cell types (Fig. 1): clear (sometimes described as secretory or serous cells), dark (or mucoid cells) and myoepithelial cells. The clear cells rest either directly on the basement membrane or on the myoepithelial cells and extend to the luminal border. The lateral borders of the clear cells are extensively plicated and the cytomembranes of two neighbouring cells form numerous slender projections (canaliculi) which become interdigitated. The membrane of the canaliculus is studded with microvilli (Fig. 2) which decrease in number when the canaliculus dilates during sweating. The luminal membrane of the clear cell is rarely exposed directly to the lumen because the luminal surface is almost totally lined by dark cells. Many mitochondria are contained in the clear cell, which closely resembles other fluid and electrolyte transporting cells and is considered to be responsible for secretion of the serous precursor of eccrine sweat. Another feature is the varying amounts of glycogen it contains; histochemically the level can be shown to become depleted after secretory activity (DOBSON 1962), but much of the glycogen is retained after "training" by heat acclimation (COLLINS et al. 1966).

The dark cells contain an abundance of mucous granules and ribosomes which stain with basic dyes and are electron-dense when visualised under the electron microscope. The granules appear to be produced in the Golgi region and as they grow larger the electron-dense material increases. During secretion there is an enhanced membrane activity associated with fusion of granules with the membrane, subsequently followed by expulsion of granules from the cell. Though it is believed that the mucosubstance in eccrine sweat is derived from the dark cell granules, there is as yet no clear biochemical proof of this.

The third cellular element of the eccrine secretory coil, the myoepithelial cells, lie on the basement membrane and about the clear cells (Fig. 3). Myoepithelial

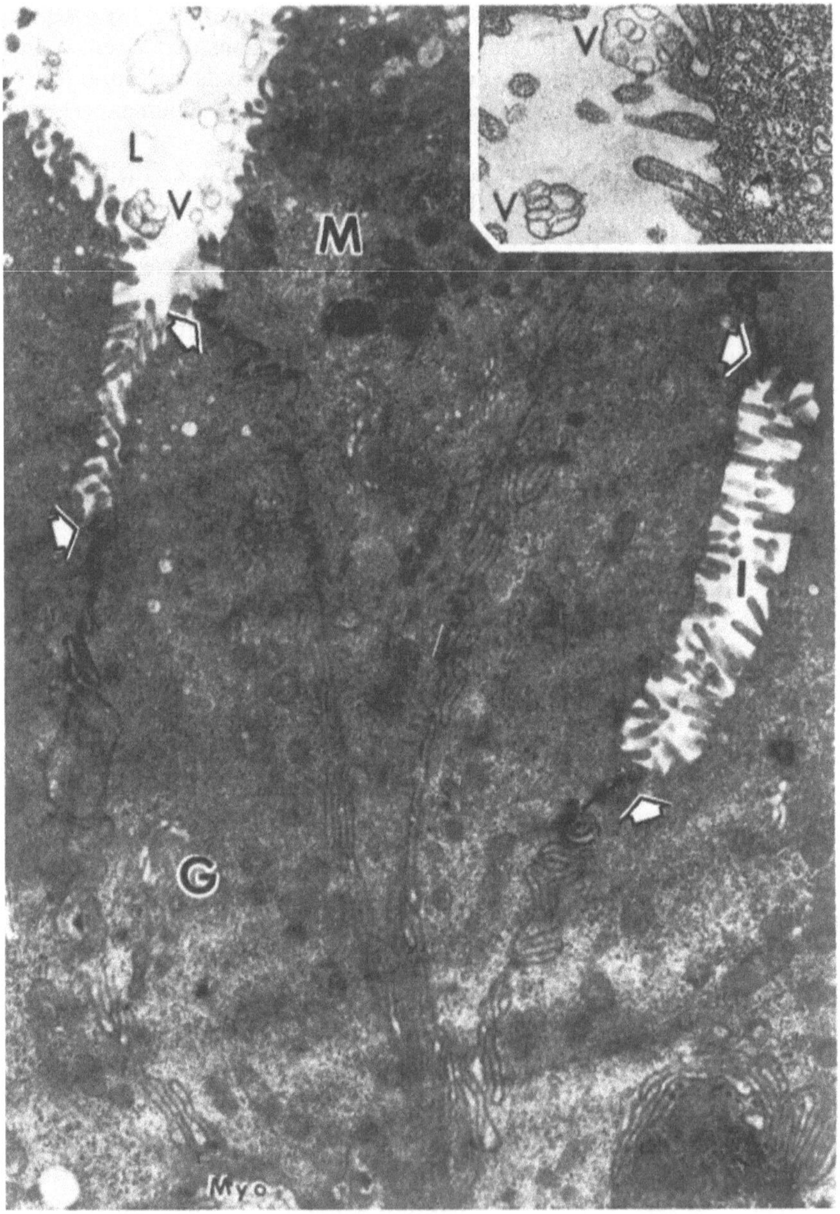

Fig. 2. Electron micrograph of eccrine gland secretory cells showing the microvilli-studded surface of an intercellular canaliculus (*I*) between two clear cells. Microvilli are also found on the surface of the lumen (*L*), of which the canaliculus is an extension. Intercellular channels leading to the canaliculus and to the lumen sealed by tight junctions (*arrows*). *V*, vesicles in the lumen; *G*, Golgi apparatus of a clear cell; *M*, dark cell; *Myo*, myoepithelial cell. (Hashimoto 1978)

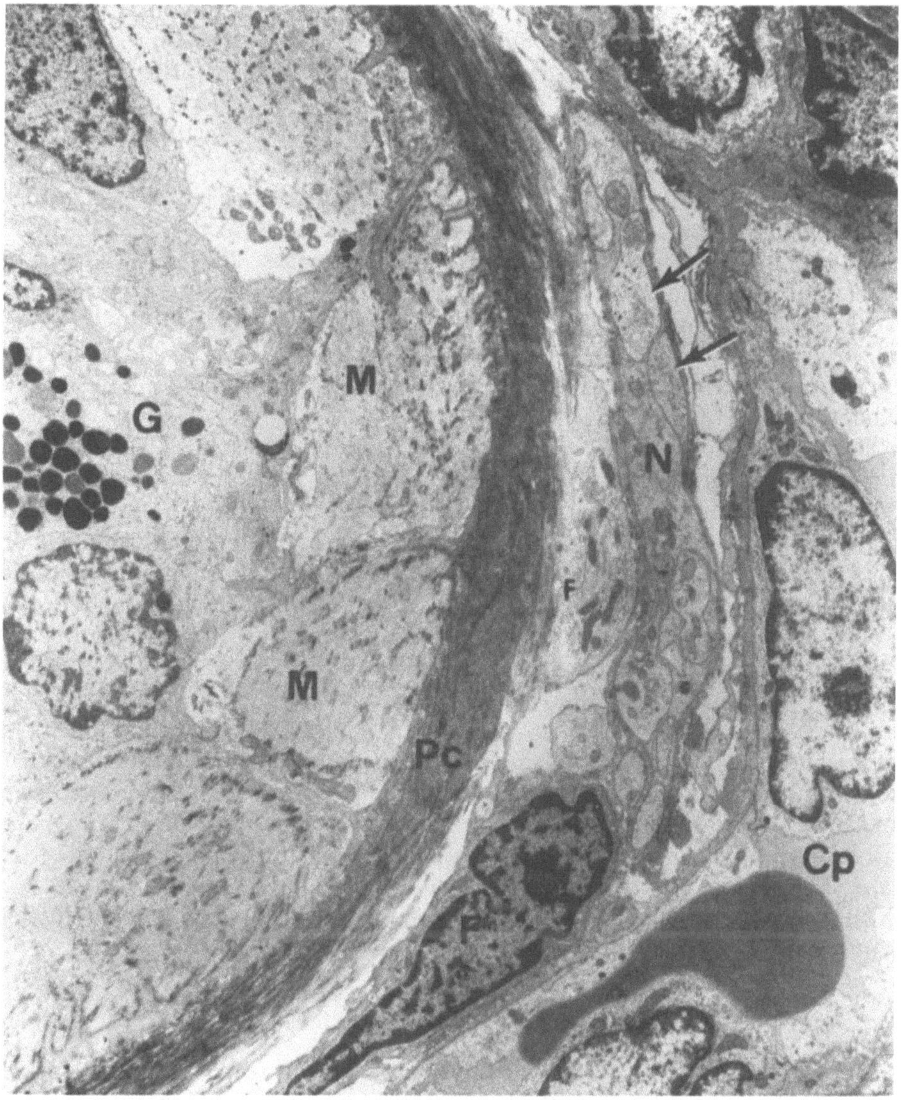

Fig. 3. Periglandular tissue of the eccrine gland shown by electron microscopy. *M*, myoepithelial cells with myofilaments; *G*, secretory cell; *Pc*, periglandular collagenous tissue; *F*, fibrocyte; *Cp*, capillary; *N*, cholinergic terminal nerve containing agranular vesicles (*arrowed*). (UNO 1977)

cells are composed of myofilaments arranged parallel to the course of the secretory tubules. The function of the myoepithelial cells is not clear. Suggested possible roles are that they act by expelling preformed sweat onto the skin by contraction, that they provide support for the sweat tubule, or that by contraction the intercellular space controlling the flow of fluid to the secretory epithelium is enlarged. SATO and co-workers claim that the amount of preformed sweat held in the luminal space is negligible and they discount the classical concept that the myoepithelial contraction contributes to eccrine secretory function by its pumping action (SATO et al. 1979; SATO 1987). They believe that the primary function is to provide structural support.

Between the coiled secretory and duct segments there is a transitional zone composed of a single layer of cylindrical luminal cells with no myoepithelial cells. The duct segment which originates from the transitional zone has two cell layers, luminal and basal cells, the luminal cells bearing numerous microvilli on the luminal surface and a cuticular border of dense tonofilaments. The lateral borders of the luminal cells are infolded and possess intercellular channels and the basal cells rest on the basement membrane, which is continuous with that of the secretory coil. The proximal coiled duct is functionally more active than the distal straight portion as judged by the Na-K-ATPase activity and the increased number of mitochondria (SATO et al. 1971).

The straight duct of the eccrine gland leads into the rete ridge and the intraepidermal segment which spirals to the eccrine pore. Here the wall is composed of an inner layer of luminal cells, the luminal border of which contains numerous microvilli, with a zone of two or three layers of outer cells and finally malpighian cells, basal cells and the basal lamina of the epidermis. The luminal cells have a large number of multivesicular bodies of various densities which are regarded as lysosomes involved in cytoplasmic digestion and lumen formation (HASHIMOTO 1978). Their presence in the mature gland indicates that lumen formation is constantly in progress as new luminal cells are produced. Luminal lysozymes may also be involved in the process of sweat reabsorption in the terminal duct. The luminal cells have, in addition, small keratohyaline granules which increase in number at the level of the lower epidermis. The eccrine pore in the epidermis is thus lined with keratinised cells or can sometimes be plugged with keratin material. Keratinisation may serve to prevent collapse of the terminal portion of the gland. It also provides the means for fluid reabsorption from the wetted skin, which, under conditions of high ambient humidity, leads to a reduction in eccrine gland secretion (COLLINS and WEINER 1962; SARKAY et al. 1965).

II. Apocrine Glands

In contrast to the eccrine secretory coil, only one type of secretory cell can be recognised in the apocrine coil. In the presecretory stage, the cells contain numerous granules which disappear after secretion. The single layer of columnar secretory cells rests on myoepithelial cells and on the basal lamina. Intercellular canaliculi occur as in eccrine glands. There is also a well-marked collagenous capsule surrounding the apocrine coil. Dense granules are formed in the secretory cells, some starting as Golgi-related vesicles and increasing in size, others being derived from mitochondria.

Different stages of merocrine and holocrine secretion can be observed in electron microscopic studies of the secretory cells. Decapitation secretion can be seen in mature cells where the luminal portion of the cells becomes swollen and devoid of large granules. At the base of the swelling, the cell membrane becomes invaginated and constriction develops (Fig. 4). Eventually the apical cap becomes de-

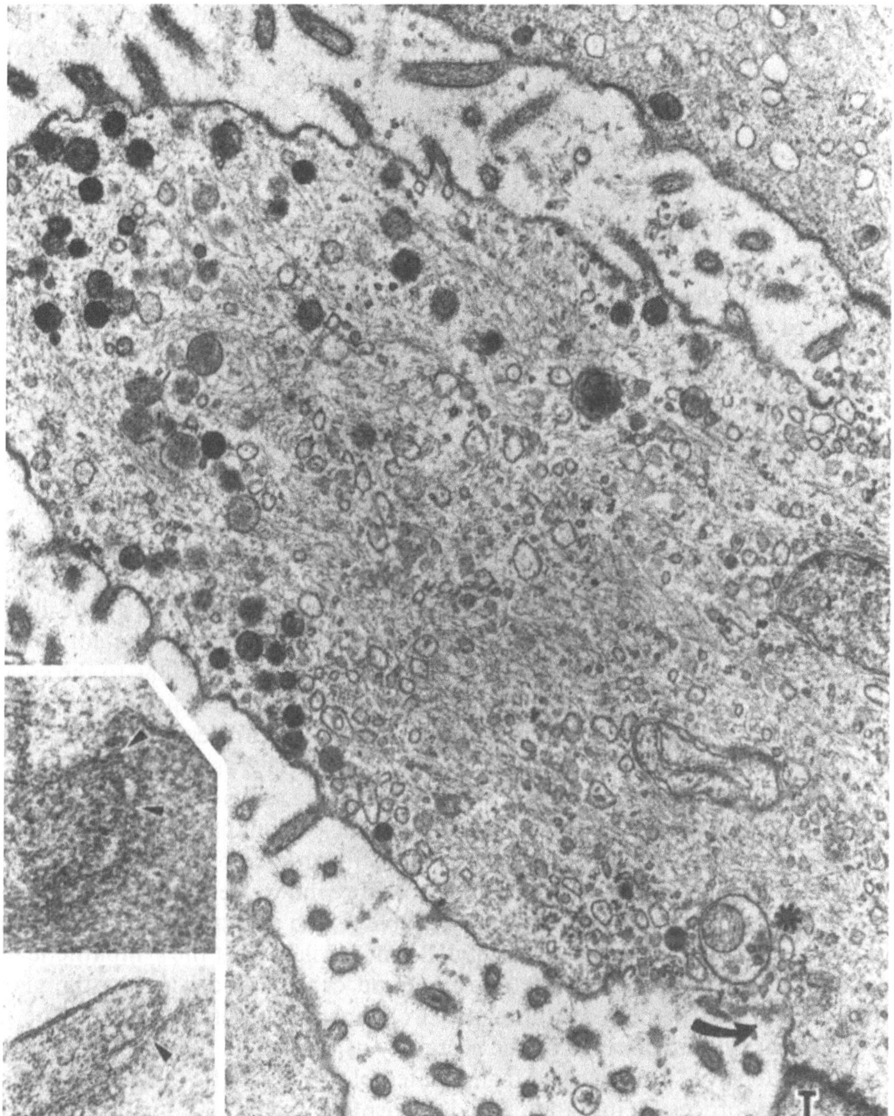

Fig. 4. Electron micrograph of decapitation secretion in an apocrine secretory cell from human axilla. At the base of the apical cap are numerous vesicles and vacuoles (*). The cell membrane shows invagination into the cytoplasm (*arrows*). *T*, luminal end of intercellular channel sealed by a tight junction. (HASHIMOTO 1978)

tached into the lumen. In some apical caps the cell membrane becomes fragmented and part of the contents of the secretory cells appears to be released.

Myoepithelial cells line the outermost layer of the human secretory apocrine tubule; the myoepithelial structure is described as being better developed than in the eccrine gland (ROTHMAN 1954). According to BLIGH (1967) and JENKINSON

(1973) this may reflect a more important role of myoepithelial contraction in the apocrine gland, although Sato (1980) maintains that though myoepithelial contraction may contribute to the initial transient expulsion, the apocrine myoepithelium, like that of the eccrine gland, primarily performs the role of giving structural support.

There are neither secretory nor myoepithelial cells in the apocrine sweat duct. Luminal cells of the duct, however, bear numerous short microvilli and in spite of the fact that reabsorption appears not to be a major function of the proximal duct, there are numerous infolded intercellular channels and mitochondria in the cells. Instead of the two-cell layers seen in the eccrine duct, the apocrine duct wall consists of three layers of cells.

C. Innervation

I. Eccrine Glands

Human eccrine glands are richly innervated by unmyelinated sympathetic nerve fibres which react strongly in tests for acetylcholinesterase (Hellmann 1952). The postganglionic sudomotor nerves are cholinergic, as demonstrated by the classical studies of Dale and Feldberg (1934). These fibres branch repeatedly in the skin and are distributed to the secretory coil of the gland as well as to the coiled portion of the duct. Free endings of the peripheral sympathetic nerves in relation to the secretory cells have rarely, if ever, been visualised. Instead, there is a dense network of very fine fibres, described by Hillarp (1959) as a "ground plexus", in the vicinity of the glands. Individual fibres exhibit a series of varicosities strung out along their length and these represent the presynaptic terminal endings (Fig. 3). Each effector cell is therefore probably innervated by many fibres arising from different sudomotor neurones and each fibre innervates many effector cells. The multiple innervation may also readily give rise to spatial and temporal summation in sweat gland responses. Acetylcholine is released at the sympathetic terminals in close proximity to the eccrine gland capsule and it enters the gland through the basal lamina and interstitial channels (Hashimoto 1978). The infolded basal cytomembranes of the secretory cells become depolarised, sodium ions and interstitial fluid derived from dilated blood vessels flow into the cell and the secretory process is initiated. Fluid absorption from the interstitial space is greatly facilitated by the large absorption surface of the secretory cells.

The coexistence of two putative transmitters, e.g. peptides and classical transmitters, in the same neurone has been demonstrated by immunohistochemical techniques in sweat gland neurones (Hokfelt et al. 1980). A dense network of vasoactive intestinal polypeptide (VIP) immunoreactive fibres has been identified around the sweat glands of the cat foot-pad, and these fibres are also rich in acetylcholinesterase. It is suggested that acetylcholine produces secretion by an atropine-sensitive action on secretory cells and that a local dilatation occurs by a synergistic atropine-resistant action of VIP. In earlier studies, Fox and Hilton (1958) demonstrated that sweat contained the vasoactive peptide bradykinin and proposed that this may increase blood flow in the neighbourhood of an active gland. VIP has a well-established vasodilatory effect and it may also influence local blood flow indirectly by activation of a kinin system.

UNO and MONTAGNA (1975) have demonstrated a loose network of catecholamine-containing nerves around the eccrine glands of the monkey paw. This observation has revived the long-standing theory of a dual adrenergic and cholinergic innervation which was first proposed when catecholamines were found to have some secretory effect on the eccrine glands (HAIMOVICI 1950). The demonstration of adrenergic nerves in the vicinity of the monkey-paw glands may, of course, be simply a species-specific difference. However, in a later histofluorescent and electron microscopic study of human eccrine glands from the skin of the back and chest, UNO (1977) described numerous cholinergic and a few adrenergic terminals around the secretory coils. Cholinergic terminals were also found along the tubules of the duct.

Of relevance to the dual innervation theory are the developmental changes in the properties of cholinergic sympathetic neurones innervating the sweat glands of the rodent foot-pad described by LANDIS and KEEFE (1983). The developing eccrine glands appear to be innervated initially by sympathetic noradrenergic axons. These lose their store of endogenous catecholamines but not their capacity for uptake and storage as the axonal plexus is elaborated. The neurones thus appear to undergo a transition from noradrenergic to cholinergic function during development.

II. Apocrine Glands

MONTAGNA (1962) found cholinesterase-containing nerve fibres around the axillary apocrine glands in man, as also did ROTHMAN (1954), though this finding cannot be considered a characteristic feature of cholinergic fibres. CAHN and SHELLEY (1955) studied the innervation of the axillary apocrine glands by methylene blue and silver staining techniques and claimed to demonstrate multiple neurofibrillae forming a meshwork about the membrane of the secretory coil. Like the eccrine gland, none of these neurofibrillae appeared to penetrate the basement membrane. SHELLEY and HURLEY (1953) also reported that human apocrine glands responded to catecholamines rather than acetylcholine and were innervated exclusively by adrenergic nerves, a conclusion which was disputed by ROTHMAN (1954). The possibility has also been raised that human apocrine glands like those of some mammals, e.g. the horse, are under the control of circulating adrenaline.

The limitations of the earlier investigative techniques have been emphasised by advances made in the last decade with electron microscopy, intracellular electrophysiology and fluorescence histochemistry. Different types of autonomic nerve can be identified according to the types of vesicle they contain. Using histofluorescence and electron microscopy, UNO (1977) observed adrenergic fibres occurring sporadically in the connective tissue of the apocrine glands of the human axilla. None formed a dense network around the glands but the evidence favoured a few adrenergic fibres innervating the apocrine secretory coil. Cholinergic terminals were also found to be present around both the secretory tubules and ducts.

The conclusion drawn by UNO (1977) was that both the eccrine and the apocrine glands of man are innervated by predominantly cholinergic terminals with a few adrenergic terminals. To some extent this arrangement of the neural network with cholinergic dominance in a dual innervation corresponds to the pharmacological responsiveness of the two types of sweat gland, both of which appear to respond to cholinergic and adrenergic agents, the former being a stronger stimulant of secretion (SATO 1977, 1980). This view, however, is not endorsed by ROBERTSHAW (1974), who maintains that apocrine glands of all species are more responsive to catecholamines and are more readily stimulated by adrenaline than by noradrenaline or acetylcholine.

D. Secretory Function of Eccrine Glands

Few species possess a mechanism of heat dissipation quite as effective as that of man, who is capable of sustaining a rate of sweat secretion as high as 1 litre per hour from the eccrine glands on the body surface for some hours. Completely evaporated, 1 litre eliminates 675 W of thermal energy. The thermoregulatory function of the sweat glands is dealt with in more detail elsewhere (Chap. 12, p. 217) but here the secretory process of the eccrine gland will be considered. Two functional types of eccrine gland are usually recognised in human skin; those on the general body surface, which are concerned with heat regulation and respond gradually in accordance with the intensity of afferent thermal stimulation, and those on the palms and soles, which respond rapidly to mental, emotional and sensory stimuli. Other areas of the skin will react to emotional stimuli but palmar and plantar glands are usually the first to become active and are the only ones to do so if the mental stimulus is weak (KUNO 1956). Eccrine glands of the axilla respond to mental stimuli, as do those of the forehead, though here the secretory activity is said to be weaker than on the palms. In the case of extreme mental excitement, fear or violent sensory stimulation, there is often a profuse and sudden outbreak of sweat on the general body surface (cold sweat) as well as on the normal "emotional" sweating areas. Palmar and plantar glands do not normally react to thermal stimulation unless the heating is sufficiently intense to cause emotional sweating. Thus there appears to be a degree of overlap in the central control and functional activity of "thermal" and "emotional" eccrine glands.

It has often been observed that the maximum sweat rate is greater in men than in women (e.g. SHAPIRO et al. 1981). This difference has sometimes been explained on the basis of differences in body weight, surface area or the degree of sweat gland training (acclimation). In order to study these differences and the underlying cause, REES and SHUSTER (1981) measured sweat rates induced by pilocarpine before and after puberty and in response to androgens and anti-androgens. The difference in sweat rate corrected to adult surface area was not observed in prepubertal boys and girls. Local administration of androgens did not increase sweat rate in the adult female nor did anti-androgen compounds decrease sweat rate in males. It is suggested that the sweat rate difference between males and females stems from endocrine changes in the skin due to gene expression during a critical stage of development, which is initiated by androgens at puberty but is not maintained by androgens in adults. This interpretation was confirmed by the finding of a low (pre-pubertal) sweat rate in males with hypogonadism of pre-pubertal origin (WALTON et al. 1983) which increased with androgen.

The assessment of thermal, fluid and electrolyte balances during sweating is or primary importance in hot environmental conditions. The measurement of sweat loss and the composition of sweat secreted onto the skin surface depends largely on satisfactory methods for collection and on the conditions under which the sweat is sampled (see also Chap. 3 in Vol. 87/II). Earlier surveys of sweat composition (e.g. S. ROBINSON and A. H. ROBINSON 1954; ROTHMAN 1954; WEINER and HELLMANN 1960) still serve as satisfactory estimates of total fluid and chemical constituents lost by the body during thermoregulatory eccrine sweating. The salient feature about sweat secreted onto the skin surface is that it is hypotonic compared to extracellular fluid; sodium and chloride ions account for about 90% of its osmotic activity. Sodium chloride concentration is usually in the range 0.1–0.4 g dl^{-1} (17–68 m*M*) and this concentration will vary according to the rate of sweating, the level of skin temperature, dietary intake of salt, adrenocortical hormone activity, sweat suppression (hidromeiosis) and the state of heat acclimation, all of which are factors that have been shown to alter the composition of sweat. Potassium concentration in sweat is usually hypertonic (4–24 m*M*) compared with plasma. The main non-electrolytes in eccrine sweat are lactate (10–40 m*M*), urea (from 1.2 to 4 times the plasma level), ammonia (0.5–8 m*M*), amino acids [usually greater than the plasma concentration but some, e.g. glutamine and cystine, less than plasma levels (GITLITZ et al. 1974)] and macromolecules (proteins, glycoproteins and mucopolysaccharides). In spite of the fact that sweat glands on the general body surface and those on the palms (and soles) respond to different stimuli, the composition of sweat collected simultaneously from the two regions does not differ substantially (COLLINS 1962).

I. The Secretory Process and Sweat Formation

Studies on the function of isolated eccrine glands during the last two decades have helped to elucidate the probable mechanism of sweat secretion. The process involves two primary stages, the formation of an isotonic precursor fluid in the secretory coil and the reabsorption of sodium in the sweat gland duct. The use of cryoscopic techniques (SLEGERS 1963; BRUSILOW and GORDES 1964), micropuncture (SCHULZ et al. 1965) and isolated sweat glands (SATO 1973a, 1977) has verified that precursor sweat secreted in the secretory coil is isotonic with respect to interstitial periglandular fluid. The formation of the precursor sweat, though not completely understood, requires the active transport of sodium from secretory cells followed by the passive flow of chloride and water to restore osmotic equilibrium. Both aerobic and anaerobic glycolysis supply the energy for this process, which is inhibited by ouabain (SATO and DOBSON 1971, 1973). Sodium- and potassium-activated adenosine triphosphatase (Na-K-ATPase) is present and highly active in the secretory coil (SATO and DOBSON 1970). The localisation of sodium transport sites in the forward-pumping secretory system of the secretory cells is thought to be in the canalicular walls between the clear cells (SLEGERS and van't HOF-GROOTENBOER 1971).

The energy metabolism of the eccrine gland has been studied by many workers. Weiner and van Heyningen (1952) suggested that there was active glycolytic metabolism in the sweat gland and that the high lactate concentration in sweat was derived not from plasma but from the metabolism of the sweat gland itself. That lactate can also be shown to be produced by a secreting isolated eccrine sweat gland in vivo and in vitro (Schulz et al. 1965; Wolfe et al. 1970; Sato and Dobson 1971, 1973) gives support to this theory. Wolfe et al. (1970) have even postulated that anaerobic metabolism of glucose to lactate is the only metabolic pathway for providing energy for the secretory function, but this has been shown by Sato (1977, p. 89) to be unlikely. The weight of evidence at present appears to favour oxidative phosphorylation of plasma glucose with ATP formation as the main route of energy production, while glycolysis and lactate production play a secondary role. Lactic acid formation may represent a more important pathway of sweat gland metabolism and energy production in relatively anoxic conditions such as may occur during mental sweating on the palms when skin blood flow is low. According to Collins (1962) the concentration of lactate in palmar sweat is higher than that in sweat from the forearm, and the concentration of lactate in thermally induced sweat increases in proportion to the severity of occlusion of the local blood supply (van Heyningen and Weiner 1952).

The absorption of sodium and chloride due to active transport of sodium ions in the luminal cells of the sweat gland duct is well established (Sato 1977; Morimoto 1978). The electrochemical potential gradient across the duct favours passive movement of chloride ions across the duct epithelium. In the isolated sweat gland from the palmar surface of the Rhesus monkey, which produces a hypotonic sweat similar to that of the human eccrine gland, sweat sampled from the secretory coil has approximately the same sodium and potassium concentrations as that of the incubation medium, while sweat from the proximal duct contains a considerably lower sodium but a higher potassium concentration than that from the secretory coil (Sato 1973 a). Ouabain has been shown to completely inhibit ductal reabsorption of sodium and chloride as well as sweat secretion in both in vivo and isolated sweat gland studies. Na-K-ATPase has also been demonstrated in both the secretory and duct segments of the eccrine gland. Slegers (1967) has introduced the concept of a two-stage sodium reabsorption process in the duct, a fast reabsorption process in the proximal (coiled) portion of the duct and a slower process in the distal (straight) part of the duct. This has an important bearing on the action of adrenocortical steroids, for Collins (1966) and subsequently others have demonstrated that aldosterone administered locally or systemically lowers sweat sodium concentration in normal subjects. In its time course and efficacy, the action of exogenous aldosterone on the sweat glands closely resembles that of the endogenous hormone secreted in response to dietary restriction (Shuster 1962) or salt and fluid imbalance produced by excessive sweating (Collins and Weiner 1968). Aldosterone appears to act at least partly on the distal duct; the cat-pad glands which lack a well-differentiated distal duct do not respond to aldosterone by sodium conservation (Collins et al. 1970). In patients suffering from cystic fibrosis it is the proximal part of the eccrine gland duct which appears to be primarily affected. Thus, an explanation is afforded for the fact that these patients secrete sweat containing an abnormally high sodium concentration but also respond to the sodium-conserving action of aldosterone (GRAND et al. 1967). Though the work of SLEGERS (1967), SATO (1977, 1987) and others has recently contributed much to the understanding of ductal function, the detailed mechanism of active sodium reabsorption has not as yet been fully explained.

E. Apocrine Gland Function

Most mammals possess apocrine glands clustered in special "apocrine" sites like man, but there are also many animals possessing functional apocrine glands distributed over most of the body surface. In a few species apocrine sweating is accredited with a specific thermoregulatory function; the best known example of this is in the horse (LOVATT EVANS and SMITH 1956), where failure of apocrine sweat secretion on the body leads to a condition of "dry coat" which bears a resemblance to the anhidrosis associated with eccrine gland failure in man. Other animals such as cattle have also been attributed with thermoregulatory apocrine sweating (BLIGH 1967). In many species, apocrine glands located in special sites, e.g. perianal or inguinal glands, are regarded as scent glands which come into activity with phases of the sexual cycle. In man, apocrine glands in the "apocrine areas" do not appear to have either a definite thermoregulatory or sexual function, though they do not develop fully until puberty and are said to involute after the menopause (ROTHMAN 1954). Cyclic activity of the glands related to the human menstrual cycle has not been established. SHELLEY et al. (1953) claim that human apocrine secretion is odourless and that odour only develops as the result of bacterial contamination. As in many other species, emotional stimuli in man can evoke a rapid discharge of the apocrine glands (SHELLEY and HURLEY 1953), but if the secretion is odourless when it is newly discharged it is difficult to attribute to it a sexual role. It was concluded by WEINER and HELLMANN (1960) that human apocrine glands are best regarded as vestigial in function.

There is equally little precise knowledge of the composition of human apocrine sweat. A technique of sampling individual glands by capillary tube is required because eccrine glands are interspersed with apocrine, and in the axilla, for example, both types of gland respond to emotional stimuli. Furthermore, the sebaceous glands also empty their oily secretion into the hair follicle, at a point just below the apocrine duct. Qualitatively, SHELLEY and HURLEY (1953) found that uncontaminated axillary apocrine sweat collected directly from follicle openings was milky, translucent (or soapy) and contained protein, reducing sugar and ammonia. The range of electrolyte concentration is not clearly documented, though the evidence is that it is hypotonic like eccrine sweat (WEINER and HELLMANN 1960). In some individuals, apocrine sweat is pigmented red or yellow (chromhidrosis); the colour has been attributed to the action of skin micro-organisms but SHELLEY and HURLEY (1954) have reported cases where no chromogenic bacteria are to be found in cultured specimens of coloured apocrine sweat. The apocrine fluid collected by SHELLEY and HURLEY (1953) was also reported to fluoresce in ultraviolet light. The material responsible for the fluorescence has not been identified though it is possible that it may have derived from sebaceous glands through their common orifice with the apocrine glands. ROTHMAN (1954, p. 186) comments that he has observed strong fluorescence of sebum fractions obtained from the hairy scalp, where apocrine glands are not present.

F. Pharmacology of Sweating

I. Eccrine Glands

The fundamental cholinergic nature of the sympathetic innervation of the eccrine glands was shown by the early classic experiments of DALE and FELDBERG (1934) using the cat's paw preparation. Many subsequent experiments have demonstrated the response of human eccrine glands to acetylcholine and allied substances and the blocking action of atropine (e.g. CHALMERS and KEELE 1952; RANDALL and KIMURA 1955; COLLINS et al. 1959). Intravenous administration of 0.5 mg atropine sulphate is sufficient to inhibit thermal sweating within 2 or 3 min (CRAIG and CUMMINGS 1965). Human eccrine glands, however, have also long been known to respond to intradermally administered catecholamines (WADA 1950; HAIMOVICI 1950) and also to intra-arterial though not intravenous infusion of adrenaline (FOSTER et al. 1970; FOSTER and WEINER 1970). The question has therefore been raised whether spontaneous sweating in thermogenic areas in patients with a phaeochromocytoma may be at least partly due to the direct stimulation of the glands by circulating adrenaline. This suggestion has been discounted by FOSTER et al. (1970), who observed that the dose of intra-arterial adrenaline required to stimulate sweating is rather high, and by PROUT and WARDLE (1969), who abolished sweating in phaeochromocytoma by hyoscine. Local adrenaline-induced sweating is quantitatively only about 10% of that produced by parasympathomimetics. It is blocked by Priscol or dibenamine and not by atropine, which suggests specific adrenergic receptors in the glands.

The use of specific α- and β-adrenergic blockers has not completely clarified the problem. α-Blockers such as phentolamine and guanethidine have been shown to inhibit adrenaline-induced sweating, but these substances are also reported to possess strong anticholinergic properties (FOSTER and WEINER 1970). HEMELS (1970) claimed that in non-atopic subjects the β-blocker propanolol increased the intensity of response to local injections of acetylcholine and suggested that an adrenergic β-receptor performed an inhibitory role in normal cholinergic sweating. This result may again be explained by a feeble anticholinergic action of propanolol. The isolated eccrine gland from the simian palm which provides a model for the human gland (SATO 1977, 1987) responds to parasympathomimetics as well as to both α-adrenergic and β-adrenergic stimulation. Acetylcholine produced the most powerful stimulant effect and the adrenergic agonists were inhibited by their respective antagonists. There was, however, some cross-inhibition between α- and β-antagonists in both cholinergic and adrenergic responses.

Though the dual innervation theory has been revived by the observation of UNO and MONTAGNA (1975) that there are both adrenergic and cholinergic periglandular nerves, the question remains as to the possible role of the adrenergic component. That the myoepithelial cells receive a specific adrenergic innervation is one possibility (ROBERTSHAW 1974). SATO et al. (1979) have, however, demonstrated that in the isolated eccrine gland the myoepithelial cells contract in response to cholinergic stimuli and not to α- or β-adrenergic agents. One further observation from the same laboratory (SATO 1973b) may be of significance. Both adrenaline and the β-adrenergic agonist isoproterenol significantly enhanced the amount of glucose metabolised via the pentose cycle and this was inhibited by propanolol. SATO (1977, 1987) tentatively suggests that the control of glandular growth (including hypertrophy) and morphological changes in gland cells during secretory activity may be a role subserved by the β-adrenergic component. An analogy may be drawn with salivary glands: parotid and submaxillary glands have a dual innervation and can be shown to hypertrophy with increased nuclear division as a result of treatment with β-adrenergic agents. The sublingual gland which lacks adrenergic innervation is unaffected (SEIFERT 1967). The fact remains that adrenergic blocking agents do not inhibit thermoregulatory or emotional eccrine sweating although complete inhibition may be produced by atropine.

II. Neurohumoral Aspects

The pharmacological properties of eccrine glands are clearly important to their functional role in thermoregulatory sweating, especially in patients undergoing therapeutic treatment with drugs acting on the autonomic nervous system. There are a number of other aspects of neurohumoral control which also throw light on the function of the glands.

1. Denervation

Eccrine glands have frequently been described as providing an exception to Cannon's rule of hypersensitivity to transmitter substances following postganglionic denervation. This has been demonstrated most consistently in human eccrine sweat glands (RANDALL and KIMURA 1955; SILVER et al. 1963), with the response to acetylcholine or adrenaline gradually diminishing after denervation. In the cat pad eccrine glands, however, there is evidence that the glands do conform to Cannon's law. REAS and TRENDELENBURG (1967) observed two phases in the changes of sensitivity to acetylcholine and pilocarpine after denervation in the cat's paw: reduced sensitivity during the first two postoperative days followed by hypersensitivity from the fourth postoperative day onward. On the basis of pre-treatment with stimulant or blocking drugs these investigators concluded that the sensitivity was inversely related to the activity of the glands: inactivity caused by blocking the nerves caused hypersensitivity, while daily pre-treatment with pilocarpine caused reduced sensitivity.

2. Axon Reflex Sweating

Nicotine injected intradermally was used by COON and ROTHMAN (1941) to demonstrate a local reflex sweat response similar to the well-known axon reflex flare of the Lewis triple response. This response can be produced in man and in the cat's paw by local injections of nicotine and acetylcholine but not with methacholine, and it is blocked by a high concentration of nicotine, by hexamethonium and by curare (COLLINS and WEINER 1961). In their response to these substances the neuroglandular units are found to resemble sensory receptor responses producing axon reflexes in the skin and mesentery. Branching terminals of the sympathetic nerve supply to the glands appear to furnish the pathways for the sweating axon reflex but there is no evidence that this local response is concerned with normal thermogenic sweating. Sweat reactions sometimes observed in the periphery of ulcers, whitlows or in other inflammatory skin conditions may, however, be due to stimulation of the axon reflex pathways.

3. Hyperhidrosis

Excessive sweat production in the emotional sweating areas (essential hyperhidrosis) is thought to be produced by an exaggerated central drive to the eccrine glands. Though usually restricted to the palms, soles and axillae, it sometimes occurs over the general body surface in areas that normally react only to thermal stimuli (ALLEN et al. 1974). In the axilla, in spite of the presence of well-developed apocrine glands, hyperhidrosis is basically due to intense activity of the eccrine glands (SHELLEY and HURLEY 1975). The use of anticholinergic agents such as atropine, scopolamine and probanthine has long been a popular form of treatment. Their local or systemic administration has, however, non-specific effects such as dry mouth, blurred vision and urinary retention which are often unacceptable. The few attempts made to use anti-adrenergic agents in the treatment of hyperhidrosis have, not surprisingly, met with little success. In the axilla, surgical excision of the well-localised hyperhidrotic area is effective, and transaxillary sympathectomy may be used for suppression of both palmar and axillary hyperhidrosis (ELLIS 1979). Hyperhidrosis is a self-limiting condition though it may last for many years, and surgical intervention may therefore be regarded as something to be avoided if possible. Topical sweat suppressants are commonly used for treatment, especially in mild to moderately severe cases. Methenamine, believed to act by slow release of formaldehyde, has been used successfully even in patients with known formaldehyde sensitivity (CULLEN 1975). Aluminium salts which block the sweat duct in the stratum corneum by protein precipitation are also effective. SHELLEY and HURLEY (1975) claim that complete control of axillary hyperhidrosis can be achieved by a careful regime involving (a) local application of 20% aluminium chloride hexahydrate in anhydrous ethyl alcohol, (b) application when the axillary glands are inactive and (c) covering the area for 6–8 h with impermeable film after application.

III. Apocrine Glands

HURLEY and SHELLEY (1954) proposed that as human axillary apocrine glands can be made to expel their preformed sweat by intradermal or intravenous injections of adrenaline and that sweat cannot be likewise produced by parasympathomimetic drugs, nor blocked by atropine, then a predominantly adrenergic control system was involved. However, AOKI (1962) was able to stimulate apocrine secretion by acetylcholine, acetyl β-methylcholine (methacholine) and carbaminoylcholine (carbachol). The parasympathomimetics caused profuse eccrine sweating also in the axilla, but in some subjects this was only slight and apocrine secretion could be easily identified and collected in capillary tubes. ROBERTSHAW (1974) is of the opinion that this is of minor importance since the apocrine sweating was only produced with a high concentration of acetylcholine and the amount produced was very small.

The apocrine gland has a well-developed myoepithelium and the secreting tubule has a large luminal space in which to retain pre-formed sweat. In the isolated apocrine gland the myoepithelium contracts in response to α-adrenergic stimulation (phenylephrine) but not to β-adrenergic stimulation (isoproterenol) (K. SATO and F. SATO 1979; SATO 1980). However, α-adrenergic stimulation was found to be the least effective stimulant of apocrine sweat secretion. Acetylcholine in the physiological range of concentration did not produce myoepithelial contraction though higher concentrations evoked a small contraction. The possibility therefore exists that secretory cell function and myoepithelial contraction may be dissociated pharmacologically (SATO 1980), with cholinergic and β-adrenergic elements controlling secretion and α-adrenergic stimulation controlling myoepithelial function.

The expulsion of sweat from apocrine glands in response to adrenaline appears to occur in one or two short bursts, which suggests myoepithelial contraction and expulsion of pre-formed material (HURLEY and SHELLEY 1954). In the eccrine glands, both adrenaline and noradrenaline cause slight but prolonged secretion lasting an hour or more. The myoepithelium is also less well developed in eccrine glands and it is less likely that here the myoepithelial cell's primary function is to expel sweat.

References

Allen JA, Armstrong JE, Roddie IC (1974) Sweat responses of a hyperhidrotic subject. Br J Dermatol 90:277–281

Aoki T (1962) Stimulation of human axillary apocrine sweat glands by cholinergic agents. J Invest Dermatol 38:41–44

Bligh J (1967) A thesis concerning the processes of secretion and discharge of sweat. Environ Res 1:28–45

Brusilow SW, Gordes EH (1964) Solute and water secretion in sweat. J Clin Invest 43:477–484

Cahn MM, Shelley WB (1955) Hyaluronidase-methylene blue staining of nerve fibres about the human axillary apocrine sweat gland. J Invest Dermatol 25:63–66

Chalmers TM, Keele CA (1952) The nervous and chemical control of sweating. Br J Dermatol 64:43–54

Collins KJ (1962) Composition of palmar and forearm sweat. J Appl Physiol 17:99–102

Collins KJ (1966) The action of exogenous aldosterone on the secretion and composition of drug-induced sweat. Clin Sci 30:207–221

Collins KJ, Weiner JS (1961) Axon reflex sweating. Clin Sci 21:333–344

Collins KJ, Weiner JS (1962) Observations on arm-bag suppression of sweating and its relationship to thermal sweat-gland fatigue. J Physiol (Lond) 161:538–556

Collins KJ, Weiner JS (1968) Endocrinological aspects of exposure to high environmental temperatures. Physiol Rev 48:785–839

Collins KJ, Sargent F, Weiner JS (1959) Excitation and depression of eccrine sweat glands by acetylcholine, acetyl β-methylcholine and adrenaline. J Physiol (Lond) 148:592–614

Collins KJ, Crockford GW, Weiner JS (1966) Local training effect of secretory activity on the response of eccrine sweat glands. J Physiol (Lond) 184:203–214
Collins KJ, Foster KG, Hubbard JL (1970) Effect of aldosterone on mammalian eccrine sweat glands. Experientia 26:1313–1314
Coon JM, Rothman S (1941) The sweat response to drugs with nicotine-like action. J Pharmacol Exp Ther 23:1–11
Craig FN, Cummings EG (1965) Speed of action of atropine on sweating. J Appl Physiol 20:311–315
Cullen SI (1975) Topical methenamine therapy for hyperhidrosis. Arch Dermatol 111:1158–1160
Dale HH, Feldberg W (1934) The chemical transmission of secretory impulses to the sweat glands of the cat. J Physiol (Lond) 82:121–128
Dobson RL (1962) The correlation of structure and function in the human eccrine sweat gland. In: Montagna W, Ellis RA, Silver AF (eds) Advances in biology of skin, vol 111. Appleton-Century-Crofts, New York
Ellis H (1979) Transaxillary sympathectomy in the treatment of hyperhidrosis of the upper limb. Ann Surg 45:546–551
Foster KG, Weiner JS (1970) Effects of cholinergic and adrenergic blocking agents on the activity of the eccrine sweat glands. J Physiol (Lond) 210:883–895
Foster KG, Ginsburg J, Weiner JS (1970) Role of circulating catecholamines in human eccrine sweat gland control. Clin Sci 39:823–832
Fox RH, Hilton SM (1958) Bradykinin formation in human skin as a factor in heat vasodilation. J Physiol (Lond) 142:219–232
Gitlitz PH, Sunderman FW, Hohnadel DC (1974) Ion-exchange chromatography of amino acids in sweat collected from healthy subjects during sauna bathing. Clin Chem 20:1305–1312
Grand RJ, di Sant' Agnese PA, Talamo RC, Pallavicini JC (1967) The effects of exogenous aldosterone on sweat electrolytes. II. Patients with cystic fibrosis of the pancreas. J Pediatr 70:357–368
Haimovici H (1950) Evidence for adrenergic sweating in man. J Appl Physiol 2:512–521
Hashimoto K (1978) In: Jarrett A (ed) The physiology and pathophysiology of the skin, vol 5. Academic, London, p 1544–1589
Hellmann K (1952) The cholinesterase of cholinergic sweat glands. Nature 169:113
Hemels HGWM (1970) The effect of propanolol on the acetylcholine-induced sweat response in atopic and non-atopic subjects. Br J Dermatol 83:312–314
Hey EN, Katz G (1969) Evaporative water loss in the new born baby. J Physiol (Lond) 200:605–619
Hillarp NA (1959) The construction and functional organisation of the autonomic innervation apparatus. Acta Physiol Scand [Suppl] 157:1–38
Hokfelt T, Johnannson O, Ljungahl A, Lundberg JM, Schultzberg M (1980) Peptidergic neurones. Nature 284:515–521
Hurley HJ, Shelley WB (1954) The role of the myoepithelium of the human apocrine sweat gland. J Invest Dermatol 22:143–155
Hurley HJ, Shelley WB (1960) The human apocrine sweat gland in health and disease. Thomas, Springfield, Ill
Jenkinson DM (1967) On the classification of sweat glands and the question of the existence of an apocrine secretory process. Br Vet J 123:311–316
Jenkinson DM (1973) Comparative physiology of sweating. Br J Dermatol 88:397–406
Kuno Y (1956) Human perspiration. Thomas, Springfield, Ill
Landis SC, Keefe D (1983) Evidence for neurotransmitter plasticity in vivo: developmental changes in properties of cholinergic sympathetic neurones. Dev Biol 98:349–372
Lovatt Evans C, Smith DFG (1956) Sweating responses in the horse. Proc R Soc Lond [Biol] 145:61–83
Montagna W (1962) The structure and function of skin. Academic, New York
Morimoto T (1978) In: Jarrett A (ed) The physiology and pathophysiology of the skin, vol 5. Academic, London, pp 1611–1620
Prout BJ, Wardell WM (1969) Sweating and peripheral blood flow in patients with phaeochromocytoma. Clin Sci 36:109–117

Randall WC, Kimura KK (1955) The pharmacology of sweating. Pharmacol Rev 7:365–387
Ranvier L (1879) Sur la structure des glandes sudoripores. CR Acad Sci [III] 89:1120–1123
Reas HW, Trendelenburg U (1967) Changes in the sensitivity of the sweat glands of the cat after denervation. J Pharmacol Exp Ther 156:126–136
Rees J, Shuster S (1981) Pubertal induction of sweat gland activity. Clin Sci 60:689–692
Robertshaw D (1974) Neural and humoral control of apocrine glands. J Invest Dermatol 63:160–167
Robinson S, Robinson AH (1954) Chemical composition of sweat. Physiol Rev 34:202–220
Rothman S (1954) Physiology and biochemistry of the skin. University of Chicago Press, Chicago
Sarkay I, Shuster S, Stammers M (1965) Occlusion of the sweat pore by hydration. Br J Dermatol 77:101–104
Sato K (1973a) Sweat induction from an isolated eccrine sweat gland. Am J Physiol 225:1147–1152
Sato K (1973b) Stimulation of pentose cycle in the eccrine sweat gland by adrenergic drugs. Am J Physiol 224:1149–1154
Sato K (1977) The physiology, pharmacology and biochemistry of the eccrine sweat gland. Rev Physiol Biochem Pharmacol 79:51–131
Sato K (1980) Pharmacological responsiveness of the myoepithelium of the isolated human axillary apocrine sweat gland. Br J Dermatol 103:235–243
Sato K (1987) Biology of eccrine sweat glands. In: Fitzpatrick TB, Eisen AZ, Wolff K, Freedberg IM, Austen KF (eds) Dermatology in general medicine, 3rd edn. McGraw-Hill, New York, pp 195–209
Sato K, Dobson RL (1970) Enzymatic basis for the active transport of sodium in the duct and secretory portion of the eccrine sweat gland. J Invest Dermatol 55:53–56
Sato K, Dobson RL (1971) Glucose metabolism of the isolated eccrine sweat gland. I. The effects of mecholyl, epinephrine and ouabain. J Invest Dermatol 56:272–280
Sato K, Dobson RL (1973) Glucose metabolism of the isolated eccrine sweat gland. II. The relation between glucose metabolism and sodium transport. J Clin Invest 52:2166–2174
Sato K, Sato F (1979) Pharmacology and function of an isolated human apocrine gland in vitro. Clin Res 27:535A
Sato K, Dobson RL, Mali JWH (1971) Enzymatic basis for the active transport of sodium in the eccrine sweat gland: localization and characterization of Na-K-ATPase. J Invest Dermatol 57:10–16
Sato K, Nishiyama A, Koboyashi M (1979) Mechanical properties and functions of the myoepithelium in the eccrine sweat gland. Am J Physiol 237:C177–184
Schiefferdecker P (1917) Die Hautdrüsen des Menschen und der Säugetiere, ihre biologische und rassenanatomische Bedeutung, sowie die Muscularis sexualis. Biol Zentralbl 37:534–562
Schulz I (1969) Micropuncture studies of the sweat formation in cystic fibrosis patients. J Clin Invest 48:1470–1477
Schulz I, Ullrich KJ, Fromter E, Holzgreve H, Frick A, Hegel U (1965) Mikropunktion und elektrische Potentialmessung an Schweißdrüsen des Menschen. Pflügers Arch 284:360–373
Seifert G (1967) Experimental sialadenosis by isoproterenol and other agents: histochemistry and electron microscopy. In: Schneyer LH, Schneyer CA (eds) Secretory mechanisms of salivary glands. Academic, New York, pp 191–208
Shapiro Y, Pandolf KB, Avellini BA, Pimental NA, Goldman RF (1981) Heat balance and heat transfer in men and women exercising in hot-dry and hot-wet conditions. Ergonomics 24:375–386
Shelley WB, Hurley HJ (1953) The physiology of the human axillary apocrine sweat gland. J Invest Dermatol 20:285–295
Shelley WB, Hurley HJ (1954) Localized chromidrosis. Arch Dermatol Syphilol 69:449–471

Shelley WB, Hurley HJ (1975) Studies on topical antiperspirant control of axillary hyperhidrosis. Acta Derm Venereol (Stockh) 55:241–260
Shelley WB, Hurley HJ, Nicholas AC (1953) Axillary odour. Experimental study of the role of bacteria, apocrine sweat and deodorants. Arch Dermatol Syphilol 68:430–446
Shuster S (1962) Adrenal control of eccrine sweat function. Proc R Soc Med 55:719–720
Silver A, Versaci A, Montagna W (1963) Studies of sweating and sensory function in cases of peripheral nerve injuries of the hand. J Invest Dermatol 40:243–258
Slegers JFG (1963) The mechanism of eccrine sweat function in normal subjects and in patients with mucoviscidosis. Dermatologica 127:242–254
Slegers JFG (1967) A mathematical approach to the two-step reabsorption hypothesis. Mod Probl Paediatr 10:74–88
Slegers JFG, van't Hof-Grootenboer (1971) The localisation of sodium transport sites in a forward pumping secretory system. Pflügers Arch 327:167–185
Uno H (1977) Sympathetic innervation of the sweat glands and piloarrector muscles of macaques and human beings. J Invest Dermatol 69:112–120
Uno H, Montagna W (1975) Catecholamine-containing nerve terminals of the eccrine sweat glands of macaques. Cell Tissue Res 158:1–13
van Heyningen RE, Weiner JS (1952) The effect of arterial occlusion on sweat composition. J Physiol (Lond) 116:404–413
Wada M (1950) Sudorific action of adrenaline on the human sweat glands and determination of their excitability. Science 111:376–377
Walton S, Chadwick L, Kendall-Taylor P, Shuster S (1983) The effect of testosterone on maximal sweat rate in eunuchs. Br J Dermatol 108:245
Weiner JS, Hellmann K (1960) The sweat glands. Biol Rev 35:141–186
Weiner JS, van Heyningen RE (1952) Observations on lactate content of sweat. J Appl Physiol 4:734–744
Wolfe S, Cage G, Epstein M, Kimberg DV (1970) Metabolic studies of isolated human eccrine sweat glands. J Clin Invest 49:1880–1884

CHAPTER 12

Thermoregulation and the Skin

W. I. Cranston

Apart from its integumentary function, one of the principal functions of the skin is in thermoregulation. Central temperature depends upon the balance between heat production and heat loss, and, apart from a normally small contribution from the respiratory system, heat loss is entirely mediated by the skin. Two mechanisms contribute to this action: control of skin blood flow and thermal sweating. These mechanisms are to some extent interdependent: alteration of blood flow alone can influence heat loss, but sweating, and increased evaporative heat loss from skin, requires vasodilatation to provide water for evaporation and increased heat flow (and hence blood flow) to the skin surface.

A. Skin Blood Flow

I. Vascular Effects on Heat Exchange

Rate of skin blood flow determines skin temperature, and the relationship between skin temperature and environmental temperature determines heat loss by radiation, conduction and convection. In temperate environments, radiation normally accounts for the greatest proportion of heat loss. In cold environments, the skin blood flow, particularly in the extremities, can fall almost to zero, whereas it can account for up to about half the cardiac output during exposure to severe heat stress.

The terminal capillary skin loops form the most effective heat exchangers, as the tissue temperature is closest to that of the environment, and the temperature gradient between blood and tissue is greater than in the deeper vessels. It has been known for many years that there are numerous arteriovenous anastomoses in the skin of the extremities (Grant 1930; Mescon et al. 1956). Opening of these anastomoses gives a large increase in local blood flow, though it is unlikely that blood passing through these anastomoses reaches temperature equilibrium with the surrounding tissues. Warm blood returning through these vessels will warm the deep venous plexus blood and its surrounding tissues, providing an additional means of increasing heat exchange.

II. Nervous Control of Cutaneous Blood Flow

Vasoconstrictor and vasodilator nerves have been described in the skin, but the importance of these two types of nerve varies in different parts of the body. In man, it is simpler to measure changes in skin blood flow in the extremities than in the head and trunk, because fairly accurate occlusion plethysmography methods can be used in the former tissues whereas more indirect and inexact methods have been employed in the latter (Blair et al. 1961; Fox et al. 1962).

In the hand and foot, nervous control of the circulation is entirely by means of sympathetic noradrenergic nerve fibres. After nerve block or sympathectomy, the blood flows in the hand and foot are raised; during exposure to thermal loads, the blood flow in a normal hand or foot does not exceed that in the opposite sympathectomised or nerve blocked extremity (Arnott and Macfie 1948; Gaskell 1956; Roddie et al. 1957a). There have been occasional claims of vasodilator nerves in the hands in patients with Raynaud's disease (Fatheree and Allen 1938; Lewis and Pickering 1931) but the bulk of the evidence strongly indicates that thermally mediated vasodilatation in this area is mediated entirely by withdrawal of sympathetic vasoconstrictor tone.

In the forearm and proximal limbs, however, there is evidence of vasodilator nerve fibres, as well as vasoconstrictor fibres. When a normal subject is exposed to heat, the blood flow to forearm skin increases in two phases. The first modest increase is contemporaneous with the increase in hand blood flow and is due to release of sympathetic vasoconstrictor tone. This phase is unaffected by atropine, as is hand blood flow (Gaskell 1956; Roddie et al. 1957b). The second increase in forearm blood flow begins as sweating starts. This increase is diminished, but not abolished, by atropine, implying that it may have a partial cholinergic component, but it is inhibited by neural blockade (Roddie et al. 1957b; Edholm et al. 1957). The synchrony between the onset of sweating and active vasodilatation suggests that the skin blood flow is modified in some way by the activation of sweat glands, a suggestion that is strengthened by the observation that nervously mediated vasodilatation is absent in patients with anhidrotic ectodermal dysplasia (Brengelmann et al. 1981). The impaired thermoregulation observed in patients with this condition probably represents failure of active vasodilatation as well as failure of sweating (Totel 1974); a similar failure of active vasodilatation has been reported in patients with diabetic neuropathy (Greeson et al. 1975). The chemical transmitters responsible for this effect are uncertain; as mentioned, acetylcholine may play a part but is certainly not the only agent involved. Fox and Hilton (1958) found that sweat collected from the hand and forearm contained bradykinin forming enzyme, and that subdermal perfusates in sweating limbs contained bradykinin: they suggested that bradykinin release might be responsible for vasodilatation, as it is, at least in part, in salivary glands (Gautvik 1970a, b). This proposition was supported by only two experiments on one subject, and appears to be unlikely for several reasons: (a) a sudden decrease of the temperature of a small area of sweating skin rapidly stops sweating, while marked cutaneous vasodilatation persists, though kinins are very rapidly broken down by tissue peptidases (Brockelhurst and Zeitlin 1967; Colman 1974); (b) bradykinin appears in sweat from peripheral areas without active vasodilatation; and (c) no bradykinin was found in forearm venous blood during thermally induced sweating (Allwood and Lewis 1964). There have been relatively unsupported suggestions that active vasodilatation might be purinergic (Burnstock 1972) or dopaminergic (Lang et al. 1976). However mediated, active vasodilatation appears to be the most important cutaneous response to central thermal stimuli, and this type of response appears to characterise trunk and head skin, with the exception of the nose, ears and lips, whose vasomotor innervations resemble that in the hand.

III. Reflex Control of Skin Blood Flow

It has been known for years that cooling of quite small areas of skin can cause an almost immediate vasoconstriction due to stimulation of cold receptors in skin (PICKERING 1933). In contrast, the effect of skin warming was only more recently recognised, probably because this reflex is only apparent if relatively large areas of skin are stimulated. Exposure of the limbs or trunk to radiant heating caused vasodilatation in the hands after 10–15 s exposure (KERSLAKE and COOPER 1950). This vasodilatation seems to be limited to the skin; muscle blood flow is not affected (BARCROFT et al. 1955). This reflex appeared to be blocked if the irradiated area had previously been sympathectomised (COOPER and KERSLAKE 1953), and it was suggested that the afferent limb of this reflex travelled in the sympathetic fibres rather than in the somatic afferents. Ordinary skin thermal sensation is apparently unaffected by sympathectomy. It is clear that the rapid skin vasoconstriction or vasodilatation caused by skin cooling or heating can be of such magnitude as to drive the central temperature in the opposite direction, a fact known as long ago as 1805 (CURRIE 1805).

IV. Central Control of Skin Blood Flow

GIBBON and LANDIS (1932) observed a slow rise in skin temperature of the fingers after the contralateral arm had been immersed in warm water for 10–15 min. This rise in skin temperature was prevented if the circulation to the immersed arm was occluded. This suggested that the warm blood from the immersed arm was being circulated to some central thermoreceptor, a suggestion strengthened by the observation that intravenous infusions of warm or cold saline could cause peripheral vasodilatation or constriction respectively (SNELL 1954). Though it had been proposed as far back as the beginning of the century (KAHN 1904) that there might be an intracranial thermoreceptor, its site was first accurately located by Ranson and his colleagues (MAGOUN et al. 1938), close to the midline, between the anterior commissure and the optic chiasma. High frequency diathermy heating of this area evoked panting in cats. Subsequently this has been confirmed extensively and it has also been shown that cooling of this area evokes shivering and vasoconstriction. A proportion of the neurones in this area have been shown to alter their firing rates in response to local temperture changes (EISENMAN and JACKSON 1967; NAKAYAMA et al. 1963; HARDY et al. 1964); though it is tempting to assume that these cells might be temperature sensors related to thermoregulation, such an assumption is not yet valid. While the hypothalamic area is the one that has been most explored, it is clear that warming and cooling areas in the midbrain (CABANAC and HARDY 1969), the medulla (CHAI and LIN 1972; LIPTON 1973) and the spinal cord (SIMON 1974) can induce appropriate thermoregulatory responses.

There is also evidence now of receptors in the posterior abdominal wall in at least two species (RAWSON and QUICK 1972; RIEDEL et al. 1972). An early suggestion that there might be thermal receptors in the walls of the great vessels is unlikely to be correct (CRANSTON et al. 1978).

In man, central thermoreceptors are extremely sensitive, and respond to changes of core temperature of less than 0.2 °C (GERBRANDY et al. 1954).

V. Interactions Between Thermal Receptors

In a system with multiple inputs, there has to be some method of integrating the thermoregulatory output, and there is good evidence that the effector mechanisms are driven by an integrated, and usually additive, output related to the inputs. For example, studies were made on blood flows through the pig's tail while hypothalamic, spinal cord and trunk temperatures were controlled, and varied independently (INGRAM and LEGGE 1971). When each site was warmed or cooled, tail blood flow rose or fell. Hypothalamic warming produced less vasodilatation when the cord was cool than when it was warm. When the trunk temperature was low, hypothalamic cooling caused more vasoconstriction in the tail than when the trunk temperture was higher. Similarly, in man, an integration of the effects of central temperature has been shown; the effect of changes in central temperature is considerably greater than that of changes in skin temperature (WYSS et al. 1975).

The effects of central receptors and integrating systems must be mediated by neurotransmitters. Many attempts have been made to identify these and to determine their importance, but this whole field remains in a very confused state (for review, see HELLON 1974).

VI. Effects of Local Temperature

The local temperature of blood vessels has direct effects upon blood flow. Other things being equal, the higher the skin temperature in the hand, the greater the blood flow, at skin temperatures above about 15 °C (SPEALMAN 1945). Below skin temperatures of about 10 °C, there is an initial vasoconstriction, followed by an increase in blood flow to well above the resting value, followed by a phasic variation in flow (GREENFIELD and SHEPHERD 1950). The mechanism may be inability of noradrenaline released at vasoconstrictor nerve endings to cause contraction of arteriolar smooth muscle at low temperatures (KEATINGE 1964): this may be regarded as a mechanism for preventing local cold injury, though it can lead to very large heat losses (GREENFIELD et al. 1951).

The tone of cutaneous veins can similarly be affected by combinations of local, central and general skin temperatures (WEBB-PEPLOE and SHEPHERD 1968). It appears that central temperature and general skin temperature may influence sympathetic adrenergic outflow to cutaneous veins, and that the local venous temperature modulates the response to this outflow (ROWELL et al. 1971 a, b).

B. Sweating

I. Neural, Humoral and Local Control

Eccrine sweat glands are normally activated by cholinergic sympathetic efferent fibres (KUNO 1956), although they can respond to intradermal injection (RANDALL and KIMURA 1955) of adrenaline or noradrenaline. Forearm sweating may also be produced by intra-arterial infusion of adrenaline (FOSTER et al. 1967).

There is little evidence to suggest that this sensitivity to locally applied catecholamines is of any physiological significance; fairly convincing evidence against any response to a naturally occurring raised level of catecholamines was obtained by PROUT and WARDELL (1969). In two patients with proven phaeochromocytomas, they iontophoresed hyoscine hydrobromide into areas of skin where the patients normally sweated during spontaneous attacks. During attacks witnessed during the succeeding few hours, suppression of sweating was observed over the iontophoresed areas of skin, while surrounding areas sweated copiously. It was shown that the effect of hyoscine was not a non-specific one.

It has been known for many years that sweat rate tends to decline with continuous sweating, and that this decline is greater in humid than in dry environments (BALDWIN and INGRAM 1968). Initially thought to be due to sweat gland fatigue, it now seems that the degree of skin wetting may be related to the phenomenon (BADE et al. 1979; ABRAMS and HAMMELL 1964) and the term hidromeiosis has been applied to it. The degree of skin wetness appears to be multiplicative, with the drive to sweating mediated by core temperature change (NADEL and STOLWIJK 1973b). The mechanism responsible for hidromeiosis is probably mechanical obstruction of the sweat duct orifices by hydration of keratin (SARKANY et al. 1965). Thus at a given central drive to sweating, hidromeiosis occurs during whole body immersion in water, but not during immersion in 15% saline (BARKER and CARPENTER 1970). Similar evidence was obtained by NADEL and STOLWIJK (1973a). A subject, after exercising in a hot environment, had both arms dried and inserted into cotton shirt sleeves. One had been exposed to saturated NaCI and then dried, the other had not. After further exercising in the heat, the sweat rate, measured by weighing the sleeves, was over 30% higher in the arm enclosed by the salt-saturated sleeve. This is probably explained by the demonstration that hydration occlusion of the sweat duct occurs with water but not with saline (SARKANY et al. 1965).

Generalised sweat rate is also influenced by local skin temperature. CRAWSHAW et al. (1975) applied cooling thermodes to different skin areas in subjects resting in an ambient temperature of 30 °C, while continuously measuring sweat rate hygrometrically over the ventral aspect of the thigh. It should be noted that in these experiments, subjects were selected who showed the greatest responses to skin cooling, and thus the results may not be generally applicable. This criticism apart, cooling of different skin areas caused a reduction in sweat rate; the reduction was proportional to the area cooled and was affected by the rate of skin cooling. The only area showing a greater than average response was the forehead.

In addition, local sweat rate is affected by local skin temperature. BULLARD et al. (1967) applied ventilated capsules to the thighs of sweating subjects, and varied the skin temperature under the capsules. At all levels of central temperature, local skin sweating increased with local temperature.

Sweat rate, like skin blood flow, is controlled by central inputs, related to the core temperature, and by peripheral ones, related to skin temperature. The influence of changes in core temperature appears to be independent of environmental temperature, except insofar as the environmental temperature will determine the basal level of sweating. Thus a given increment of core temperature results in a similar increases in sweat rate at different ambient temperatures (LIND 1964).

The relationship between central and peripheral drives to sweating has been extensively investigated. In general terms, granted a roughly linear increment in the relationship between central temperature and sweat rate, changes in peripheral inputs [one of which, at lest during exercise, might arise from muscle (NIELSON 1969)] usually affect the slope of the line relating core temperature to sweat rate, without seriously affecting the intercept (NADEL and STOLWIJK 1973 b).

II. Acclimation and Fatigue of Sweat Glands

Repeated exposure to heat stresses over weeks improves the ability of subjects to tolerate heat (LADELL 1964). The maximal sweat rate increases, and the proportional increment of sweat rate at lower (but elevated) core temperatures is greater (BELDING and HATCH 1963; WYNDHAM 1967). The line relating central temperature to sweat rate is shifted laterally, but its slope is unaffected (NADEL and STOLWIJK 1973 b). This probably represents a peripheral effect – the sweat glands producing more sweat for a given neural input than they would do in normal circumstances. Repeated local injection of stimulant drugs can result in an increased local sweat secretion in response to general stimuli (COLLINS et al. 1965).

It has been suggested that continued stimulation of sweat gland activity may result in fatigue of the glands. The earlier work of this type was done at fairly high sweat rates, in which sweat samples were obtained from arm bags worn continuously, and where hidromeiosis almost certainly played a part (BULMER and FORWELL 1956). Over long continued experiments at high sweat rates, fluid and electrolyte deprivation may affect observed sweat rates. In long exposures to dry environments where fluid balance is maintained, central temperature rise does not change with time, as one would expect if sweat glands were subject to fatigue. There is no convincing evidence of peripheral fatigue of sweat glands.

References

Abrams R, Hammel HT (1964) Hypothalamic temperature in unanaesthetized albino rats during feeding and sleeping. Am J Physiol 206:641–646

Allwood MJ, Lewis GP (1964) Bradykinin and forearm blood flow. J Physiol (Lond) 170:571–581

Arnott WM, Macfie JM (1948) Effect of ulnar nerve block on blood flow in the reflexly vasodilated digit. J Physiol (Lond) 107:233–238

Bade H, Braun HA, Hensel H (1979) Parameters of the static burst discharge of lingual cold receptors in the cat. Pflügers Arch 382:1–5

Baldwin BA, Ingram DL (1968) The influence of hypothalamic temperature and ambient temperature on the thermoregulatory mechanisms in the pig. J Physiol (Lond) 198:517–529

Barcroft H, Bock KD, Hensel H, Kitchin AH (1955) Die Muskeldurchblutung des Menschen bei indirekter Erwärmung und Abkühlung. Pflügers Arch Ges Physiol 261:199–210
Barker JL, Carpenter DO (1970) Thermosensitivity of neurons in sensorimotor cortex of the cat. Science 169:597–598
Belding HS, Hatch TF (1963) Relation of skin temperature to acclimation and tolerance to heat. Fed Proc 22:881–883
Blair DA, Glover WE, Roddie IC (1961) Cutaneous vasomotor nerves to the head and trunk. J Appl Physiol 16:119–122
Brengelmann GL, Freund PR, Rowell LB, Olerud JE, Kraning KK (1981) Absence of active cutaneous vasodilation associated with congenital absence of sweat glands in man. Am J Physiol 240(4):H571–575
Brockelhurst WE, Zeitlin IJ (1967) Determination of plasma kinin and kininogen levels in man. J Physiol (Lond) 191:417–426
Bullard RW, Banerjee MR, MacIntyre BA (1967) The role of skin in negative feedback regulation of eccrine sweating. Int J Bioclim Biomet 11:93–104
Bulmer MG, Forwell GD (1956) The concentration of sodium in thermal sweat. J Physiol (Lond) 132:115–122
Burnstock G (1972) Purinergic nerves. Pharmacol Rev 24:509–581
Cabanac M, Hardy JD (1969) Résponses unitaires et thermorégulatrices lors de rechauffements et refroidissements localisés de la région préoptique et du mésencéphale chez le lapin. J Physiol (Paris) 61:331–347
Chai CY, Lin MT (1972) Effects of heating and cooling the spinal cord and medulla oblongata on thermoregulation in monkeys. J Physiol (Lond) 225:297–308
Collins KJ, Crockford GW, Weiner JS (1965) Sweat-gland training by drugs and thermal stress. Arch Environ Health 11:407–420
Colman RW (1974) Formation of human plasma kinin. N Engl J Med 291:509–515
Cooper KE, Kerslake D McK (1953) Abolition of nervous reflex vasodilatation by sympathectomy of the heated area. J Physiol (Lond) 119:1829
Cranston WI, Hellon RF, Townsend Y (1978) Thermal stimulation of abdominal veins in conscious rabbits. J Physiol (Lond) 277:49–52
Crawshaw LI, Nadel ER, Stolwijk JAJ, Stamford BA (1975) Effect of local cooling on sweat rate and cold sensation. Pflügers Arch 354:19–27
Currie J (1805) Medical reports on the effects of water cold and warm as a remedy in fever and other diseases. Cadell and Davies, London, p 216
Edholm OG, Fox RH, MacPherson RK (1957) Vasomotor control of the cutaneous blood vessels in the human forearm. J Physiol (Lond) 139:455–465
Eisenman JS, Jackson DC (1967) Thermal response patterns of septal and preoptic neurons in cats. Exp Neurol 19:33–45
Fatheree FJ, Allen EV (1938) Sympathetic vasodilator fibres in the upper and lower extremities. Arch Intern Med 62:1015–1028
Foster KG, Ginsburg J, Weiner JS (1967) Adrenaline and sweating in man. J Physiol (Lond) 191:131
Fox RH, Hilton SM (1958) Bradykinin formation in human skin as a factor in heat vasodilatation. J Physiol (Lond) 142:210–232
Fox RH, Goldsmith R, Kidd DJ (1962) Cutaneous vasomotor control of the human head, neck and chest. J Physiol (Lond) 161:298–312
Gaskell P (1956) Are there sympathetic vasodilator nerves to the vessels of the hands? J Physiol (Lond) 131:647–656
Gautvik K (1970a) Studies on kinin formation in functional vasodilatation of the submandibular salivary gland in cats. Acta Physiol Scand 79:174–187
Gautvik K (1970b) The interaction of two different vasodilator mechanisms in the chorda-tympani activated sub-mandibular salivary gland. Acta Physiol Scand 79:188–203
Gerbrandy J, Snell ES, Cranston WI (1954) Oral, rectal and oesophageal temperatures in relation to central temperature control in man. Clin Sci 13:615–624
Gibbon JHH, Landis EM (1932) Vasodilatation in the lower extremities in response to immersing the forearms in warm water. J Clin Invest 11:1019–1036

Grant RT (1930) Observations on direct communications between arteries and veins in the rabbit's ear. Heart 15:281–303
Greenfield ADM, Shepherd JT (1950) A quantitative study of the response to cold of the circulation through the fingers of normal subjects. Clin Sci 9:323–347
Greenfield ADM, Shepherd JT, Whelan RF (1951) The loss of heat from the hands and from the fingers immersed in cold water. J Physiol (Lond) 112:459–475
Greeson TP, Freedman RI, Levan NE, Wong WH (1975) Cutaneous vascular responses in diabetics. Microvasc Res 10:8–16
Hardy JD, Hellon RF, Sutherland K (1964) Temperature-sensitive neurones in the dog's hypothalamus. J Physiol (Lond) 175:242–253
Hellon RF (1974) Monoamines pyrogens and cations: their actions on the central control of body temperature. Pharmacol Rev 26:289–321
Ingram DL, Legge KF (1971) The influence of deep body temperatures and skin temperatures on peripheral blood flow in the pig. J Physiol (Lond) 215:693–707
Keatinge WR (1964) Mechanism of adrenergic stimulation of mammalial arteries and its failure at low temperatures. J Physiol (Lond) 174:184–205
Kerslake DMcK, Cooper KE (1950) Vasodilatation in the hand in response to heating the skin elsewhere. Clin Sci 9:31–47
Khan RH (1904) Über die Erwärmung des Carotidenblutes. Arch Anat Physiol Physiol Abt (Suppl):81–134
Kuno Y (1956) Human perspiration. Thomas, Springfield, Ill
Ladell WSS (1964) Terrestrial animals in humid heat. In: Dill DB (ed) Man. Handbook of physiology, sect. 4. American Physiological Society, Washington, pp 625–659
Lang WJ, Bell C, Conway EL, Padanyi R (1976) Cutaneous and muscular vasodilation in the canine hindlimb evoked by central stimulation. Circ Res 38:560–566
Lewis T, Pickering GW (1931) Vasodilatation in the limbs in response to warming the body: with evidence for sympathetic vasodilator nerves in man. Heart 16:33–51
Lind AR, Leithead CS (1964) Heat stress and heat disorders. Churchill, London
Lipton JM (1973) Thermosensitivity of medulla oblongata in control of body temperature. Am J Physiol 224:890–897
Magoun HW, Harnson F, Brobeck JR, Ranson SW (1938) Activation of heat loss mechanisms by local heating of the brain. J Neurophysiol 1:101–114
Mescon H, Hurley HJ, Moretti G (1956) Anatomy and histochemistry of the arteriovenous anastomosis in digital skin. J Invest Dermatol 27:133–145
Nadel ER, Stolwijk JAJ (1973 a) Effect of skin wettedness on sweat gland response. J Appl Physiol 35(5):689–694
Nadel ER, Stolwijk JAJ (1973 b) Sweat gland response to the efferent thermoregulatory signal. Arch Sci Physiol 27:A67–A77
Nakayama T, Hammel HT, Hardy JD, Eisenman JS (1963) Thermal stimulation of electrical activity of single units of the preoptic region. Am J Physiol 204:1122–1126
Nielsen B (1969) Thermoregulation in rest and exercise. Acta Physiol Scand [Suppl] 323:1–74
Pickering GW (1933) The vasomotor regulation of heat loss from the human skin in relation to external temperature. Heart 16:115–135
Prout BJ, Wardell WM (1969) Sweating and peripheral blood flow in patients with phaeochromocytoma. Clin Sci 36:109–117
Randall WC, Kimura KK (1955) The pharmacology of sweating. Pharmacol Rev 7:365–380
Rawson RO, Quick KP (1972) Localization of the intraabdominal thermoreceptors in the ewe. J Physiol (Lond) 222:665–677
Riedel W, Iriki M, Simon E (1972) Regional differentiation of sympathetic activity during peripheral heating and cooling anaesthetized rabbits. Pflügers Arch 332:239–247
Roddie IC, Shepherd JT, Whelan RF (1957 a) A comparison of the heat elimination from the normal and nerve-blocked finger during body heating. J Physiol (Lond) 138:445–448
Roddie IC, Shepherd JT, Whelan RF (1957 b) The contribution of constrictor and dilator nerves to the skin vasodilation during body heating. J Physiol (Lond) 136:489–497

Rowell LB, Brengelmann GL, Detry JM, Wyss C (1971 a) Venomotor responses to rapid changes in skin temperature in exercising men. J Appl Physiol 30:64–71
Rowell LB, Brengelmann GL, Detry JM, Wyss C (1971 b) Venomotor responses to local and remote thermal stimuli to skin in exercising men. J Appl Physiol 30:72–77
Sarkany I, Shuster S, Stammers MC (1965) Occlusion of the sweat pore by hydration. Br J Dermatol 77:101–104
Simon E (1974) Temperature regulation: the spinal cord as a site of extra-hypothalamic thermoregulatory functions. Rev Physiol Biochem Pharmacol 71:1–76
Snell ES (1954) The relationship between the vasomotor response in the hand and heat changes in the body induced by intravenous infusions of hot or cold saline. J Physiol (Lond) 125:361–372
Spealman GR (1945) Effect of ambient air temperature on blood flow in the hands. Am J Physiol 145:218–222
Totel GL (1974) Physiological responses to heat of resting man with impaired sweating capacity. J Appl Physiol 37:346–352
Webb-Peploe MM, Shepherd JT (1968) Responses of dog's cutaneous veins to local and central temperature changes. Circ Res 23:693–699
Wyndham CH (1967) Effect of acclimatization on the sweat rate rectal temperature relationship. J Appl Physiol 22:27–30
Wyss CR, Brengelmann GL, Johnson JM, Rowell LB, Silverstein D (1975) Altered control of skin blood flow at high skin and core temperatures. J Appl Physiol 38:839–845

CHAPTER 13

Hair and Nail

R. P. R. Dawber

Since hair and nail are derived embryologically from the same tissue – the primitive epidermis – it is logical to consider them together. The definitive hair follicle complement and the nail apparatus are both fully developed by 20 weeks in utero. It is of interest that in many congenital and acquired ectodermal and integumentary diseases hair and nail changes are commonly seen though in acquired diseases predominantly affecting the epidermis, nail changes are frequently present whereas the hair follicle may remain normal. This may in part reflect greater evolutionary and embryological development away from the epidermis: as will be discussed later, the epidermis and nail apparatus grow continuously throughout life whilst hair is produced intermittently in the hair cycle.

A. Hair Follicle and Hair Shaft

Hair is not necessary for survival in man, though it still has many important functions, including: (a) sexual attractiveness, by its appearance and as a "carrier" of apocrine secretion; (b) as a tactile organ – hair follicles have a liberal supply of fine nerve endings responsive to light touch; and (c) as an organ of recognition.

In other primates and moulting animals a coat of exact length and colour is necessary for survival. Also the central medulla is a heat conserver in certain species, e.g. the polar bear and porcupine, in which medullary air spaces may take up more that 50% of the hair diameter.

I. Follicle Structure

The fully mature hair follicle during which the hair shaft is being continuously produced is the stage known as metanagen (Fig. 1). All the layers within the outer root sheath (ORS) are products of the hair matrix in the hair bulb. The ORS is in continuity with the interfollicular epidermis and is functionally similar (Rook and Dawber 1982).

The hair bulb matrix cells are of uniform type, having a large spherical nucleus and very little cytoplasm. In differentiation these cells soon transform into six concentric cylinders of ascending cells (Fig. 2); these layers are clearly discernible at the level of the upper bulb. The inner three layers are the presumptive hair, from within outwards, the medulla, the cortex and the cuticle. In many follicles medullary cells are absent or are only produced intermittently. The three cell layers outside the presumptive hair are those of the inner root sheath (IRS): the

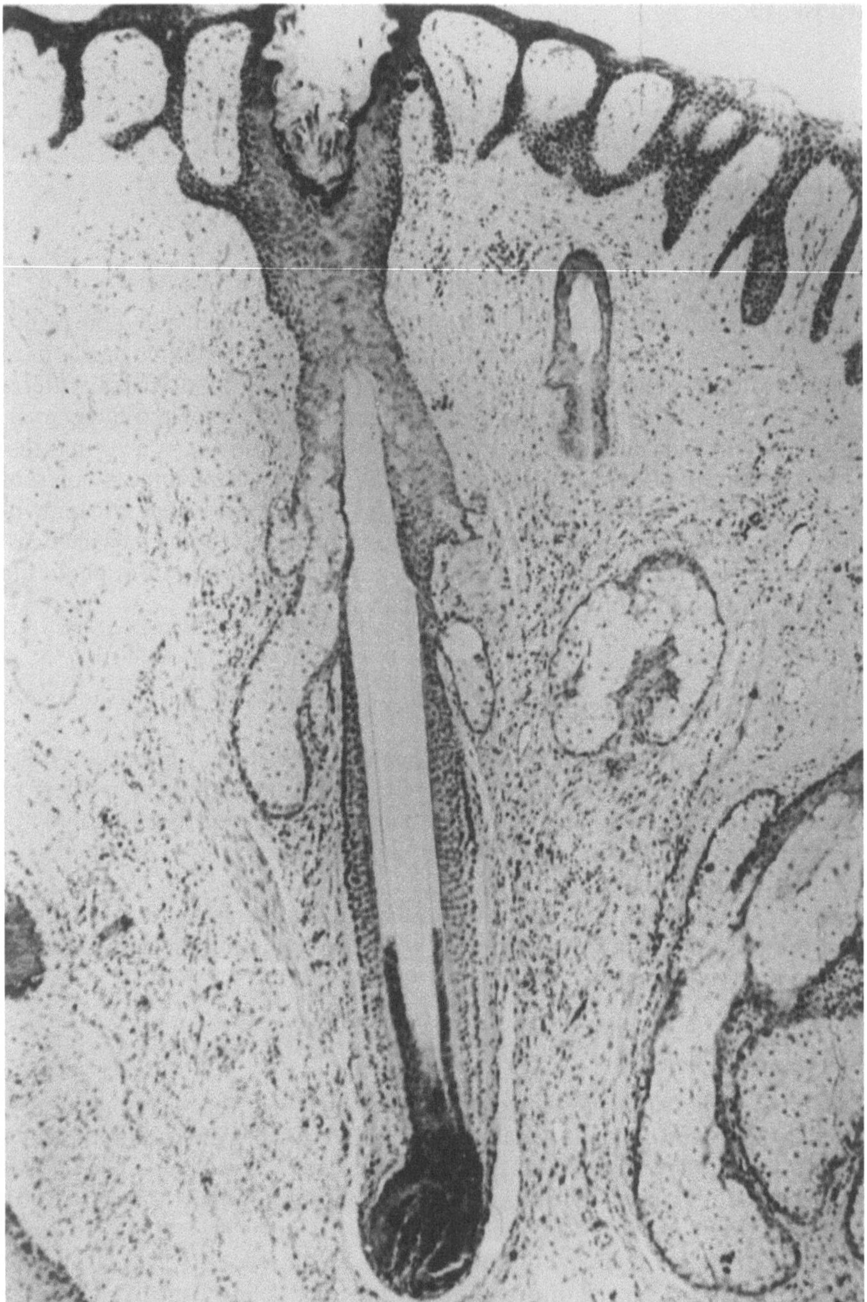

Fig. 1. Structure of hair follicle in growing phase (metanagen)

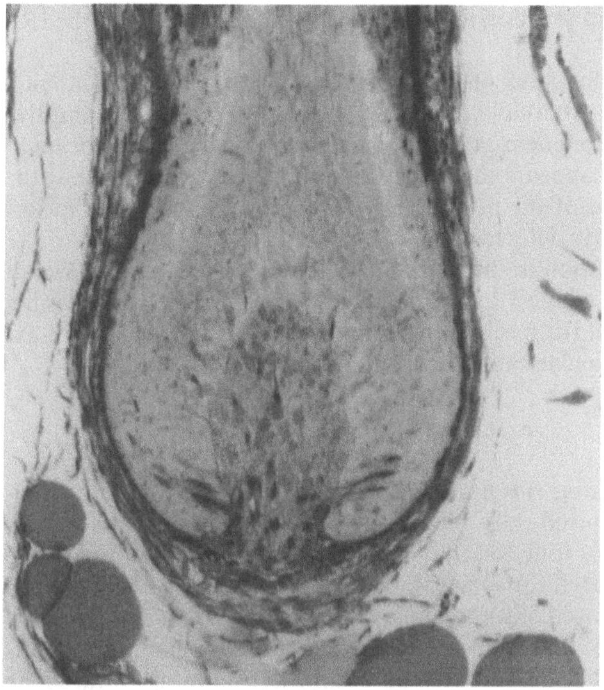

Fig. 2. Hair bulb surrounding the central dermal papilla in the suprabulbar region; distinct layers of differentiating cells are visible (light micrograph; 1-μm section)

IRS cuticle (outside the hair cuticle), the Huxley layer (the most prominent, containing large cytoplasmic trichohyalin granules), and the outer Henle layer, which is one cell thick.

The ORS surrounds the IRS and the developing hair within it. It is divided into two parts: (a) a short section surrounding the outer part of the hair bulb; this is one or two cells thick; and (b) an upper part from the neck of the hair bulb to the level of the sebaceous duct. The ORS possesses a basal layer in continuity with the basal layer of the surface epidermis. Differentiation occurs in a centripetal direction towards the IRS; the cells enlarge, flatten and become vacuolated with large quantities of cytoplasmic glycogen of unknown function. The exact fate of cells adjacent to the Henle layer is not known but it is likely that they move vertically with the inner cell layers and are shed into the follicular infundibulum with the IRS at the level of the sebaceous duct (EPSTEIN and MAIBACH 1969). Below the level of the sebaceous duct, the ORS contains no functioning melanocytes.

The exact function of the IRS and ORS is not known but it is generally considered that they control the rate of movement of the cells of the developing hair within them and also shape the hair (STRAILLE 1965; STROUD 1980). The latter function is proposed because the layers of the IRS harden before the developing hair keratinises. It has been postulated that many genetic hair dystrophies in which hair shape is abnormal, e.g. Monilethrix and Marie-Unna syndrome (ROOK and DAWBER 1982), possess an IRS "mould" that hardens abnormally, the hair within it assuming the shape of the IRS. This is quite likely in view of the intimate desmosomal and gap junction links between the hair cuticle and the IRS cuticle.

Within the hair bulb at the apex of the dermal papilla, melanocytes donate pigment to the cells that will eventually form the hair cortex; rarely, presumptive hair cuticle cells also receive pigment granules. The mechanism of melanocyte to matrix cell pigment transfer appears to be the same as in the epidermis, but the complex spatial arrangement of the many cell lines within the hair bulb means that this has never been clearly defined.

By mid-follicle, keratinisation in the hair and IRS is complete. At the level of the sebaceous duct the cells of the IRS are shed from the hair surface, possibly in response to enzyme action from sebum, leaving the hair cuticular surface exposed by the level of the follicular canal (or infundibulum).

II. Hair Shaft Structure

The fully formed hair shaft (Fig. 3) is thus composed essentially of a highly organised long cylinder of keratinised cells, biochemically built to withstand the rigours of the environment. This function may not be important in man but in the coat of most animals it is crucial for survival.

The outer cuticle is composed of six to ten layers of flat cells 0.2–0.5 μm thick. Surface cells overlap like roof tiles (imbrication), the exposed free margin pointing towards the tip. Each cuticle cell is made up of lamellar pointing towards the tip. Each cuticle cell is made up of lamellar components within the cell membranes

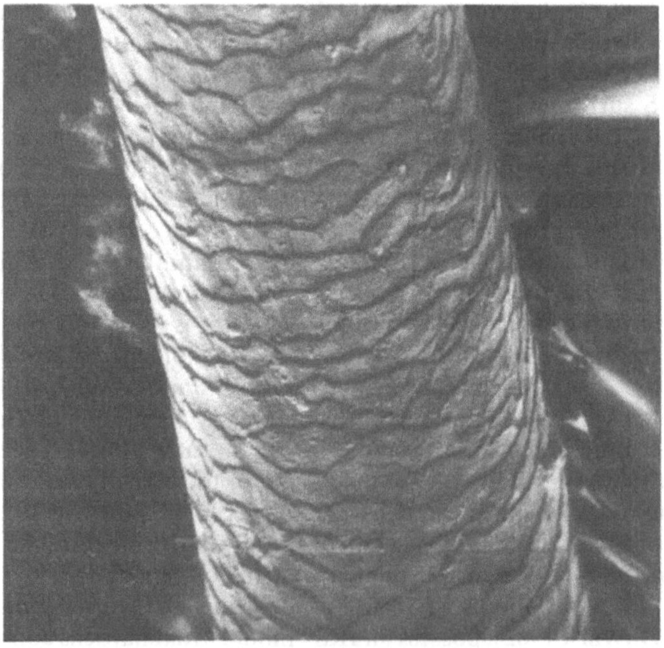

Fig. 3. Normal hair at root end (scanning electron micrograph)

– an outer exocuticle, rich in high sulphur protein, particularly within its A-layer (SWIFT 1977), and the endocuticle, composed of cytoplasmic debris.

The main bulk of the hair shaft is the cortex. Within its long fusiform cells, orientated along the axis of the hair, are α-keratin fibrils embedded in a matrix of high sulphur protein. The colour of hair is mainly dependent on the number, shape and size of pigment granules within these cells.

The medulla may be absent, intermittent or continuous within the cortex; it is composed of a cortex-like framework of "spongy" keratin supporting thin shells of amorphous material bounding air spaces of variable size.

III. Hair Cycle

Unlike the pepdermis, which divides and differentiates continuously throughout life, the follicle only produces hair intermittently. The metanagen phase lasts 3–6 years on the normal scalp; after this period, the follicles involute for 3–6 weeks (catagen) and then enter an inactive (telogen) stage of several months during which, in man, the hair is shed. The dermal papillae do not disappear during telogen but contracts into a ball of apparently inert cells underlying the telogen "club" root. The mechanism by which the dermal papillae reactivate and stimulate new follicular growth (anagen) is not known (JOHNSON 1977). In humans, removal of the papillae generally prevents follicular growth and transplanted dermal papillae can produce new follicle formation from the overlying epidermis (OLIVER 1970). It is likely that the matrix cells of the new anagen are formed from the outer root sheath.

The length attained by hair mainly depends on the duration of the anagen phase; this is longest on the scalp in both sexes and the beard area in men. Other factors governing hair length include linear growth and the natural breakdown ("weathering") of hair due to physical (e.g. friction) and chemical (e.g. permanent waving) forces as the fibres grows away from the body. After puberty, anagen length is greater in the female on the scalp and shorter on the beard. Racial factors also play a part; for example women from Iran and some parts of Northern India not infrequently have hair up to 120 cm in length (SAITOH et al. 1970).

On the normal scalp, 80%–95% of follicles on the vertex are in anagen, i.e. telogen counts as judged by root microscopy of plucked hairs may be up to 15% in normal individuals; occipital telogen counts are consistently lower in all age groups after puberty – less than 10%. Since under normal circumstances telogen hairs are shed, between 120 and 130 hairs may be shed per day after puberty.

Three morphological types of hair are described:

1. Lanugo hair is formed between 20 and 32 weeks in utero on all body sites bar palms, soles and the palmar and plantar surfaces of fingers and toes respectively. By 32 weeks of foetal life, this lightly pigmented or apigmented hair is shed and replaced by two types of hair:
2. Coarse pigmented, often medullated hair on the scalp, eyebrows and eyelashes – terminal hair.
3. Vellus hair: fine, short, apigmented or hypopigmented hair on all sites bar those with terminal hair.

IV. Endocrine Control Factors

At puberty, secondary sexual hair develops in response to stimulation by ovarian and adrenal androgens in the female and testicular androgens in the male. The latter are said to give the greater prominence of secondary sexual hair seen in men. Oestrogens have a limiting effect on follicular androgen activity (EBLING 1981; LESHIN and WILSON 1981). Testicular testosterone is the principal androgen circulating in the plasma in men, whereas in women the less potent adrenal and ovarian steroid androstenedione is the major circulating androgen. The androgenic activity of the latter is probably dependent on enzymic conversion to testosterone in peripheral tissues by 17β-hydroxysteroid dehydrogenase (LESHIN and WILSON 1981). Testosterone and androstenedione also undergo 5α-reduction in peripheral tissues to form dihydrotestosterone and 5α-androstenedione respectively.

The limiting effect of oestrogens on androgenic activity in target tissues is important in assessing androgen-mediated hair growth; for example, the increased facial hair at the menopause particularly relates to decreased oestrogen production, ovarian androgen secretion not decreasing concurrently. This oestrogen–androgen balance also governs the normal vertex pattern in both sexes, i.e. the extent to which men and women bald is genetically controlled, there being a different response to androgens at tissue level in different scalp sites (HAMILTON 1951; LUDWIG 1977). The female pattern of common baldness (diffuse) is less pronounced partly because of oestrogens; post-menopausally the female may, with increasing age, equate in pattern to that attained by her male relatives in the third and fourth decades of life. Also, female virilising syndromes, with increased androgen secretion, may be associated with vertex "patterning" – as in male common baldness – even before the menopause. Thus, decreased oestrogen production and increased ovarian androgen production may have similar effects on the hair follicle. Despite these facts, the main control of the vertex pattern of hair is at follicular receptor level (LESHIN and WILSON 1981).

B. Nail Apparatus

The component parts of the nail apparatus are shown in Fig. 4 (BARAN and DAWBER 1984). The primary function of these structures is to produce a strong, relatively inflexible keratinous nail plate over the dorsal surface of the end of each digit. The nail plate gives a protective covering for the fingertip; by exerting counter-pressure over the volar skin and pulp when the fingers are in use, the flat plate adds to the precision and delicacy of touch, the ability to pick up small objects and other subtle functions. Finger-nails cover approximately one-seventh of the dorsal surface, whilst on the great toe the nail may cover up to 50% of the dorsum of the digit.

The rectangular nail plate is the largest structure, resting on and being firmly attached to the nail bed. Underlying the proximal part of the nail is the white lunula; the nail bed distal to the lunula is typically pink because its translucency allows the redness of the vascular bed to be seen through it. The "free" terminal part of the nail is white to grey because of the intrinsic colour of the nail plate.

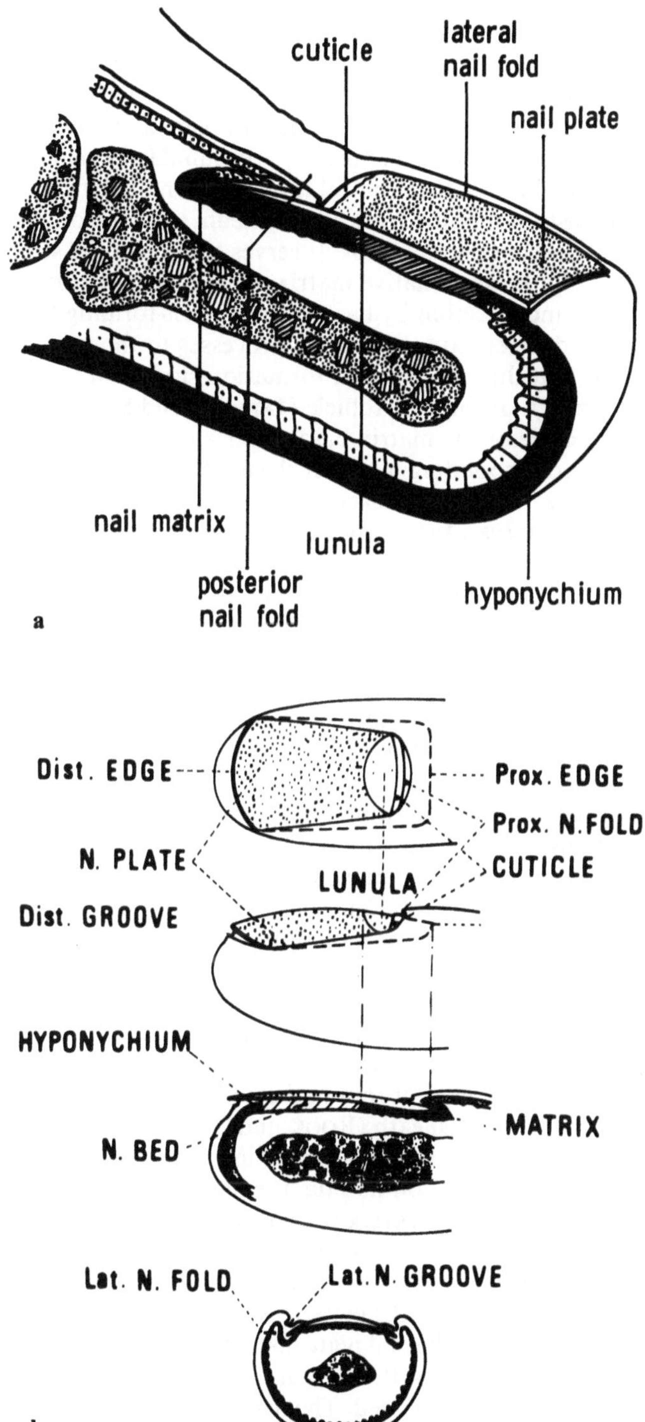

Fig. 4 a, b. Component parts of the nail apparatus

In longitudinal cross-section, the epidermis of the proximal nail fold is seen to be in continuity with the skin over the dorsum of the terminal phalanx; the epidermis of this fold continues round the roof of the nail as the germinative nail matrix. The matrix surrounds the base of the nail plate and continues under the proximal part of the nail to the distal border of the lunula; the nail bed epithelium (ventral matrix) contributes a thin ventral layer to the nail plate.

The cuticle consists of modified stratum corneum reflected onto the nail plate from the proximal and lateral nail folds. It serves to protect the structures at the nail base, particularly the germinative matrix, from inflammatory processes due to wetting, solvents and infection by bacteria and yeast-forming fungi.

The nail matrix differentiates by similar processes to those in the epidermis, but keratinises without the keratohyalin formation similar to that which occurs in the inner root sheath of the hair follicle (JARRETT and SPEARMAN 1966). This apart, the ultrastructure of the matrix basal layer and differentiating cells is the same as that of the corresponding layers of the epidermis.

Why the lunula is white is not exactly known but there are many possible structural reasons (BURROWS 1919; LEWIN 1965):

1. Matrix epithelial cells are nucleated, appearing parakeratotic.
2. The nail surface is smoother and shinier than the nail beyond the lunula.
3. The nail plate is thinner, and the underlying matrix thicker than the red nail bed distal to the lunula.
4. The nail plate/nail bed attachment is less firm.
5. The nail plate over the matrix is more opaque.
6. The capillary network is less dense.
7. Dermal connective tissue is less compact.

The lunula appears to have no functional significance, being absent on some nails, and on all nails in some individuals.

The nail matrix epithelium contains melanocytes in the lower two cell layers. These donate pigment to differentiating cells as in the epidermis. In Caucasoids, little or no pigment is evident in the nail plate; in Negroid individuals it is commonly seen as longitudinal linear streaks.

The nail bed epithelium (ventral matrix) and underlying dermis are closely apposed to the phalangeal periosteum, there being no subcutaneous fat layer in this site. The epithelium is typically only two layers thick; the transformation from living cell to fully keratinised cell is abrupt in the nail bed, similar to the Henle layer of the hair follicle inner root sheath (ROOK and DAWBER 1982). ACHTEN (1968) has suggested that the nail apparatus is similar in many respects to half a hair follicle cut longitudinally and laid on its side, the hair bulb being equivalent to the nail "root" matrix and the hair cortex the equivalent of the nail plate. Nail bed epithelial cells differentiate vertically and then distally as they approach the nail plate, moving horizontally towards the tip of the digit.

The nail bed dermo-epidermal junction is thrown into regular longitudinal folds, giving firm adhesion of the nail plate distal to the lunula. The nail bed capillary network is orientated along these folds, leading to longitudinal "splinter" haemorrhages when they are damaged. The nail bed dermal connective tissue network radiates vertically from the phalangeal periosteum to the epithelium, providing firm adhesion of the nail plate to the bone.

The nail bed is richly supplied with glomus bodies (93–510 per cm^2), neurovascular bodies which act as arteriovenous anastomoses. Each glomus body is an oval encapsulated organ 300 μm long, composed of a tortuous vessel uniting an artery and vein, and a nerve supply; surrounding these are many modified muscle cells resembling epithelioid cells (RYAN 1973). The nerve fibres within the bodies are cholinergic. Glomus bodies lie parallel to the capillary reservoirs; they are able to contract rhythmically at a slower rate than the associated arterioles and asynchronously with them. In response to cold, arterioles constrict whilst glomus bodies dilate. They are particularly important in the preservation of blood supply to the peripheries in cold conditions.

At the junction of the distal border of the nail bed, in some nails the proximal hyponychial area thickens, producing hyperkeratosis in a narrow band, the solehorn (PINKUS 1927); this may contribute soft keratin to the nail plate ventrally. It has been suggested that in diseases such as pityriasis rubra pilaris, this structure is the seat of the subungual hyperkeratosis characteristic of the disease.

The nail plate is made up of three layers, dorsal (thin), intermediate (thickest) and ventral (thinnest) plates. The actual flatness or curvature of the nail depends on the shape of the terminal phalanx.

Microscopically the plate is composed of closely knit horny squamous cells arranged in lamellae (PARENT et al. 1985). Acidophilic masses may be seen in older subjects, the pertinax bodies of LEWIS and MONTGOMERY (1955). The plasma membranes of the flattened squamous cells are tortuous, the many folds interlocking. The keratin pattern within these cells is granular, similar to the hair cuticle.

Comparison of the keratin biochemistry between nail and hair shows essentially the same fractions – α-fibrillar, low sulphur protein; globular, high sulphur matrix protein; and high glycine–tyrosine matrix protein. Amino acid analysis shows more cysteine, glutamic acid and serine than in hair, and less tyrosine; most cysteine is in the intermediate nail plate.

Calcium is the chief element in the nail plate; in the keratinized cells it is present as the phosphate in hydroxyapatite crystals bound to phospholipid. FORSLIND et al. (1976) suggest that calcium contributes little to the hardness of the nail plate in man.

References

Achten G (1968) Normale Histologie und Histochemie des Nagels. In: Jadassohn J (ed) Handbuch der Haut- und Geschlechtskrankheiten. Springer, Berlin Heidelberg New York, pp 339–376

Baran R, Dawber RPR (1984) Diseases of the nails and their management, 1st edn. Blackwell, Oxford

Burrows MT (1919) The significance of the lunula of the nail. Johns Hopkins Hosp Rep 18:357–361

Ebling FJ (1981) Hormonal control of hair growth. In: Orfanos CE, Montagna W, Stuttgen G (eds) Hair research. Springer, Berlin Heidelberg New York, pp 195–204

Epstein W, Maibach HI (1969) Cell proliferation and movement in human hair bulbs. In: Montagna W, Dobson RL (eds) Advances in biology of skin. Pergamon, New York

Forslind B, Wroblewski R, Alzelius BA (1976) Calcium and sulphur location in human nail. J Invest Dermatol 67:223

Hamilton JB (1951) Patterned long hair in man; types and incidence. Ann NY Acad Sci 53:708
Jarrett A, Spearman RI (1966) The histochemistry of the human nail. Arch Dermatol 94:652
Johnson E (1977) The control of hair growth. In: Jarrett A (ed) Physiology and pathophysiology of the skin, vol 4. The hair follicle. Academic, London, p 1351
Leshin M, Wilson JD (1981) Mechanisms of androgen-mediated hair growth. In: Orfanos CE, Montagna W, Stuttgen G (eds) Hair research. Springer, Berlin Heidelberg New York, pp 205–209
Lewin K (1965) The normal fingernail. Br J Dermatol 77:421
Lewis BL, Montgomery H (1955) The senile nail. J Invest Dermatol 24:11–18
Ludwig E (1977) Classification of the types of androgenic alopecia arising in the female sex. Br J Dermatol 97:249
Oliver RF (1970) The induction of follicle formation in the hooded rat by vibrissae dermal papillae. J Embryol Exp Morphol 23:219
Parent D, Achten G, Stouffs-Vanhoof F (1985) Ultrastructure of the normal human nail. Am J Dermato-pathol 1(6):529–535
Pinkus F (1927) In: Jadassohn J (ed) Handbuch der Haut- und Geschlechtskrankheiten. Springer, Berlin, pp 267–289
Rook RA, Dawber RPR (1982) Diseases of the hair and scalp. Blackwell Scientific, Oxford, pp 10–14
Ryan TJ (1973) The arterio-venous anastomoses. In: Jarrett A (ed) The physiology and pathophysiology of the skin, vol 2. Academic, London, p 612
Saitoh M, Uzuka M, Sakamoto M (1970) Human hair cycle. J Invest Dermatol 54:65
Straille WE (1965) Root sheath – dermal papillary relationships in the control of hair growth. In: Lyne AC, Shorth BF (eds) Biology of skin and hair growth. Angus and Robertson, Sydney, pp 35–37
Stroud JD (1980) Complementation of the inner root sheath of human hair. In: Brown AC, Crounse RG (eds) Hair, trace elements and human illness. Praeger, New York, pp 163–168
Swift JA (1977) The histology of keratin fibres. In: Asquith RS (ed) The chemistry of natural protein fibres. Wiley, London, pp 81–146

CHAPTER 14

The Sebaceous Glands

A. J. THODY and S. SHUSTER

A. Introduction

The sebaceous glands are lipid-secreting glands found in the dermis of mammals. Most are associated with hair follicles as *pilosebaceous glands* found on most of the body surface: in man they are largest and most numerous on the face, forehead, scalp and front and back of the upper chest. The *free sebaceous glands* are not associated with hairs and occur in the transitional zones between the skin and mucous membranes. They are particularly prevalent in the anogenital and periareolar skin and the buccal mucosal membranes, where they increase in size with age. In some mammals free sebaceous glands have become highly specialised and are concerned with the secretion of pheromones, e.g. the preputial glands of rodents, the ventral glands of the gerbil, the costovertebral glands of hamsters and the large brachial glands of lemurs.

All sebaceous glands are similar in structure. In man they consist of a collection of lobules of various size and shape that open into a system of ducts; these form the main excretory duct, which opens into the pilary canal. Free sebaceous glands open directly onto the skin's surface. There are at least two types of cell: the lipid-producing cells of the fundus and the stratified squamous epithelial cells of the duct. The germinative or basal cells of the fundus are in contact with the basement membrane and after division the cells differentiate and synthesise and store lipid. The mature lipid-filled cells are moved towards the duct as more cells are produced by division; there they die and disintegrate, releasing lipid and cell debris which reaches the skin surface as *sebum*.

B. Development

Like the hair follicle and the sweat glands, the pilosebaceous glands are derived embryologically from the epidermis. In the human embryo they first appear at around 9–10 weeks as a small bulge on the side of the developing hair follicle. Differentiation into sebaceous tissue occurs between 13 and 15 weeks, and by 17 weeks the glands are fully developed. Sebaceous glands are active in the foetus and much of the vernix caseosa is composed of sebum. The sebaceous glands of new-born infants are large but soon after birth they regress and remain small until puberty, when they enlarge under the influence of various hormones. Free sebaceous glands develop later than pilosebaceous glands and at times which vary widely in different regions; e.g. the mammary and anal sebaceous glands develop prenatally, whereas those in the genitalia and oral mucosa appear after birth, which suggests that unlike the sweat glands they can differentiate from epithelium that is already fully developed. Differential development and control (see below) suggests that glands in different regions may serve different functions. In addition there are considerable species variations (see SHUSTER and THODY 1974; STRAUSS et al. 1983).

C. Sebum

I. Formation

Sebum is a holocrine secretion. Basal cells in the periphery of the sebaceous lobule proliferate and as they are moved passively towards the centre of the lobule they differentiate and synthesise lipids. In man the turnover time is of the order of 21–25 days (PLEWIG and CHRISTOPHERS 1974). Lipid droplets are found in basal cells (ELLIS and HENDRIKSON 1963), indicating that differentiation occurs before the cells commence movement. There is a progressive increase in the size of these lipid droplets as the cells migrate centrally. A range of different lipids are formed (see below) depending on the state of maturity of the cells (WHEATLEY et al. 1979; ALVES et al. 1983). Immature preputial cells predominantly contain triglycerides but as they approach maturity they are rich in wax esters. By the time the cells have reached the centre of the lobule and are close to the excretory duct they are fully mature: at this stage they are large, with a small nucleus, little cytoplasm and as many lipid vacuoles as 100 per cell (ELLIS 1967). At the excretory duct the cells disrupt and are discharged as sebum.

Although the sebocytes will incorporate a variety of substrates into sebaceous lipid de novo lipogenesis occurs from glucose (COOPER et al. 1974) and it is thought that sebaceous lipogenesis in situ does not require the delivery of complex substrates that occur in other sites. In most species the fatty acids that are found in sebum are distinct from those found in dietary or circulating lipids. In humans, for example, most of the unsaturated fatty acids in surface sebum are $\Delta 6$ compounds whereas dietary lipids are $\Delta 9$ compounds. It appears that sebum is composed almost entirely of lipids that are produced within the sebaceous gland and that dietary fats have little effect upon its composition (see later). In human sebaceous glands two major biosynthetic pathways exist, one giving rise to triglycerides, wax and sterol esters and the other involving the synthesis of squalene. An unusual feature of the sebaceous glands is that there is no conversion of squalene to cholesterol and as a result squalene accumulates. This explains the abundance of squalene in human sebum. A similar situation may exist in other species, e.g. the beaver, where squalene can account for 80% of the skin surface lipids (LINDHOLM and DOWNING 1980). The sebaceous glands also synthesise phospholipids. These are required for the considerable increase in plasma membrane synthesis that occurs as the sebaceous cells grow in size during development. However, phospholipids are not found in surface sebum (see below) and it appears that as the cells disintegrate to form sebum the phospholipids are completely degraded. The resulting fatty acids may be incorporated into wax esters (DOWNING et al. 1977). For further details of the synthesis of sebum see STRAUSS et al. (1983).

II. Composition

The main lipids in sebum are fatty acids, triglycerides, wax esters and squalene. Sebum is rapidly contaminated by other substances, including keratinocyte lipids and other products, eccrine and apocrine sweat and exogenous material. Sebum is rapidly exposed to micro-organisms (see Vol. II) which convert triglycerides to free fatty acids, and virtually all of the free fatty acids in sebum are formed in this way (NICOLAIDES and WELLS 1957).

Many factors affect the composition of sebum, e.g. sex (KIRK 1948), age (COTTERILL et al. 1972), menstrual cycle (MAC DONALD and CLARKE 1970), race (NICOLAIDES and ROTHMAN 1953), calorie intake (POCHI et al. 1970), disease (RUST et al. 1970) and skin region (RUST et al. 1968; GREENE et al. 1970). Another factor is rate of sebum production because at low rates the epidermal components (normally $<5\%$ and predominantly cholesterol and its esters) appear to increase proportionally.

III. Function

Despite suggestions and speculations, e.g. antimicrobial and emollient actions, the function of sebum is unknown. The primary function of pilosebaceous sebum is likely to be associated with hair function and is therefore probably concerned with thermoregulation in the hairy mammals. In man hair has little or no thermoregulatory importance and because no clear function has been described for human sebaceous glands they are often considered to be vestigial. Although specialised glands serve as pheromone producers in many of the lower mammals (STODDART 1980), olfactory communication is relatively unimportant in man with the possible exception of the peri-areolar sebaceous glands, which may release odours that attract the infant (RUSSELL 1976). The fact that human sebaceous glands have been maintained in association with vestigial hairs and have a complex endocrine control does, however, suggest a function (SHUSTER 1976). What this function is remains to be seen.

IV. Factors Affecting the Rate of Sebum Production (see Vol. II)

1. *Sebaceous gland density, size, cell turnover and lipogenic rate:* The amount of sebum that is secreted in a given area of skin will depend on all these characteristics. Sebaceous gland density shows much regional variation and in human adults the highest rates of sebum excretion occur on the forehead.
2. *Skin temperature:* Sebum excretion increases with skin temperature but this may be a collection artefact due to decreased viscosity of sebum (BURTON 1970).
3. *Circadian rhythm:* Sebum secretion shows a circadian rhythm (BURTON et al. 1970) which is not due to temperature.
4. *Menses:* There is a small but significant variation in sebum secretion during the menstrual cycle (BURTON et al. 1972a).
5. *Stress:* Although some workers have reported changes in sebum secretion during stress (WOLFF et al. 1951; ROBIN and KEPECS 1953), this has not been confirmed (IKAI 1959).

6. *Diet:* Diet has little effect upon sebum secretion (FULTON et al. 1969) although POCHI et al. (1970) reported a decrease during extreme fasting.
7. *Acne:* Probably the most important normal correlate of sebum excretion is whether or not the subject has had acne (CUNLIFFE and SHUSTER 1969) (see Vol. II).
8. *Drugs:* Various drugs, e.g. anti-androgens, retinoids, oral contraceptives and phenothiazines, will alter the rate of sebum excretion. Although certain drugs are able to exert direct actions on the sebaceous glands, they can in many cases also act by blocking the actions of or by affecting the secretion of various hormones and in particular those of the pituitary and gonads (see later).
9. *Diseases:* A number of diseases are characterised by changes in sebum excretion, e.g. parkinsonism and acromegaly (see later).

D. Control

Sebum secretion depends upon (a) gland size, (b) proliferative activity (cell turnover) and (c) lipogenic capacity. It is clear, therefore, that rate of sebum production cannot be inferred from gland size alone. Although it seems likely that androgens are important in regulating the proliferative activity of sebaceous cells, very little is known about this aspect of sebaceous gland control (see PLEWIG and CHRISTOPHERS 1974). Much of what is known about sebaceous gland control has come from studies that have measured the rate of sebum secretion and, more recently, sebaceous lipogenesis, although the biochemical aspects of the latter process are still poorly understood.

In man virtually all of our knowledge is confined to pilosebaceous glands, the control of which is likely to differ from the free sebaceous glands. As has been discussed previously (SHUSTER and THODY 1974; SHUSTER 1982), there is considerable disagreement on certain aspects of sebaceous function. We consider that the differences are mostly due to technical factors and will only be resolved when more work is done. We recommend that the reader also considers the accounts by STRAUSS et al. (1983) and the different views on androgen action summarised below and in Vol. II.

I. Non-endocrine Control

In man there is no evidence of a non-endocrine control of pilosebaceous glands (see SHUSTER and THODY 1974). The increase in the sebum excretion rate measured below the neurological lesion in paraplegic patients (THOMAS et al. 1983, 1985) is probably explained by an increase in the lipid reservoir of the gland that is likely to occur during immobility. In the rat repeated removal of surface lipid stimulates secretion but the mechanism is not understood (SHUSTER and THODY 1974). The speed of appearance of odours from the free sebaceous glands suggests that release mechanisms may operate but whether these are neural is not yet clear.

II. Endocrine Control

1. **Androgens** (see Vol. II)

Androgens are the best known stimulators of sebaceous gland activity. Testosterone increases sebaceous gland growth and sebum production in man (STRAUSS and POCHI 1963) and in experimental animals (EBLING 1963; EBLING and SKINNER 1967; ARCHIBALD and SHUSTER 1967; THODY and SHUSTER 1970). Numerous other androgens are effective in stimulating sebum secretion in the rat but to varying degrees (ARCHIBALD and SHUSTER 1969; NIKKARI and VALAVAARA 1970; EBLING et al. 1971). The most active are those with a 17β-hydroxy group (testosterone, 5α-dihydrotestosterone, 5α-androstene-3β,17β-diol). The 17-oxosteroids such as dehydroepiandrosterone (DHA), DHA sulphate, androstenedione and androsterone appear to be less potent.

5α-Dihydrotestosterone (DHT) is generally considered to be the active androgen at the tissue level. This was first proposed by BRUCHOVSKY and WILSON (1968) for the rat prostate when they observed that nuclei from these glands rapidly converted testosterone to DHT. The enzyme responsible for this, 5α-reductase, is also present in skin (GOMEZ and HSIA 1968; WILSON and WALKER 1969; FLAMIGNI et al. 1971), and the conversion of testosterone to DHT could represent an important step in the stimulation of sebum secretion. EBLING et al. (1971) suggested that this conversion may be under pituitary control. They observed that while testosterone had no effect on sebum secretion after hypophysectomy, an effect was seen with DHT. However, this view is untenable for a number of reasons:

1. Testosterone is active in stimulating sebum secretion after hypophysectomy (NIKKARI and VALAVAARA 1970; THODY and SHUSTER 1970, 1971 a) and this has since been confirmed by EBLING et al. (1975).
2. The sebotrophic activity of DHT is not greater than that of testosterone even after hypophysectomy.
3. The activity of sebaceous 5α-reductase is not decreased after hypophysectomy; on the contrary, it is probably increased (HAY et al. 1976, 1982).

There is evidence which suggests that DHT is not the main stimulator of sebaceous activity. Thus intradermal DHT failed to stimulate sebaceous lipogenesis in human subjects and in some actually decreased it (COOPER et al. 1979). Moreover, the 5α-reduction of testosterone and sebaceous lipogenesis showed no correlation in the sampe pieces of skin from male subjects and in females there was actually a negative correlation (COOPER et al. 1977). On the other hand, a clear correlation was found between sebaceous lipogenesis and the metabolism of DHA (HAY et al. 1977). Similar findings were made in a female patient with clonal acne. Thus sebaceous lipogenesis and DHA metabolism were increased on the side of her body that showed the more severe acne, yet conversion to DHT was similar on both sides (COOPER et al. 1976 a). The current view that DHT has a role in human sebaceous gland function (MAUVAIS-JARVIS et al. 1981; see Vol. II) must therefore be questioned. Its importance in sebaceous control in the rat is also open to question. As mentioned above, 5α-reduction is not always related to sebaceous activity. Furthermore inhibition of 5α-reductase with 4-androstan-3-one,17β-carboxylic acid (HSIA and VOIGT 1974) produced a small stimulation of sebum secretion when given with testosterone in the rat (THODY and SHUSTER 1975, unpublished observations). DHT would therefore appear to be a weak androgen agonist on sebaceous cells.

2. Oestrogens

Oestradiol at a dose of 2–4 µg/day has been shown to decrease sebaceous gland size in intact, hypophysectomised and adrenalectomised female rats (EBLING 1963). A similar dose also reduced the rate of sebum production (EBLING and SKINNER 1967) although others (SHUSTER and THODY 1974) were unable to confirm this and in ovariectomised rats found only a small decrease in sebum secretion after a very high dose (40 µg ethynyl oestradiol/day for 3 weeks). There are reports that high doses of ethynyl oestradiol (0.1–0.25 mg/day) will suppress sebum secretion in the human male and it is thought that this may be indirect through an action on the pituitary (STRAUSS and POCHI 1963).

There is, however, little doubt that oestrogens can stimulate the sebaceous glands but this action seems to be restricted to specialised glands. Thus there are reports that oestrogens will increase the size of the ventral gland in gerbils (GLENN and GRAY 1965) and the preputial gland in the rat (GLENN et al. 1959). Although others have found no effects on preputial gland size (THODY and DIJKSTRA 1978), oestrogens have been shown to stimulate preputial lipogenesis (ALVES et al. 1983) and it is likely that this is related to the production of preputial gland pheromones (THODY and DIJKSTRA 1978; LUCAS et al. 1982). These effects on preputial lipogenesis are sex related, for whilst occurring in both ovariectomised and intact female rats they were not seen in castrated or intact males (ALVES et al. unpublished observations).

There is also little doubt that oestrogens can act directly upon the preputial glands and receptors for oestradiol have been demonstrated (PIETRAS 1981). In addition in vitro effects on lipogenesis have been demonstrated (ALVES et al. 1983).

Whether other sebaceous glands can respond in the same way to oestrogens or whether these effects are specific to pheromone-producing glands is not yet known. If the latter is the case then the sebaceous glands in the human breast could be under an oestrogen-stimulated control.

3. Progesterone

Whether or not progesterone stimulates sebaceous gland activity has been a matter of some debate. HASKIN et al. (1953) reported an increase in sebaceous gland size in the rat. EBLING (1948, 1961), on the other hand, found no effect with 1 mg progesterone/day, and although a higher dose (10 mg/day) produced a small increase in gland size there was no change in sebum production (EBLING et al. 1969 b). Progesterone has also been reported to have no effect on sebaceous gland activity in man (STRAUSS and KLIGMAN 1961). Recent evidence suggests that the response to progesterone depends upon the sex of the animal and also changes in the early endocrine environment. Thus progesterone stimulated sebum secretion in adult female rats but in the male a similar effect was only seen after castration. In intact males progesterone inhibited sebum secretion, presumably through an anti-androgen action. Moreover, the degree of stimulation in castrated and ovariectomised females was found to be related to the timing of gonadectomy and was most pronounced when this was carried out pre-pubertally (SHUSTER et al. 1977).

4. Glucocorticoids

There is little information on the effects of glucocorticoids on sebaceous gland activity. Sebum excretion is decreased in patients with adrenal insufficiency (POCHI et al. 1963; GOOLAMALI et al. 1974). However, replacement therapy with glucocorticoids had no effect on sebum excretion, which suggests that the decreases result from the loss of adrenal androgen rather than glucocorticoids. Similar findings have been reported in rats (THODY and SHUSTER 1971 b; SHUSTER and THODY 1974).

5. Thyroid Hormones

The thyroid is involved in the control of the sebaceous glands. In the rat thyroidectomy decreases sebum secretion and thyroxine will reverse this by an action which is independent of testosterone (THODY and SHUSTER 1972 a). In man it appears that thyroid hormones are important for sebaceous gland function but may only operate in the hypothyroid and euthyroid ranges (GOOLAMALI et al. 1976).

6. Insulin

Although it seems likely that insulin may be necessary for the control of sebaceous lipogenesis, there is little evidence to support this. However, TOH (1982) has recently shown that sebum secretion is decreased in rats with streptozotocin-induced diabetes.

7. The Pituitary

The overall control of the sebaceous glands is by the pituitary. Hypophysectomy decreases sebaceous volume and the rate of sebum secretion in the rat to a level which is lower than that seen after castration (HASKIN et al. 1953; EBLING et al. 1969 a, b; NIKKARI and VALAVAARA 1969, 1970; THODY and SHUSTER 1970). Patients with hypopituitarism also show a low rate of sebum excretion (GOOLAMALI et al. 1973). The pituitary exerts its control either through its various target glands or via the direct action of its hormones on the sebaceous glands.

a) Gonadotrophic Hormones

Gonadotrophic hormones act by stimulating the secretion of gonadal steroids and have no effect in rats that have been castrated and hypophysectomised (SHUSTER and THODY 1974). Gonadotrophic hormones have also been reported to stimulate sebum secretion in the human male (STRAUSS and POCHI 1963).

b) Thyrotrophic Hormone

Thyrotrophic Hormone (TSH) increases sebum secretion in hypophysectomised rats (THODY and SHUSTER 1972a). This action involves stimulation of the thyroid and not a direct action on the sebaceous glands (EBLING et al. 1970) since no effect was seen after thyroidectomy (THODY and SHUSTER 1972a).

c) Adrenocorticotrophic Hormone

The sebotrophic effect of adrenocorticotrophic hormone (ACTH), like that of the gonadotrophins and TSH, is mediated via its target gland (THODY and SHUSTER 1971b). This is presumably through stimulation of adrenal androgens and, possibly, progesterone, for glucocorticoids would appear to have little effect on sebum secretion in the rat (see above).

d) Growth Hormone and Prolactin

Growth hormone appears to increase sebum secretion and sebaceous lipogenesis through its general somatotrophic action (NIKKARI and VALAVAARA 1969; EBLING et al. 1969a; SHUSTER and THODY 1974). It therefore has no specific effect on sebaceous glands. Similarly, prolactin is not sebotrophic.

It has been suggested that growth hormone and prolactin increase the effect of testosterone on the sebaceous glands (EBLING et al. 1969a), but this has not been confirmed (NIKKARI and VALAVAARA 1970; SHUSTER and THODY 1974). There is evidence that growth hormone and α-melanocyte stimulating hormone (α-MSH) may act together to increase sebaceous lipogenesis in both cutaneous sebaceous glands and preputial glands (COOPER et al. 1976b). Growth hormone may also have a stimulatory effect in man and may account for the enhanced sebum excretion rate in acromegalics (BURTON et al. 1972b).

e) α-Melanocyte Stimulating Hormone

The decrease in sebum secretion and preputial gland weight that follows posterior hypophysectomy (THODY and SHUSTER 1972b) has been shown to be due to the removal of the pars intermedia, the source of MSH (THODY and SHUSTER 1973). α-MSH has been shown to stimulate sebum secretion in hypophysectomised rats and to completely restore it to normal levels in posterior hypophysectomised rats (THODY and SHUSTER 1973). Skin lipid biosynthesis, which is decreased after posterior hypophysectomy, was also restored to normal after administration of α-MSH (COOPER et al. 1974).

Although completely hypophysectomised rats showed an increase in sebum secretion after α-MSH, in contrast to the response seen in posterior hypophysectomised animals, the level of sebum secretion did not return to normal (THODY and SHUSTER 1973, 1975). This implies that the anterior pituitary is necessary for α-MSH to exert its full effect. As discussed above, the anterior pituitary has its main effect on the sebaceous glands through the control of androgen production. When hypophysectomised rats received a physiological dose of testosterone together with α-MSH a synergistic increase in sebum secretion occurred, with a restoration to normal levels (SHUSTER and THODY 1974; THODY and SHUSTER 1975).

EBLING et al. (1975) have since confirmed that α-MSH and testosterone act together in a synergistic manner to stimulate sebum secretion. Modified sebaceous glands such as the preputial glands also show this synergism (KRAHENBUHL and DESAULLES 1969; THODY and SHUSTER 1975).

The mechanism of this synergism between α-MSH and testosterone is not yet clear. Since α-MSH also synergises with DHT and 5α-androstanediol (BOWDEN et al. 1976; COOPER et al. 1975), it is unlikely that the synergism is due to an enhanced conversion of testosterone on its 5α-metabolites (see above). Furthermore, α-MSH does not alter testosterone metabolism in skin (THODY and SHUSTER 1976). While testosterone may affect both cellular proliferation and lipogenesis, α-MSH acts predominantly on the latter and, in particular, on wax ester biosynthesis (COOPER et al. 1974; THODY et al. 1976). These two different sites of action could account for the synergism of α-MSH and testosterone on the sebaceous glands. α-MSH also acts synergistically with progesterone to stimulate sebum secretion and preputial gland size (KRAHENBUHL and DESAULLES 1969; THODY and SHUSTER 1975) but has no such effect with the oestrogens. Thus although oestrogens may stimulate lipogenesis in preputial cells (ALVES et al. 1983, 1986) their mode of action is different to that of progesterone and androgens. α-MSH and oestrogens may nevertheless act synergistically to stimulate the secretion of preputial gland odours (DONOHOE et al. 1981).

f) Pituitary Peptides and Human Sebaceous Glands

There is no evidence to suggest that MSH and related peptides are sebotrophic in man. Although β-lipotrophin (β-LPH) is present in the human and is released together with ACTH, no correlation has been found between circulating levels and sebum excretion (SHUSTER et al. 1976). L-Dopa decreases the seborrhoea of parkinsonism but has no effect in normal subjects (BURTON and SHUSTER 1970) and it was therefore postulated that the seborrhoea was due to a pituitary hormone under an inhibitory control such as an MSH peptide. There are a number of other situations such as late pregnancy, suckling and administration of phenothiazines (BURTON and SHUSTER 1970; BURTON et al 1971, 1973, 1975; SHUSTER et al. 1973) where the increase in sebum excretion could be due to the action of such a hormone, possibly analogous to that of α-MSH in the rat. This particular peptide is found in the human circulation, but with the possible exception of an increase in late pregnancy it is normally present at very low concentrations (CLARK et al. 1978). Recent evidence suggests that it may also be raised in Cushing's syndrome (THODY et al. 1985), a condition in which there is a decrease in sebum excretion (GOOLAMALI et al. 1974). α-MSH is present in increased amounts in the foetal human pituitary (SILMAN et al. 1976) and this could explain the enhanced sebum production in late foetal and neonatal life.

g) Sebotrophic Activity of Peptides Related to MSH

It is now known that MSH and related peptides are derived from a precursor protein pro-opiomelanocortin (NAKANISHI et al. 1979). Of these peptides ACTH, β-LPH and α-MSH are sebotrophic in the rat. Although part of the effect of ACTH is undoubtedly mediated by the adrenals (see above), the possibility that it also has a direct effect has not been excluded. β-LPH, which is the immediate precursor of β-MSH, is a potent stimulator of sebum secretion (THODY and SHUSTER 1971 c). All of these peptides contain a common heptapeptide sequence (ACTH 4-10) which is responsible for their similar biological properties (melanocyte stimulating activity, central actions). However, unlike α-MSH this particular sequence, as well as ACTH 1-10, failed to increase preputial gland weight (MEDDIS et al. 1978). The sequence responsible for the sebotrophic activity of MSH and related peptides is therefore unknown.

8. The Early Endocrine Environment

There is now much evidence to suggest that hormones exert permanent effects during foetal and neonatal life. This was first demonstrated when THODY and SHUSTER (1971 d) observed that administration of androgen during the first week of life produced an increase in sebum secretion in female rats that was still evident during adult life. Removal of the tests at birth, on the other hand, reduced adult sebum secretion to a level approaching that of females (TOH 1980 a). Although castration at 4 weeks of age produced a smaller reduction (TOH 1980 a), sebum secretion during adult life in rats castrated at 3 weeks of age was lower than in those castrated during adult life (ARCHIBALD 1973). The ovaries do not seem to exert the same long-lasting influence, for their removal has no effect on adult sebum secretion rates (TOH 1980 b). Other sebotrophic hormones are also important in early life and α-MSH given in utero and to a lesser extent during neonatal life permanently increases sebum secretion (THODY et al. 1978). Neonatal thyroxine, on the other hand, decreases sebum secretion in adult rats, but this effect may be related to a decreased thyroid function (TOH 1979).

As well as affecting the intrinsic basal activity of the sebaceous glands, the early endocrine environment also affects the way they respond to sebotrophic hormones in adult life. Thus gonadectomy at different stages produces a totally different response to progesterone, with the greatest response being seen in rats gonadectomised at 3 weeks of age (SHUSTER et al. 1977). The reasons for this are not yet clear, but it seems possible that endocrine ablation at different stages will permanently induce different enzymic steps in androgen metabolism or changes in steroid receptors. Whatever the mechanism, it appears that the sebaceous gland response can be conditioned throughout life by changes in the early endocrine environment. These various effects of hormones from direct modulation to modulation and regulation of responses and finally to irreversible gene expression are discussed by SHUSTER (1982) and may be of importance in disorders such as primary cutaneous virilism (idiopathic hirsutism).

References

Alves A, Thody A, Shepherd L, Shuster S (1983) Estrogen stimulates preputial lipogenesis. J Invest Dermatol 80:358

Alves A, Thody A, Fisher C, Shuster S (1986) Measurement of lipogenesis in isolated preputial gland cells of the rat and the effect of oestrogen. J Endocrinol 109:1–7

Archibald A (1973) A study of factors influencing sebaceous gland activity in the rat. PhD Thesis, University of Newcastle upon Tyne

Archibald A, Shuster S (1967) Bioassay of androgen using the rat sebaceous gland. J Endocrinol 37:22
Archibald A, Shuster S (1969) The bioassay of androgens and antiandrogens using sebum secretion in the rat. Proc R Soc Med 62:887–888
Bowden PE, Meddis D, Cooper MF, Thody AJ, Shuster S (1976) Effects of 5α-reduced androgens on preputial gland size and lipogenic activity. Biochem Soc Trans 4:795–797
Bruchovsky N, Wilson JD (1968) The conversion of testosterone to 5α-androstan-17β-ol-one by rat prostate in vivo and in vitro. J Biol Chem 243:5953–5960
Burton JL (1970) The physical properties of sebum in acne vulgaris. Clin Sci 39:757
Burton JL, Shuster S (1970) Effect of L-dopa on seborrhoea of parkinsonism. Lancet II:19–20
Burton JL, Cunliffe WJ, Shuster S (1970) Circadian rhythm in sebum excretion. Br J Dermatol 82:497–501
Burton JL, Libman LJ, Hall R, Shuster S (1971) Laevo-dopa in acne vulgaris. Lancet II:370
Burton JL, Cartlidge M, Shuster S (1972a) Variation in sebum secretion during the menstrual cycle. Acta Derm Venereol (Stockh) 53:81–84
Burton JL, Libman LJ, Cunliffe WJ, Wilkinson R, Hall R, Shuster S (1972b) Sebum excretion in acromegaly. Br Med J 1:406–408
Burton JL, Shuster S, Cartlidge M, Libman LJ, Martell U (1973) Lactation, sebum excretion and melanocyte stimulating hormone. Nature 243:349–350
Burton JL, Shuster S, Cartlidge M (1975) The sebotrophic effect of pregnancy. Acta Derm Venereol (Stockh) 55:11–13
Clark D, Thody AJ, Bowers H, Shuster S (1978) Immunoreactive α-MSH in human plasma in pregnancy. Nature 274:163–164
Cooper MF, Thody AJ, Shuster S (1974) Hormonal regulation of cutaneous lipogenesis: effects of hypophysectomy, posterior hypophysectomy and α-melanocyte stimulating hormone treatment. Biochim Biophys Acta 360:193–204
Cooper MF, Bowden PE, Meddis D, Thody AJ, Shuster S (1975) The effect of α-MSH and various androgens on preputial gland activity. Acta Endocrinol [Suppl] 119:250
Cooper MF, Hay JB, McGibbon D, Shuster S (1976a) Androgen metabolism and sebaceous activity in clonal acne. J Invest Dermatol 66:261
Cooper MF, Meddis D, Bowden PE, Thody AJ, Shuster S (1976b) A comparison of the effects of growth hormone and α-MSH on the sebaceous gland. J Endocrinol 72:29P–30P
Cooper MF, McGibbon D, Shuster S (1977) Response of sebaceous lipogenesis to testosterone in acne. J Invest Dermatol 68:255
Cooper MF, McGibbon D, Wilson PD, Shuster S (1979) Androgenic control of the human sebaceous gland. J Invest Dermatol 72:267
Cotterill JA, Cunliffe WJ, Williamson B, Bulusu L (1972) Age and sex variation in skin surface lipid composition and sebum excretion rate. Br J Dermatol 87:333–340
Cunliffe WJ, Shuster S (1969) Pathogenesis of acne. Lancet I:685–687
Donohoe SM, Thody AJ, Shuster S (1981) Effect of α-melanocyte stimulating hormone and ovarian steroids on preputial gland function in the female rat. J Endocrinol 90:53–58
Downing DT, Strauss JS, Norton LA, Pochi PE, Stewart ME (1977) The time course of lipid formation in human sebaceous glands. J Invest Dermatol 69:407–412
Ebling FJ (1948) Sebaceous glands. I. The effect of sex hormones on the sebaceous glands of the female albino rat. J Endocrinol 5:297–302
Ebling FJ (1961) Failure of progesterone to enlarge sebaceous glands in the female rat. Br J Dermatol 73:65–68
Ebling FJ (1963) Hormonal control of sebaceous glands in experimental animals. In: Montagna W, Ellis RA, Silvers AF (eds) Advances in biology of skin, vol 4, The sebaceous glands. Pergamon, Oxford, pp 200–210
Ebling FJ, Skinner J (1967) The measurement of sebum production in rats treated with testosterone and oestradiol. Br J Dermatol 79:386–393

Ebling FJ, Ebling E, Skinner J (1969a) The influence of pituitary hormones on the response of the sebaceous glands of the rat to testosterone. J Endocrinol 45:245–256
Ebling FJ, Ebling E, Skinner J (1969b) The influence of the pituitary on the response of the sebaceous and preputial glands of the rat to progesterone. J Endocrinol 45:257–263
Ebling FJ, Ebling E, Skinner J (1970) The effects of thyrotrophic hormone and of thyroxine on the response of the sebaceous glands of the rat to testosterone. J Endocrinol 48:83–90
Ebling FJ, Ebling E, McCaffery V, Skinner J (1971) The response of the sebaceous glands of the hypophysectomised castrated male rat to 5α-dihydrotestosterone, androstenedione, dehydroepiandrosterone and androsterone. J Endocrinol 51:181–190
Ebling FJ, Ebling E, Randall V, Skinner J (1975) The synergistic action of α-melanocyte stimulating hormone and testosterone on the sebaceous, prostate, preputial, Harderian and lachrymal glands, seminal vesicles and brown adipose tissue in the hypophysectomised castrated rat. J Endocrinol 66:407–412
Ellis RA (1967) Eccrine, sebaceous, and apocrine glands. In: Zelickson AS (ed) Ultrastructure of normal and abnormal skin. Lea & Febiger, Philadelphia
Ellis RA, Hendrikson RC (1963) The ultrastructure of the sebaceous glands of man. In: Montagna W, Ellis RA, Silver AF (eds) Advances in biology of skin, vol 4, The sebaceous glands. Pergamon, Oxford
Flamigni CA, Collins WP, Koullapis EN, Craft I, Dewhurst C, Sommerville IF (1971) Androgen metabolism in human skin. J Clin Endocrinol Metab 32:737–743
Fulton JE, Plewig G, Kligman AM (1969) Effect of chocolate on acne vulgaris. JAMA 210:2071–2074
Glenn EM, Gray J (1965) Effect of various hormones on the growth and histology of the gerbil (*Meriones unguiculatus*) abdominal sebaceous gland pad. Endocrinology 76:1115–1123
Glenn EM, Richardson SL, Bowman BJ (1959) A method of assay of anti-tumour activity using a rat mammary fibroadenoma. Endocrinology 64:379–389
Gomez EC, Hsia SL (1968) In vitro metabolism of testosterone-4-^{14}C and Δ^4-androstene-3, 17-dione-4-^{14}C in human skin. Biochemistry 7:24–32
Goolamali SK, Burton JL, Shuster S (1973) Sebum excretion in hypopituitarism. Br J Dermatol 89:21–24
Goolamali SK, Plummer N, Burton JL, Shuster S, Thody AJ (1974) Sebum excretion and melanocyte stimulating hormone in hypoadrenalism. J Invest Dermatol 63:253–255
Goolamali SK, Evered D, Shuster S (1976) Thyroid disease and sebaceous function. Br Med J 1:432–433
Greene RS, Downing DT, Pochi PE, Strauss JS (1970) Anatomical variation in the amount and composition of human skin surface lipids. J Invest Dermatol 54:240–247
Haskin D, Lasher N, Rothman S (1953) Some effects of ACTH, cortisone, progesterone and testosterone on sebaceous glands in the white rat. J Invest Dermatol 20:207–211
Hay JB, Meddis D, Thody AJ, Shuster S (1976) Androgen metabolism in preputial glands of hypophysectomised rats. J Endocrinol 71:96
Hay JB, Cooper MF, McGibbon D, Shuster S (1977) Comparison between sebaceous lipogenesis and androgen metabolism in skin from acne patietns. J Invest Dermatol 68:253
Hay JB, Meddis D, Thody AJ, Shuster S (1982) Mechanism of action of α-melanocyte stimulating hormone in rat preputial glands: the role of androgen metabolism. J Endocrinol 94:289–294
Hsia SL, Voigt V (1974) Inhibition of DHT production: an effective means of blocking androgen action in the hamster sebaceous gland. J Invest Dermatol 62:224–227
Ikai K (1959) Rate of sebum excretion from the glands onto the skin surface. J Invest Dermatol 32:27–33
Kellum RE, Strangfield K, Ray LF (1970) Acne vulgaris. Studies in pathogenesis: triglyceride hydrolysis by *Corynebacterium acnes* in vitro. Arch Dermatol 101:41–47
Kirk E (1948) Quantitative determination of the skin lipid secretion in middle aged and old individuals. J Gerontol 3:251–266

Krahenbuhl C, Desaulles PA (1969) Interactions between α-MSH and sex steroids on the preputial glands of female rats. Experientia 25:1193–1195
Lindholm JS, Downing DT (1980) Occurrence of squalene in skin surface lipids of the otter, the beaver and the kinkajou. Lipids 15:1062–1063
Lucas PD, Donohoe SM, Thody AJ (1982) The role of estrogen and progesterone in the control of preputial gland sex attractant odors in the female rat. Physiol Behav 28:601–607
Mauvais-Jarvis P, Kuttenn F, Mowszowicz I (1981) Hirsutism. Springer, Berlin Heidelberg New York
McDonald I, Clarke G (1970) Variations in the levels of cholesterol and triglyceride in the skin surface fat during the menstrual cycle. Br J Dermatol 83:473–476
Meddis D, Thody AJ, Shuster S, Greven HM (1978) The effect of α-melanocyte stimulating hormone and various analogues on preputial gland weight in the rat. IRCS Medical Science 6:433
Nakanishi S, Inone A, Kita T, Nakamura M, Chang ACY, Cohen SN, Numa S (1979) Nucleotide sequence of cloned cDNA for bovine corticotropin-β-lipotrophin precursor. Nature 278:423–427
Nicolaides N, Rothman S (1953) Studies on the chemical composition of human hair fat. J Invest Dermatol 21:9–14
Nicolaides N, Wells GC (1957) On the biogenesis of the free fatty acids in human skin surface fat. J Invest Dermatol 29:423–433
Nikkari T, Valavaara M (1969) The production of sebum in young rats. Effects of age, sex, hypophysectomy and treatment with somatotrophic hormone and sex hormones. J Endocrinol 43:113–118
Nikkari T, Valavaara M (1970) The influence of age, sex, hypophysectomy and various hormones on the composition of the skin surface lipids of the rat. Br J Dermatol 183:459–472
Pietras RJ (1981) Sex pheromone production by preputial gland: the regulatory role of estrogen. Chem Sens 6:391–408
Plewig G, Christophers E (1974) Renewal rate of human sebaceous glands. Acta Derm Venereol (Stockh) 54:177–182
Pochi PE, Strauss JS, Mescon H (1963) The role of adrenocortical steroids in the control of human sebaceous gland activity. J Invest Dermatol 41:391–399
Pochi PE, Downing DT, Strauss JS (1970) Sebaceous gland response in man to prolonged total caloric deprivation. J Invest Dermatol 55:303–309
Robin M, Kepecs JC (1953) The relationship between certain emotional states and the rates of secretion of sebum. J Invest Dermatol 20:373–384
Russell MJ (1976) Human olfactory communication. Nature 260:520–522
Rust S, Harth P, Herrmann F (1968) Untersuchungen der freien Fettsäuren im Hautoberflächenfett von Hautgesunden. Arch Klin Exp Dermatol 231:300–310
Rust S, Harth P, Herrmann F (1970) Untersuchungen der freien Fettsäuren in Hautoberflächenfett von Psoriatikern. Arch Klin Exp Dermatol 238:207–216
Shuster S (1976) Hypothesis: the biological purpose of acne. Lancet I:1328
Shuster S (1982) The sebaceous glands and primary cutaneous virilism. In: Jeffcoate SL (ed) Androgens and antiandrogen therapy. Wiley, Chichester
Shuster S, Thody AJ (1974) The control and measurement of sebum secretion. J Invest Dermatol 62:172–190
Shuster S, Thody AJ, Goolamali SK, Burton JL, Plummer NA, Bates D (1973) Melanocyte stimulating hormone and parkinsonism. Lancet I:463–465
Shuster S, Goolamali SK, Smith AG, Thody AJ, Alvarez-Ude F, Kerr DNS (1976) Decreased sebum excretion in chronic renal failure. Br Med J 1:23–24
Shuster S, Hinks WM, Thody AJ (1977) Effect of sex and age at gonadectomy on the sebaceous response to progesterone. J Endocrinol 73:67–70
Silman RE, Chard T, Lowry PJ, Smith I, Young IM (1976) Human foetal pituitary peptides and parturition. Nature 260:716–718
Stoddart DM (1980) Aspects of the evolutionary biology of mammalian olfaction. In: Stoddart DM (ed) Olfaction in mammals. Academic, London

Strauss JS, Kligman AM (1961) Effect of progesterone and progesterone like compounds on the human sebaceous glands. J Invest Dermatol 36:309–318
Strauss JS, Pochi PE (1963) The hormonal control of sebaceous glands. In: Montagna W, Ellis RA, Silvers AF (eds) Advances in biology of skin, vol 4. Sebaceous glands. Pergamon, Oxford
Strauss JS, Downing DT, Ebling FJ (1983) Sebaceous glands. In: Goldsmith LA (ed) Biochemistry and physiology of the skin, vol II. Oxford University Press, New York
Thody AJ, Dijkstra H (1978) Effect of ovarian steroids on preputial gland odours in the female rat. J Endocrinol 77:397–403
Thody AJ, Shuster S (1970) The effect of hypophysectomy and testosterone on the activity of the sebaceous glands of castrated rats. J Endocrinol 47:219–224
Thody AJ, Shuster S (1971a) The effect of hypophysectomy on the response of the sebaceous gland to testosterone propionate. J Endocrinol 49:329–333
Thody AJ, Shuster S (1971b) The effect of adrenalectomy and adrenocorticotrophic hormone on sebum secretion in the rat. J Endocrinol 49:325–328
Thody AJ, Shuster S (1971c) Sebotrophic activity of β-lipotrophin. J Endocrinol 50:533–534
Thody AJ, Shuster S (1971d) Increased sebum secretion in adult female rats after neonatal treatment with testosterone propionate. J Endocrinol 49:677–681
Thody AJ, Shuster S (1972a) A study of the relationship between the thyroid gland and sebum secretion in the rat. J Endocrinol 54:239–244
Thody AJ, Shuster S (1972b) The control of sebum secretion by the posterior pituitary. Nature 237:346–347
Thody AJ, Shuster S (1973) A possible role of MSH in the mammal. Nature 245:207–209
Thody AJ, Shuster S (1975) Control of sebaceous gland function in the rat by α-melanocyte stimulating hormone. J Endocrinol 64:504–510
Thody AJ, Shuster S (1976) α-MSH and the metabolism of testosterone in the skin of the rat. J Invest Dermatol 66:264
Thody AJ, Cooper MF, Bowden PE, Meddis D, Shuster S (1976) Effect of α-melanocyte stimulating hormone and testosterone on cutaneous and modified sebaceous glands in the rat. J Endocrinol 71:279–288
Thody AJ, Meddis D, Shuster S (1978) Increased sebaceous gland activity in the adult rat after α-melanocyte stimulating hormone treatment during early life. J Invest Dermatol 70:328–330
Thody AJ, Fisher C, Kendal-Taylor P, Jones MT, Price J, Abraham RR (1985) The measurement of immunoreactive α-melanocyte stimulating hormone in human plasma. Acta Endocrinol 110:313–318
Thomas SE, Harrington CI, Ebling FJ (1983) An investigation into the sebum excretion rate above and below the lesion in paraplegic patients. Br J Dermatol 109:696
Thomas SE, Conway J, Ebling FJ, Harrington CI (1985) Measurement of sebum excretion rate and skin temperature above and below the neurological lesion in paraplegic patients. Br J Dermatol 112:569–573
Toh YC (1979) Effect of neonatal administration of thyroxine on the rate of sebum production in rats. J Endocrinol 83:199–203
Toh YC (1980a) Role of the gonads in the regulation of sebaceous glands in rats: comparison of the effects of castration at birth and after birth. J Endocrinol 85:261–265
Toh YC (1980b) Effect of ovariectomy at birth on the regulation of sebaceous glands in rats. J Endocrinol 86:179–182
Toh YC (1982) Effect of streptozotocin-induced diabetes on the activity of the sebaceous glands in rats. Endokrinologie 80:56–59
Wheatley VR, Potter JER, Lew G (1979) Sebaceous gland differentiation. II. The isolation, separation and characterization of cells from mouse preputial gland. J Invest Dermatol 73:291–296
Wilson JD, Walker JD (1969) The conversion of testosterone to 5α-androstan-17β-ol-3-one (dihydrotestosterone) by skin slices of man. J Clin Invest 48:371–379
Wolff HG, Lorenz TH, Graham DT (1951) Stress, emotions and human sebum: their relevance to acne vulgaris. Trans Assoc Am Physicians 64:435–444

CHAPTER 15

Metabolism of Sex Steroids

F. WRIGHT and P. MAUVAIS-JARVIS

A. Introduction

In recent years, human skin has been recognised as a target organ for several steroid hormones, among them sex steroids, i.e. androgens, oestrogens and progesterone. Human skin participates not only in the catabolism of these steroids but also in the synthesis of active hormones from steroid precursors supplied by the blood vessels.

The androgen model in this respect is unique. It has long been observed that abnormally excessive androgen action on human skin is responsible for undesirable effects such as acne and hirsutism in women and alopecia in both sexes whereas abnormally low androgen levels, in eunuchs for example, lead to a lack of beard growth and absence of acne in men together with regression of the more classical target organs for androgens such as the prostate.

B. Hydroxysteroid Dehydrogenase Activities

The interconversion of hydroxyl and keto groups in cutaneous steroid metabolism is a most frequent phenomenon. These reactions take place at various positions of the steroid molecule and are very often associated with the microsomal fraction of the cell-free preparations of human skin. They are linked to the $NAD^+/NADH,H^+$ rather than to the $NADP^+/NADPH,H^+$ co-factor system. Soluble enzymes are also present, however, requiring preferentially NADPH as a co-factor.

I. Androgens

1. Dehydroepiandrosterone

Dehydroepiandrosterone [Δ^5-Androstene-3β-ol-20-one (DHA)] metabolism has been studied in skin from shoulder and thigh. Shoulder skin metabolises DHA more actively than does thigh skin. CHAKRABORTY et al. (1970) related these findings to the fact that shoulder skin is an acne area whereas thigh skin is not. They also concluded that variations in Δ^5-3β-hydroxysteroid dehydrogenase may interfere with hair growth of sexual areas of human skin.

2. Δ^4-Androstene-3,17-dione

The enzyme 17β-hydroxysteroid dehydrogenase converts Δ^4-androstene-3,17-dione (Δ^4) to testosterone. In human skin homogenates this activity was first demonstrated by GOMEZ and HSIA (1968) and subsequently by FLAMIGNI et al. (1971).

The transformation of Δ^4 to testosterone is a reversible reaction. The activity of 17β-hydroxysteroid dehydrogenase has been investigated in both growing and resting hair follicles by ADACHI (1973). In the growing hair follicles (anagen phase) testosterone, when incubated, is transformed to Δ^4 as the major catabolic product. However, in parallel, the growing hair follicle produces more dihydrotestosterone (DHT), the active androgen, than the resting follicle (telagen phase).

II. Oestrogens

The formation of oestrone (E_1) from 17β-oestradiol (E_2) and not the reverse reaction predominates in human skin from both sexes (FROST et al. 1966). However, in vaginal mucosa, the formation of E_2 is favoured over that of E_1. Since the latter tissue is a primary target tissue for oestrogen action and since E_2 rather than E_1 is considered as the active oestrogen, this shift in equilibrium correlates with the local action of oestrogens.

A specific receptor protein which has a high affinity for E_2 and not E_1 has been inferred from the above observed shift of equilibrium. This E_2 receptor protein has been reported in human skin (MOWSZOWICZ et al. 1981 b).

It is interesting to point out that the 17β-hydroxysteroid dehydrogenase which transforms E_2 to E_1 is the same enzyme as is responsible for Δ^4-androstenedione metabolism to testosterone.

III. Progesterone

A 20α-hydroxysteroid dehydrogenase transforms progesterone, the luteal hormone, into 20α-derivatives in human skin (FROST et al. 1969).

C. 5α Reduction of Testosterone and Progesterone

I. Testosterone

The possibility that testosterone might be a circulating pre-hormone was first envisaged by BRUCHOVSKY and WILSON (1968 a, b). Evidence has since accumulated and DHT is now recognised beyond any doubt as the active androgen at the level of most target tissues.

In human skin the 5α reduction of testosterone to DHT is a very active pathway, as first demonstrated by VOIGT et al. (1970). This stereospecific 5α reduction requires a minimal structure whose characteristics are a 3-keto group, a double bond in the C4 position and a 17β derivative (VOIGT and HSIA 1973). This 5α reduction is associated with the microsomal fraction of human skin preparations (Fig. 1) and requires NADPH as co-factor.

WILSON and WALKER (1969) were the first to report in detail the regional variations in the 5α-reductase activity of human skin. The rate of formation of DHT in skin specimens from non-perineal sites (thigh, back, leg, breast etc.) is low

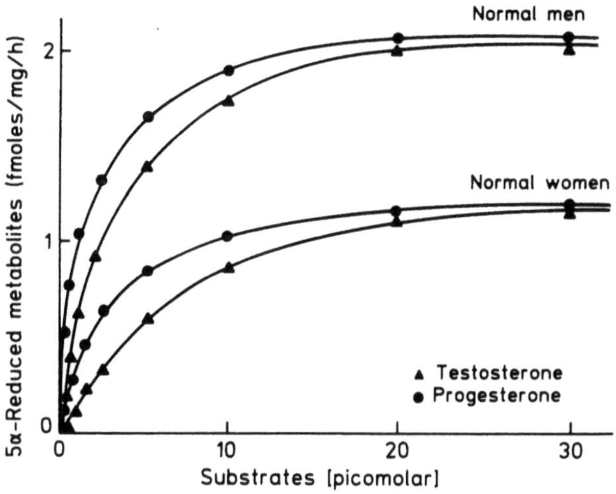

Fig. 1. Kinetic studies of 5α-reductase as a function of substrate concentration in human skin microsomes (1 mg proteins) supplemented with NADPH 2×10^{-5} *M*. (Wright et al. 1980)

(50 pmol/h for 100 mg tissue) and contrasts with that observed in perineal tissues (foreskin, scrotum, labia mojora and clitoris), which reaches a maximum of 1070 pmol/h for 100 mg tissue.

After tritiated testosterone incubations together with skin homogenates the mean recovery of 5α-reduced metabolites observed by our group was low in prepubertal subjects of both sexes (Mauvais-Jarvis et al. 1975, 1976; Mauvais-Jarvis 1977; Kuttenn and Mauvais-Jarvis 1975). It increased dramatically at puberty in boys to reach 150 ± 30 fmol/mg skin. It also increased in women but remained lower (50 ± 10 fmol/mg skin). These results suggest the presence of a sex-dependent enzyme in this anatomical region where the growth of hair is known to be androgenically controlled.

II. Progesterone

Frost et al. (1969) were the first to report progesterone 5α reduction of skin slices and Mauvais-Jarvis et al. (1969b) confirmed these results in vivo.

In human skin microsomes (Wright and Giacomini 1980) the 5α reduction of ^{3}H-progesterone was studied together with that of ^{3}H-testosterone in men and women (Fig. 2). These results show that progesterone is a better substrate for the enzyme than testosterone itself. Moreover, the apparent maximal velocities are identical, a fact which is in favour of a unique enzyme for both substrates. The mechanism of this enzymic inhibition was foreseen by Voigt and Hsia (1973). Progesterone, indeed, is an active anti-androgen at this key enzyme level when applied in vivo (Fig. 2).

To summarise here, human skin is capable of retaining circulating androgens and of metabolising them to DHT, thus producing the active androgen necessary to suit its local needs as an androgen target organ. The regional variations in 5α-reductase activity measured in human skin also suggest a predominant role of DHT in the primary sexual tissues when compared with other body sites. By con-

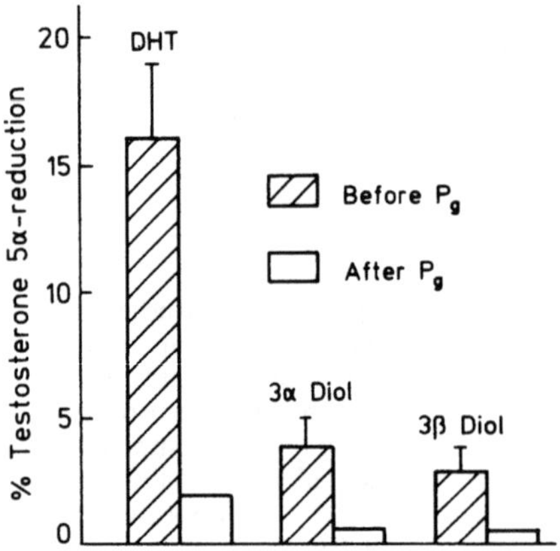

Fig. 2. Conversion rate of testosterone to DHT and androstanediols by skin homogenates from five normal men volunteers before and after topical application of 100 mg progesterone (mean ± SE). (MAUVAIS-JARVIS et al. 1975)

trast, the differentiation of the secondary sexual skin (pubis) is under the control of an androgen-dependent 5α-reductase.

D. 3α- and 3β-Hydroxysteroid Dehydrogenases

I. Androgens

In human skin, as in all androgen target organs, DHT does not remain as such but is further metabolised to 3α- and 3β-diols. Both 3α- and 3β-hydroxysteroid dehydrogenases (3α- and 3β-HSD) are present in human skin cytosol and microsomes (WRIGHT and GIACOMINI 1980). The 3α-HSD is predominantly NADPH linked and more active in the soluble fraction than in the microsomes. The 3β-HSD is preferably NADH linked and more active in the microsomes than in the cytosol. These very active enzymes therefore take part in the control of DHT availability in the androgen target cells. Moreover, some data are now available suggesting that these two metabolites of DHT could exert certain specific androgenic activities and may be mediators per se. This was first demonstrated in prostate organ culture (ROBEL et al. 1971). Neither 3α- nor 3β-HSD is reversible in human skin (WRIGHT et al. 1980) whereas 3α-HSD is reversible and 3β-HSD nonreversible in prostate (BRUCHOVSKY and LIESKOVSKY 1979).

Human scalp and back skin, two regions containing large sebaceous glands, metabolise testosterone to diols (STEWART et al. 1977). 3β-diol is a major metabolite in vitro. This result is in agreement with the data of EBLING (1963) in rodents suggesting that this metabolite could be a key molecule in the stimulation of sebum production.

II. Progesterone

Both progesterone and its 5α-reduced metabolite 5α-pregnane-3,20-dione (DHP) inhibit 3α- and 3β-HSD activities in human skin cytosol. Progesterone does not act as a classical competitive inhibitor of the enzymes whereas DHP does (GIACOMINI and WRIGHT 1980).

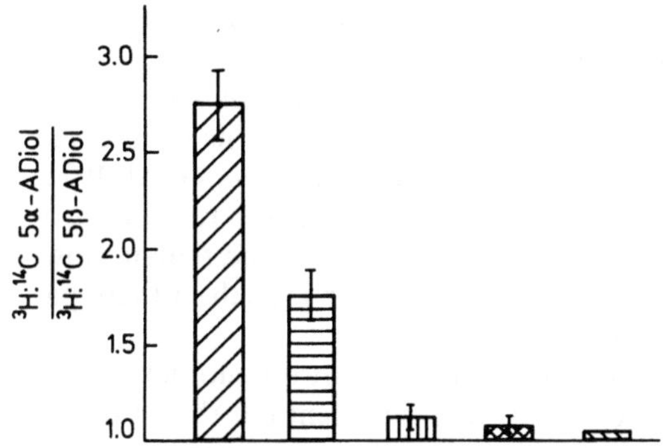

Fig. 3. Cutaneous coefficient of testosterone 5α-reduction (mean ±SD) as expressed by the ratio of the labelled 5α-reduced metabolites over the ratio of the labelled 5β-reduced metabolites of ^{14}C-testosterone administered i.v. and ^{3}H-testosterone administered percutaneously. ▨, normal men; ▤, normal women; ▥, testicular feminisation syndrome; ⊠, hypogonadal men; ▧, one patient with 5α-reductase deficiency. (KUTTENN et al. 1980)

E. Normal Control of Androgen Metabolism in Human Skin

Steroid hormone metabolism can be studied after percutaneous administration. Around 10% of the applied dose penetrates the skin (FELDMAN and MAIBACH 1969; MAUVAIS-JARVIS et al. 1969a).

Percutaneously administered testosterone leads to the production of 5α-reduced metabolites whereas intravenously administered testosterone is mainly 5β-reduced (MAUVAIS-JARVIS et al. 1969a). These in vivo results are in agreement with in vitro results on the metabolism of testosterone in human skin (RONGONE 1966; GOMEZ and HSIA 1968). More 5α-reduced metabolites are derived from percutaneously administered testosterone in men than in women, attesting to a more active 5α-reductase in normal male skin as a whole than in skin of normal women or even hypogonadal men (Fig. 3). MAUVAIS-JARVIS et al. reported that in three hypogonadal men percutaneous DHT for 3 months (50 mg daily) induced the production of 5α metabolites as measured by the above technique followed by normalisation of libido and secondary sex differentiation (pubic hair growth, deepening of the voice etc.).

In fact, there seem to be two different 5α-reductases. First, in external genital areas 5α-reductase seems to be acquired early in embryogenesis (SIITERI and WILSON 1974) and is always very active. However, it awaits androgen production and in the absence of testosterone it remains quiescent for lack of substrate. Second, in pubic skin 5α-reductase is absent or very low in both sexes before puberty and increases sharply at puberty. These results point to an androgen-dependent 5α-reductase activity in pubic skin, a region where secondary sex characteristics are expressed.

F. Abnormal Control of Androgen Metabolism in Human Skin

The complex mechanism of action of androgens is not yet fully understood. However, in the past decade, the initial steps of androgen action have been identified (BRUCHOVSKY and WILSON 1968a, b; BAULIEU and JUNG 1970; MAINWARING 1975). The presence of a specific receptor protein for androgens was thus ascertained in the classical androgen target organs and also in human skin (MOWSZOwicz and Wright 1979; Mowszowicz et al. 1980). This protein is necessary for the androgen target cell to express its identity.

The lack of this protein leads to male pseudo-hermaphroditism where circulating androgens are high but the external genitalia are female. This is observed in the testicular feminisation syndrome in its complete form (Keenan et al. 1975; Mowszowicz et al. 1981a). A consequence of this lack of androgen receptor protein is that the androgen-dependent 5α-reductase is not induced and remains low, which explains the typical absence of pubic or axillary hair growth in these patients (Mauvais-Jarvis et al. 1969a, 1970).

Imperato McGinley et al. (1974) were the first to report on another type of pseudo-hermaphroditism where the receptor protein is present but the 5α-reductase systems are deficient (Kuttenn et al. 1979). Pseudovaginal perineoscrotal hypospadias consists of XY individuals with predominantly female external genitalia at birth who develop a partially normal male virilisation at puberty as evidenced by a deepening of the voice, enlargement of the phallus, descent of the testes etc. Testosterone is within the normal range whereas DHT is very low. This plasma DHT/T ratio, which is already very low under basal conditions, decreases after HCG stimulation, attesting to the defect (Kuttenn et al. 1980). Direct ev-

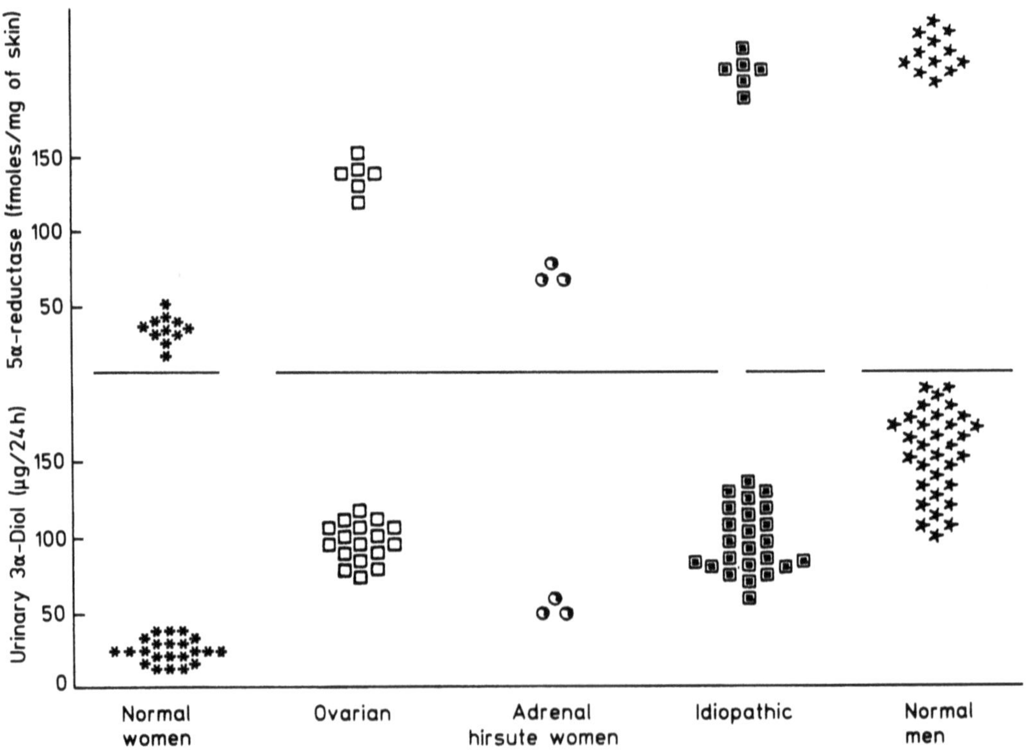

Fig. 4. *TOP:* Percentage of testosterone converted to 5α-reduced metabolites (dihydrotestosterone + 3α- and 3β-androstanediols) in homogenates of pubic skin. *BOTTOM:* Excretion rates of urinary 5α-androstane-3α,17β-diol (3α-diol) in the same patients or groups of patients

idence of a lack of 5α-reductase was observed after in vitro incubations of pubic and perineal skin homogenates. No 5α-reduced metabolites were obtained in spite of a normal level of receptor protein for androgens (KUTTENN et al. 1979).

Hirsutism in women can arise from increased production of androgens by the adrenals or the ovaries (MAHOUDEAU et al. 1971). However, increased utilisation of circulating androgens by the skin also results in hirsutism. This parameter has been investigated in our group by two different techniques. First, in vivo, by measuring urinary 3α-diol, a unique index of androgen utilisation by its target organs (WRIGHT et al. 1978). Second, in vitro, by measuring pubic skin 5α reduction of testosterone on homogenates. In all patients with hirsutism but especially in those with idiopathic hirsutism, where androgen production presents a minimal increase, both 3α-diol and cutaneous 5α-reduction are high (Fig. 4).

G. Conclusion

Recently a large body of evidence has accumulated showing the important role of steroid metabolism in human skin. Human skin is capable of retaining weak circulating androgens (DHEAS, DHA, Δ^4) or testosterone and of metabolising them into a more potent molecule, DHT, which is the active mediator in the development of certain local sex characteristics such as pilosebaceous gland activity via its androgen-dependent 5α-reductase activity. Another metabolite, 3β-diol, could be responsible for excessive sebum production leading to acne. In addition, human skin is a readily accessible androgen target organ which can be used to investigate the action of anti-androgens and particularly those which could be used in topical form.

One can speculate on the role of oestrogen metabolism in human skin. The question remains, however, of whether this metabolism of oestrogens (17β-dehydrogenase) represents a potentiation of their biological activity on specifically oestrogen-sensitive structures or merely catabolism necessary to inactivate 17β-oestradiol.

Human skin is sensitive to progesterone, a natural antagonistic hormone which is both anti-androgenic and anti-oestrogenic. This property has already been used therapeutically in the treatment of hyperseborrhoea (CHERIF-CHEIKH and DE LIGNIERES 1974) in both men and women via the percutaneous route.

Much has been unveiled on steroid metabolism and action in human skin; however, a lot more of "the iceberg" remains "underwater".

References

Adachi K (1973) The metabolism and control mechanism of human hair follicles. Curr Probl Dermatol 5:37–78

Baulieu EE, Jung I (1970) A prostatic cytosol receptor. Biochem Biophys Res Commun 38:599–606

Bruchovsky N, Lieskovsky G (1979) Increased ratio of 5α-reductase:3 α (β) hydroxysteroid dehydrogenase activities in the hyperplastic human prostate. J Endocrinol 80:289–301

Bruchovsky N, Wilson JD (1968 a) The conversion of testosterone to 5 α-androstane-17β-ol-3-one by rat prostate in vivo and in vitro. J Biol Chem 243:2012–2021

Bruchovsky N, Wilson JD (1968 b) The intranuclear binding of testosterone and 5 α-androstane-17β-ol-3-one by rat prostate. J Biol Chem 243:5953–5960

Chakraborty J, Thomson J, Mac Swenn MP, Muir AV, Calman KC, Grant JK, Milne JA (1970) The in vitro metabolism of 7α-^{3}H-dehydroepiandrosterone by human skin. Br J Dermatol 83:477–482

Cherif-Cheikh JL, de Lignieres B (1974) Traitement de la séborrhée du cuir chevelu par la progestérone percutanée. Sem Hôp Ther 50:489–506

Ebling FJ (1963) Hormonal control of the sebaceous gland in experimental animals. In: Montagna W, Ellis RA, Silver AF (eds) Sebaceous glands, vol IV. Pergamon, Oxford, pp 200–211

Feldman RJ, Maibach HI (1969) Percutaneous penetration of steroids in man. J Invest Dermatol 52:89–94

Flamigni C, Collins WP, Koullapis EN, Craft I, Dewhurst CJ, Sommerville IF (1971) Androgen metabolism in human skin. J Clin Endocrinol Metab 32:737–743

Frost P, Weinstein GD, Hsia SL (1966) Metabolism of estradiol-17β and estrone in human skin. J Invest Dermatol 46:584–585

Frost P, Gomez EC, Weinstein GD, Lamas J, Hsia SL (1969) Metabolism of progesterone-4-^{14}C in vitro in human skin and vaginal mucosa. Biochemistry 8:948–952

Giacomini M, Wright F (1980) The effects of progesterone and pregnanedione on the reductive metabolism of dihydrotestosterone in human skin. J Steroid Biochem 13:645–651

Gomez EC, Hsia SL (1968) In vitro metabolism of testosterone-4-^{14}C and Δ^{4}-androstene-3,17-dione-4-^{14}C in human skin. Biochemistry 7:24–32

Imperato McGinley J, Guerrero L, Gautier T, Peterson RE (1974) Steroid 5α-reductase deficiency in man: an inherited form of male pseudohermaphrodism. Science 186:1213–1215

Keenan BS, Meyer WJ, Hadjian AJ, Migeon CJ (1975) Androgen receptor in human skin fibroblasts. Characterization of a specific 17β-hydroxy-5α-androstan-3-one-protein complex in cell sonicates and nuclei. Steroids 25:535–552

Kuttenn F, Mauvais-Jarvis P (1975) Testosterone 5 α-reduction in the skin of normal subjects and of patients with abnormal sex development. Acta Endocrinol 79:164–176

Kuttenn F, Mowszowicz I, Wright F, Baudot N, Jaffiol C, Robin M, Mauvais-Jarvis P (1979) Male pseudohermaphroditism: a comparative study of one patient with 5 α-reductase deficiency and three patients with the complete form of testicular feminization. J Clin Endocrinol Metab 49:861–865

Kuttenn F, Mowszowicz I, Mauvais-Jarvis P (1980) Androgen metabolism in human skin. In: Mauvais-Jarvis P, Vickers CFH, Wepierre J (eds) Percutaneous absorption of steroids. Academic, London, pp 99–121

Mahoudeau JA, Bardin CW, Lipsett MB (1971) The metabolic clearance rate and origin of plasma dihydrotestosterone in man and its conversion to the 5 α-androstanediols. J Clin Invest 50:1338–1344

Mainwaring WIP (1975) Steroid hormone receptors: a survey. Vitam Horm 33:223–246

Mauvais-Jarvis P (1977) Androgen metabolism in human skin: mechanisms of control. In: Martini L, Motta M (eds) Androgens and antiandrogens. Raven, New York, pp 229–245

Mauvais-Jarvis P, Bercovici JP, Gauthier F (1969 a) In vivo studies on testosterone metabolism by skin of normal males and patients with the syndrome of testicular feminization. J Clin Endocrinol Metab 29:417–419

Mauvais-Jarvis P, Baudot N, Bercovici JP (1969 b) In vitro studies on progesterone metabolism by human skin. J Clin Endocrinol Metab 29:1580–1585

Mauvais-Jarvis P, Bercovici JP, Crepy O, Gauthier F (1970) Studies on testosterone metabolism in subjects with testicular feminization syndrome. J Clin Invest 49:31–40

Mauvais-Jarvis P, Kuttenn F, Baudot N (1974) Inhibition of testosterone conversion to dihydrotestosterone in men treated percutaneously by progesterone. J Clin Endocrinol Metab 38:142–147

Mauvais-Jarvis P, Kuttenn F, Wright F (1975) La progesterone administrée par voie percutanée. Un antiandrogène à action locale. Ann Endocrinol 36:55–62

Mauvais-Jarvis P, Kuttenn F, Gauthier-Wright F (1976) Testosterone 5α-reduction in human skin as an index of androgenicity. In: James VHT, Serio M, Giusti G (eds) The endocrine function of the human ovary. Academic, New York, pp 481–494
Mowszowicz I, Wright F (1979) A simple and reliable technique for separating the androgen receptor from testosterone binding globulin in human tissues. Anal Biochem 92:164–169
Mowszowicz I, Kirchhoffer MO, Kuttenn F, Mauvais-Jarvis P (1980) Testosterone 5α-reductase activity of skin fibroblasts increase with serial subcultures. Mol Cell Endocrinol 17:41–50
Mowszowicz I, Riahi M, Wright F, Bouchard P, Kuttenn F, Mauvais-Jarvis P (1981 a) Androgen receptor in human skin cytosol. J Clin Endocrinol Metab 52:338–344
Mowszowicz I, Kopp F, Martin PM (1981 b) Multiple steroid binding sites in human skin cytosol. IInd CIRD international symposium, Sophia Antipolis France, October 1981
Robel P, Lasnitzki I, Baulieu EE (1971) Hormone metabolism and action: testosterone and metabolites in prostate organ culture. Biochimie 53:81–96
Rongone EL (1966) Testosterone metabolism by human male mammary skin. Steroids 7:489–504
Siiteri PK, Wilson JD (1974) Testosterone formation and metabolism during male sexual differentiation in the human embryo. J Clin Endocrinol Metab 38:113–125
Stewart ME, Pochi PE, Strauss JS, Wotiz HH, Clark SJ (1977) In vitro metabolism of ^{3}H-testosterone by scalp and back skin: conversion of testosterone to 5α-androstane-3β,17β-diol. J Endocrinol 72:385–390
Voigt W, Hsia SL (1973) Further studies on testosterone 5α-reductase of human skin. Structural features of steroid inhibitors. J Biol Chem 248:4280–4285
Voigt W, Fernandez EP, Hsia SL (1970) Transformation of testosterone into 17β-hydroxy-5α-androstane-3-one by microsomal preparation of human skin. J Biol Chem 245:5594–5599
Wilson JD, Walker JD (1969) The conversion of testosterone to 5α-androstan-17β-ol-3-one (dihydrotestosterone) by skin slices of man. J Clin Invest 48:371–378
Wright F, Giacomini M (1980) Reduction of dihydrotestosterone to androstanediols by human female skin in vitro. J Steroid Biochem 13:639–643
Wright F, Mowszowicz I, Mauvais-Jarvis P (1978) Urinary 5α-androstane-3α,17β-diol radioimmunoassay: a new clinical evaluation. J Clin Endocrinol Metab 47:850–854
Wright F, Kirchhoffer MO, Giacomini M (1980) Anti-androgen activity of progesterone in human skin. In: Mauvais-Jarvis P, Vickers CFH, Wepierre J (eds) Percutaneous absorption of steroids. Academic, London, pp 123–137

CHAPTER 16

Melanophores, Melanocytes and Melanin: Endocrinology and Pharmacology

A. J. THODY and S. SHUSTER

Many animals undergo changes in skin colour. In lower vertebrates these changes can occur rapidly, involving the movement of pigment granules within specialised cells known as chromatophores. Such *physiological* or *chromomotor* colour changes are normally transitory and allow the animal to adapt to its environment. *Morphological* or *chromogenic* colour change are much slower and longer lasting processes since they depend upon amount of pigment synthesised or number of pigment cells. This is the most important type of colour change in homoiotherms and is the main determinant of skin and hair colour in mammals.

A. Pigment Cells

There are four types of pigment cell or chromatophore:

1. Melanophores which contain melanin granules or melanosomes
2. Iridophores which contain guanine-rich reflecting platelets
3. Xanthophores containing carotenoid vesicles and pteridine-rich pterinosomes
4. Erythrophores which also contain carotenoid vesicles and pterinosomes

All four types are found in cold-blooded vertebrates but the best known and only type to be found in mammals and other homoiotherms is the melanophore. Melanophores can be further divided on the basis of their appearance and location in the skin.

B. Dermal Melanophores

Dermal melanophores are large flat or basket-shaped cells with processes that radiate upwards and outwards from the main body of the cell. They are especially prevalent in cold-blooded vertebrates and it is these pigment cells which are responsible for the rapid, chromomotor colour changes that are characteristic of these animals. The melanophores that are found in other sites, e.g. CNS, eye and adrenal medulla, are similar in appearance and are probably closely related to dermal melanophores.

C. Epidermal Melanophores

Epidermal melanophores are situated at the dermo-epidermal junction and their long, thin dendritic processes ramify among neighbouring epidermal keratinocytes. This arrangement allows the transfer of melanosomes into the keratinocytes and in mammals it is this deposition of melanin that determines the pigmentation of the skin and hair. Epidermal melanophores are not involved in rapid chromomotor colour changes and because of their limited ability to mobilise their pigment they are referred to as *melanocytes*. The melanocytes responsible for hair pigmentation are situated in the bulb of the hair follicle and normally are active only when the hairs are growing (anagen); this allows the transfer of melanin into the developing cortical and medullary cells. This explains why changes in coat colour occur only when animals are moulting, which is when the old coat is lost and new hairs are growing. It is not clear what happens to the melanocytes during the resting phase of the hair cycle (telogen). The degree of pigmentation of the hair may be entirely different to that of the surrounding skin, which suggests that the pigmentary processes in follicular and epidermal melanocytes can occur independently.

D. Melanosome Dispersion

The rapid chromomotor colour changes that are characteristic of lower vertebrates occur as a result of melanosome dispersion within dermal melanophores. Mechanisms proposed for melanosome movement are:

1. A change in the intracellular current flow resulting from changes in membrane potential (electrophoretic theory; KINOSITA 1963)
2. The ionic release or exchange of membrane-bound ions (ion-exchange theory; NOVALES and NOVALES 1965)
3. Sol and gel transformation which produces changes in the hydrostatic pressure of the cytoplasm (sol–gel transformation theory, MARSLAND and MEISNER 1967)
4. Channels surrounded by microtubules (microtubule theory; GREEN 1968)

MOELLMANN et al. (1973) extended the microtubule theory, suggesting a relationship between microtubules and 100-Å filaments: when the skin is pale and the melanosomes are aggregated, micotubules predominate, but during the darkening response, when the melanosomes become dispersed, there is an abundance of 100-Å filaments. Similar filaments have been found in human melanocytes (JIMBOW et al. 1976) but there is no evidence of an association between melanosomes and microtubules, which are only rarely seen in the dendritic processes of the melanocyte.

E. Melanogenesis

All melanophores produce melanin, of which there are two types: the eumelanins, which are brownish black, and the phaeomelanins, which range from yellow to reddish brown. Eumelanins are insoluble, complex polymers of indole quinone

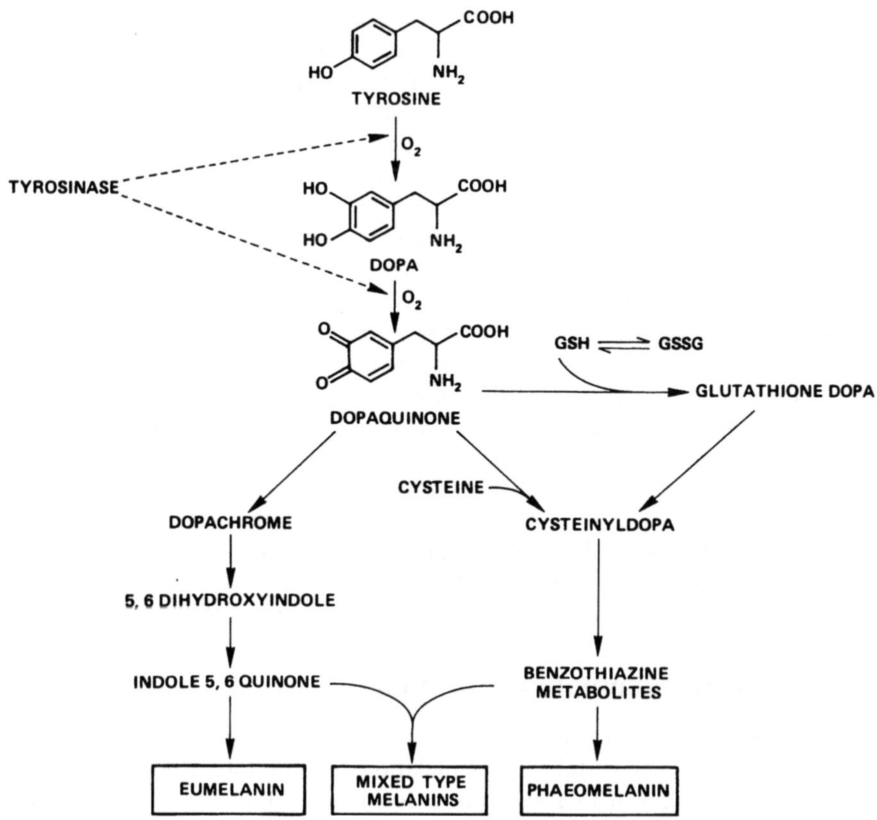

Fig. 1. Melanogenesis pathway

and in mammals are always attached to a protein. The phaeomelanins have an even more complex structure and differ drom the eumelanins in their high sulphur content and solubility in dilute alkali (see PROTA 1980). These pigments are formed in the melanosome. According to the Raper-Mason scheme (see Fig. 1), tyrosine is first converted into dopa and this is then oxidised to dopaquinone. Both reactions are catalysed by the enzyme tyrosinase. It has been suggested that mammalian tyrosinase has only dopa oxidase activity and that the conversion of tyrosine to dopa is catalysed by peroxidase (OKUN et al. 1970). However, HOLSTEIN et al. (1973) and HEARING and EKEL (1975) have shown that a true tyrosinase is present in mammalian melanocytes and melanoma cells and at least three and possibly four isozymic forms of this enzyme have been identified (BURNETT 1971; HEARING et al. 1978, 1981).

Once dopaquinone is formed a series of complex reactions follow, involving cyclisation and oxidative polymerisation, and this results in the formation of eumelanin (see PROTA 1980 and Fig. 1). These reactions are thought to occur spontaneously although recent evidence suggests that certain of them have a regulatory control and PAWELEK et al. (1980) identified three factors from Cloudman mouse melanoma cells, the first of which stimulates the conversion of dopachrome to 5,6-dihydroxyindole and the second the conversion of 5,6-dihydroxyindole to melanin. The third factor acts to retard the formation of melanin from 5,6-dihydroxyindole. These activities appear to be associated with the different isozymes of tyrosinase. Thus the isozymes T_1, T_2, and T_3 contain blocking factor activity and the isozyme T_4 contains activity that accelerates the formation of eumelanin from dopachrome (HEARING et al. 1982). These workers suggest that while the latter (T_4) is active within the eumelanosome and regulates the conversion to eumelanin, blocking activity is associated with tyrosinase outside of the melanosome and this prevents the accumulation of melanin and its toxic precursors within the cytoplasm.

According to PROTA (1980) all that is required to switch the pathway from eumelanin synthesis to that of phaeomelanin is the presence of –SH containing cysteine residues. These reactive –SH groups combine with dopaquinone to give cysteinyldopa compounds which are quickly oxidised into the benzothiazines and thence into the phaeomelanins. The cysteinyldopas could also be formed from glutathionedopa that results from the reaction between dopaquinone and glutathione. The relative amounts of the latter depend upon the activity of glutathione reductase. Glutathione could have a regulatory role in melanogenesis because unless the glutathionedopa is hydrolysed to cysteinyldopa its formation would result in the side tracking of the dopaquinone (PROTA 1980 and Fig. 1).

F. Melanocyte Stimulating Hormone

Probably the most well known stimulator of melanosome dispersion in the dermal melanophores of lower vertebrates is melanocyte stimulating hormone (MSH). From studies carried out in *Rana* it now seems clear that the α-MSH molecule contains two active sequences, His-Phe-Arg-Trp and Lys-Pro-Val-NH_2 (SCHWYZER and EBERLE 1976; EBERLE 1980). These two sites seem to act in synergy to stimulate melanosome dispersion. The N terminal sequence of α-MSH, Ser-Tyr-Ser, is also important since it acts as a potentiating sequence and enhances the effect of the two active sequences. This structure-activity relationship may not hold for all pigment cells and there is evidence that the MSH receptors on the *Anolis* melanophore may recognise different sequences within the α-MSH molecule (LUCAS et al. 1987a). There is little information on the structure-activity relationships of other MSH peptides, e.g. β-MSH and γ-MSH. It is, however, interesting that other peptides such as β-endorphin, which arises together with the MSH peptides from the 31-K precursor pro-opiomelanocortin (POMC) (Fig. 2), may enhance the action of MSH. Thus β-endorphin potentiates the action of α-MSH and β-MSH in the *Anolis* melanophore (CARTER and SHUSTER 1979). The 88-91 sequence of β-endorphin, Lys-Lys-Gly-Glu was found to be responsible for this action and was thus named melanotrophin potentiating factor or MPF (CARTER et al. 1979). More recently, it has been shown that the same peptide will potentiate the action of α-MSH in tyrosinase activity and melanogenesis in the hair follicle of the Siberian hamster (LOGAN et al. 1981) although it fails to affect the action of MSH on mouse melanoma cells (LANDE et al. 1981; MACNEIL 1982). Different MSH peptides also interact. Thus des-acetyl MSH and γ-MSH act as partial agonists to α-MSH on its action on *Anolis* melanophore (MCCORMACK et

There has been controversy as to whether internalisation of the MSH receptor complex is necessary for the initiation of melanogenesis. VARGA et al. (1976) showed that in mouse melanoma cells ^{125}I-labelled MSH is internalised to the Golgi region of the cell and suggested that MSH receptor-activated adenylate cyclase complexes are transported by an endocytotic process to the Golgi region, where they fuse with the vesicles that contain inactive tyrosinase. More recently, however, EBERLE (1982) showed that doubly labelled α-MSH is not internalised into melanoma cells in an undegraded form. Furthermore, when covalently bound to large proteins such as albumin, thyroglobulin or tobacco mosaic virus, α-MSH was still capable of eliciting a biological response despite its inability to enter the cell (EBERLE 1982). It seems likely that internalisation may take place but while it is not essential for the biological response it may be important for degradation of the MSH receptor complex.

Not all melanocytes respond to MSH. In mice, for instance, while α-MSH has an action in hair follicular melanocytes (GESCHWIND et al. 1972; BURCHILL and THODY 1986a; BURCHILL et al. 1986), MSH has no effect on the proliferation (NORDLUND et al. 1986) or the tyrosinase activity (SEECHURN and THODY 1986) of mouse epidermal melanocytes. However, there are reports that the peptide has a role in the differentiation of epidermal melanocytes during early life (HIROBE and TAKEUCHI 1977). It has also been reported that MSH will activate epidermal melanocytes in guinea-pigs (SNELL 1962). Skin darkening has been observed in man following the administration of either α- or β-MSH (LERNER and MCGUIRE 1961), although cultured human melanocytes fail to respond to MSH despite their sensitivity to cAMP (HALABAN et al. 1983; FRIEDMANN and GILCHREST 1987). This lack of effect of MSH on human melanocytes raises the question as to the significance of the MSH peptides as pigmentary hormones in man.

I. Physiological Significance of MSH Peptides as a Pigmentary Hormone in Man

It was originally considered that the main pigmentary hormone in man was a 22 amino acid β-MSH but it is now generally agreed that this particular peptide does not exist and the immunoreactive "β-MSH" found in man is likely to be due to either β- or γ-LPH (see THODY 1980). Under normal conditions the human pituitary contains very little α-MSH, and apart from a possible increase in patients with Cushing's disease, circulating plasma immunoreactive α-MSH levels are undetectable (THODY et al. 1985). The inability of the human pituitary to produce large amounts of MSH peptides is related to its poorly developed intermediate lobe and it would appear that the main pigmentary peptides to be produced by the human pituitary are ACTH and LPH. These peptides are derived from the same precursor protein, POMC, as the MSH peptides (Fig. 2). However, while ACTH and LPH are produced in both the anterior and the intermediate lobe of the pituitary, further cleavage of these peptides to α- and β-MSH seems only to occur in the intermediate lobe. While there is evidence that α-MSH is secreted in patients with adrenal insufficiency (THODY et al. 1985), it seems likely that in these conditions it is the ACTH and LPH that have the major pigmentary effect. The hyperpigmentation seen in chronic renal failure may be the result of an impaired metabolism of LPH by the kidney (SMITH et al. 1976). An intermediate lobe is, nevertheless, present in foetal human pituitary and at this stage the human pituitary, like the foetal pituitaries of other mammals, is capable of producing true MSH peptides (see THODY 1980). The MSH peptides may therefore play an important role in the development of the pigmentary system in both man and other mammals (QUEVEDO and FLEISCHMAN 1980).

G. Catecholamines

It has been known for some time that catecholamines may either stimulate or inhibit melanosome dispersion in the melanophores of lower vertebrates (for reviews see HADLEY and BAGNARA 1975; EBERLE 1980). The stimulatory effect is mediated by β-receptors and their stimulation increases cAMP levels and the dispersion of the melanosomes. The inhibitory effect, on the other hand, is mediated by α-receptors. Not all melanophores possess α-receptors. They appear to be absent on the melanophores of certain *Rana* and *Xenopus laevis* and in this case the effect of the catecholamines is to stimulate melanosome dispersion (β-effect). However, when α-receptors are present, as in *Anolis carolinensis*, they tend to dominate and the overall effect of the catecholamines is then an inhibitory one. Recent studies suggest that in the lizard *Anolis carolinensis*, these α-receptors are of the α_2 subtype (CARTER and SHUSTER 1982). α_2-Adrenoceptors are normally associated with a decrease in cAMP and this would therefore explain their inhibitory effect on melanosome dispersion. Dopamine appears to act through the α_2-receptor in *Anolis* (CARTER and SHUSTER 1982). Catecholamines can therefore either increase (β-effect) or decrease (α_2-effect) melanosome dispersion. By use of agonists and antagonists it has been shown that this interaction between MSH and catecholamines in the control of melanosome is not on the receptor and is therefore likely to be at the second messenger level.

There is evidence that the β-agonist isoprenaline is able to stimulate tyrosinase activity and melanogenesis in mouse hair follicular melanocytes (BURCHILL and THODY 1986a). This is likely to involve an activation of adenylate cyclase as occurs with α-MSH (see above). Conversely, bromocriptine and the specific D_2 agonist LY 171555 decrease tyrosinase activity in mouse hair follicular melanocytes, presumably as a result of an inhibition of cAMP production (BURCHILL and THODY 1986b).

H. Melatonin

Melatonin has a lightening effect on amphibian melanophores (MORI and LERNER 1960). This indoleamine also inhibits pigmentation of the hair in a number of mammals (RUST and MEYER 1969; LOGAN and WEATHERHEAD 1980). In the case of the Siberian hamster it appears that inhibitory mechanisms, activated by melatonin, are more important than MSH in the control of coat colour (LOGAN and WEATHERHEAD 1980). In this species melatonin inhibits melanogenesis in both normal and MSH-stimulated hair follicular melanocytes (LOGAN and WEATHERHEAD 1980). This inhibition was found to occur at some post-tyrosinase step in the melanogenic pathway and recent experiments suggest that this effect might be mediated by cGMP (WEATHERHEAD and LOGAN 1981). However, not all mammals respond to melatonin in this way (THODY et al. 1984). It is also not clear whether melatonin inhibits melanogenesis in man although it has been reported to produce a slight lightening of skin when administered in large doses to hyperpigmented patients (NORDLUND and LERNER 1977).

J. Steroids

Ovarian steroids have been shown to increase skin pigmentation in ovariectomised guinea-pigs (SNELL 1967). Others, however, have been unable to demonstrate any pigmentary effects with either oestrogens or progesterone in the hamster although increases in pigmentation occurred after testosterone administration. Similar findings have been reported in mice, rats and humans but in the guinea-pig testosterone has produced either no effect or a slight reduction in skin pigmentation. The effects of adrenal steroids have been less well studied but they would appear to produce no pigmentary changes in male guinea-pigs (SNELL 1967); however, more potent glucocorticoids may inhibit the function of tyrosinase in mouse melanocytes (NORDLUND et al. 1981). Increased pigmentation of the nipples, areola, face, abdominal skin and genitalia is often seen during pregnancy. These changes are similar to those produced by oestrogens and progesterone in the guinea-pig and SNELL (1967) suggested that these hormones are responsible for the pigmentary changes in pregnancy. The chloasma that sometimes occurs in women taking oral contraceptives is probably due to the steroid content of the pill and there are no changes in circulating β-LPH in chloasma (SMITH et al. 1977). In the idiopathic hirsutes or primary cutaneous virilism the hair is dark and a major effect of anti-androgens is to make the hair less visible by lightening its colour (personal observation).

How gonadal steroids increase pigmentation is not clear. In guinea-pigs the pigmentary changes were associated with an increase in melanocyte number and similar findings were obtained in abdominal skin from pregnant women (SNELL 1967). Whether these steroids affect melanogenesis is not yet known.

It is possible that gonadal steroids and MSH peptides may interact to affect pigmentation. There are reports that MSH and progesterone will act synergistically to stimulate melanosome dispersion in *Rana pipiens* (HIMES and HADLEY 1971) and it is interesting to recall that MSH and gonadal steroids also interact synergistically on other target sites, e.g. sebaceous gland (THODY and SHUSTER 1975). In a more recent study, in *Anolis carolinensis*, however, gonadal steroids had no effect upon the pigmentary action of MSH and it would seem that this lizard shows no pigmentary response to gonadal steroids (LUCAS 1984, unpublished observations). The possibility should also be considered that in some mammals part of the pigmentary effect of gonadal steroids could be explained by an increase in MSH release. Ovarian steroids have been shown to stimulate the pars intermedia and MSH secretion in a number of mammals (see THODY 1980).

K. Prostaglandins

Prostaglandins have been shown to stimulate melanosome dispersion in the melanophores of lower vertebrates and it has even been suggested that they mediate the effect of MSH (VAN DER VEERDONK and BROUWER 1973). However, this seems unlikely in view of the recent finding that indomethacin, which blocks prostaglandin synthesis, has no effect on MSH action (LUCAS et al. 1985). It is not yet known how the prostaglandins exert their effect on melanosome dispersion and whether they have any other effects on pigment cells, i.e. melanogenesis.

There is evidence that prostaglandins E_1, E_2, and D_2 stimulate the proliferation of mouse epidermal melanocytes although no effects were seen with prostaglandin $F_{2\alpha}$, thromboxane and prostacyclin (NORDLUND et al. 1986). Arachidonic acid also stimulates melanocyte proliferation and part of its action was blocked by indomethacin, which inhibits cyclo-oxygenase and prostaglandin formation (NORDLUND and ACKLES 1985).

References

Abe K, Butcher KW, Nicholson WE, Baird CE, Liddle RA, Liddle GW (1969) Adenosine 3′,5′ monophosphate (cAMP) as the mediator of the actions of melanocyte-stimulating hormone (MSH) and norepinephrine on the frog skin. Endocrinology 84:362–368

Bitensky MW, Burnstein SR (1965) Effects of cyclic adenosine monophosphate and melanocyte stimulating hormone on frog skin in vitro. Nature 208:1282–1284

Burchill SA, Thody AJ (1986a) Melanocyte stimulating hormone and the regulation of tyrosinase activity in hair follicular melanocytes of the mouse. J Endocrinol 111:225–232

Burchill SA, Thody AJ (1986b) Dopaminergic inhibition of tyrosinase activity in hair follicular melanocytes of the mouse. J Endocrinol 111:233–237

Burchill SA, Thody AJ, Ito S (1986) Melanocyte stimulating hormone, tyrosinase activity and the regulation of eumelanogenesis and phaeomelanogenesis in the hair follicular melanocytes of the mouse. J Endocrinol 109:15–21

Burchill SA, Virden R, Fuller B, Thody AJ (1987) Regulation of tyrosinase synthesis by α-melanocyte stimulating hormone in mouse hair follicular melanocytes. J Endocrinol 115:1–7

Burnett JB (1971) The tyrosinase of mouse melanoma. Isolation and molecular properties. J Biol Chem 246:3079–3091

Burnett JB, Birch DE, Fuller BB, Hadley ME (1981) Cyclic AMP-dependent protein kinase activity in Cloudman mouse melanoma cell cultures. In: Beiji M (ed) Pigment cell. University of Tokyo Press, Tokyo

Carter RJ, Shuster S (1979) β-Endorphin potentiates melanocyte stimulating activity on the skin of *Anolis carolinensis*. J Endocrinol 80:7

Carter RJ, Shuster S (1982) The association between the melanocyte-stimulating hormone receptor and the α_2-adrenoceptor in the *Anolis* melanophore. Br J Pharmacol 75:169–176

Carter RJ, Shuster S, Morley JS (1979) Melanotropin potentiating factor is the C-terminal tetrapeptide of human β-lipotropin. Nature 279:74–75

De Graan PNE, Eberle AN, van de Veerdonk ECG (1981) Photoaffinity labelling of MSH receptors reveals a dual role of calcium in melanophore stimulation. FEBS Lett 129:113–116

Eberle A (1980) MSH receptors. In: Schulster D, Levitzki A (eds) Cellular receptors for hormones and neurotransmitters. Wiley, Chichester

Eberle A (1982) MSH receptors. Br J Dermatol 106:111–113

Friedmann PS, Gilchrest BA (1987) Ultra violet radiation directly induces pigment production by cultured hormone melanocytes. J Cell Physiol 133:88–94
Geschwind II, Huseby RA, Nishioka R (1972) The effect of melanocyte-stimulating hormone on coat color in the mouse. Rec Prog Horm Res 28:91–130
Green L (1968) Mechanism of movements of granules in melanocytes in *Fundulus heterochitus*. Proc Natl Acad Sci USA 59:1179–1186
Hadley ME, Bagnara JT (1975) Regulation of release and mechanism of action of MSH. Am Zool 15 (Suppl 1):81–104
Hadley ME, Anderson B, Heward CB, Sawyer TK, Hruby VJ (1981) Calcium dependent prolonged effects on melanophores of [4-norleucine, 7-D-phenylalanine] α-melanotropin. Science 213:1025–1027
Halaban R, Pomerantz SH, Lambert DT, Lerner AB (1983) Regulation of tyrosinase in human melanocytes grown in culture. J Cell Biol 97:480–488
Halaban R, Pomerantz SH, Marshall S, Lerner AB (1984) Tyrosinase activity and abundance in Cloudman melanoma cells. Arch Biochem Biophys 1230:383–387
Hearing VJ, Ekel TM (1975) Involvement of tyrosinase in melanin formation in murine melanoma. J Invest Dermatol 64:80–83
Hearing VJ, Nicholson JM, Montague PM, Ekel TM, Tomecki KJ (1978) Mammalian tyrosinase: structural and functional interrelationship of enzymes. Biochim Biophys Acta 522:327–339
Hearing VJ, Ekel TM, Montague PM (1981) Mammalian tyrosinase. Isoenzymic form of the enzyme. Int J Biochem 13:99–103
Hearing VJ, Korner AM, Pawelek JM (1982) New regulators of melanogenesis are associated with purified tyrosinase isoenzymes. J Invest Dermatol 79:16–18
Himes PJ, Hadley ME (1971) In vitro effects of steroid hormones on frog melanophores. J Invest Dermatol 57:337–342
Hirobe T, Takeuchi T (1977) Induction of melanogenensis in the epidermal melanoblasts of newborn mouse skin by MSH. J Embryol Exp Morphol 37:79–90
Holstein TJ, Stowell CP, Quevedo WC Jr, Zarcaro RM, Blenieki TC (1973) Peroxidase, protyrosinase and multiple forms of tyrosinase in mice. Yale J Biol Med 46:560–571
Jimbow K, Davison PF, Pathak MA, Fitzpatrick TB (1976) Cytoplasmic filaments in melanocytes, their nature and the role in melanin pigmentation. In: Riley V (ed) Pigment cell. Karger, Basel, pp 13–32
Kinosita H (1963) Electrophoretic theory of pigment migration within fish melanophore. Ann NY Acad Sci 100:992–1004
Lande S, Pawelek J, Lerner AB, Emanuel JR (1981) Assay of melanotropic peptides in an in vitro mammalian system. J Invest Dermatol 77:244–245
Lerner AB, McGuire JS (1961) Effect of α and β-melanocyte stimulating hormone on the skin colour of man. Nature 189:176–179
Logan A, Weatherhead B (1980) Post-tyrosinase inhibition of melanogenesis by melatonin in hair follicles in vitro. J Invest Dermatol 74:47–50
Logan A, Carter RJ, Shuster S, Thody AJ, Weatherhead B (1981) Melanotrophin potentiating factor (MPF) potentiates MSH induced melanogenesis in hair follicle melanocytes. Peptides 2:121–123
Lucas AM, Shuster S, Thody AJ (1983 a) Calcium has several points of action on MSH-stimulated melanosome dispersion. J Invest Dermatol 80:367
Lucas AM, Thody AJ, Shuster S (1983 b) MSH stimulated melanosome dispersion – the role of Ca^{2+} and cAMP. Regul Pept 7:293
Lucas AM, Thody AJ, Shuster S (1985) The stimulatory effects of prostaglandins on the melanophores of the lizard, *Anolis carolinensis*. Life Sci 36:835–840
Lucas AM, Shuster S, Thody AJ, Eberle A, Girard J (1987 a) A comparison of structure activity relationships within α-MSH on melanophores of *Anolis carolinensis* and *Rana pipiens*. Regul Pept 18:43–50
Lucas AM, Thody AJ, Shuster S (1987 b) Role of protein kinase C in the pigment cell of the lizard (*Anolis carolinensis*). J Endocrinol 112:283–287
MacNeil S (1982) Stimulation of B16 melanoma adenylate cyclase activity by MSH related peptides. Br J Dermatol 106:115–116

MacNeil S, Johnson SK, Bleehen SS, Brown BL, Tomlinson S (1981) Stimulation of the adenylate cyclase of a B16 melanoma cell line by pro-opiocortin-related peptides – a structure activity study. Regul Pept 2:193–200
Marsland D, Meisner D (1967) Effects of D_2O or the mechanism of pigment dispersal in the melanocytes of *Fundulus heterochitus:* a pressure temperature analysis. J Cell Physiol 70:209–215
McCormack AM, Carter RJ, Thody AJ, Shuster S (1982) Des-acetyl MSH and γ-MSH act as partial agonists to α-MSH on the *Anolis* melanophore. Peptides 3:13–16
Moellmann G, McGuire J, Lerner AB (1973) Ultrastructure and cell biology of pigment cells. Intracellular dynamics and the fine structure of melanocytes (with special reference to the effects of MSH and cyclic AMP in microtubules and 10 nm filaments). Yale J Biol Med 46:337–360
Mori W, Lerner AB (1960) A microscopic bioassay for melatonin. Endocrinology 67:443–450
Nordlund JJ, Ackles AE (1985) Melanocytoxins and fatty acids stimulate proliferation of melanocytes in murine pinnal epidermis: a possible role for arachidonic acid metabolism in regulation of melanocyte growth. In: Bagnara J, Klaus S, Paul E, Schartel M (eds) Biological, molecular and clinical aspects of pigmentation, pigment cell. University of Tokyo Press, Tokyo, pp 501–513
Nordlund JJ, Lerner AB (1977) The effects of oral melatonin on skin color and on the release of pituitary hormones. J Clin Endocrinol Metab 45:768–774
Nordlund JJ, Ackles AA, Traynor FF (1981) The proliferative and toxic effects of ultraviolet light and inflammation on epidermal pigment cells. J Invest Dermatol 77:361–368
Nordlund JJ, Collins CE, Rheins LA (1986) Prostaglandin E_2 and D_2 but not MSH stimulate the proliferation of pigment cells in the pinnal epidermis of the DBA/2 mouse. J Invest Dermatol 86:433–437
Novales RR, Davis WJ (1967) Melanin dispersing effect of adenosine 3′,5′-monophosphate on amphibian melanophores. Endocrinology 81:283–290
Novales RR, Novales BJ (1965) The effects of osmotic pressure and calcium deficiency on the response of tissue cultured melanophores to melanocyte stimulating hormone. Gen Comp Endocrinol 5:568–576
Okun M, Edelstein L, Or N, Hamada G, Donnellan B (1970) The role of peroxidase versus the role of tyrosinase in enzymatic conversion of tyrosine to melanin in melanocytes, mast cells and eosinophils: an autoradiographic-histochemical study. J Invest Dermatol 55:1–12
Pawelek J (1976) Factors regulating growth and pigmentation of melanoma cells. J Invest Dermatol 66:201–209
Pawelek J, Korner AM, Bergstrom A, Bologna S (1980) New regulators of melanin biosynthesis and the autodestruction of melanoma cells. Nature 286:617–619
Prota G (1980) Recent advances in the chemistry of melanogenesis in mammals. J Invest Dermatol 75:122–127
Quevedo WC Jr, Fleischman RD (1980) Developmental biology of mammalian melanocytes. J Invest Dermatol 75:116–120
Rust CC, Meyer R (1969) Hair color, molt and testis size in male, short-tailed weasels treated with melatonin. Science 165:921–922
Schwyzer R, Eberle A (1977) On the molecular mechanism of α-MSH receptor interactions. In: Tilders FJH, Swaab DF, van Wimersma Greidanus TJB (eds) Melanocyte stimulating hormone. Control, chemistry and effects. Karger, Basel, pp 18–25 (Frontiers of hormone research, vol 4)
Seechurn P, Thody AJ (1986) α-MSH and the differentiation of epidermal melanocytes in the C57BL mouse. J Invest Dermatol 87:167
Smith AG, Shuster S, Thody AJ, Alvarez-Ude F, Kerr DNS (1976) Role of the kidney in regulating plasma immunoreactive β-melanocyte stimulating hormone. Br Med J 1:874–876
Smith AG, Shuster S, Thody AJ, Peberdy M (1977) Chloasma, oral contraceptives and plasma immunoreactive β-melanocyte stimulating hormone. J Invest Dermatol 68:169–170

Snell RS (1962) Effect of the melanocyte stimulating hormone of the pituitary in melanocytes and melanoma in the skin of guinea pigs. J Endocrinol 25:249–258
Snell RS (1967) Hormonal control of pigmentation in man and other mammals. In: Montagna W, Hu F (eds) The pigmentary system. Pergamon, New York, pp 447–466
Thody AJ (1980) The MSH peptides. Academic, London
Thody AJ, Shuster S (1975) Control of sebaceous gland function in the rat by α-melanocyte stimulating hormone. J Endocrinol 64:503–510
Thody AJ, Ridley K, Carter RJ, Lucas AM, Shuster S (1984) α-MSH and coat color changes in the mouse. Peptides 5:1031–1036
Thody AJ, Fisher C, Kendal Taylor P, Jones MT, Price J, Abraham RR (1985) The measurement and characterisation by high pressure liquid chromatography of immunoreactive α-melanocyte stimulating hormone in human plasma. Acta Endocrinol (Copenh) 110:313–318
Varga JM, Di Pasquale A, Pawelek J, McGuire JS, Lerner AB (1974) Regulation of melanocyte stimulating hormone action at the receptor level: discontinuous binding of hormone to synchronized mouse melanoma cells during the cell cycle. Proc Natl Acad Sci USA 71:1590–1593
Varga JM, Moellman G, Fritsch P, Godawska E, Lerner AB (1976) Association of cell surface receptors for melanotropin with the Golgi region in mouse melanoma cells. Proc Natl Acad Sci USA 73:559–562
Veerdonk van der FCG, Brouwer E (1973) Role of calcium and prostaglandin (PGE_1) in the MSH induced activation of adenylate cyclase in *Xenopus laevis*. Biochem Biophys Res Commun 52:130–136
Vesely DL, Hadley ME (1976) Receptor-specific calcium requirements for melanophore stimulating hormone. In: Riley V (ed) Pigment cell, vol 3. Karger, Basel, pp 265–274
Weatherhead B, Logan A (1981) Interaction of α-melanocyte stimulating hormone, melatonin, cyclic AMP and cyclic CMP in the control of melanogenesis in hair follicle melanocytes in vitro. J Endocrinol 90:89–96
Wong G, Pawelek J (1975) Melanocyte stimulating hormone promotes activation of preexisting tyrosinase molecules in Cloudman S91 melanoma cells. Nature 255:644–646
Wong G, Pawelek J, Sansone M, Morowitz J (1974) Response of mouse melanoma cells to melanocyte stimulating hormone. Nature 248:351–354

CHAPTER 17

Cytokines in Relation to Inflammatory Skin Disease

K. A. Brown, B. A. Ellis, and D. C. Dumonde

A. Introduction

Cytokines are soluble, non-antibody proteins that orchestrate many immunological reactions including cell-mediated immune responses; lymphocyte activation, proliferation and differentiation; and haemopoietic mechanisms. The term encompasses the mediators derived from activated T-lymphocytes, i.e. lymphokines (LKs) (Dumonde et al. 1969); monocytes/macrophages, i.e. monokines (MKs) (Waksman and Namba 1976); and those mediators whose cellular origin is known to be more diverse, e.g. interferons. All cytokines were originally described in terms of their biological activity(ies) with little knowledge of their chemical structure. From the application of more precise biological assays, immunoassays, and more sensitive biochemical techniques, it became apparent that different lymphokine or cytokine activities could actually be mediated by the same molecule(s). Consequently, at the 2nd International Workshop on Lymphokines in 1979, the generic term "interleukin" ("between leucocytes") was introduced to bring together some of these mediators (Aarden et al. 1979). Historically, the concept of special mediators of immunological regulation, rather than direct cellular or antibody-mediated mechanisms, evolved from studies of delayed-type hypersensitivity and inflammation, and has been extensively documented (Adelman et al. 1979; Rocklin et al. 1980; Watson 1981; Hanson et al. 1982; Dumonde and Hamblin 1983).

This review shall be confined to those interleukins whose production and function have been associated with certain dermatological disorders.

B. Interleukins: Structure and Physicochemical Properties

Physicochemical characterisation of interleukins progressed relatively slowly even with the development of cell lines (Rosenstreich and Wahl 1979; Gillis and Watson 1980; Gillis et al. 1980, Dumonde and Hamblin 1983) and T-T cell hybridomas (Schrader and Clark-Lewis 1981; Gillis and Stull 1982; Kobayashi et al. 1982). This was due to the relatively low amounts of protein produced and the observation that many cell lines elaborated multiple activities. In general terms, interleukins are described as proteins which have varying degrees of glycosylation and exhibit isoelectric point (pI) values between 4.5 and 7.0. They have a relative molecular mass (M_r) between 12 K daltons and 100 K daltons and a variable susceptibility to proteolytic enzymes and temperature.

Genetic engineering techniques have been applied to interleukin activities. In addition to identifying more precisely the chemical composition of the active moieties they have provided information regarding gene structure and location, structural and functional interrelationships, and evolutionary conservation of the molecule. This approach has facilitated an appreciation of the hormone-like mechanism of action of cytokines at both a molecular and cellular level. Interleukins produced by bacterial recombinant DNA techniques, although not glycosylated, are usually biologically active in vivo (Lotze et al. 1985) and exhibit biological activities in vitro that often parallel those of natural products (Rosenberg et al. 1984).

In man, two distinct cDNA's have been identified for interleukin-1 (IL-1) (March et al. 1985); these genes code for precursor molecules of 271 and 269 amino acid residues respectively, with a M_r of approximately 30 K daltons. Neither gene codes for a recognised N-terminal signal sequence although the biological activity is associated with the C-terminal 159 and 153 amino acid residues, designated α and β respectively. This adds weight to the proposal that IL-1 is not actively secreted but is released by damaged cells (Gery et al. 1981). The α and β forms show only 26% homology whereas the α form shows 62% homology with the equivalent mouse IL-1 for which similar data is available (Lomedico et al. 1984).

In contrast, only a single gene, located on chromosome 4q and composed of four exons, has been identified for human IL-2 (Holbrook et al. 1984; Seigel et al. 1984). This gene codes for a polypeptide chain of 153 amino acid residues of which the first 20 residues account for an N-terminal signal peptide. The resulting mature product has a predicted M_r of 15.42 K daltons – which correlates well with values established for the natural product – and also has a neutral isoelectric point, the differences in molecular weight and isoelectric point being accounted for by the variable glycosylation of the natural product (Robb and Smith 1981). The amino acid sequence shows no site for N glycosylation but one site (a threonine residue at position 3) is suitable for O glycosylation. Of the three cysteine residues predicted, a disulphide bridge between those at positions 58 and 105 is required to maintain an active conformation (Robb et al. 1984; Wang et al. 1984; Liang et al. 1986), whilst that at position 125 may be substituted without loss of activity (Liang et al. 1986). Nucleotide sequence analysis of the cDNA for mouse IL-2 shows that the three cysteine residues are conserved and that no N glycosylation site is available (Taniguchi et al. 1983). Interspecies homology has been estimated as being 76% at the nucleotide level but only 63% at the protein level.

Sequence analysis of the cloned cDNA (derived from the WEHI-3 myelomonocytic cell line) for IL-3 shows that it codes for a polypeptide of 166 amino acids, including a signal peptide of (possibly) 27 amino acid residues (Fung et al. 1984). Protein sequence analysis of mature processed IL-3 indicated a M_r of 15.10 K daltons and four potential N glycosylation sites. There is no significant sequence homology with human IL-2 (Taniguchi et al. 1983) or interferon-γ (Gray et al. 1982). From studies using a chemically synthesised protein, it has been shown that a disulphide bond incorporating the cysteine residue at position 17 is essential for IL-3 activity (Clark-Lewis et al. 1986).

Similar structural analysis is rapidly becoming available for the other interleukins – IL-4, IL-5 and IL-6 (HIRANO et al. 1986; KINASHI et al. 1986; YOKOTA et al. 1986) – as well as for those cytokines whose activities overlap (GRAY et al. 1982; PENNICA et al. 1984).

The two forms of IL-1 appear to bind to a single receptor on the target cell, e.g. T- and B-lymphocytes, dermal and synovial fibroblasts, chondrocytes and endothelial cells (DOWER and URDEL 1987). More detailed analysis of the heterodimeric IL-2 receptor has enabled discrimination between low- and high-affinity receptors, although to date only one gene (consisting of eight exons and located on chromosome 20) for the α-chain has been identified (LEONARD et al. 1984; NIKAIDO et al. 1984). This gene codes for the 55 K dalton glycoprotein observed on the membranes of normal, activated, peripheral blood T cells. The β-chain (75 K daltons in man), which is as yet of undefined composition, may be non-covalently linked to the α-chain to give the high-affinity IL-2 receptor (SMITH 1987).

C. Biological Activities of Interleukins

Interleukin-1, previously referred to as lymphocyte activating factor, endogenous pyrogen and leucocytic endogenous mediator (GERY et al. 1971; KAMPSCHMIDT et al. 1978; ROSENWASSER et al. 1979), is an essential signal in T-cell activation. Originally thought to be produced only by macrophages, IL-1 is now believed to be secreted by most nucleated cells. It possesses a wide range of biological activities in vitro, involving a host of target cells – summarised in Table 1 – that have been reviewed extensively elsewhere (DINARELLO 1984a, b; DURUM et al. 1985; OPPENHEIM et al. 1986). Interleukin-1 acts on the thermoregulatory centre of the

Table 1. Biological properties of interleukin-1 in vitro

Target cell	Response
Polymorphonuclear cells	Release from bone marrow Chemotaxin for neutrophils Enhanced adherence to cultured endothelium Degranulation
Lymphocytes	Stimulation of interleukin (IL-2, IL-3) production Induction of IL-2 receptors on T-cells Chemotaxin for lymphocytes Cofactor for the proliferation/differentiation of B-cells
Fibroblasts	Proliferation Production and secretion of collagenase Release of plasminogen activator Release of prostaglandins
Endothelium	Release of prostaglandins Proliferation Induction of procoagulant activity

For further details of these and other biological activities, see DINARELLO (1984a, b), DARUM et al. (1985), and OPPENHEIM et al. (1986).

hypothalamus, causing pyrexia, and exerts effects on bone marrow, inducing granulocytosis. It also acts on hepatocytes causing alterations associated with the acute phase response, i.e. a decrease in the production of albumin with a concomitant increase in the production of acute phase proteins (C-reactive protein, fibrinogen, serum amyloid A and haptoglobin). Furthermore, it regulates the serum concentration of trace metals, causing a reduction in zinc and iron (reflecting a liberation of lactoferrin from neutrophils) and an increase in the level of copper (as a result of increased synthesis of ceruloplasmin (DINARELLO 1984 a, b; DURUM et al. 1985; OPPENHEIM et al. 1986).

Originally described as a blastogenic factor in the culture supernatants of mixed lymphocyte cultures (KASAKURA and LOWENSTEIN 1965; GORDON and MACLEAN 1965), and subsequently as a product of antigen-specific stimulation of sensitised lymphocytes (DUMONDE et al. 1969; MAINI et al. 1969), IL-2 is considered primarily as a T-cell growth factor. The biological activities of IL-2 are almost entirely restricted to lymphocyte populations, e.g. enhancement of natural killer cell activity, activation of killer cells, and a growth and differentiation of B cells; however, IL-2 also activates macrophages (ROBB 1984; SMITH 1984).

Interleukin-3, also referred to as multi-colony stimulating and mast cell growth factor, is a haemopoietic stimulating factor (IHLE et al. 1982; CLARK and KAMEN 1987). It is a T-cell derived factor thought to play a key role in the regulation of haemopoiesis by the immune system, and a role in the oncogenesis of haemopoietic cells has been proposed (METCALF 1985).

B-cell stimulating factor 1, or B-cell growth factor (HOWARD et al. 1982), now reclassified as IL-4, acts on (mouse) B cells before they enter the cell cycle, causing cell proliferation and enhanced synthesis and secretion of IgG_1 and IgE. Interleukin-4 also promotes the differentiation and proliferation of T cells (WIDMER and GRABSTEIN 1987) – an example of the overlap, and, in some instances, the synergy exhibited by the interleukins.

Interleukin-5, or eosinophil differentiation factor (LOPEZ et al. 1986), is a 45 K dalton glycoprotein that induces the production of eosinophils in bone marrow cultures. It is identical to B-cell growth factor II (SWAIN et al. 1983) and T-cell replacing factor (SCHIMPL and WECKER 1972).

Interferon-β_2, now proposed as IL-6, was first detected on the basis of its interferon-like antiviral activity (SAGAR et al. 1982) and has subsequently been found to be identical to a T cell differentiation factor or B-cell stimulating factor 2 (HIRANO et al. 1986). It is structurally similar to a fibroblast-derived 26 K dalton potent growth factor, which is capable of replacing feeder cells in the maintenance of certain hybridoma and plasmacytoma cell lines (BILLIAU 1986).

D. Interleukins and Normal Skin

For a considerable period many immunologists looked upon the skin as fulfilling two functions: firstly as the main physical barrier in restricting the entry of infectious organisms and secondly as an appropriate site for eliciting hypersensitivity responses to intradermally applied antigens. This view was to change following the demonstration that Langerhans' cells, a subgroup of dendritic cells (TEW et al. 1981), are very efficient at presenting antigen to lymphoid cells; accordingly, the skin is now regarded as an important immunological organ.

Accessory cells provide two signals for T-cell activation: the presentation of antigen associated with class II MHC molecules, and the production of IL-1. The IL-1 acts as a maturation signal by inducing T cells to generate IL-2 and to express IL-2 receptors which, in turn, promote clonal expansion. Both Langerhans' cells and keratinocytes produce epidermal thymocyte-activating factor (ETAF), which is now acknowledged to be physicochemically and functionally identical to IL-1 (LUGER et al. 1981; ANSEL et al. 1983; LUGER and OPPENHEIM 1983; LUGER et al. 1983a; SAUDER et al. 1984) and to have the same mRNA expression (ANSEL et al. 1986; KUPPER et al. 1986). However, keratinocytes have also been shown to express an additional IL-1-like mRNA that is different from that found in macrophages, introducing the possibility that these cells may also encode a new IL-1-like species (BELL et al. 1987). In addition to Langerhans' cells, antigen presentation may also be undertaken by resident macrophages, endothelial cells of the local vasculature, and fibroblasts. Although interferon treatment of dermal fibroblasts induces these cells to express surface class II molecules and to present antigen (UMETSU et al. 1985), the addition of IL-2 is required in order to generate the proliferation of unactivated resting T cells from sensitised donors.

Both IL-1 and its precursor are present in normal human epidermis in vivo (HAUSER et al. 1986; SCHMITT et al. 1986) at concentrations 100–1000 greater than those found in other tissues. Such high resting values of IL-1 in normal tissue suggest that either it is continuously being non-specifically released by the constituent cells or that it is being produced in response to persistent external immunological stimuli of a subclinical nature.

E. Dermatological Disorders and IL-1

The impaired cell-mediated immunity of patients with atopic dermatitis (HANIFIN and LOBITZ 1977) may underlie the increased susceptibility of these patients to cutaneous bacterial, fungal and viral infections. Dendritic cells are powerful stimulators of the allogeneic and autologous mixed leucocyte reaction (AMLR) (KUNTZ-CROW and KUNKEL 1982), both of which are believed to have a significant role in the regeneration of helper and suppressor/cytotoxic lymphocytes (SMOLEN et al. 1982). In patients with atopic dermatitis a defective AMLR, using Langerhans' cells as the stimulator population, may be due to an impaired generation of cytokines, as shown by the inability of epidermal cells from these patients to produce normal quantities of IL-1 (RASANEN et al. 1987a). Restoration of the AMLR response in atopic dermatitis has been achieved following the addition of exogenous IL-1 or IL-2 (LEUNG et al. 1983). The failure to produce adequate quantities of IL-1 in atopic dermatitis may be a defect common to other accessory cells, as demonstrated by the decreased production of IL-1 by monocytes from patients with this disorder (MIZOGUCHI et al. 1985; RASANEN et al. 1987b).

The clinical effectiveness of retinoids in the treatment of a number of dermatoses may be related to their IL-1-inducing activity (DAYER 1982; TRECHSEL et al.

1985; WALSH et al. 1985). The systemic administration of detretinate, a synthetic retinoid, to rats produced an increase of epidermal IL-1 whose activity has been proposed to prevent the hyperproliferation of keratinocytes (SCHMITT et al. 1987). However, in vitro, IL-1 stimulates the growth of these cells (GILCHRIST and SAUDER 1984).

The inflammatory skin changes and cutaneous malignancies induced by ultraviolet (UV) radiation (HARBER and BICKERS 1981; PARRISH et al. 1984), together with the systemic sequelae of nausea, headache and fever associated with severe sunburn, are thought to arise from the action of soluble inflammatory factors (e.g. 5-hydroxytryptamine, histamine and prostaglandins) released from UV-damaged cells of the epidermis and dermis (HAWK et al. 1983; VENINGA and DE DOER 1986). Recent work has shown that animals exposed to acute UV irradiation, which develop the systemic disorders associated with acute phototoxicity, show an increase in serum IL-1 activity (ANSEL et al. 1987). Since IL-1 is currently regarded as an effector molecule central to the induction of the acute phase response, these results suggest that this molecule could play a significant role in the manifestations of acute phototoxicity, particularly as exposure of epidermal cells to UV irradiation either in vivo or in vitro enhances their production of IL-1 (ANSEL et al. 1984; GAHRING et al. 1984; KUPPER and MCGUIRE 1986; MCGUIRE et al. 1986). However, this view is complicated by the finding that low doses of ultraviolet B radiation have a profound inhibitory effect on the release of IL-1 from human epidermal cells used as accessory cells in a T-cell mitogenic assay (ELMENTS et al. 1986).

Ultraviolet irradiation also induces immunosuppression as shown by the inhibition of contact and delayed-type hypersensitivity in treated mice, an effect that is believed to be mediated either by impaired antigen presentation (GREENE et al. 1979; NOONAN et al. 1981) or by the induction of suppressor cells (DREBIN et al. 1983). Although there is little evidence to implicate IL-1 in these particular mechanisms of immunosuppression, pretreatment of mice with high doses of this interleukin impairs their capacity to elicit contact hypersensitivity responses (DAYNE et al. 1986). Such an inhibitory effect is not seen in mice receiving indomethacin, an inhibitor of prostaglandin production, suggesting that prostaglandins, which have also been proposed to control production of IL-2 (KUNKELL and CHENSUE 1985), mediate the IL-1-induced depression of contact hypersensitivity (ROBERTSON et al. 1987). In aged animals the hyporesponsiveness to contact sensitisation is fully restored by the administration of IL-2 (BELSITO et al. 1987).

Attempts to implicate IL-1 with the UV-induced impairment of contact hypersensitivity are complicated by the finding that UV irradiation of mouse epidermal cells induces the synthesis and release of a specific inhibitor of IL-1 activity (SCHWARZ et al. 1987). This inhibitor's molecular weight (40 K daltons) is similar to that of a factor which is released from UV-treated murine keratinocytes, and which inhibits the induction of contact hypersensitivity (SCHWARZ et al. 1986). However, continued exposure to UV irradiation can indeed produce a state of hyposensitisation in which the animal becomes refractory to further challenge with the contact sensitiser. It has been suggested that the UV-induced desensitisation in mice arises from the inability of IL-1 generated within the skin to reach its target, or that the changes in the structure of skin of chronically UV-exposed animals prevent induction of IL-1 release on appropriate stimulation (GAHRING and DAYNES 1986).

Scleroderma is a chronic inflammatory disorder characterised by excessive collagen production and fibroblast proliferation. Since IL-1 increases the synthesis of glycosaminoglycans and the proliferation of fibroblasts in scleroderma (WHITESIDE et al. 1987), studies have been performed to determine if there is an impairment in the regulation of IL-1 production in this disorder. In one such investigation monocytes from one-third of patients spontaneously released IL-1, in contrast to control monocytes whose supernatants were devoid of any activity (ALCOCER-VARELA et al. 1985). A similar finding was reported by another group where only diluted supernatants obtained from unstimulated blood mononuclear cells of scleroderma patients revealed enhanced IL-1 activity (SANDBORG et al. 1985). The undiluted supernatants were later shown to contain an inhibitor of IL-1 which also possessed fibroblast stimulatory activity (SANDBORG et al. 1986). Further experiments are needed to identify the cellular source of this inhibitor and to define its pathological implication for scleroderma.

F. Dermatological Disorders and IL-2

Pemphigus vulgaris is an autoimmune disease involving skin and mucous membranes. Sera from patients with this disorder contain an antibody which reacts with an intercellular autoantigen in both the epidermis and mucous membrane. These patients also have defects in IL-1 and IL-2 production and in the expression of IL-2 receptors on blood mononuclear cells following mitogen stimulation (KERMANI-AROB et al. 1984; POLITSTEIN-WILLINGER 1985). As an immunoregulation disorder is believed to underlie the enhanced expression of autoantibodies, (ALLISON et al. 1971) it is of interest that an impairment of IL-2 production is also seen in patients with systemic lupus erythematosus (SLE) (ALCOCER-VARELA and ALARCON-SEGOVIA 1982; LINKER-ISRAELI et al. 1983), and in strains of mice that develop SLE-like autoimmune disease (DAUPHINE et al. 1981; WOLFSY et al. 1981).

A failure of T-lymphocytes to respond to IL-2 may be an immunological aberration common to a number of inflammatory dermatological diseases which do not necessarily share an identifiable autoimmune component. Patients with early active Behçet's disease have a decreased number of IL-2 receptor-bearing T cells, whereas patients with chronic active or inactive disease possess cells with a low expression of IL-2 receptors (SAKANE et al. 1986). Such defects in IL-2 immunoregulation may be overcome by appropriate therapy. For example, the inability of lymphocytes from patients with visceral leishmaniasis to produce IL-2 during periods of active disease is restored by successful chemotherapy (CARVALHO et al. 1985) whilst the IL-2 defect in Pemphigus vulgaris (see above) is corrected following palliative treatment with gold salts (POLITSTEIN-WILLINGER 1985). There is experimental evidence that IL-2 itself may have an application in virus infections of the skin. Thus, administration of IL-2 to mice infected with herpes simplex virus increases the antiviral activity of adoptively transferred T-lymphocytes (SANDER and KATZ 1982), and in vitro IL-2 restores the defective lymphocyte transformation of patients with recurrent herpes simplex virus infection (Wainberg et al. 1985). Whereas in certain dermatological disorders there may be a need to suppress IL-2 production, in others the reverse may well apply. For example, an important mode of action of cyclosporin A is believed to be its inhibition of IL-2 mRNA production (Elliot et al. 1984). This may well underlie its ability to suppress delayed hypersensitivity and contact sensitisation in animals (Shidani et al. 1984; Borel et al. 1977; Rullan et al. 1984), and its therapeutic effects in psoriasis (Harper et al. 1984; Van Hooff et al. 1985; Van Joost et al. 1986; Wentzell et al. 1987) and atopic dermatitis (Van Joost et al. 1987). Suppression of IL-2 production and activity might well be of therapeutic value in cutaneous T-cell lymphomas, and it is of interest that IL-2 has been used in the laboratory to maintain a cultured T-cell line which originated from a patient with Sezary's syndrome (Namiuchi et al. 1986).

G. Dermatological Disorders, Interleukins, and Leucocyte-Endothelial Interactions

A number of "idiopathic" dermatological diseases (e.g. psoriasis) are characterised by an intense infiltration of leucocytes into the dermal or epidermal layers, and it is likely that interleukins may play an important role in the recruitment of such leucocytes.

In health, vascular endothelium appears to regulate the compartmentation of the different leucocyte classes between blood and tissue, whereas in inflammatory disease (such as psoriasis) a breakdown of endothelial homeostasis may well underlie the pathogenesis of extravascular leucocyte infiltration and activation. Indeed, changes in the morphology and function of capillaries within the dermis appear to precede the epidermal cell proliferation associated with this disorder (Majwski et al. 1987). In psoriatic lesions, polymorphonuclear cells (PMNs) migrate from the enlarged capillaries in the dermal papillae into the upper regions of the epidermis where they form microabscesses (Pincus and Mehregan 1966; Tagami and Ofiyi 1976). In the lower region of the epidermis (Ternowitz 1986) and the upper region of the dermis the leucocyte infiltrate is predominantly composed of mononuclear cells. The continued infiltration of blood leucocytes into psoriatic skin may be due to a disorder of leucocyte-endothelial interaction, in which interleukins play an important role (Bos and Krieg 1985; Boker et al. 1984).

At first sight, IL-1 appears to have all the necessary credentials to be considered as a regulatory molecule central to the control of leucocyte extravasation into inflamed skin. It is secreted by endothelial cells (Miossec et al. 1986) and is a chemotactic agent for leucocytes (Luger et al. 1983; Miossec et al. 1984); a chemotaxin that biochemically resembles IL-1 has been identified in psoriatic scales (Schroder and Christophers 1985; Camp et al. 1986). Interdermal injection of IL-1 results in the accumulation of neutrophils (Cybulsky et al. 1985). Although extracts of psoriatic lesions contain a number of chemoattractants, e.g. platelet activating factor (PAF) and leukotriene B4 (Brain et al. 1984; Mallet and CUNNINGHAM 1985), to date these mediators do not appear to have the multiplicity of biological effects on endothelial cells that is seen with IL-1. Pretreatment of cultured endothelium with low concentrations of IL-1 increases the adhesion of polymorphonuclear cells (BEVILACQUA et al. 1985; SCHLEIMER and RUTLEDGE 1986), monocytes (BEVILACQUA et al. 1985) and lymphocytes (CAVENDER et al. 1986). This effect is believed to be due to the induction on the endothelial cell surface of adhesion proteins whose optimum expression is seen after 4-h incubation (POBER et al. 1986; POHLMAN et al. 1986). An IL-1-induced augmentation of leucocyte infiltration may also be mediated through its activation of endothelial cells to release other inflammatory factors such as prostaglandin E2 (ALBRIGHTSON et al. 1985), and granulocyte-macrophage colony stimulating factor (SIEFF et al. 1987).

Another prominent feature of psoriasis is the hyperproliferation of epidermal keratinocytes. Since ETAF is released from keratinocytes and stimulates their growth in vitro (GILCHRIST and SAUDER 1984), this molecule could be associated with the hyperproliferation of the psoriatic epidermis. However, the attractive proposal that IL-1 is a dominant factor in promoting leucocyte extravasation and keratinocyte proliferation in psoriasis is questioned by reports showing that there is no intrinsic abnormality of ETAF production in psoriatic keratinocytes (KRAGBALLE et al. 1987) and that the IL-1-like activity present in crude horny tissue extracts from inflammatory skin of psoriatic patients is far lower than that seen in similar extracts from normal skin and patients with other inflammatory dermatoses (TAKEMATSU et al. 1986). These decreased levels of IL-1 could be due to the activity of an inhibitor of IL-1 which is present in psoriatic skin (CAMP et al. 1986). Alternatively, low concentrations of IL-1 in aqueous extracts from psoriatic scale could result from the continued consumption of IL-1 by infiltrating leucocytes, or by cells which promote leucocyte recruitment, such as the local vascular endothelium.

H. Comment

Few studies have investigated a possible role for interleukins, other than IL-1 and IL-2, in the pathology of dermatological diseases. In culture, normal and malignant human epidermal cells spontaneously release factors with IL-3-like activity (LUGER et al. 1985; DANNER and LUGER 1987), suggesting that the epidermis may have the capacity to recruit leucocytes from the bone marrow and induce their differentiation. There is growing interest in seeking associations between other interleukins and dermatological disorders, e.g. the eosinophilic activities of IL-5 and the features of allergic skin disease. With a growing interaction between immunology and dermatology and a growing appreciation of synergy, cascade and pleiotropic interactions between interleukins and other cytokines, the stage seems set for a burgeoning of research activity in this field.

Acknowledgements. We thank the Trustees and Research (Endowments) Committee of St. Thomas' Hospital for research support.

References

Aarden LA, Brunner TK, Cerottini JC et al. (1979) Letter to the editor: revised nomenclature for antigen non-specific T-cell proliferation and helper factors. J Immunol 123:2928–2929
Adelman NE, Hammond ME, Cohen S, Dvorak HF (1979) Lymphokines as inflammatory mediators. In: Cohen S, Pick E, Oppenheim JJ (eds) Biology of the lymphokines. Academic, New York, pp 13–58
Albrightson CR, Baenziger NL, Needleman P (1985) Exaggerated human vascular cell prostaglandin biosynthesis mediated by monocytes: role of monokines and interleukin 1. J Immunol 135:1872–1877
Alcocer-Varela J, Alarcon-Segovia D (1982) Decreased production of and response to interleukin-2 by cultured lymphocytes from patients with systemic lupus erythematosus. J Clin Invest 69:1388–1396
Alcocer-Varela J, Martinez-Cordero E, Alarcon-Segovia D (1985) Spontaneous production of, and defective response to, interleukin-1 by peripheral blood mononuclear cells from patients with scleroderma. Clin Exp Immunol 59:666–672
Allison AC, Denman AM, Barnes RD (1971) Cooperating and controlling functions of thymus-derived lymphocytes in relation to autoimmunity. Lancet II:135–140
Ansel J, Luger TA, Green I (1983) The effect in vitro and in vivo of UV irradiation on the production of ETAF activity by human and murine keratinocytes. J Invest Dermatol 81:519–523
Ansel J, Luger TA, Klock A, Hochstein D, Green I (1984) The effect of in vivo UV irradiation on the production of IL-1 by murine macrophages and P388D1 cells. J Immunol 133:1350–1355
Ansel JC, Luger TA, Lowry DR, Mountz JD (1986) Expression of IL-1 in murine keratinocytes. J Invest Dermatol 87:127
Ansel JC, Luger TA, Green I (1987) Fever and increased serum IL-1 activity as a systemic manifestation of acute phototoxicity in New Zealand white rabbits. J Invest Dermatol 89:32–37
Bell TV, Harley CB, Stetsko D, Sander DN (1987) Expression of mRNA homologous to interleukin 1 in human epidermal cells. J Invest Dermatol 88:375–379
Belsito DV, Dersarkissian RM, Thorbecke GJ, Baer RL (1987) Reversal by lymphokines of the age-related hyperresponsiveness to contact sensitization and reduced Ia expression on Langerhans cells. Arch Dermatol Res 279:576–580
Bevilacqua MP, Pober JS, Wheeler MC, Contran RS, Gimbrone MA Jr (1985) Interleukin 1 acts on cultured human vascular endothelium to increase the adhesion of polymorphonuclear leukocytes, monocytes and related leukocyte lines. J Clin Invest 76:2003–2011
Billiau A (1986) BSF-2 is not just a differentiation factor. Nature 324:415
Boker BS, Swain AF, Fry L, Valdimarsson H (1984) Epidermal T lymphocytes and HLA-DR expression in psoriasis. Br J Dermatol 110:555–564
Borel JF, Feurer C, Magnee C, Stahelin H (1977) Effects of the new antilymphocyte peptide cyclosporin A in animals. Immunology 32:1017–1025
Bos JD, Krieg SR (1985) Psoriasis infiltrating cell immunophenotype: changes induced by PUVA or corticosteroid treatment in T-cell subsets, Langerhans cells and interdigitating cells. Acta Derm Venereol (Stockh) 65:390–397
Brain S, Camp R, Dowd P, Kobza Black A, Greaves MW (1984) The release of leukotriene B4-like material in biologically active amounts from the lesional skin of patients with psoriasis. J Invest Dermatol 83:70–77
Camp RDR, Fincham NJ, Cunningham FM, Greaves MW, Morris J, Chu A (1986) Psoriatic skin lesions contain biologically active amounts of an interleukin 1-like compound. J Immunol 137:3469–3474
Carvalho EM, Badaro R, Reed SG, Jones TC, Johnson WD (1985) Absence of gamma interferon and interleukin 2 production during active visceral leishmaniasis. J Clin Invest 76:2066–2069
Cavender DE, Haskard DO, Joseph B, Ziff M (1986) Interleukin 1 increases the binding of human B and T lymphocytes to endothelial cell monolayers. J Immunol 136:203–207

Clark SC, Kamen R (1987) The human hematopoietic colony stimulating factors. Science 236:1229–1237
Clark-Lewis I, Aebersold R, Ziltener H, Schrader JW, Hood LE, Kent SBH (1986) Automated chemical synthesis of a protein growth factor for hemopoietic cells, interleukin-3. Science 231:134–139
Cybulsky MI, Colditz IG, Movat HZ (1985) Interleukin 1 activity in the local recruitment of PMNs: its potential role in endotoxin induced acute inflammation. Fed Proc 44:1260 (abstr)
Danner M, Luger TA (1987) Human keratinocytes and epidermoid carcinoma cell lines produce a cytokine with interleukin 3-like activity. J Invest Dermatol 88:353–361
Dauphine MJ, Kipper BS, Wolfsy D, Talal N (1981) Interleukin-2 deficiency is a common feature of autoimmune mice. J Immunol 127:2483–2487
Dayer JM (1982) Effects of retinoids on prostaglandin E2 (PGE2) and mononuclear cell factor (MCF) production by human monocyte cultures. Clin Res 30:469 (abstr)
Dayne R, Samlowski WE, Burnhan DK, Gahring LC, Roberts LK (1986) Immunobiological consequences of acute and chronic UV exposure. Curr Probl Dermatol 15:177–194
Dinarello CA (1984a) Interleukin 1. Rev Infect Dis 6:51–55
Dinarello CA (1984b) Interleukin 1 and the pathogenesis of the acute-phase response. N Engl J Med 311:1413–1418
Dower SK, Urdel DL (1987) The interleukin-1 receptor. Immunol Today 8:46–51
Drebin AJ, Schatten S, Tominga A (1983) Effect of ultraviolet radiation-induced impairment of antigen-presenting cell function at the cellular and molecular level. In: Parrish JA (ed) The effect of ultraviolet radiation on the immune system. Johnson and Johnson, New York, p 123
Dumonde DC, Wolstencroft RA, Panayi GS, Matthew M, Morley J, Howson WT (1969) Lymphokines: non-antibody mediators of cellular immunity generated by lymphocyte activation. Nature 224:38–42
Dumonde DC, Hamblin AS (1983) Lymphokines. In: Holborow EJ, Reeves WG (eds) Immunology in medicine, 2nd edn. Academic, New York, pp 121–150
Durum SK, Schmidt JA, Oppenheim JJ (1985) Interleukin 1: an immunological perspective. Ann Rev Immunol 3:263–287
Elliot YF, Lin Y, Mizel SB, Blackley RC, Harnish DG, Paetkau V (1984) Induction of interleukin 2 messenger RNA inhibited by cyclosporin A. Science 226:1439–1441
Elments C, Rick E, Urda G, Fujiwara H, Ellner J (1986) UVB radiation inhibits accessory cell signals required for mitogenic responses of human T lymphocytes. J Invest Dermatol 86:473 (abstr)
Fung MC, Hapel AJ, Ymer S, et al. (1984) Molecular cloning of cDNA for murine interleukin 3. Nature 307:233–237
Gahring L, Baltz M, Pepys MB, Daynes R (1984) Effect of ultraviolet radiation on the production of epidermal cell thymocyte-activating factor/interleukin 1 in vivo and in vitro. Proc Natl Acad Sci USA 81:1198–1202
Gahring LC, Daynes RA (1986) Desensitization of animals to the inflammatory effects of ultraviolet radiation is mediated through mechanisms which are distinct from those responsible for endotoxin tolerance. J Immunol 136:2868–2874
Gery I, Gershon RK, Waksman BH (1971) Potentiation of cultured mouse thymocyte responses by factors released by peripheral leukocytes. J Immunol 107:1778
Gery I, Davies P, Derr J, Krett N, Barranger JA (1981) Relationship between production and release of lymphocyte-activating factor (interleukin 1) by murine macrophages. I. Effects of various agents. Cell Immunol 64:293–303
Gilchrist BA, Sauder DN (1984a) Autocrine growth stimulation of human keratinocytes by epidermal cell-derived thymocyte activating factor (ETAF): implications for cellular ageing. J Invest Dermatol 82:439 (abstr)
Gilchrist BA, Sauder DN (1984b) Autocrine growth stimulation of human keratinocytes by epidermal cell-derived thymocyte activating factor (ETAF): implications for cellular ageing. Clin Res 32:585A
Gillis S, Stull DD (1982) Tumor and hybridoma cell line interleukin-2 production. In: Pick E (ed) Lymphokines, vol 5. Academic, New York, pp 371–385

Gillis S, Watson J (1980) Biochemical and biological characterization of lymphocyte regulatory molecules. V. Identification of an interleukin-2 producer human leukemic T-cell line. J Exp Med 152:1709–1719

Gillis S, Scheid M, Watson J (1980) Biochemical and biological characterization of lymphocyte regulatory molecules. III. The isolation and phenotypic characterization of interleukin-2 producing T-cell lymphomas. J Immunol 125:2570–2578

Gordon J, MacLean LD (1965) A lymphocyte-stimulating factor produced in vitro. Nature 208:795–796

Gray PW, Leung DW, Pennica D, et al. (1982) Expression of human immune interferon cDNA in *E. coli* and monkey cells. Nature 295:503–508

Greene MI, Sy MS, Kripke ML, Benacerraf B (1979) Impairment of antigen-presenting function by ultraviolet radiation. Proc Natl Acad Sci USA 76:6591–6595

Hanifin JM, Lobitz WC Jr (1977) Newer concepts of atopic dermatitis. Arch Dermatol 113:667–670

Hanson JM, Rumjanek VM, Morley J (1982) Mediators of cellular immune reactions. Pharmacol Ther 17:165–198

Harber LC, Bickers DR (1981) Photosensitivity diseases. Principles of diagnosis and management. Saunders, Philadelphia

Harper JI, Keat ACS, Staughton RCD (1984) Cyclosporin for psoriasis. Lancet II:981–988

Hauser C, Saurat JH, Schmitt A, et al. (1986) Interleukin 1 is present in normal human epidermis. J Immunol 136:3317–3322

Hawk JLM, Black AK, Jaenicke KF, et al. (1983) Increased concentrations of arachidonic acid, prostaglandin E2, D2 and 6-oxo-F1α and histamine in human skin following UVA irradiation. J Invest Dermatol 80:476–498

Hirano T, Yasukawa K, Harada H, et al. (1986) Complementary DNA for a novel human interleukin (BSF-2) that induces B lymphocytes to produce immunoglobulin. Nature 324:73–76

Holbrook NJ, Smith KA, Fornace AJ, Comeau CC, Wiskocil RL, Crabtree GR (1984) T-cell growth factor: complete nucleotide sequence and organisation of the gene in normal and malignant cells. Proc Natl Acad Sci USA 81:1634–1638

Howard M, Farrar J, Hilfiker M, et al. (1982) Identification of a T-cell derived B-cell growth factor distinct from interleukin 2. J Exp Med 155:914–923

Ihle JN, Rebar L, Keller J, Lee JC, Hapel AJ (1982) Interleukin 3: possible roles in the regulation of lymphocyte differentiation and growth. Immunol Rev 63:11–32

Kampschmidt RF, Pulliam LA, Merriman CR (1978) Further similarities of endogenous pyrogen and leukocytic endogenous mediator. Am J Physiol 243:E332–E337

Kasakura S, Lowenstein L (1965) A factor stimulating DNA synthesis derived from the medium of leucocyte cultures. Nature 208:794–795

Kermani-Arob V, Hirji K, Ahmed AR, Fahey JL (1984) Deficiency of interleukin-2 production and interleukin-2 receptor expression on peripheral blood leukocytes after phytohaemagglutinin stimulation in pemphigus. J Invest Dermatol 83:101–104

Kinashi T, Harada N, Severinson E, et al. (1986) Cloning of complementary DNA encoding T-cell replacing factor and identity with B-cell growth factor II. Nature 324:70–73

Kobayashi Y, Asada M, Higuchi M, Osawa T (1982) Human T-cell hybridomas producing lymphotoxin and migration inhibitory factor. J Immunol 128:417–422

Kragballe K, Marcelo CL, Voorhees JJ, Sauder DN (1987) Formation of epidermal cell thymocyte-activating factor (ETAF) from cultured human keratinocytes from normal and uninvolved psoriatic skin. J Invest Dermatol 88:8–10

Kunkell SL, Chensue SW (1985) Arachidonic acid metabolites regulate interleukin-1 production. Biochem Biophys Res Commun 128:892–897

Kuntz-Crow M, Kunkel MG (1982) Human dendritic cells: major stimulators of the autologous and allogeneic mixed leucocyte reactions. Clin Exp Immunol 49:338–346

Kupper TS, McGuire J (1986) Hydrocortisone reduces both constitutive and UV-elicted release of epidermal thymocyte activating factor (ETAF) by cultured keratinocytes. J Invest Dermatol 87:570–573

Kupper T, Gubler U, Ballard D, et al. (1986) Identification of interleukin 1 mRNA in human and murine keratinocyte. Clin Res 34:640–647

Leonard WJ, Depper JM, Crabtree GR, et al. (1984) Molecular cloning and expression of cDNAs for the human interleukin-2 receptor. Nature 311:626–631
Leung DYM, Saryan JA, Frankel R, Lareau M, Geha RS (1983) Impairment of the autologous mixed lymphocyte reaction in atopic dermatitis. J Clin Invest 72:1482–1486
Liang S-M, Thatcher DR, Wang C-M, Allet B (1986) Studies of structure-activity relationships of human interleukin-2. J Biol Chem 261:334–337
Linker-Israeli M, Bahke AC, Kitridou RC, Gendler S, Gillis S, Horwitz DA (1983) Defective production of interleukin-1 and interleukin-2 in patients with systemic lupus erythematosus (SLE). J Immunol 130:2651–2655
Lomedico PT, Gubler U, Hellmann CP, et al. (1984) Cloning and expression of murine interleukin-1 cDNA in *Escherichia coli.* Nature 312:458–462
Lopez AF, Begley CG, Williamson DJ, et al. (1986) Murine eosinophil differentiation factor. An eosinophil-specific colony-stimulating factor with activity for human cells. J Exp Med 163:1085–1099
Lotze MT, Matory YL, Ettinghausen SE, et al. (1985) In vivo administration of purified human interleukin 2. II. Half-life, immunologic effects, and expansion of peripheral lymphoid cells in vivo with recombinant IL-2. J Immunol 135:2865–2875
Luger TA, Oppenheim JJ (1983) Characteristics of interleukin 1 and epidermal cell-derived thymocyte activating factor. Adv Inflam Res 5:1–25
Luger TA, Stadler BM, Katz SI, et al. (1981) Epidermal cell (keratinocyte)-derived thymocyte-activating factor (ETAF). J Immunol 127:1493–1498
Luger TA, Sztein MB, Schmidt J, et al. (1983 a) Properties of murine and human epidermal cell-derived thymocyte-activating factor. Fed Proc 42:2772–2776
Luger TA, Charon JA, Colot M, et al. (1983 b) Chemotactic properties of partially purified human epidermal cell-derived thymocyte-activating factor (ETAF) for polymorphonuclear and mononuclear cells. J Immunol 131:816–820
Luger TA, Wirth U, Koeck A (1985) Epidermal cells synthesize a cytokine with interleukin 3 like properties. J Immunol 134:915–919
Maini RN, Bryceson ADM, Wolstencroft RA, Dumonde DC (1969) Lymphocyte mitogenic factor in man. Nature 224:43–44
Majwski S, Tigalonowa M, Jablonska S, Polakowski I, Janczura E (1987) Serum samples from patients with active psoriasis enhance lymphocyte-induced angiogenesis and modulate endothelial cell proliferation. Arch Dermatol 123:221–225
Mallet AI, Cunningham FM (1985) Structural determination of platelet activating factor in psoriasis. Biochem Biophys Res Commun 126:192–199
March CJ, Mosley B, Larsen A, et al. (1985) Cloning, sequence and expression of two distinct human interleukin-1 complementary DNAs. Nature 315:641–647
McGuire J, Longdon R, Kupper T (1986) Interleukin-1 mRNA increases following irradiation with ultraviolet-B. J Invest Dermatol 87:155 (abstr)
Metcalf D (1985) The granulocyte-macrophage colony stimulating factors. Science 229:16–22
Miossec P, Yu C-L, Ziff M (1984) Lymphocyte chemotactic activity of human interleukin 1. J Immunol 133:2007–2011
Miossec P, Cavender D, Ziff M (1986) Production of interleukin 1 by human endothelial cells. J Immunol 136:2486–2491
Mizoguchi M, Furusawa S, Okitsu S, Yoshino K (1985) Macrophage-derived interleukin 1 activity in atopic dermatitis. J Invest Dermatol 84:303–311
Namiuchi S, Kumagai S, Sano H, et al. (1986) A human T cell line established from a patient with Sezary syndrome. J Immunol Methods 94:215–224
Nikaido T, Shimizu A, Ishida N, et al. (1984) Molecular cloning of cDNA encoding human interleukin-2 receptor. Nature 311:631–635
Noonan FP, Kripke ML, Pederson GM, Green MI (1981) Suppression of contact sensitivity of mice by ultraviolet radiation is associated with defective antigen presentation. Immunology 43:527–533
Oppenheim JJ, Kovacs EJ, Matsushima K, Durum SK (1986) There is more than one interleukin 1. Immunol Today 7:45–56
Parrish JA, Kripki ML, Warwick ML (1984) Photoimmunology. Plenum, New York

Pennica D, Nedwin GE, Haylick JS, et al. (1984) Human tumour necrosis factor: precursor structure, expression and homology to lymphotoxin. Nature 312:724–729
Pincus H, Mehregan AM (1966) The primary histological lesion of seborrhaeic dermatitis and psoriasis. J Invest Dermatol 46:109–116
Pober JS, Bevilacqua MP, Mendrick DL, Lapierre LA, Fiers W, Gimbrone MA Jr (1986) Two distinct monokines, interleukin 1 and tumor necrosis factor, each independently induce biosynthesis and transient expression of the same antigen on the surface of cultured human vascular endothelial cells. J Immunol 136:1680–1687
Pohlman TH, Stanness KA, Beatty PG, Ochs MD, Harlan JM (1986) An endothelial cell surface factor(s) induced in vitro by lipopolysaccharide, interleukin 1, and tumour necrosis factor – increases neutrophil adherence by a CDW 18-dependent mechanism. J Immunol 136:4548–4553
Politstein-Willinger E (1985) Normalization of defective interleukin 1 and interleukin 2 production in patients with pemphigus vulgaris following chrysotherapy. Clin Exp Immunol 62:705–714
Rasanen L, Lehto M, Reunola T, Jansen C, Lehtinen M, Leinikki P (1987 a) Langerhans cell- and T-lymphocyte functions in patients with atopic dermatitis with disseminated cutaneous herpes simplex virus infection. J Invest Dermatol 89:15–18
Rasanen L, Lehto M, Reunala T, Jansen C, Leinikki P (1987 b) Decreased monocyte production of interleukin-1 and impaired lymphocyte proliferation in atopic dermatitis. Arch Dermatol Res 279:215–218
Robb RJ, Smith KA (1981) Heterogeneity of human T-cell growth factor(s) due to variable glycosylation. Mol Immunol 18:1087–1094
Robb RJ (1984) Interleukin 2: the molecule and its function. Immunol Today 5:203–209
Robb RJ, Kutny RM, Panico M, Morris HR, Chowdrey V (1984) Amino acid sequence and post-translational modification of human interleukin-2. Proc Natl Acad Sci USA 81:6456–6490
Robertson B, Gahring L, Newton R, Daynes R (1987) In vivo administration of interleukin 1 to normal mice depresses their capacity to elicit contact hypersensitivity responses: prostaglandins are involved in this modification of immune function. J Invest Dermatol 88:380–387
Rocklin RE, Bendtzen K, Greineder D (1980) Mediators of immunity: lymphokines and monokines. Adv Immunol 29:55–136
Rosenberg SA, Grimm EA, McGrogan M, et al. (1984) Biological activity of recombinant human interleukin-2 produced in *Escherichia coli*. Science 223:1412–1415
Rosenstreich DL, Wahl SM (1979) Cellular sources of lymphokines. In: Cohen S, Pick E, Oppenheim JJ (eds) Biology of the lymphokines. Academic, New York, pp 210–242
Rosenwasser LJ, Dinarello CA, Rosenthal AS (1979) Adherent cell function in murine T-lymphocyte antigen recognition. IV. Enhancement of murine T cell antigen recognition by human leukocytic pyrogen. J Exp Med 150:709–714
Rullan PP, Barr RJ, Cole GW (1984) Cyclosporine and murine allergic contact dermatitis. Arch Dermatol 120:1179–1186
Sagar AD, Sehgal PB, Slate DL, Ruddle FH (1982) Multiple human interferon genes. J Exp Med 156:744–755
Sakane T, Suzuki N, Ueda Y, et al. (1986) Analysis of interleukin-2 activity in patients with Behcet's disease. Arthritis Rheum 29:371–378
Sandborg CI, Berman MA, Andrews BS, Friou GJ (1985) Interleukin-1 production by mononuclear cells from patients with scleroderma. Clin Exp Immunol 60:294–302
Sandborg CI, Berman MA, Andrews BS, Mirick GR, Friou GJ (1986) Increased production of an interleukin 1 (IL-1) inhibitor with fibroblast stimulating activity by mononuclear cells from patients with scleroderma. Clin Exp Immunol 66:312–319
Sauder DN, Katz SI (1982) Immune modulation by epidermal cell products: possible role of ETAF in inflammatory and neoplastic skin diseases. J Am Acad Dermatol 7:651–654
Sauder DN, Dinarello CA, Morhenn VB (1984) Langerhans cell production of interleukin-1. J Invest Dermatol 82:605–607
Schimpl A, Wecker E (1972) Replacement of T-cell function by a T-cell product. Nature 237:15–17

Schleimer RP, Rutledge BK (1986) Cultured human vascular endothelial cells acquire adhesiveness for neutrophils after stimulation with interleukin 1, and tumour-promoting phorbol diesters. J Immunol 136:649–654

Schmitt A, Hauser C, Janin F, Dayer JM, Saurat JH (1986) Normal rat epidermis contains high amounts of natural tissue IL-1. Biochemical analysis by HPLC identifies a MW17 Kd form with a pI 5.7 and a MW 30 Kd form. Lymphokine Res 5:105–117

Schmitt A, Hauser C, Didierjean L, Merot Y, Dayer JM, Saurat JH (1987) Systemic administration of detretinate increases epidermal interleukin 1 in the rat. Br J Dermatol 116:615–622

Schrader JW, Clark-Lewis I (1981) T-cell hybridoma-derived regulatory factors. I. Production of T-cell growth factor following stimulation with concanavalin A. J Immunol 126:1101–1105

Schroder JM, Christophers E (1985) Identification of a new and potent chemotaxin in psoriatic scales. J Invest Dermatol 84:444–451

Schwarz T, Urbanska A, Gschnait F, Luger TA (1986) Inhibition of the induction of contact hypersensitivity by an UV mediated epidermal cytokine. J Invest Dermatol 87:289

Schwarz T, Urbanska A, Gschnait F, Luger T (1987) Biological and biochemical properties of an UVB induced cell derived inhibitor of CHS. J Immunol 138:1457–1464

Seigel LJ, Harper ME, Wong-Staal F, Gallo RC, Nash WG, O'Brien SJ (1984) Gene for T-cell growth factor: location on human chromosome 4q and feline chromosome B1. Science 223:175–178

Shidani B, Milon G, Marchal G, Truffa-Bachi P (1984) Cyclosporin A inhibits the delayed-type hypersensitivity reaction: impaired production of early pro-inflammatory mediator(s). Eur J Immunol 14:314–318

Sieff CA, Tsai S, Faller DV (1987) Interleukin 1 induces cultured human endothelial cell production of granulocyte-macrophage colony-stimulating factor. J Clin Invest 79:48–51

Smith KA (1984) Interleukin 2. Ann Rev Immunol 2:319–333

Smith KA (1987) The two-chain structure of high affinity IL-2 receptors. Immunol Today 8:11–13

Smolen JS, Chused TM, Novotny EA, Steinberg AD (1982) The human autologous mixed lymphocyte reaction. J Immunol 129:1050–1053

Swain SL, Howard M, Kappler J, et al. (1983) Evidence for two distinct classes of murine B cell growth factors which have activities in different functional assays. J Exp Med 158:822–835

Tagami H, Ofiyi S (1976) Leukotactic properties of soluble substances in psoriasis scale. Br J Dermatol 95:1–8

Takematsu M, Suzuki R, Tagami H, Kumagai K (1986) Interleukin-1 like activity in horny layer extracts: decreased activity in scale extracts of psoriasis and sterile pustular dermatosis. Dermatologica 172:236–240

Taniguchi T, Matsui H, Fujita T, et al. (1983) Structure and expression of a cloned cDNA for human interleukin-2. Nature 302:305–310

Ternowitz T (1986) Monocyte and neutrophil chemotaxis in psoriasis. Relation to the clinical status and the type of psoriasis. J Am Acad Dermatol 15:1191–1199

Tew J-Y, Thorbecke GJ, Steinman RM (1981) Dendritic cells in the immune response: characteristics and recommended nomenclature. J Reticuloendothel Soc 31:371–380

Trechsel U, Evequoz V, Fleisch H (1985) Stimulation of interleukin 1 and 3 production by retinoic acid in vitro. Biochem J 230:339–344

Umetsu DT, Pober JS, Jabova HH, et al. (1985) Human dermal fibroblasts present tetanus toxoid antigen to antigen-specific T cell clones. J Clin Invest 76:254–260

Van Hooff JP, Leunissen RML, Staak WVD (1985) Cyclosporin and psoriasis. Lancet I:335–342

Van Joost T, Heule F, Stolz E, et al. (1986) Short-term use of cyclosporin A in severe psoriasis. Br J Dermatol 114:615–620

Van Joost T, Stolz E, Heule F (1987) Efficacy of low-dose cyclosporine in severe atopic skin disease. Arch Dermatol 123:166–167

Veninga TS, de Doer JC (1986) Urinary excretion pattern of serotonin and 5-hydroxyindole acetic acid in ultraviolet induced erythema. J Invest Dermatol 50:1–8
Wainberg MA, Portnoy JD, Clecner B, et al. (1985) Viral inhibition of lymphocyte proliferative responsiveness in patients suffering from recurrent lesions caused by Herpes Simplex virus. J Infect Dis 152:441–448
Waksman BH, Namba Y (1976) On soluble mediators of immunologic regulation. Cell Immunol 21:161–176
Walsh LJ, Seymour GJ, Powel RN (1985) The in vitro effect of retinol on human gingival epithelium. II. Modulation of Langerhans cell markers and interleukin-1 production. J Invest Dermatol 85:501–506
Wang A, Lu S-D, Mark DF (1984) Site-specific mutagenesis of the human interleukin-2 gene: structure-function analysis of the cysteine residues. Science 224:1421–1433
Watson JS (1981) Lymphokines and the induction of immune responses. Transplantation 31:313–317
Wentzell JM, Baughman RD, O'Connor GT, Bernier GM (1987) Cyclosporine in the treatment of psoriasis. Arch Dermatol 123:163–165
Whiteside TL, Worrall JG, Buckingham RB, Rodnan GP (1987) Soluble mediators from mononuclear cells increase the synthesis of glycosaminoglycan by dermal fibroblast cultures derived from normal subjects and progressive systemic sclerosis patients. Arthritis Rheum 28:188–197
Widmer MB, Grabstein KH (1987) Regulation of cytolytic T-lymphocyte generation by B cell stimulatory factor. Nature 326:795–798
Wolfsy D, Murphy ED, Roths JB, Dauphine MJ, Kipper SB, Talal N (1981) Deficient IL-2 activity in MRL/MP and C57B2/6J mice bearing the lpr gene. J Exp Med 154:1671–1680
Yokota T, Otsuka T, Mosmann T, et al. (1986) Isolation and characterization of a human interleukin cDNA clone, homologous to mouse B-cell stimulatory factor 1, that expresses B-cell- and T-cell-stimulating activities. Proc Natl Acad Sci USA 83:5894–5898

Section B: Autocoids in Normal and Inflamed Skin

CHAPTER 18

Histamine, Histamine Antagonists and Cromones

J. C. FOREMAN

A. Introduction

Histamine was discovered independently from two sources in the first decade of this century by KUTSCHER (1910) and by WINDAUS and VOGT (1907). WINDAUS and VOGT prepared histamine synthetically by the decarboxylation of histidine whereas Kutscher identified histamine as a base in ergot. Histamine is so called because the Greek word *histos* means tissue and BARGER and DALE (1910) isolated the amine from guinea-pig intestinal tissue. Following the discovery of histamine in mammalian tissue, there was an extensive study of its biological effects by DALE and LAIDLAW (1910, 1911), who noted the parallel between the actions of histamine in animals and the response of an animal to a foreign protein, normally inert, but to which the animal had been sensitised by prior injection. DALE and LAIDLAW (1919) also demonstrated that it was possible to produce shock in animals by the injection of histamine. DALE (1913) and SCHULTZ (1910) both independently demonstrated an anaphylactic reaction in isolated smooth muscle, following the original description in whole animals of the anaphylactic reaction to foreign protein by PORTIER and RICHET (1902).

The link between histamine and the anaphylactic reaction was established by the work of BARTOSCH et al. (1932) in which it was shown that an anaphylactic reaction in isolated lung was accompanied by the release of histamine. At about the same time, LEWIS (1927) injected histamine into human skin and noted that the response was virtually identical to that observed following an injury to a point in the skin. Lewis described the "triple response" to injury of the skin and suggested that histamine release in the skin could be responsible for producing the triple response.

Histamine thus became a candidate for a mediator of acute inflammatory reactions, and in the sections which follow, the evidence for the mediator and modulator role of histamine in immune and inflammatory responses in the skin will be examined. A valuable way of assessing the role of a putative mediator of inflammation is to judge the evidence against a set of criteria developed by DALE. It is ironic that these criteria have more often been applied to putative neurotransmitters than to mediators of inflammation since DALE did so much of the fundamental work on histamine as a mediator of anaphylaxis.

The criteria are as follows:

1. The substance, when given at appropriate doses in vivo and in vitro, should produce the effects seen in the inflammatory reaction.
2. The inflammatory reaction should lead to the formation and/or release of the putative mediator.

3. The enzymes necessary for the production of the mediator should be present at the site of its formation, and such enzyme activity should increase when the inflammatory stimulus causes increased turnover of the mediator.
4. A mechanism such as metabolism, uptake or desensitisation must be available to terminate the actions of the mediator so that its effects do not persist indefinitely.
5. Pharmacological interference with release, metabolism, storage, synthesis or action of the mediator should give rise to predictable changes in the inflammatory reaction.
6. Clinical or experimental conditions involving deficiencies or overproduction of the mediator or its metabolising enzymes should give rise to appropriate alterations in the inflammatory reaction.
7. Receptors or other recognition–transduction systems should be demonstrable on the relevant cells by means of pharmacological techniques or binding experiments.

These criteria will now form the basis for examining the experimental data relating to the role of histamine in the skin.

B. Histamine Content of Skin

An understanding of the content and localisation of histamine in the skin is of major importance in assessing the role of this amine in the physiology and pathophysiology of the skin. There remains considerable uncertainty about the precise localisation of histamine in skin largely because there is no *specific* histochemical method of detecting histamine with a sufficiently high resolution. There is no doubt that much of the histamine contained in skin is located within mast cells, but non-mast cell sources of histamine within the skin have to be given serious consideration. The problem is further complicated by the fact that the mast cells, including skin mast cells, appear to be a heterogeneous population of cells, both morphologically and functionally (PEARCE 1983; DAMAS and LECOMTE 1983).

One of the first estimates of the histamine content of human skin came, not surprisingly, from the laboratory of Thomas Lewis, where HARRIS (1927) showed that human skin contained about 10 μg histamine per gram of skin (wet weight). HARRIS (1927) also noted that there was regional variation in the histamine content of human skin. Since that early observation, many investigators have determined the skin content of histamine employing samples from different sites and using different methods of histamine assay. Despite this, quite good agreement exists on the actual content and regional variation of the histamine content of skin. RILEY and WEST (1956) confirmed the value of HARRIS (1927) and found that breast skin contained 10 μg/g wet weight in the dermo-epidermal region. A lower value of 3 μg histamine per gram of tissue was obtained for the deeper dermis. FELDBERG and LOESER (1954) obtained values of about 9 μg histamine/g wet weight for human breast and abdominal wall skin and had some evidence that the skin from the neck contained higher levels of histamine. JOHNSON (1957) confirmed the regional variation of histamine content of skin, showing that the content of skin from the head and neck about 30 μg histamine per gram of skin (wet weight) whereas the skin from the chest wall and abdomen had a content of about 8 µg/g. In a detailed profile of the histamine content of human skin, ZACHARIAE (1964) observed the following distribution:

Head > neck > dorsum of hand > forearm = leg > chest = dorsum of foot > back = palm > abdomen = upper arm

The absolute value given for the forearm was 30 µg/g but this value can be expected to be higher than those reported by other workers because the skin samples were both defatted and dried before the estimation of histamine. ZACHARIAE (1964) also noted that the histamine content of skin fell with increasing age. JUHLIN (1967) obtained a value of about 4 µg/g wet weight for skin from human forearm, which would correspond to a value of about 12 µg/g dry weight. More recently, EADY et al. (1979) have demonstrated regional variation of histamine content within the skin of the arm, the values being 14 µg/g dry weight for the upper arm and 19 µg/g for the forearm. HORNER and WINKELMANN (1968) also found similar values for the histamine content of forearm skin of 8 µg/g wet weight, corresponding to about 24 µg/g dry weight. An unusual source of skin has recently been used for studies of histamine release, where a value of 18 µg/g wet weight was obtained for human foreskin (THARP et al. 1983).

Apart from the studies of histamine content in human skin, a variety of animal studies have been conducted. RILEY and WEST (1953) found that the skin from cat and mouse contained 20 and 45 µg histamine/g wet weight respectively. In the paper by RILEY and WEST (1953) evidence was provided to show that histamine in tissues is located within mast cells. Using normal tissues, including skin, and tissues treated with histamine-liberating compounds, they demonstrated a very high degree of correlation between mast cell content and histamine content of the tissues. RILEY and WEST (1953) also showed that in mast cell tumours there was a corresponding increase of histamine within the tumour. Furthermore, painting of skin with carcinogenic hydrocarbons resulted in an increase in the number of mast cells and also the tissue histamine content. In addition, in patients with urticaria pigmentosa, who have increased numbers of mast cells within the skin, there was an increase in the skin content of histamine which correlated with the increased numbers of mast cells. In a subsequent paper, RILEY and WEST (1956) determined the histamine content of skin from seven different mammalian species. They observed that the histamine was concentrated at the dermo-epidermal junction in all but one species, the rat, and that this was also the region of greatest mast cell density. It is worth noting at this point that in man also, mast cell density is greatest at the dermo-epidermal junction (COWEN et al. 1979). The values they obtained for histamine content were: man,10; dog,10; cat,20; mouse,45; rat,8; guinea-pig,2; rabbit,2 (all µg/g wet weight in the dermo-epidermal region). For the rat, a concentration of 35 µg/g was obtained for deeper dermal tissue whereas in the other species the deeper dermal content of histamine was markedly lower than that in the dermo-epidermal region. FELDBERG and TALESNIK (1953) obtained values of about 30 µg/g wet weight for the histamine content of skin from rat abdomen and upper leg and DIXON (1959) found that the histamine content of rat skin varied with development. From 18.5 days before birth to 7 days after birth, there was a good correlation between histamine content of skin and the mast cell content. Also, after birth, between days 3 and 168, the histamine content of skin rose from 9 to 17 µg/g wet weight. MONGAR and SCHILD (1952) obtained a value of the histamine content of guinea-pig skin of 3 µg/g wet weight, and JOHNSON (1956) demonstrated that the regional variation of histamine content of human skin, described above, is also seen in guinea-pig, with skin from the head and neck region having greater concentrations than skin from the abdomen. JOHNSON (1956) also noted that as guinea-pigs grew older and larger the histamine content of skin fell, animals less than 2.5 months old having about twice as much histamine in skin than older animals. This is again similar to human skin, where histamine content appears to fall with age (ZACHARAIE 1964), and it is worth noting the mast cell density in skin also decreased with age (HELLSTRÖM and HOLMGREN 1950). GRAHAM et al. (1955) estimated that dog skin contains 18 µg histamine/g dry weight, which is in reasonable agreement with the figure produced by RILEY and WEST (1956), and these authors also noted a good correlation between the mast cell content of skin and the histamine content. Using the correlation between histamine and mast cell content, GRAHAM et al. (1955) derived a value of about 7 pg/cell for the histamine content of dog skin mast cells. In a similar exercise, RILEY and WEST (1956) derived a value of 1 pg/cell for cat skin mast cells.

The studies cited above provide general agreement about the histamine content of skin in different species and in different locations within a single species. The good correlation between histamine content of skin and the number of mast cells is the basis of the belief that skin histamine is located within mast cells. Indeed, the values of 1 pg/cell and 7 pg/cell for the histamine content of skin mast cells would support the view that most skin histamine is contained within mast cells. If there was a large amount of histamine not contained in mast cells these values would be expected to be much larger. There is really only one source of pure mast cells where sound quantitative data can be obtained and this is the rat mast cell from the peritoneal cavity. HELANDER and BLOOM (1974) quote values for rat peritoneal mast cells of between 9 and 40 pg/cell. Despite a measure of agreement between the values for the histamine content of mast cells, the possibility of the existence of non-mast cell histamine remains and there is some evidence in support of this. However, it must also be noted that historically, tissue mast cells have been identified by their staining reaction of metachromasia with toluidine blue, a stain which binds to the glycosaminoglycan of the mast cell granule. ENNERBÄCK (1966a, bc) has shown that there is a population of mast cells which require special fixation and staining in order that they may be visualised. In fact, as already stated, it is now recognised that mast cells are a heterogeneous population of cells both morphologically and functionally (PEARCE 1983), and hence it is possible that in early studies the total number of mast cells was underestimated since only some of them were stained. It is very likely that this is the case for lung tissue but no evidence has yet emerged with regard to skin (GREENWOOD 1985).

Even with these uncertainties on the relationship between mast cell numbers and histamine content, other lines of evidence suggest the presence of non-mast cell histamine. The histochemical study by JUHLIN (1967) identified histamine in the vascular wall but it is well established that mast cells, particularly in skin, are concentrated around blood vessels (EADY et al. 1979), although the variability of mast cell distribution is great. Using sections of large arteries and veins from a variety of species, including man, EL-ACKAD and BRODY (1975) found a discrepancy between mast cell numbers and histamine content and concluded that in vessel walls there were significant quantities of non-mast cell histamine. SCHAYER (1962) found that blood vessels, probably in their endothelium, could synthesise histamine and that this synthesis did not appear to be by mast cells. KAHLSON et al. (1960b), studying histamine synthesis in foetal tissue, also came to the conclusion that histamine was not synthesised by mast cells in skin and other tissues. It has also been reported that embryonic tissue contains histamine not stored in mast cells (KAHLSON et al. 1963). It must be remembered, however, that interpretation of this work is complicated by current knowledge of the heterogeneous staining properties of mast cells. In normal human skin, it has been reported that there is a good correlation between mast cell numbers, histamine-forming capacity and histamine content (SØNDERGAARD and GLICK 1971). Similarly, a histochemical study of skin demonstrated the coexistence of material reacting with *o*-phthalaldehyde (i.e. histamine) and with toluidine blue (glycosaminoglycan) in cells presumed to be mast cells. These cells were increased in number in urticaria pigmentosa and in keloids (HÅKANSON et al. 1969). More recently, it has been shown that W/W^v mice, which are deficient in mast cells, have virtually no histamine in skin compared with normal mice (YAMATODANI et al. 1982). Furthermore, the irreversible inhibitor of histidine decarboxylase, α-fluoromethylhistidine, did not cause skin histamine levels to fall, whereas it did produce a reduction of brain and stomach histamine (MAEYAMA et al. 1982). Thus, skin histamine appears to be stored in mast cells and the turnover is low: a finding also reported by SCHAYER (1956b). Nevertheless, despite the weight of evidence compatible with the view that skin histamine is located in mast cells, the hypothesis of SCHAYER (1963) that some histamine is not stored but newly formed ("induced" or "nascent") when required, by cells other than mast cells, cannot be excluded. A recent study identifying histamine and a histamine-metabolising system in purified, isolated endothelial cells from the microvasculature of rats and guinea-pigs (ROBINSON-WHITE and BEAVEN 1982) supports the idea that some histamine in blood vessel walls is not of mast cell origin. These endothelial cells also failed to release histamine in response to compound 48/80, adding to the evidence that the histamine is not in mast cells.

C. Histamine-Forming Capacity in Skin

Histamine is formed from the amino acid L-histidine by the action of enzyme activity referred to as histidine decarboxylase. In fact, there are two enzymes which catalyse the decarboxylation of histidine, both of which require pyridoxal phosphate as a co-factor. One enzyme is non-specific and catalyses the decarboxylation of all naturally occurring aromatic L-amino acids whereas the other enzyme is specific for L-histidine. The enzyme specific for L-histidine is inhibited by α-methylhistidine (KAHLSON et al. 1962) and it is present in rat peritoneal mast cells (SCHAYER 1956a). SCHAYER (1957) has shown that the specific histidine decarboxylase is present in skin and a variety of other mammalian tissues. The low turnover of skin histamine has already been referred to above, and it has been shown that while histamine formation can be inhibited by feeding a pyridoxine-free diet and using the histidine decarboxylase inhibitor, semicarbazide, skin histamine levels remain unchanged (KAHLSON 1960). A number of inflammatory processes within skin are associated with increased histidine decarboxylase activity. SCHAYER (1960) found that histamine itself, as well as 5-hydroxytryptamine and noradrenaline, stimulated histamine formation in mouse skin. In urticaria pigmentosa, where increased numbers of mast cells are present in the skin, it has been shown that the skin has increased histidine decarboxylase activity which is about 10 times greater in the lesion than in the unaffected skin (DEMIS and BROWN 1961). No increased histamine-forming capacity was found in skin from the sites of chronic urticaria. Increased histamine formation, but reduced total levels of histamine in rat skin, have been reported at the site of the classical cell-mediated immune response initiated by the injection of tuberculin into the skin of sensitive rats (GRAHAM and SCHILD 1967). The histamine-forming capacity of human skin increases at the site of allergic contact dermatitis when antigen is applied in a "patch test" to a sensitive individual (SØNDERGAARD and GLICK 1972). The change in histamine-forming capacity was found to be in the upper dermis. Another cell-mediated immune reaction, graft rejection, is also associated with increased histamine formation and reduced total levels of histamine in the region of the skin where the graft is made (FAN and LEWIS 1982). Furthermore, the immunosuppressive agent cyclosporin A prevented the changes in both histamine-forming capacity and histamine content of the skin at the same time as prolonging allograft survival. MOORE and SCHAYER (1969) found greater stimulation of histamine formation at the site of an allograft than at the side of an autograft. The co-carcinogen 12-*O*-tetradecanoylphorbol-13-acetate, which stimulates histamine release from the mast cells of the rat (SAGI-EISENBERG et al. 1985), has been shown to increase histidine decarboxylase activity in the skin of W/W^v mice which are genetically deficient in mast cells (TAGUCHI et al. 1979). This finding again emphasises the possibility that cells other than mast cells in skin can synthesise and store histamine. Histamine formation in skin is also increased at the site of a healing wound (KAHLSON et al. 1960a) and suppression of this histamine-forming capacity by pyridoxine-free diet and semicarbazide prolongs wound healing.

Semicarbazide has been referred to above as an inhibitor of histidine decarboxylase but it is both non-specific and of low potency in this respect, and other more potent and specific inhibitors of the enzyme have been employed to determine its role in various systems. The compound 4-bromo-3-hydroxybenzyl oxyamine (NSD 1055 or brocresine) inhibits histidine decarboxylase in rat skin (JOHNSON and KAHLSON 1967). LEVINE (1966) has shown that this compound reduces urinary output of histamine after a histidine load. In patients with mastocytosis, the compound partially prevented symptoms such as flushing, itching and abdominal cramps, and also prevented the exacerbation of the symptoms induced by a histidine load. However, ZACHARIAE et al. (1969) found no effect of brocresine in chronic urticaria and GREAVES (1971) reported no change in the dermographic response or urinary histamine excretion in patients with urticaria pigmentosa who were treated with the compound. The data must be considered inconclusive since the lack of clinical effect of the drug goes hand in hand with the absence of effect on histamine formation in vivo. The compound α-fluoromethylhistidine produces a profound fall in skin histamine-forming capacity (MAEYAMA et al. 1982) in animals and it would be interesting to know the clinical effects of an irreversible inhibitor of histidine decarboxylase.

D. Histamine Catabolism in Skin

Skin from guinea-pig, rabbit, cat and human was shown to have the ability to destroy histamine by GRANROTH and NILZEN (1948). Histamine catabolism can occur by two major routes: oxidative deamination and *N*-methylation. In guinea-pig, both diamine oxidase and *N*-methyltransferase are responsible for histamine catabolism (YAMAMOTO et al. 1976a). Diamine oxidase converts histamine to imidazole acetic acid, with imidazole acetaldehyde being formed as an intermediate. *N*-methyltransferase requires *S*-adenosyl methionine as a co-factor and it converts histamine into *N*-methyl histamine, which can then be oxidised to methyl imidazole acetic acid. In man, the proportions of the various metabolites which appear in the urine are: free histamine 2–3; methyl histamine 4–8; methyl imidazole acetic acid 42–47; imidazole acetic acid and its riboside conjugate 25–34 (each value is a percentage of total) (SCHAYER and COOPER 1956).

In human skin there appears to be no diamine oxidase and the principal route of histamine catabolism is through *N*-methylation (FRANCIS et al. 1977). The *N*-methyltransferase is located *intracellularly* in the upper part of the dermis and has a K_m for histamine of 4.2 μ*M*. The levels of the enzyme are low and, interestingly, histamine 25–100 μ*M* acts as a non-competitive inhibitor of the enzyme in vitro (FRANCIS et al. 1980). The enzyme is also inhibited by the antagonist of histamine at H_2-receptors for histamine, burimamide (YAMAMOTO et al. 1976b).

The lack of non-toxic, selective inhibitors of the enzymes of histamine catabolism makes it difficult to assess whether catabolism is important for limiting the actions of histamine in skin. Aminoguanidine, a diamine oxidase inhibitor, potentiates the passive cutaneous anaphylactic reaction in guinea-pig skin (YAMAMOTO et al. 1976a) but apparently there is no effect on such reactions of blocking *N*-methyltransferase in human skin with chloroquine (FRANCIS and GREAVES 1979). The intracellular localisation of the *N*-methyltransferase is difficult to interpret but there is some evidence that histamine is taken up by cells of the skin (THOMPSON et al. 1981). It may well be that histamine catabolism in skin is not important for terminating the action of histamine and this view is supported by the observation that large amounts of histamine released from skin mast cells by an anaphylactic reaction can be detected in venous blood draining from the area of skin (IND et al. 1984). Also, the kinetics and capacity of the *N*-methyltransferase in skin probably cannot account for the removal of the quantities of histamine likely to be released at the site of an inflammatory reaction in the skin. However, it is noteworthy that patients with urticaria pigmentosa have increased levels of methyl imidazole acetic acid in urine (GRANERUS et al. 1983).

E. The Release of Histamine from Skin

Having considered the content, formation and catabolism of histamine in skin, it is appropriate to survey the experimental data on the release of histamine from skin stimulated in a variety of ways.

In some very elegant experiments in which patients sensitive to ragweed antigen were challenged intradermally and an exposed blister base rinsed with saline (KATZ 1942), it was shown that an immediate hypersensitivity reaction in human skin caused histamine release. At about the same time, ROSE (1941) reported that the levels of histamine in venous blood draining from the skin increased when a patient with cold urticaria was given a low temperature challenge. In animal experiments, studies involving perfused skin of cat and dog have demonstrated histamine release in response to antigen challenge in sensitive tissue, compound 48/80, *d*-tubocurarine and morphine (FELDBERG and PATON 1951; FELDBERG and SCHACHTER 1952). There is some evidence that an opiate receptor is involved in histamine release from human skin (CASALE et al. 1984). MONGAR and SCHILD (1952) also demonstrated the release of histamine from chopped guinea-pig skin in vitro in response to an antigen–antibody reaction, compound 48/80 and *d*-tubocurarine.

The technique of dermal perfusion (WINKELMANN 1966) has been widely used to study histamine release from skin. Using this technique, HORNER and WINKELMANN (1968) have shown that compound 48/80 causes histamine to be released into human skin. However, intradermal challenge of sensitive subjects with appropriate antigen did not produce any histamine release detectable by this method, although kinin release was measured (MICHEL et al. 1970). Also, GREAVES and SØNDERGAARD (1970a) detected histamine release by this method in only 3 of 22 patients with atopic eczema. However, using dermal perfusion, histamine and other spasmogenic substances were detected in the perfusate from patients with urticaria pigmentosa when they were appropriately stimulated (GREAVES and SØNDERGAARD 1970b).

The original experiments of KATZ (1942) in which an exposed blister base was perfused with saline have been developed in two directions. DUNSKY and ZWEIMAN (1978) abraded the skin and then applied a chamber to the abraded area; the saline in the chamber could then be sampled and assayed for histamine. By this method, intradermal challenge of ragweed-sensitive patients produced histamine release into the chamber, although the release was slow compared with morphological changes in skin mast cells at the site of challenge (TING et al. 1980). Another development of Katz's blister base method employs the application of suction to the skin to form a blister from which exudate can be sampled by hypodermic needle for assay of histamine (KOBZA BLACK et al. 1977). With this technique, histamine release into the blister fluid has been observed following UVA irradiation of the skin, intradermal challenge of allergic individuals and cold challenge of patients with cold urticaria (MISCH et al. 1983; HAWK et al. 1983; DORSCH et al. 1982). Chemically induced inflammation failed to cause histamine release into the blister fluid but it did cause the formation of prostaglandins (KOBZA BLACK et al. 1976).

In support of the data obtained using skin perfusion and blister techniques, there are a number of studies reporting histamine release into the blood following a variety of stimuli applied to the skin. DUNER et al. (1960) detected increases in blood histamine following cold challenge of patients with cold urticaria and this has been confirmed by others (KAPLAN et al. 1975; SOTER et al. 1976).

Intradermal challenge of allergic individuals with antigen causes an elevation of plasma histamine (HEAVEY et al. 1983), as does intradermal injection of the neuropeptide substance P (HEAVEY et al. 1984) and also codeine (IND et al. 1984).

Histamine release from skin discussed above has been by direct measurement of histamine, but in some cases injection of a substance into the skin has been assumed to release histamine on the grounds that plasma extravasation into the skin in response to the injection is prevented by anti-histamine pre-treatment of the subject or animal. Thus, the complement fragment C5a releases histamine from skin (REGAL et al. 1983). Similar studies suggest that substance P also releases histamine from mast cells of human skin (FOREMAN et al. 1983).

It is clear from the data quoted above that a variety of immunological and non-immunological inflammatory stimuli promote histamine release from skin in humans and experimental animals. It is assumed that most of this released histamine comes from mast cells in the skin. Details of the immunopharmacology of materials released from mast cells are discussed in Chap. 8.

F. Application of Histamine to the Skin and the Pharmacological Modification of Its Effects

Histamine is classically considered to be a mediator of acute inflammation which is characterised by "rubor, calor, dolor and turgor". However, other actions of histamine on cells in the skin need to be considered, including cell proliferation, wound healing and chemotaxis. It must also be borne in mind that the application of histamine to skin can produce indirect as well as direct effects and important considerations in this respect are the axon reflex and release of other mediators of inflammation.

I. Vascular Effects

Intradermal injection of histamine into human skin produces the triple response described by LEWIS and GRANT (1924), namely an intense red spot at the site of the injection surrounded by a rapidly developing area of redness of lesser intensity (flare) which spreads several centimetres from the site of the injection. The third component is an area of oedema (weal) in the skin local to the injection site; this area is smaller than the area of flare. Both flare and weal areas are dependent on the amount of histamine injected (HERXHEIMER and SCHACHTER 1959). The red spot at the site of the injection is thought to be a direct vasodilator effect of histamine, probably on pre-capillary arterioles (DIETZEL et al. 1969). The weal also appears to be a direct effect of histamine on the permeability of venules (MAJNO and PALADE 1961; MAJNO et al. 1961). The oedema formation cannot totally be explained by an increase in transcapillary pressure (DIETZEL et al. 1969; BAKER 1979) but is the result of an increase in the permeability of the venule wall induced by histamine. The flare is not a direct effect of histamine but is the result of an axon reflex initiated by histamine (LEWIS 1927). BAYLISS (1901) demonstrated that antidromic stimulation of primary sensory neurones produced vasodilatation in the skin and LEWIS (1927) proposed that histamine stimulated nerves within the skin, generating impulses passing antidromically through the arborisations of these nerves to produce vasodilatation (flare) at distances of several centimetres from the original point of stimulation. The nature of the substance released from the nerves to produce flare is unknown but it is possible that it may be a neuropeptide. Such a view is compatible with the knowledge that the neuropeptide substance P and other neuropeptides are present in sensory nerves, and that intradermal injection of these neuropeptides produces weal and flare responses (FOREMAN and JORDAN 1984; PIOTROWSKI and FOREMAN 1985). Flare, then, may be due to neuropeptides acting on blood vessels or the neuropeptides may themselves stimulate mast cells to release histamine which in turn is responsible for the vasodilatation. In this context, it is important to note that vasodilatation resulting from direct nerve stimulation is blocked by anti-histamines (GRAHAM and LIOY 1973; POWELL and BRODY 1976).

Use of selected agonist and antagonists for histamine receptors has shown that, at low doses of histamine, the flare response is mediated solely by an H_1-receptor, but at higher doses of histamine, or where histamine is released in situ from mast cells by compound 48/80, both H_1- and to a lesser extent H_2-receptors are involved in the production of flare (DUNER and PERNOW 1952; FERMONT et al. 1976; MARKS and GREAVES 1977; ROBERTSON and GREAVES 1978). The newer anti-histamines, e.g. astemizole, totally inhibit flare induced by histamine compound 48/80 and antigen in antigen-sensitive individuals (HUMPHREYS and SHUSTER 1987).

The weal produced by injection of histamine into human skin is due to increased vascular permeability and this can be monitored in experimental animals by measuring leakage into the tissues of dye placed in the vascular compartment. The increase in vascular permeability induced by histamine is mimicked in both animals and man by inducing histamine release from mast cells with compound 48/80 or with antigen in sensitised subjects (MILES and MILES 1952; FELDBERG and MILES 1953; LAST and LOEW 1947). H_1-antagonists suppress histamine-induced, compound 48/80-induced and antigen-induced increases in vascular permeability (SPECTOR and WILLOUGHBY 1963; BAIN 1949; LAST and LOEW 1947; COOK et al. 1973; HUMPHREYS and SHUSTER 1987). In these and other early studies the inhibition by H_1-antagonists of histamine-mediated increases in vascular permeability was only weak (GREAVES and SHUSTER 1967). Although the relatively selective H_2-agonist 4-methyl histamine produces weal in human skin (ROBERTSON and GREAVES 1978), in rat skin the selective H_2-agonist impromidine did not cause increased vascular permeability (OWEN et al. 1984). Furthermore, selective H_2-antagonists either alone or in combination with the then available H_1-antagonists did not themselves inhibit vascular permeability due to histamine (OWEN et al. 1984; MARKS and GREAVES 1977; NATHAN et al. 1981; ROBERTSON and GREAVES 1978; HARVEY and SHOCKET 1980). These studies have now to be re-interpreted because the newer H_1-receptor antagonists terfenadine and astemizole will cause almost complete inhibition of histamine weal and flare and partial inhibition of the response to compound 48/80 and house dust mite (see Chap. 8, 29, Vol. I and Chap. 18, 31, Vol. II and HUMPHREYS and SHUSTER 1987). However, measurements of the rate of formation and resorption of weals induced by histamine suggest that mediators other than histamine are released secondarily (COOK and SHUSTER 1979), and it is worth pointing out that neuropeptides such as substance P are capable of producing a direct, histamine-independent increase in vascular permeability in human skin (FOREMAN et al. 1983; JORRIZO et al. 1983).

II. Sensory Effects

Pain is a notoriously difficult sensory modality to assess objectively. Intradermal injection of histamine into human skin produces a very variable response as assessed by reports from the subjects. "Pricking" and "itching" are used to describe the sensations produced by intradermal injection of histamine but "pain" is certainly too strong a word to describe the sensation. ROSENTHAL (1977) has presented some evidence that histamine is a mediator of cutaneous nociception and DAVIES and GREAVES (1980) have shown that itch in response to histamine is mediated by an H_1-receptor in human skin, with no contribution from H_2-receptors. The evidence suggests that mediators other than histamine itself are responsible for the nociceptive effect of histamine in human skin (DAVIES and GREAVES 1980), and prostaglandins appear to be candidates for mediators of histamine-induced nociception (JUAN 1981). The effect of anti-histamines on pruritus is discussed elsewhere (Chap. 18, Vol. II).

III. Other Effects

Repair and regeneration is a response secondary to an inflammatory reaction which destroys tissue. Reference has already been made above to the role of histamine formation in wound healing. Depletion of histamine in tissue by use of compound 48/80 has been shown to delay wound healing, and exogenous histamine appears to accelerate wound healing (BOYD and SMITH 1959; FITZPATRICK and FISHER 1982). Secretion from mast cells in rat skin induced by compound 48/80 has been found to be associated with mitogenesis (NORRBY 1983) but histamine appears to be without effect on epidermal proliferation in skin with normal turnover (MARKS et al. 1982). In fact, in in vitro experiments, histamine seems to inhibit epidermal cell proliferation (HALPRIN et al. 1984). The data relating to histamine in wound healing remain, therefore, to be explained.

Acute inflammatory reactions are associated with some degree of cellular infiltration and the delayed reaction is associated with a more marked cellular infiltration. This raises the question as to whether or not histamine is chemotactic. Repeated injection of histamine into human skin produces a slight perivenular accumulation of neutrophils, eosinophils and mononuclear cells (JAMES et al. 1981). In contrast, compound 48/80 produces more marked cellular accumulation, and it is likely that mast cell-derived products other than histamine are responsible for any chemotaxis resulting from mast cell stimulation. Indeed, whilst histamine increases neutrophil chemokinesis, it inhibits chemotaxis (ANDERSON et al. 1977). Nevertheless, H_1 histamine antagonists do appear to suppress cell accumulation in late phase allergic reactions in skin (LEMANSKE et al. 1983) although it is not clear whether histamine may have some weak chemotactic effect or is chemotactic by virtue of the release of secondary mediators.

IV. Anti-allergic Drugs

Cromoglycate and related drugs inhibit histamine release from some populations of mast cells (ASSEM and MONGAR 1970; CHURCH and GRADIDGE 1980; GARLAND 1973; BIERMAN et al. 1979). In human skin, cromoglycate neither inhibits antigen-induced histamine release nor suppresses an antigen-evoked skin response in sensitive individuals in vitro (PEARCE et al. 1974; Assem and MONGAR 1970; PHILLIPS et al. 1983; TING et al. 1983a) nor in vivo (HUMPHREYS and SHUSTER 1987). Ketotifen and oxatomide both inhibit histamine release from mast cells (BIERMAN et al. 1979; CHURCH and GRADIDGE 1980), but both of these compounds are also anti-histamines at H_1-receptors. Ketotifen inhibits antigen-evoked skin responses in sensitive individuals (PHILLIPS et al. 1983) and oxatomide also prevents passive cutaneous anaphylactic reactions (AWOUTERS et al. 1977). Oxatomide does not produce significant inhibition of histamine release from human skin (GATTI et al. 1980).

It appears likely that in skin anti-allergic drugs do not inhibit mediator release from mast cells and, where they have anti-allergic activity in the skin, this is more likely to be due to antagonism of mediator effects.

G. Clinical Conditions Associated with Histamine in Skin

Urticaria, the condition most closely associated with histamine in the skin, is often linked with angioedema and has a number of underlying pathological mechanisms. The characteristic wealing, erythema and itching are, of course, very similar to the effects of histamine injection into the skin. Urticaria pigmentosa is essentially a proliferation of mast cells within the skin and is associated with increased amounts of histamine and histamine release into the skin. Blocking histamine formation in this condition has produced equivocal results (see above) and total H_1-receptor blockade with astemizole does little for the dermographism of the condition (KRAUSE and SHUSTER 1985a, b). Almost certainly, mediators other than histamine are generated in urticaria pigmentosa.

In cold urticaria there is histamine release into the circulation on cold challenge (see above) and some relief of symptoms is provided by treatment with anti-histamines (WANDERER et al. 1977; KAUR et al. 1981). However, histamine release in this condition may be inhibited by prednisone without clinical improvement in the patients (KOBZA BLACK et al. 1981). Exposure to ultraviolet radiation causes an increase in circulating histamine and histamine is released into skin (HAWK et al. 1980, 1983) but anti-histamines are of relatively little effect in patients with solar urticaria and, again, other mediators such as prostaglandins appear to be involved (HAWK et al. 1983).

Factitious urticaria (dermographism) is associated with normal mast cell numbers and histamine content of the skin. The increased release of histamine on mechanical challenge (GREAVES and SØNDERGAARD 1970a) appears to result from mast cell fragility, though some neurogenic mechanism has not been excluded. The dermographism is best controlled by H_1-anti-histamines, H_2-receptor antagonists having a small additional effect (MATTHEWS et al. 1979; COOK and SHUSTER 1983). However, even the newer H_1-receptor antagonists terfenadine and astemizole, which will almost completely inhibit the histamine weal, only partially inhibit dermographic wealing (KRAUSE and SHUSTER 1984b, 1985b), suggesting that mediators in addition to histamine are involved in this form of urticaria.

In chronic urticaria and atopic skin lesions, there is evidence (see above) for a role for histamine and the weal and flare response to histamine is increased (KRAUSE and SHUSTER 1985c) but, once again, it seems likely that other mediators are also involved because complete inhibition of the response to histamine with astemizole does not totally alleviate the condition (KRAUSE and SHUSTER 1984a).

Whilst there are conditions associated with increased numbers of mast cells or histamine in skin there appears to be no knowledge of the effect of a total lack of histamine. Mice genetically deficient in mast cells are capable of survival but these mice do seem to have histamine in cells not identifiable as mast cells (YAMATODAMI et al. 1982). It is interesting that vascular effects similar to those seen in Raynaud's disease can be induced with histamine antagonists (LAFFERTY et al. 1983) and the role of histamine in this condition would seem to be worthy of further consideration.

H. Concluding Remarks

In the context of Dale's criteria outlined in the Introduction, it is clear that histamine satisfies many of them as a mediator of inflammation in skin. Histamine is present in skin; there is a specific synthesising mechanism in skin which is simulated in some inflammatory conditions; there is a histamine metabolising system in skin, though its role in terminating histamine's action is unclear; and application of histamine mimics certain types of inflammation in the skin. Disappointingly, data available on the pharmacological effects of modifying histamine synthesis, release, metabolism and action all tend to suggest that mediators other than histamine are involved in most inflammatory processes within skin, and the need remains for more potent and selective drugs for interfering with some of these processes. Finally, although excess histamine plays a part in certain clinical conditions, there is no information on the clinical effect of total absence of histamine: perhaps such a situation is incompatible with life.

Acknowledgement. I should like to thank Mr. N.A. HAYES for his help in preparing this manuscript.

References

Anderson R, Glover A, Rabson AR (1977) The in vitro effects of histamine and metiamide on neutrophil motility and their relationship to intracellular cyclic nucleotide levels. J Immunol 118:1690–1696

Assem ESK, Mongar JL (1970) Inhibition of allergic reactions in man and other species by cromoglycate. Int Arch Allergy Appl Immunol 38:68–77
Awouters F, Niemegeers CJE, Van den Berk J et al. (1977) Oxatomide, a new orally active drug which inhibits both the release and the effects of allergic mediators. Experientia 33:1657–1659
Bain WA (1949) The quantitative comparison of histamine antagonists in man. Proc R Soc Med 42:615–623
Baker CH (1979) Nonhemodynamic effects of histamine on gracilis muscle capillary permeability. J Pharmacol Exp Ther 211:672–677
Barger G, Dale HH (1910) Chemical structure and sympathomimetic action of amines. J Physiol (Lond) 41:19–59
Bartosch R, Feldberg W, Nagel E (1932) Das Freiwerden eines histaminähnlichen Stoffes bei der Anaphylaxie des Meerschweinchens. Pflügers Arch 230:129–153
Bayliss WM (1901) On the origin from the spinal cord of the vasodilator fibres of the hind limb, and on the nature of these fibres. J Physiol (Lond) 26:173–209
Bierman CW, Assem ESK, Mongar JL (1979) Inhibition and stimulation of histamine release by oxatomide. Int J Immunopharmacol 1:227–231
Boyd JF, Smith AN (1959) The effect of histamine and histamine releasing agent (compound 48/80) on wound healing. J Pathol Bacteriol 78:379–388
Casale TB, Bowman S, Kaliner M (1984) Induction of human cutaneous mast cell degranulation by opiates and endogenous opioid peptides: evidence for opiate and nonopiate receptor paticipation. J Allergy Clin Immunol 73:775–781
Church MK, Gradidge CF (1980) Inhibition of histamine release from human lung in vitro by antihistamines and related drugs. Br J Pharmacol 69:663–667
Commens CA, Greaves MW (1978) Cimetidine in chronic ideopathic urticaria: a randomised double blind study. Br J Dermatol 99:675–679
Cook J, Shuster S (1979) The measurement and mechanism of histamine wealing. J Invest Dermatol 72:283
Cook J, Shuster S (1983) The effect of an H_1 and H_2 receptor antagonist on the dermographic response. Acta Derm Venereol (Stockh) 63:260–262
Cook TJ, MacQueen DM, Wittig HJ, Thornby JI, Lantos RL, Virtue CM (1973) Degree and duration of skin test suppression and side effects with antihistamine. J Allergy Clin Immunol 51:71–77
Cowen T, Trigg P, Eady RAJ (1979) Distribution of mast cells in human dermis: development of a mapping technique. Br J Dermatol 100:635–640
Dale HH (1913) The anaphylactic reaction of plain muscle in the guinea pig. J Pharmacol Exp Ther 4:167–223
Dale HH, Laidlaw PP (1910) The physiological action of β-imidazolylethylamine. J Physiol (Lond) 41:318–344
Dale HH, Laidlaw PP (1911) Further observations on the action of β-imidazolylethamine. J Physiol (Lond) 43:182–195
Dale HH, Laidlaw PP (1919) Histamine shock. J Physiol (Lond) 52:355–390
Damas J, Lecomte J (1983) Mast cell heterogeneity in the rat. Experientia 39:1311–1312
Davies MG, Greaves MW (1980) Sensory responses of human skin to synthetic histamine analogues and histamine. Br J Clin Pharmacol 9:461–465
Demis DJ, Brown DD (1961) Histidine metabolism in urticaria pigmentosa. J Invest Dermatol 36:253–257
Dietzel W, Massion WH, Hinshaw LB (1969) The mechanism of histamine-induced transcapillary fluid movement. Pflügers Arch 309:99–106
Dixon JB (1959) Histamine, 5-hydroxytryptamine and serum globulins in the foetal and neonatal rat. J Physiol (Lond) 147:144–152
Dorsch W, Ring J, Reimann HJ, Geiger R (1982) Mediator studies in skin blister fluid from patients with dual skin reactions after intradermal antigen injection. J Allergy Clin Immunol 70:236–242
Duner H, Pernow B (1952) Cutaneous reactions produced by local administration of acetylcholine, acetyl-β-methylcholine (methacholine), piperidine and histamine. Acta Physiol Scand 25:38–48

Duner H, Pernow B, Sterky G (1960) The histamine concentration in the blood on exposure to cold and heat. A study in the healthy subject and patients with cold allergy. Allergy 15:417–424
Dunsky EH, Zweiman B (1978) The direct demonstration of histamine release in allergic reactions in the skin using a skin chamber technique. J Allergy Clin Immunol 62:127–130
Eady RAJ, Cowen T, Marshall TF, Plummer V, Greaves MW (1979) Mast cell population density, blood vessel density and histamine content in normal human skin. Br J Dermatol 100:623–633
El-Ackad TM, Brody MJ (1975) Evidence of non-mast cell histamine in the vascular wall. Blood Vessels 12:181–191
Ennerbäck L (1966a) Mast cells in the gastrointestinal tract. 1. Effect of fixation. Acta Pathol Microbiol Immunol Scand [B] 66:289–302
Ennerbäck L (1966b) Mast cells in rat gastrointestinal mucosa. 2. Dye-binding and metachromatic properties. Acta Pathol Microbiol Immunol Scand [B] 66:303–312
Ennerbäck L (1966c) Mast cells in the gastrointestinal tract. 3. Reactivity towards compound 48/80. Acta Pathol Microbiol Immunol Scand 66:313–322
Fan TPD, Lewis GP (1982) Blood flow, histamine content and histadine decarboxylase activity in rat skin grafts and their modification by cyclosporin A. Br J Pharmacol 76:491–497
Feldberg W, Loeser AA (1954) Histamine content of human skin in different clinical disorders. J Physiol (Lond) 126:286–292
Feldberg W, Miles AA (1953) Regional variations of increased permeability of skin capillaries induced by a histamine liberator and their relation to the histamine content of the skin. J Physiol (Lond) 120:203–213
Feldberg W, Paton WDM (1951) Release of histamine from skin and muscle in the cat by opium alkaloids and other histamine liberators. J Physiol (Lond) 114:490–509
Feldberg W, Schachter M (1952) Histamine release by horse serum from skin of the sensitized dog and non-sensitized cat. J Physiol (Lond) 118:124–134
Feldberg W, Talesnik J (1953) Reduction of tissue histamine by compound 48/80. J Physiol (Lond) 120:550–568
Fermont DC, Haggie SJ, Wyllie JH (1976) Histamine receptors in human skin. Br J Surg 63:160
Fitzpatrick DW, Fisher H (1982) Histamine synthesis, imidazole dipeptides and wound healing. Surgery 91:430–434
Foreman JC, Jordan CC (1984) Neurogenic inflammation. Trends Pharmacol Sci 5:116–119
Foreman JC et al (1983) Structure-activity relationships for some substance P-related peptides that cause weal and flare reactions in human skin. J Physiol (Lond) 335:449–465
Francis D, Greaves MW (1979) Histamine-*N*-methyltransferase of human skin: evaluation of its role in the regulation of histamine-mediated reactions. J Invest Dermatol 72:282–283
Francis D, Greaves MW, Yamamoto S (1977) Enzymatic histamine degradation by human skin. Br J Pharmacol 60:583–587
Francis DM, Thompson MF, Greaves MW (1980) The kinetic properties and reaction mechanism of histamine methyltransferase from human skin. Biochem J 187:819–828
Galant SP, Bullock J, Wong D, Maibach HI (1973) The inhibitory effect of antiallergy drugs on allergic and histamine induced weal and flare response. J Allergy Clin Immunol 51:11–21
Garland LG (1973) Effect of cromoglycate on anaphylactic histamine release from rat peritoneal mast cells. Br J Pharmacol 49:128–130
Gatti S, Coutts A, Francis D, Greaves MW (1980) Oxatomide: in vitro assessment of antagonistic activity, and effects on histamine release and enzymatic histamine degradation in skin. Br J Dermatol 103:671–677
Graham BH, Lioy F (1973) Histaminergic vasodilation in the hind limb of the dog. Pflügers Arch 342:307–318

Graham HT, Lowry OH, Wahl N, Preibat MIT (1955) Mast cells as sources of skin histamine. J Exp Med 102:307–318
Graham P, Schild HO (1967) Histamine formation in the tuberculin reaction of the rat. Immunology 12:727–727
Granerus G, Olafsson JH, Roupe G (1983) Studies on histamine metabolism in mastocytosis. J Invest Dermatol 80:410–416
Granroth T, Nilzen A (1948) On the histaminolytic activity of skin extracts. Acta Physiol Scand 15:188–192
Greaves MW (1971) Histamine excretion and demographism in urticaria pigmentosa before and after administration of a specific histidine decarboxylase inhibitor. Br J Dermatol 85:467–470
Greaves MW, Shuster S (1967) Responses of skin blood vessels to bradykinin, histamine and 5-hydroxytryptamine. J Physiol (Lond) 193:255–267
Greaves MW, Søndergaard T (1970 a) A new pharmacological finding in human allergic contact eczema. Arch Dermatol 101:659–661
Greaves MW, Søndergaard T (1970 b) Urticaria pigmentosa and factitious urticaria: direct evidence for release of histamine and other smooth muscle contracting agents in dermographic skin. Arch Dermatol 101:418–425
Greenwood B (1985) The histology of mast cells. In: Engstrom I, Lindholm NB (eds) Current views on bronchial asthma 1. Fisons, Stockholm, pp 143–149
Håkanson R, Owman CH, Sjöberg N-O, Sporrong B (1969) Direct histochemical demonstration of histamine in cutaneous mast cells: urticaria pigmentosa on keloids. Experimentia 25:854–855
Halprin KM, Taylor JR, Comerford M (1984) Control of epidermal cell proliferation in vitro. Br J Dermatol 111 (Suppl 27):13–26
Harris KE (1927) Observations upon a histamine-like substance in skin extracts. Heart 14:161–176
Harvey RP, Schocket AL (1980) The effect of H_1 and H_2 blockade on cutaneous histamine response in man. J Allergy Clin Immunol 65:136–139
Hawk JLM, Eady RAJ, Challoner AVJ, Kozba Black A, Keahey TM, Greaves MW (1980) Elevated blood histamine levels and mast cell degranulation in solar urticaria. Br J Clin Pharmacol 9:183–186
Hawk JLM, Kozba Black A, Jaenicke KF et al. (1983) Increased concentrations of arachidonic acid, prostaglandins E, D and 6-oxo-F histamine in human skin following UVA irradiation. J Invest Dermatol 80:496–499
Heavey DJ, Ind PW, Miyatake A, Dollery CT (1983) Local histamine release in man following intradermal antigen challenge. Br J Clin Pharmacol 106:214
Heavey DJ, Fuller RW, Barnes PJ, Ind PW, Brown MJ, Dollery CJ (1984) Release of histamine by intradermal injection of substance P in man. Proceedings 9th International Congress Pharmacol. London, July 1983, p 1067
Helander HF, Bloom GD (1974) Quantitative analysis of mast cell structure. J Microsc 100:315–321
Hellstrom B, Holmgren HJ (1950) Numerical distribution of mast cells in human skin and heart. Acta Anat (Basel) 10:82–107
Herxheimer A, Schachter M (1959) Weal and flare in human skin produced by histamine and other substances. Nature 183:1510–1511
Horner FA, Winkelmann RH (1968) Histamine release produced in human skin by compound 48/80. Ann Allergy 26:107–116
Humphreys F, Shuster S (1987) The effect of nedocromil on weal reactions in human skin. Br J Clin Pharmacol 24:405–408
Humphreys F et al. (1983) The effect of astemizole and indomethacin on weal and flare reactions to histamine 48/80 and house dust mite antigen. Br J Dermatol 116:435
Ind PW, Miyatake A, Heavey DJ, Dollery CT (1984) Local histamine release after immunological and non-immunological mast cell degradation in vivo. Agents Actions 14:417–419
James MP, Kennedy AR, Eady RAJ (1981) A microscopic study of inflammatory reactions in human skin induced by histamine and compound 48/80. J Invest Dermatol 78:406–413

Johnson HH (1956) Variations in histamine levels in guinea pig skin related to skin region; age (or weight); and time after death of the animal. J Invest Dermatol 27:159–163
Johnson HH (1957) Histamine levels in human skin. Arch Dermatol Venereol (Stockh) 76:726–730
Johnson M, Kahlson G (1967) Experiments on the inhibition of histamine formation in the rat. Br J Pharmacol 30:274–282
Jorizzo JL, Coutts AA, Eady RAJ, Greaves MW (1983) Vascular responses of human skin to injection of substance P and mechanism of action. Eur J Pharmacol 87:67–76
Juan H (1981) Dependence of histamine-evoked nociception on prostaglandin release. Agents Actions 11:706–710
Juhlin L (1967) Localization and content of histamine in normal and diseased skin. Acta Dermatol Venereol 47:383–391
Kahlson G (1960) A place for histamine in normal physiology. Lancet I:67–71
Kahlson G, Nilsson K, Rosengren E, Zederfeldt B (1960a) Wound healing as dependent on rate of histamine formation. Lancet II:230–234
Kahlson G, Rosengren E, White T (1960b) The formation of histamine in the rat foetus. J Physiol (Lond) 151:131–138
Kahlson G, Rosengren E, Svensson SC (1962) Inhibition of histamine formation in vivo. Nature 194:876
Kahlson G, Rosengren E, Thunberg R (1963) Observations on the inhibition of histamine formation. J Physiol (Lond) 169:467–486
Kaplan AP, Gray L, Shaff RE, Horakova Z, Beaven MA (1975) In vivo studies of mediator release in cold urticaria. J Allergy Clin Immunol 55:394–402
Katz G (1942) Histamine release in the allergic skin reaction. Proc Soc Exp Biol Med 49:272–277
Kaur S, Greaves MW, Eftekhari N (1981) Factitious urticaria (dermographism): treatment by cimetidine and chlorpheniramine in a randomised double-blind study. Br J Dermatol 104:185–190
Kobza-Black, Greaves MW, Hensby CN, Plummer NA (1976) A new method for obtaining human skin inflammatory exudate for pharmacological analysis. Br J Pharmacol 58:317
Kobza-Black A, Greaves MW, Hensby CN, Plummer NA, Eady RAJ (1977) A new method for recovery of exudates from normal and inflamed human skin. Clin Exp Dermatol 2:209–216
Kobza Black A, Keahey TM, Eady RAJ, Greaves MW (1981) Dissociation of histamine release and clinical improvement following treatment of acquired cold urticaria by prednisone. Br J Clin Pharmacol 12:327–331
Krause L, Shuster S (1984a) H_1 receptor active histamine not sole cause of chronic idiopathic urticaria. Lancet II:929–930
Krause L, Shuster S (1984b) The effect of terfenadine on dermographic wealing. Br J Dermatol 110:73–80
Krause L, Shuster S (1985a) Minimal effect of complete H_1 receptor blockade in urticaria pigmentosa. Acta Derm Venereol (Stockh) 65:338–340
Krause L, Shuster S (1985b) A comparison of astemizole and chlorpheniramine in dermographism urticaria. Br J Dermatol 112:447–453
Krause L, Shuster S (1985c) Enhanced weal and flare response to histamine in chronic idiopathic urticaria. Br J Clin Pharmacol 20:486–488
Kutscher F von (1910) Die physiologische Wirkung einer Secalbase und des Imidazolyäthylamins. Zentralbl Physiol 24:163–165
Lafferty K, De Trafford JC, Roberts VC, Cotten LT (1983) On the nature of Raynaud's phenomenon: the role of histamine. Lancet II:313–315
Last MR, Loew ER (1947) Effect of antihistamine drugs in increased capillary permeability following intradermal injection of histamine, horse serum and other agents in rabbits. J Pharmacol Exp Ther 89:81–91
Lemanske RF, Barr L, Guthman DA, Kaliner M (1983) The biologic activity of mast cell granules. V. The effects of antihistamine treatment on rat cutaneous early- and late-phase allergic reactions. J Allergy Clin Immunol 71:94–99

Levine RJ (1966) Histamine synthesis in man: inhibition by 4-bromo-3-hydroxybenzyloxy-amine. Science 154:1017–1019
Lewis T (1927) The blood vessels of the human skin and their responses. Shaw, London
Lewis T, Grant RT (1924) Vascular reactions of the skin to injury. Heart 11:209–265
Maeyama K, Watanabe T, Taguchi Y, Yamatodani A, Wada H (1982) Effect of α fluoromethylhistidine, a suicide inhibitor of histidine decarboxylase on histamine levels in mouse tissues. Biochem Pharmacol 31:2367–2370
Majno G, Palade GE (1961) Studies on inflammation. I. The effect of histamine and serotonin on vascular permeability: an electron microscope study. J Biophys Biochem Cytol 11:571–605
Majno G, Palade GE, Schoefl GI (1961) Studies on inflammation. II. The site of action of histamine and serotonin along the vascular tree: a topographical study. J Biophys Biochem Cytol 11:607–626
Marks R, Greaves MW (1977) Vascular reactions to histamine and compound 48/80 in human skin: suppression by a histamine H_2-receptor blocking agent. Br J Clin Pharmacol 4:367–369
Marks R, Dykes PJ, Tan CY (1982) Histamine and epidermal proliferation. Br J Dermatol 107:15–20
Matthews CNA, Boss JM, Warin RP, Storari F (1979) The effect of H_1 and H_2 histamine antagonists on symptomatic dermographism. Br J Dermatol 101:57–61
Michel B, Russell TH, Winkelmann RK, Gleich GJ (1970) Release of kinins from site of wheal-and-flare allergic skin reactions. Int Arch Allergy Appl Immunol 39:616–624
Miles AA, Miles EM (1952) Vascular reactions to histamine, histamine liberator and leukotaxine in the skin of guinea-pigs. J Physiol (Lond) 118:228–257
Misch KJ, Greaves MW, Kobza-Black A (1983) Histamine and the skin. Br J Dermatol 109 (Suppl 25):10–13
Mongar JL, Schild HO (1952) A comparison of the effects of anaphylactic shock and of chemical histamine releasers. J Physiol (Lond) 118:461–478
Moore TC, Schayer RW (1969) Histidine decarboxylase activity of autografted and allografted rat skin. Transplantation 7:99–104
Nathan RA, Segall N, Schocket AL (1981) A comparison of the actions of H_1 and H_2 antihistamines on histamine-induced bronchoconstriction and cutaneous wheal response in arthritic patients. J Allergy Clin Immunol 67:171–177
Norby K (1983) Intradermal mast cell secretion causing cutaneous mitogenesis. Virchows Arch [B] 42:263–269
Owen DAA, Pipkin MA, Woodward DF (1984) Studies on cutaneous vascular permeability in the rat: increases caused by histamine and histamine-like agents. Agents Actions 14:40–42
Pearce CA, Greaves MW, Plummer VM, Yamamoto S (1974) Effect of disodium cromoglycate on antigen-induced histamine release from human skin. Clin Exp Immunol 17:437–440
Pearce FL (1983) Mast cell heterogeneity. Trends Pharmacol Sci 4:165–167
Phillips MJ, Meyrick TRM, Moodley I, Davies RJ (1983) A comparison of the in vivo effects of ketotifen, clemastine, chlorpheniramine and sodium cromoglycate on histamine and allergen induced wheal in human skin. Br J Clin Pharmacol 15:277–286
Piotrowski W, Foreman JC (1985) Some effects of calcitonin gene-related peptide in human skin and on histamine release. Br J Dermatol 114:37–46
Portier P, Richet C (1902) Bull Soc Biol 170
Powell JR, Brody MJ (1976) Participation of H_1- and H_2-histamine receptors in physiological vasodilator responses. Am J Physiol 231:1002–1009
Regal JF, Hardy TM, Casey FB, Chakrin LW (1983) C5a-induced histamine release. Int Arch Allergy Appl Immunol 72:362–365
Riley JF, West GB (1953) The presence of histamine in tissue mast cells. J Physiol (Lond) 120:528–537
Riley JF, West GB (1956) Skin histamine. Its location in the tissue mast cells. Arch Dermatol 74:471–478
Robertson I, Greaves MW (1978) Responses of human skin blood vessels to synthetic histamine analogues. Br J Clin Pharmacol 5:319–322

Robinson-White A, Bevan MA (1982) Presence of histamine and histamine-metabolising enzyme in rat and guinea pig microvascular endothelial cells. J Pharmacol Exp Ther 223:440–445
Rose B (1941) Studies on blood histamine in cases of allergy. J Allergy 12:327
Rosenthal SR (1977) Histamine as the chemical mediator for cutaneous pain. J Invest Dermatol 69:98–105
Sagi-Eisenberg R, Foreman JC, Shelly R (1985) Histamine release induced by histone and phorbol ester from rat peritoneal mast cells. Eur J Pharmacol 113:11–17
Schayer RW (1956a) Formation and binding of histamine by free mast cell of rat peritoneal fluid. Am J Physiol 108:199–202
Schayer RW (1956b) Formation and binding of histamine by rat tissues in vitro. Am J Physiol 187:63–65
Schayer RW (1957) Histidine decarboxylase of rat stomach and other mammalian tissues. Am J Physiol 189:533–536
Schayer RW (1960) Relationship of induced histidine decarboxylase activity and histamine synthesis to shock from stress and from endotoxin. Am J Physiol 198:1187–1192
Schayer RW (1962) Evidence that induced histamine is an intrinsic regulator of the microcirculatory system. Am J Physiol 202:66–72
Schayer RW (1963) Induced synthesis of histamine, microcirculatory regulation and the mechanism of action of the adrenal glucocorticoid hormones. Prog Allergy 7:187–212
Schayer RW, Cooper JAD (1956) Metabolism of ^{14}C histamine in man. J Appl Physiol 9:481–483
Schultz WM (1910) Physiological studies in anaphylaxis. 1. The reaction of smooth muscle of the guinea pig sensitized with horse serum. J Pharmacol Exp Ther 1:549–567
Søndergaard J, Glick D (1971) Quantitative histochemistry of histamine and histidine decarboxylase activity in the normal human skin. J Invest Dermatol 56:231–234
Søndergaard J, Glick D (1972) Histidine decarboxylase activity in human allergic contact dermatitis. J Invest Dermatol 59:247–250
Soter NA, Wasserman SI, Austin FF (1976) Cold urticaria: release into the circulation of histamine and eosinophil chemotactic factor of anaphylaxis during cold challenge. N Engl J Med 294:687–690
Spector WG, Willoughby DA (1963) The antagonism of substances that increase vascular permeability in the rat. J Pathol 86:487–496
Taguchi Y, Tsuyama K, Watanabe T, Wada H, Kitamura Y (1979) Increase in histidine decarboxylase activity in skin of genetically mast cell-deficient W/W^v mice after application of phorbol-12-myristate-13-acetate: evidence for the presence of histamine-producing cells without basophilic granules. Proc Natl Acad Sci USA 79:6837–6841
Tharp MD, Suvunrungsi RT, Sullivan TJ (1983) IgE-mediated release of histamine from human cutaneous mast cells. J Immunol 130:1896–1901
Thompson MF, Isaacs JL, Greaves MW (1981) Histamine penetration of human epidermal cells. J Invest Dermatol 76:421
Ting S, Dunsky EH, Lavker R, Zweiman B (1980) Patterns of mast cell alterations and in vivo mediator release in human allergic reactions. J Allergy Clin Immunol 66:417–423
Ting S, Zweiman B, Lavker RM (1983a) Cromoglycate does not modulate human allergic skin reactions in vivo. J Allergy Clin Immunol 71:12–17
Ting S, Zweiman B, Lavker RM, Dunsky EH (1983b) Effect of cimetidine on exogenous histamine inhibition of histamine release in vivo. Allergy 38:11–17
Wanderer AA, St Pierre JP, Ellis EF (1977) Primary acquired urticaria: a double-blind comparative study of treatment with cyproheptadine, chlorpheniramine and placebo. Arch Dermatol 113:1375–1377
Windaus A, Vogt W (1907) Synthese des Imidazolyläthylamins. Ber Dtsch Chem Ges 40:3691–3695
Winkelmann RK (1966) Technique of dermal perfusion. J Invest Dermatol 46:220–223
Yamamoto S, Francis D, Greaves MW (1976a) Enzymic histamine metabolism in guinea pig skin and its role in immediate hypersensitivity reactions. Clin Exp Immunol 26:583–589

Yamamoto S, Francis D, Greaves MW (1976b) In vitro anaphylaxis in guinea pig skin: amplification by burimamide. J Invest Dermatol 67:696–699
Yamatodani A, Maeyama K, Watanabe K, Wada T, Kitamura Y (1982) Tissue distribution of histamine in mutant mouse deficient in mast cells. Biochem Pharmacol 31:305–309
Zachariae H (1964) Histamine in human skin. Acta Derm Venereol (Stockh) 44:219–222
Zachariae H, Brodthagen H, Søndergaard J (1969) Brocresine, a histidine decarboxylase inhibitor in chronic urticaria. J Invest Dermatol 53:341–343

CHAPTER 19

Kallikreins and Kinins

V. EISEN

A. Introduction

This article has two aims:

1. To describe briefly the general aspects of kinins and the factors concerned with the formation and removal of kinins, so as to enable the reader to evaluate the role kinins may play in healthy and diseased skin. Comprehensive accounts of older and recent work on the kinin system with extensive lists of references may be found in two volumes of this Handbook (ERDÖS 1970, 1979 a) and in numerous monographs, reviews and proceedings of learned meetings (FREY et al. 1968; SCHACHTER 1969, 1980; ERDÖS 1976; HABERLAND et al. 1973–1977; EISEN 1970, 1980; PISANO and AUSTEN 1976; HABERLAND 1978; FUJII et al. 1979; REGOLI and BARABE 1980; FRITZ et al. 1982, 1983; GREENBAUM and MARGOLIUS 1986).
2. To discuss more fully the literature dealing specifically with kinins as factors in the physiology and pathology of the skin.

Kinins is the name given to peptides which are generated, and sometimes stored, in human and animal organisms, and which exert the following biological actions:

1. A potent dilating or constricting effect on small blood vessels.
2. An increase in the permeability of vascular walls. This is usually described as increased "capillary permeability", but develops more readily at the level of post-capillary venules.
3. Stimulation of the nerve endings of pain-conducting fibres.
4. Actions on non-vascular smooth muscles. A large number are contracted. Some, for example the rat duodenum, are relaxed by kinins. Preparations of isolated smooth muscle organs were of great importance in the early work on kinins. The term bradykinin (*βϱαδυς* = slow; *κινεῖν* = to move) was coined for the nonapeptide to reflect the characteristic slow contraction produced on the guinea-pig ileum (ROCHA e SILVA et al. 1949). Unfortunately, this term refers to effects which, though extremely potent in vitro, are far less clear-cut in vivo.

Subsequent work has shown that in addition to the actions listed under 1–4, kinins produce numerous other effects (cf. p. 315).

The basic term *kinins* has been used to name several associated factors. *Kininogens* are protein substrates from which kinins can be released. The enzymes capable of this action are collectively called *kininogenases*. As will be seen later, both enzymes and substrates may play biological roles which are not related to kinins; they are therefore also known by other names. This is equally true of the factors which oppose the formation of kinins and ensure their removal. All proteolytic inhibitors which restrain kinin-forming enzymes also inhibit other proteinases. The most important *kininases* (kinin-degrading enzymes) attack other biologically active peptides as well.

B. Chemistry and Biological Activities of Kinins

I. Principal Types of Kinins

Peptides which occur naturally in the animal organism and which can be regarded as kinins may be divided into five groups:

1. The plasma kinins, which usually circulate incorporated into plasma kininogens from which they are released by proteolytic action. The same peptides are found in urine (urinary kinins). The three main plasma kinins are peptides of 9, 10, and 11 amino acids (Table 1). All three contain the nonapeptide sequence of bradykinin. Extensive studies with a variety of physicochemical methods suggest that in aqueous solutions these peptide chains assume a random coil formation. A hydrogen bond involving the carboxyl group of Ser^{6} may form across Pro^{7}. As in many other biologically active peptides, the constitution of the C-terminal end

Table 1. Representative kinins

Occurs in	Name	Structure
Mammalian plasma and urine	Bradykinin	Arg^{1}–Pro^{2}–Pro^{3}–Gly^{4}–Phe^{5}–Ser^{6}–Pro^{7}–Phe^{8}–Arg^{9} (1 060 daltons)
	Kallidin	Lys–bradykinin
	Met–Lys–bradykinin	Met–Lys–bradykinin
Rat plasma	T-kinin bradykinin	Ile–Ser–bradykinin
Rat plasma	T-kinin	Ile–Ser–bradykinin
Colostrum	Colostrokinin	Not known
Amphibians and reptiles		
Frog skin (numerous species)	Bradykinin	Bradykinin
	Phyllokinin	Bradykinin–Ile–Tyr(SO_3H)
	Val^{1}, Thr^{6}bradykinin	Val–Pro–Pro–Gly–Phe–Thr–Pro–Phe–Arg
	Ranakinin N	Bradykinin–Val–Ala–Pro–Ala–Ser
	Physalaemin	<Glu–Ala–Asp–Pro–Asn–Lys–Phe–Tyr–Gly–Leu–Met–NH_2
Turtle plasma	Thr^{6}bradykinin	Arg–Pro–Pro–Gly–Phe–Thr–Pro–Phe–Arg
Insects		
Venom sac of wasp (several species)	Vespakinin X	Als–bradykinin–Ile–Val
	Polisteskinin	<Glu–Thr–Asn–Lys–Lys–Lys–Leu–Arg–Gly–bradykinin
	Vespulakinin 1	Thr–Ala–Thr—Thr–Arg–Arg–Arg–Gly–bradykinin CHO CHO

CHO = N–acetylgalactosamine or galactose residues

of the chain is critical for full biological activity. There are only a few exceptions to the rule that Arg^9 must be terminal and its carboxyl group free. The large basic side chain of Arg^1 is also important, but it may be modified and further amino acids may be added without total loss of activity. The three proline residues restrain the flexibility of the molecule. The rigidity provided by Pro^2 and Pro^7 facilitates binding to receptors (STEWART 1979).

2. Complement kinin: a large kinin derived from component C2 (see Chap. 23).

3. Colostrokinin: a kinin of about 2000 daltons released by kallikreins, trypsin and chymotrypsin from human colostrum.

4. Leukokinins: a group of large kinins which lysosomal cathepsin D-like enzymes release from globulins (leukokininogens) found in ascites, burn blisters and other exudates and in calf and rat skin, but not in normal untreated plasma. The precursor of leukokininogen may be activated in plasma by serine proteinases. The leukokinins detected in mouse and human ascites consist of 20 and 23 amino acids, respectively; those generated by cathepsin D-like enzymes from rabbit polymorphonuclear cells and from macrophages contain 21 and 25 amino acids, respectively. None of them contain the bradykinin sequence, as only one phenylalanine residue is present (GREENBAUM 1979).

5. Kinins stored in free form in the venom glands and/or skin of insects, amphibia and reptiles. These kinins serve purposes of defence and/or aggression. The common integument of man and other mammals is therefore occasionally exposed to these substances, and their actions are obviously relevant to skin pharmacology.

The kinins found so far in non-mammalian organisms show a remarkable degree of homology with bradykinin, across a wide phylogenetic range (Table 1). Amongst the kinins present in the venom sacs of wasps and related species, some (*Polistes* and *Vespula* kinin) are characterised by their numerous basic residues and consequently release histamine, in addition to acting as kinins. The importance of the carbohydrate residues in *Vespula* kinin is not known (PISANO 1979). Amongst the amphibian and reptilian kinins are close bradykinin analogues, with only one or two amino acids replaced. Some have remained very potent although the C-terminal end is extended.

II. Mechanisms of Biological Actions

Kinins achieve their principal pharmacological effects in concentrations which are low both in absolute terms and in relation to the available potential of forming enzymes and substrates. Only actions of specific relevance to the skin will be discussed here. Detailed discussions of the general pharmacology of kinins may be found in the references listed on p. 309.

In recent years, considerable progress has been made in our understanding of the intimate mechanisms by which kinins act on target cells. REGOLI and his colleagues (REGOLI and BARABÉ 1980) have accumulated extensive evidence that at least two types of receptor can be distinguished by the order of potency of kinin agonists, by the affinity of competitive antagonists and by selective desensitisation.

Table 2. Preparations containing B_1 and B_1+B_2 bradykinin receptors (REGOLI and BARABÉ 1980)

Type B_1	
	Rabbit: aorta, renal artery, pulmonary artery, carotid artery, anterior mesenteric vein, portal vein, jejunum
	Cat: aorta, carotid artery
Types B_1 and B_2	
	Rabbit: posterior vena cava, renal vein, trachea
	Rat: anterior mesenteric vein, stomach, ileum, urinary bladder
	Dog: jugular vein, mesenteric vein, superior vena cava, subclavian artery
	Man: circular muscle of the colon

The type of receptor present in the isolated rabbit aorta was denoted as B_1 (REGOLI et al. 1977). The B_1 receptors are sensitive to the natural kinins bradykinin and kallidin, which, however, act even more potently when their C-terminal arginine is removed; des-Arg^{11}-Met-Lys-bradykinin is also highly effective. Thus, B_1 receptors respond to bradykinin to which a positively charged side chain has been either added at the N-terminal end or removed from the C-terminal end. Leu^8-des-Arg^9-bradykinin is a potent antagonist. The sensitivity of isolated tissues containing B_1 receptors (Table 2) increases with time, and the specific binding of ^{3}H-des-Arg^9-bradykinin grows in parallel. This phenomenon is prevented when protein synthesis is abolished by early application of cycloheximide or actinomycin D. Such apparent de novo synthesis of B_1 receptors could be induced in vitro or in vivo by various types of damage. For example, 5 h after receiving *E. coli* endotoxin i.v., rabbits responded to des-Arg^9-bradykinin with pronounced hypotension, and their isolated veins and arteries with increased sensitivity. From the findings it was calculated that in quiescent conditions, tissues contain few if any B_1 receptors, but promptly form them in response to noxious stimuli. This phenomenon may play an important role in tissue damage, as the B_1 agonist des-Arg^9-bradykinin is the normal product of kininase I activity (see p. 321).

The receptors gathered under the term B_2 are present in numerous tissues (Table 2), and may in fact comprise several types. B_2 receptors are stable components of cell membranes. Observations on the rat uterus and guinea-pig ileum suggest that the concentration and/or affinity of B_2 receptors may be increased by oestrogens and androgens. Bradykinin is the most potent agonist, but veins are more sensitive to the constricting effect of kallidin.

The mechanisms by which the message of the kinin–receptor complex is transmitted to the responding apparatus of the target cell are not yet fully understood. At least some kinin actions appear to be associated with changes in the production of cyclic nucleotides and/or in intracellular calcium levels (REGOLI and BARABÉ 1980). A regulatory role of calcium ions is supported by the finding that 5 mM Ca^{2+} depresses by 50% binding of tritiated bradykinin to receptors in mammalian tissue membranes; a similar degree of depression is produced by 80 mM Na^+ or Li^+ (INNIS et al. 1981).

In some instances, the effects on cyclic nucleotide levels may be the result of the influence kinins exert on arachidonic acid metabolism which leads to the synthesis of prostaglandins and related lipids. Activation of phospholipase A_2, the enzyme which releases arachidonic acid from phospholipids in cell membranes, has been reported in several tissues (McGIFF 1980). SHAYMAN and MORRISON (1985) report that bradykinin reversibly increases phosphatidyl inositol turnover in renal collecting duct cells.

III. Actions on Blood Vessels

The potent effects of kinins on the calibre of blood vessels have been investigated in a vast variety of isolated arteries and veins, vascular regions and whole animals. The effects differ according to species and the type of vessel. Dilatation is by no means the rule, but is sufficiently widespread amongst critical vascular regions (heart, splanchnic bed, liver, kidney, intestine, spleen, pancreas, brain, skin, muscles) to lower total peripheral resistance and thus lead to arterial hypotension. Veins are more often constricted. The main site of the dilating action are the smooth muscles of the arterioles which provide the pre-capillary resistance. As a consequence, kinins increase the number of patent capillaries in most tissues (JOHNSON 1979; REGOLI and BARABÉ 1980).

Animal and later human skin was used extensively in establishing and analysing kinin effects on vessel permeability. On a molar basis, the potency of bradykinin exceeds that of histamine by 1–2 orders of magnitude. Threshold doses lie in the range of 0.05–1.0 μg bradykinin.

Light and electron microscopy, often combined with monitoring of tracer materials, suggest that kinins increase the permeability of microvasculature by acting both on the calibre and on the structure of the vessel walls. The prompt dilatation of arterioles distends capillaries and post-capillary venules, an effect often reinforced by constriction of the small veins which collect the blood from the venules. Simultaneously, kinins stimulate contractile elements inside endothelial cells, particularly in the distended venules. Four types of deformation of endothelial nuclei are seen (notches, folds, closing folds and pinches); the cytoplasm accumulates in the subnuclear zone, where it is thrown into folds. This changes the shape of the cells and widens the junctions from an estimated 3–100 nm radius to about 400 nm (MAJNO et al. 1969; JOHNSON 1979).

Facilitation of water and macromolecular transport by one or more of the transcellular pathways has been suggested as an additional or even alternative mechanism by which kinins enhance permeability (JOYNER et al. 1974; CARTER et al. 1974). It is not yet certain whether kinins promote transport by macro- or micropinocytosis, through small or large pore junctions or through transitory channels across cells.

Experiments in animal skin have shown that the kinin effects on blood vessels are greatly influenced by prostaglandins of the E series and prostacyclins (see Chap. 21). These arachidonate metabolites are vasodilators in their own right and may also potentiate the dilating effect of kinins. In this way, prostaglandins play a crucial role in the effects of kinins and other permeability-enhancing endogenous agents (WILLIAMS and MORLEY 1973; MONCADA et al. 1973; IKEDA et al. 1975; WILLIAMS and PECK 1977). This represents an important amplifying mechanism, since a vast variety of noxious and other stimuli – including bradykinin – increase the formation of prostaglandins which then sensitise target tissues.

IV. Pain-Producing Effects

A great deal of the essential early information on the pain-producing actions of kinins was obtained in experiments on the skin. A simple convenient method of administering substances to the denuded base of a skin blister was developed by C. A. KEELE and his colleagues (KEELE and ARMSTRONG 1964). In this method, skin blisters are raised by applying cantharidin plasters to the forearm. The separated epidermis is cut off and the exudate removed. Substances are applied dissolved in a balanced salt solution, so that the subject is not aware of their nature. The intensity of pain is recorded graphically as no (0), mild (1), moderate (2) and severe (3) pain. Bradykinin 0.1–1 μM produces pain of a burning character and slower in onset than that caused by acetylcholine or potassium. Lower concentrations of bradykinin produce pain within 30 s, and higher concentrations in 7–10 s. Repeated doses tend to reduce the sensitivity of the area to bradykinin, but not to other pain-producing compounds, such as 5-hydroxytryptamine or acetylcholine.

Kallidin and Met-Lys-bradykinin are effective in μM concentrations. Pain of a similar character and intensity is caused by plasma whose clotting factor XII and therefore kinin-forming system has been activated by exposure to glass or by some other procedure. Such activation could also be induced in blister fluid, synovial exudate and other body fluids containing plasma proteins.

Intradermal injections of bradykinin (FERREIRA 1972) produced the same type of pain as application to the blister base.

A fuller appreciation of the pain-producing role of kinins in the skin may be gained when the potentiating influence of prostaglandins is recognised (FERREIRA 1980). FERREIRA (1972) observed that in man subdermal infusions of combinations of bradykinin (150 ng/min) plus prostaglandin E_1 (30 ng/min) produced definite pain, whereas none was felt when either compound was infused in the same concentration by itself. The sensitising effects of prostaglandins were still detected 5–15 min after their infusion had been stopped. In rats, 0.5–1.0 μg bradykinin injected into the femoral artery elicited afferent impulses in the saphenous nerve only during infusions of prostaglandin E_1 (CHAHL and IGGO 1977).

Kinins excite all categories of primary afferent fibres. Application to afferent fibres from the skin of anaesthetised cats showed that bradykinin excited particularly high percentages of (a) slow-conducting ($\leqq$2 m/s) thin non-myelinated C fibres which receive mechanical, thermal and chemical noxious stimuli and are responsible for poorly localised burning pain, and (b) slightly faster (6–76 m/s), thin myelinated Aδ fibres, some of which are concerned with mechanical and thermal nociception and transmit prompt well-localised pain (IGGO 1976; BECK and HANDWERKER 1974).

Detailed analyses of human skin responses to i.d. injections of bradykinin were carried out by GREAVES and SHUSTER (1967) and GREAVES and SØNDERGAARD (1970b). After single injections of 1.0 μg bradykinin, subjects experienced burning pain without itching, lasting 10–45 s. Weals were visible for up to 5 h, and their size was dose dependent in the range of 0.1–10.0 μg. Spreading of the pronounced erythema could be prevented by compressing the skin vessels; the erythema was therefore due to a direct action on the blood vessels and not to an axon reflex. Unlike histamine and serotonin, bradykinin did not induce tachyphylaxis: a first i.d. injection during arterial occlusion did not produce a weal; when the circulation was restored after 30 min, a second injection was fully effective.

If a kinin is produced in sweating skin (see p. 319), it may significantly influence the sodium concentration and therefore osmolality of sweat in the early phase of sweating. After iontophoretic application of bradykinin, the normal pattern of sweat osmolality, which is a gradual decrease during the first 3–5 min, changed to an abrupt fall (GORDON and SCHWARZ 1971).

V. Other Actions of Kinins

The actions of kinins on blood vessels and sensory nerve endings may clearly have a considerable impact on skin functions. Several other kinin actions must be briefly mentioned. Effects on *renal haemodynamics* and *tubular functions* lead to increased excretion of water and sodium. More importantly, kinins form part of an intricate network which adapts renal functions to the requirements of homeostasis. The other participants in this network include prostaglandins, vasopressin, angiotensins II and III, aldosterone and catecholamines. The precision, speed, subtlety and flexibility of this network are ensured by the fact that each one of these agonists influences the generation and/or the activities of most, if not all, other participating agonists.

Bronchoconstrictory effects are pronounced in some species, but their role in normal human respiration and in functional asthma is not clearly established.

Release of catecholamines from the adrenal medulla and superior cervical ganglion requires fairly high doses (FELDBERG and LEWIS 1964; LEWIS and REIT 1965).

Kinins may increase the intestinal absorption of carbohydrates and amino acids. The utilisation of glucose by muscles is promoted by 1 nM bradykinin and depressed by higher concentrations (HABERLAND et al. 1973–1977; HABERLAND 1978).

C. Kinin Formation in Mammals

I. Specific and Non-specific Kinin-Forming Enzymes

Plasma kinins can be released from kininogens by extra- and intracellular proteolytic enzymes which attack many substrates. Such enzymes are plasmin, trypsin and acrosin.

More selective proteinases which act only on a few proteins besides kininogens, are present in plasma and in body fluids derived from it, and in the pancreas, salivary, tear and sweat glands, kidney and urine, intestinal wall and neutrophil and basophil leucocytes. Although there are considerable differences between the selective proteinases from different tissues, they are usually called by the common name kallikreins [from kallikreas (καλλίκρεας) the term used by GALEN for pancreas; coined because plasma kallikrein was erroneously believed to be secreted by the pancreas].

Both the specific and the non-specific kinin-forming enzymes split peptide bonds formed by the carboxyl groups of lysine or arginine, contain a functionally essential serine in their catalytic centres (serine proteinases), and attack the same kininogens from which they release the same peptides, i.e. bradykinin and kallidin. Methionyl-lysyl-bradykinin is mainly released by pepsin-like enzymes.

Cathepsin D type aspartic proteinases in the lysosomes of white blood cells, macrophages, spleen cells and other normal or malignant cells (GREENBAUM 1979) form leukokinins from different substrates (leukokininogens) with acid pH optima (see p. 311).

II. Kinin Formation by Plasma Kallikreins and Other Plasma Enzymes

The kinin-forming system in human and most other mammalian plasmas forms part of the intrinsic blood clotting pathway (Fig. 1). Both processes are initiated by clotting factor XII (Hageman factor), which also activates intrinsic fibrinolysis (see Chap. 22, Vol. II).

In man factor XII is a single chain β-glycoprotein of 75000–80000 daltons (COCHRANE and GRIFFIN 1979). Its principal mode of activation is by contact with surfaces which carry dense negative charges. Surfaces of this type used ex vivo include glass and other silicates; when their charges are neutralised by positively charged polycations, their ability to activate factor XII is reduced or lost (EISEN 1964). In vivo, a number of pathological processes give rise to activating structures such as monosodium urate and calcium pyrophosphate crystals, lipopolysaccharides of bacterial cell walls, and altered collagen and vascular basement membrane (COCHRANE and GRIFFIN 1979). On its activation, factor XII initiates

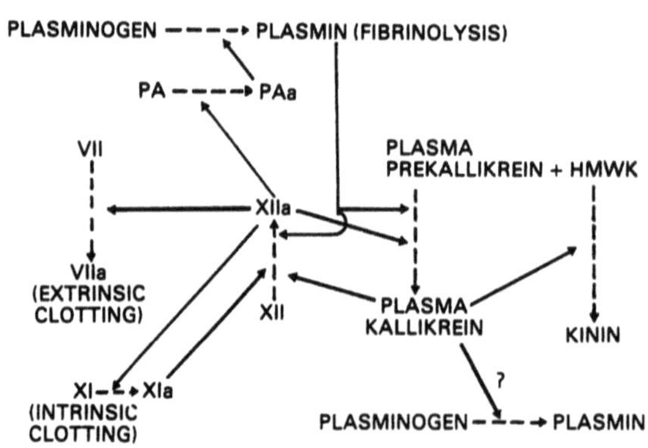

Fig. 1. Triggering of kinin formation, blood clotting and fibrinolysis in plasma. Contact with negatively charged surfaces converts Hageman factor (clotting factor XII) to its active form (XIIa). The active enzyme XIIa then initiates intrinsic and extrinsic blood clotting, fibrinolysis and plasma kinin formation by activating clotting factors XI and VII, plasminogen activator (*PA*) and pre-kallikrein, respectively. The active factor XIa, plasmin and plasma kallikrein then convert more factor XII to XIIa. High molecular weight kininogen (*HMWK*) facilitates the interaction between XIIa and kallikrein, and also serves as kinin-yielding substrate. Pre-kallikrein is also activated by plasmin. Kallikrein may act as a plasminogen activator

kinin formation both at the site of contact (solid phase reaction) and in circulating blood (fluid phase); blood clotting is triggered almost exclusively in the solid phase. The molecular mechanism by which negative charges activate factor XII is still under investigation. Provided no other plasma factors are present, contact only splits a single peptide bond which is inside a disulphide loop, so that the resulting 54000- and 26000-dalton peptides remain linked. Contrary to earlier views, the binding and cleavage appear to enhance the catalytic activity of factor XII only slightly or not at all. Rather, factor XII is rendered more susceptible to proteolytic attack by plasma kallikrein (500-fold), plasmin (100-fold) and clotting factor XI (30-fold; GRIFFIN 1978).

Additional or even alternative mechanisms have been proposed. Experiments with purified rabbit factors (WIGGINS and COCHRANE 1979) suggest that contact induces traces of catalytic activity which then activate autocatalytically the bulk of bound factor XII. It is also possible that non-activated factor XII possesses weak enzymic activity.

Negatively charged surfaces also bind pre-kallikrein, which mostly circulates as a complex with its substrate high molecular weight kininogen (HMWK). Pre-kallikrein is a single chain of 619 amino acids in which clotting factor XIIa splits a single Arg-Ileu bond. The activated kallikrein consists of a heavy chain of 371 amino acids with four tandem repeats, and a light chain of 248 amino acids which contains the catalytic centre (CHUNG et al. 1986). Human HMWK is a single peptide chain of 110000 daltons, with one disulphide bridge. The bradykinin sequence is slightly nearer to the C-terminal end. This end therefore forms, after excision of bradykinin, the light chain which contains binding sites for pre-kallikrein and factor XI. Its high affinity for negative surface charges, due to its histidine-rich region, ensures close apposition of factor XII and pre-kallikrein. As the two enzymes activate each other's precursors, the rates of their activation rapidly accelerate. Similar but less effective rate amplification by reciprocal activation exists between factors XII and XI, and between factor XII and plasminogen activator. Factor XII and its three substrates, as well as HMWK, show a high degree of homology. They are all single chain peptides of 80000–110000 daltons (factor XI may circulate as a dimer). From each of the enzymes activated by factor XII a light chain of about 30000 daltons is released. These carry the catalytic sites. The light chains of kallikrein and plasminogen activator, but not of activated factor XI, enter the circulation, where they disseminate fluid phase kinin formation and fibrinolysis (COCHRANE and REVAK 1980). They also hydrolyse factor XII outside the disulphide loop, so that the smaller fragment is separated and can act in the circulation.

The light chain of activated kallikrein may remain linked to the heavy chain by a disulphide loop (MOVAT 1979; COCHRANE and GRIFFIN 1979). In either form, plasma kallikrein may act not only on factor XII and HMWK but also on plasminogen, prorenin (LECKIE 1981) and complement component C5 (WIGGINS et al. 1981).

III. Kinin Formation by Blood Cells

Certain white blood cells also contain specific kinin-forming enzymes. Neutrophil polymorphonuclear cells engaged in phagocytosis release from their azurophil granules an enzyme similar to, but not identical with elastase (WENDT and BLÜMEL 1979) which splits from HMWK methionyl-lysyl-bradykinin or a related kinin, at neutral pH.

A kallikrein-like enzyme has been described in human basophil leucocytes (NEWBALL et al. 1979). When purified by gel filtration, chromatography on DEAE-Sepharose and Sepharose 6B, and electrophoresis in polyacrylamide gel, the enzyme activity was associated with a main peak of 1 200 000 daltons, and with smaller peaks of 400 000 and 100 000 daltons.

The basophil kallikrein differed from plasma kallikrein not only by its higher molecular weight, but also by its responses to some inhibitors. However, it activated and cleaved factor XII in similar fashion as plasma kallikrein (see p. 317). Challenge of sensitised basophils by specific antigen or by anti-human IgE led to variable release (average 23% of the total content) in a time- and temperature-dependent reaction, requiring calcium and metabolic energy. The percentage released was mostly, but not always, smaller than that of histamine, and there was no correlation between the two values; these findings suggested that the two mediators may be contained in two distinct populations of basophil granules. The release was depressed by procedures which raised intracellular cAMP levels.

On interaction with IgE, mast cells also release a neutral tryptase of 144 000 daltons, which generates kinin predominantly from HMWK (SCHWARTZ et al. 1986).

IV. Glandular Kallikreins

The kallikreins found in the exocrine pancreas, salivary and tear glands, kidneys and urine show close mutual resemblance. They differ from plasma kallikrein by their smaller molecular weight (23 000–43 000 daltons), by their susceptibility to inhibitors and by the fact that they act on low molecular weight kininogen (LMWK) as well as or better than on HMWK. It was generally assumed, though not established, that the kallikrein secreted by sweat glands is very similar to the other glandular kallikreins. If this is the case, the kallikrein-like enzyme described in human skin by TOKI and YAMURA (1978, 1979) is clearly a different factor.

Estimates of the molecular weight of the single chain glycoprotein LMWK range from 50 000 to 70 000 daltons (MOVAT 1979); it tends to aggregate in purified form.

The total capacity of the kinin-forming system in man is amply sufficient to generate rapidly effective concentrations of plasma kinin. The kininogens in 1 ml of human plasma can yield 3–10 μg bradykinin (FREY et al. 1968; EISEN 1963); of this 20%–30% is derived from HMWK, the rest from LMWK.

V. Kinin Formation in Human Skin

In addition to the kinin-forming factors in cells and extracellular fluids which skin shares with other tissues, there are activators and enzymes producing kinins which are specific to the skin.

The surface of healthy unbroken skin activates clotting factor XII in plasma; it also triggers the extrinsic clotting pathway, as it accelerates clotting of Hageman trait plasma (NOSSEL 1966; OGSTON et al. 1969). These actions are at least partly due to the secretions of sebaceous glands. Another activating constituent is collagen. WILNER et al. (1968) found that acid-soluble collagen from human foetal skin and acid-dispersable polymerised collagen from adult skin had a clot-promoting effect which was mostly due to activation of clotting factor XII. When the triple helical conformation of collagen was changed into a random coil formation by exposure to 35 °C at pH 2.5, most of the clot-promoting activity was lost.

Considerable kinin-forming activity is found in human sweat induced by heating of the forearm at 38 °C (FOX and HILTON 1958). A peak of the sweating rate was reached after 10–15 min, and of the kinin-forming activity at 20–30 min. Perfusates of cool subcutaneous tissue (skin temperature 25 °–27 °C) contained kinin (1–5 µg/ml) which increased up to 5 times during body heating (skin temperature up to 34 °C). Atropine (0.3 mg into the brachial artery) slightly delayed but did not diminish the vasodilatory response to heat, whilst the sweating response was considerably delayed and reduced.

This evidence that sweat glands secrete a kallikrein was the basis of FOX and HILTON's (1958) suggestion that kinins mediate periglandular functional vasodilatation in the skin, in a similar manner as proposed (HILTON 1970) for the kinin-forming apparatus in the pancreas and salivary and tear glands. The role of kinins in the regulation of blood flow in the latter three organs has been the subject of conflicting views (SCHACHTER 1969). Doubt was cast on the role of kinin formation in sweating and cutaneous vasodilatation by FREWIN et al. (1973), who were unable to detect kinin or kinin-forming activity in sweat obtained in volunteers through exercise or heating of limbs at 43 °C.

It is not known whether the observed kinin-forming activities were due to the kallikrein-like enzyme purified and characterised by TOKI and YAMURA (1978, 1979, 1980). Skin from amputated breasts was extracted with buffered 1.5 M NaCl; after dialysis, chromatography on DEAE-cellulose and hydroxyl-apatite-cellulose and Sephadex 4B gel filtration, the enzyme migrated as a homogeneous β-globulin in cellulose acetate electrophoresis. It also proved homogeneous on sodium dodecyl sulphate polyacrylamide gel electrophoresis and ultracentrifugal analysis. The molecular weight was 104000 daltons. The ratio of kinin-forming

Table 3. Biological (kinin-forming) and esterolytic activities of kallikreins and the kinin-forming enzyme isolated from human skin (PKF enzyme) (TOKI and YAMURA 1979)

Enzyme preparation	Biological activity (ng/mg)	Esterolytic activity (TAMe) (µmol/min/mg)	Biological/Esterolytic ratio
Human plasma kallikrein	476	6.38	74.6
Human urinary kallikrein	126	1.87	67.4
Hog pancreas kallikrein	16	0.23	69.6
PKF enzyme	229	0.35	654.3

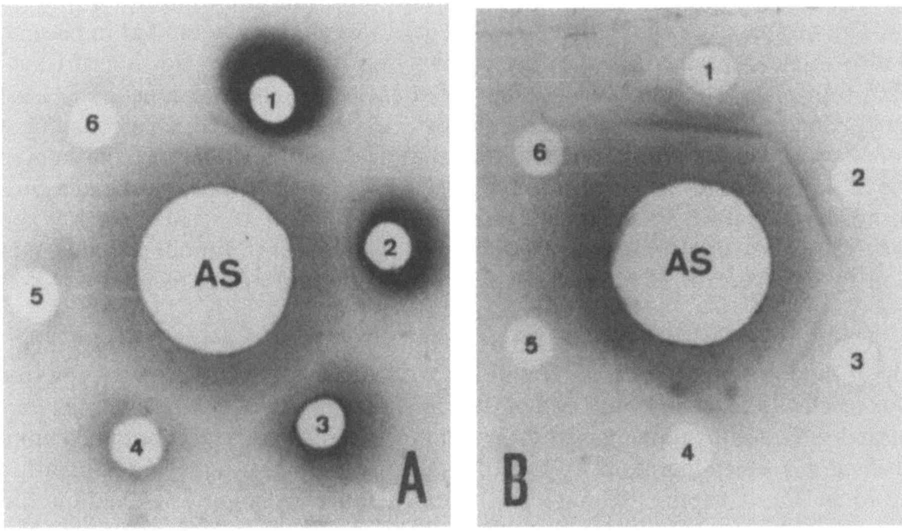

Fig. 2 A, B. Differentiation by double diffusion immunoprecipitation of kinin-forming enzyme purified from human skin (KFES). **A** Centre well (*AS*): 200 μl undiluted rabbit antiserum against KFES. Wells 1–6 contained the following in 20 μl: KFES, 10 μg (*1*) and 5 μg (*2*); human plasma kallikrein, 10 μg (*3*); human urinary kallikrein, 10 μg (*4*); human pancreatic kallikrein, 10 μg (*5*); hog pancreatic kallikrein, 10 μg (*6*). **B** Centre well (*AS*): 200 μg undiluted rabbit antiserum against human plasma kallikrein. Wells 1–6 contained the following in 20 μl; human plasma kallikrein, 10 μg (*1*) and 5 μg (*2*); human serum diluted 15-fold (*3*) and 30-fold (*4*); KFES, 10 μg (*5*) and 5 μg (*6*). Note that KFES is immunologically distinct from human plasma, urinary and pancreatic kallikrein. (TOKI and YAMURA 1979)

activity to arginine ester-splitting activity was much higher than found in other human kallikreins (Table 3). The skin kallikrein was inhibited by tosyl-lysine-chloromethyl-ketone, soya bean trypsin inhibitor, α_1-antitrypsin, but not by α_2-macroglobulin. This combination of responses distinguished the skin enzyme from the kallikreins in urine, plasma and other tissues. The distinct nature of the skin kallikrein was confirmed by immunoprecipitation tests (Fig. 2).

D. Inhibitors of Kinin Formation

The capacity of the potent and varied proteolytic inhibitors in human plasma is reflected by the fact that about 10% of the total plasma proteins are associated with this activity. When their concentrations and quality are unimpaired, these inhibitors mostly succeed in limiting the duration and intensity of plasma kinin formation to tolerable levels. Most of them inhibit several enzymes, and in this way often constitute indirect functional links between several proteolytic processes. Both factor XII and plasma kallikrein share inhibitors with proteinases of the complement, clotting and fibrinolytic systems (see Chaps. 23, Vol. I and Chap. 22, Vol. II) and with cellular or secreted proteinases.

The bulk of the kinin-forming potential in human plasma is controlled by C1 inhibitor (see Chap. 23) and α_2-macroglobulin; their relative importance is not yet generally agreed. α_1-Antitrypsin and α_2-antiplasmin check kinin formation mainly by inhibiting plasmin; α_1-antitrypsin also has a weak and slow effect on plasma kallikrein (FRITZ et al. 1979). α_2-Macroglobulin inhibits numerous serine, thiol, aspartic and metallo-proteinases by "entrapping" the enzymes after they induce a conformational change in the inhibitor (BARRET and STARKEY 1973). The kinin-forming activity of entrapped plasma kallikrein is reduced but also protected from other inhibitors, so that the complex remains active much longer than free kallikrein (VOGT and DUGAL 1976).

α_1-Antitrypsin is the only one amongst the proteolytic inhibitors in plasma for which inhibitory activity against glandular kallikreins has been definitely established (FRITZ et al. 1979).

Tissues seem to contain several proteolytic inhibitors, but most of them are not yet fully characterised. Detailed information is available on the polyvalent proteolytic inhibitor described in 1929 by FREY et al. (1968), who named it kallikrein inactivator, and by KUNITZ and NORTHROP (1936), who described it as a trypsin inhibitor. This small protein of 58 amino acids and a molecular weight of 6511 daltons is present in many bovine tissues, mostly in mast cells (FRITZ et al. 1979). It binds several serine proteinases with a very high affinity, and potently inhibits trypsin ($K_I = 6 10^{-14}$ *M* at pH 8.0). Glandular kallikreins are more effectively inhibited than human plasma kallikrein (EISEN and GLANVILLE 1969; FRITZ et al. 1979). Highly purified commercial preparations of the inhibitor (aprotinin, Trasylol) are used clinically.

Proteolytic inhibitors can be extracted from human skin with buffer solutions of physiological pH and osmolality. JUNNILA et al. (1971) separated two, one which resembled α_1-antitrypsin and one which they did not identify. TOKI and YAMURA (1978, 1980) described an inhibitor of kinin-forming activity.

E. Fate in the Body of Formed Plasma Kinins

The peptide chain of plasma kinins can be cleaved by enzymes at several sites (Fig. 3). Whilst the removal of methionine and lysine and the C-terminal arginine leads to shorter, but still active peptides, other cleavages yield inactive fragments.

Not all of the enzymic actions listed in Fig. 3 occur with sufficient frequency and intensity to be of biological significance. In some instances, enzymes and kinins meet in vivo only in rare, mostly pathological circumstances. The activities of those enzymes to which kinins are regularly exposed – in body fluids and on cell surfaces – are such that the half-life of circulating bradykinin is very short, in man about 30 s (SAAMELI and ESKES 1962) and in other species even shorter (VANE 1969).

Aminopeptidases which act on the N-terminal end of kinins are present in plasma and urine and in various tissues. They require divalent metal cations as co-factors.

Prolidases (imidopeptidases) which inactivate kinins by splitting the bond between Arg^1 and Pro^2 are also found in several tissues, notably the renal cortex, brain and red blood cells (ERDÖS 1979 a, b).

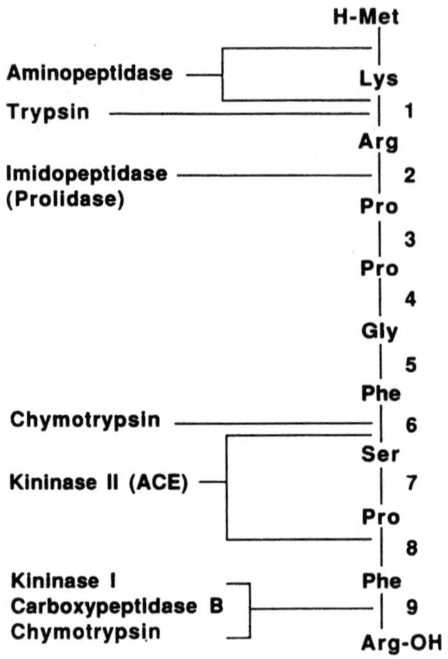

Fig. 3. Enzymes which cleave plasma kinins. Numbered amino acids from bradykinin sequence

The life span of kinins is mainly determined by enzymes which act on the C-terminal end. Kininase I (EC 3.4.17.3) resembles pancreatic carboxypeptidase B in that it removes from peptides C-terminal arginine or lysine. It is secreted, probably by the liver, into blood plasma, where it accounts for about 90% of bradykinin inactivation (ZACEST et al. 1974). The plasma enzyme is a molecule of 280000 daltons built of several subunits; the catalytic centre resides in a subunit of 90000 daltons (ERDÖS 1979b). Contrary to earlier views, the removal of Arg^9 from bradykinin does not necessarily inactivate it, but may preserve or even enhance the action on B_1 receptors (see p. 312). A similar pattern may be discerned when kininase I removes the C-terminal arginine from the complement fragments C3a and C5a (see Chap. 23), which alters, but does not abolish their activities.

Kininase II (angiotensin I converting enzyme; peptidyl dipeptidase; EC 3.4.15.1) is found in many cells as a membrane-bound enzyme (ERDÖS 1970, 1976, 1979a, b). The pulmonary blood vessels are the most important site, because of their strategic position in the circulation; high concentrations of kininase II have been demonstrated immunocytochemically near the luminal surface of the vascular endothelium. The enzyme can thus act on the total cardiac output spread over an area of some 70 m^2, so that 1 ml blood occupies over 10 miles of capillary wall. Kininase II and other endothelial enzymes split five bonds in the bradykinin molecule (RYAN and RYAN 1974). It is not surprising that up to 90% of circulating bradykinin is inactivated during one passage through the lung (VANE 1969). High concentrations of kininase II are also present in the brush border of epithelia in the proximal renal tubule and intestine, and in the choroid plexus. Kininase II in plasma is probably derived from these sources; its level is raised in sarcoidosis, leprosy and Gaucher's disease, and lowered in chronic asthma and some other pulmonary conditions.

Kininase II is a glycoprotein which contains Zn^{2+} and requires Cl^- for its activity. Purification from various sources has led to estimates of its molecular weight ranging from 90000 to 480000 daltons (ERDÖS 1979b).

Kininase II inactivates plasma kinins by removing the Phe^8-Arg^9 dipeptide with high affinity (Km $1–9 \times 10^{-7}$ *M*; ZACEST et al. 1974; DORER et al. 1974); the Ser^6-Pro^7 fragment may then also be separated. In contrast, removal of the C-terminal His^9-Leu^{10} from angiotensin I generates the active hypertensive peptide angiotensin II. Leu- and Met-enkephalin are also hydrolysed by kininase II (ERDÖS et al. 1978).

F. Kinins in Experimental and Clinical Damage of Human Skin

I. Assessment of the Role of Kinins

As some kinin-forming factors are normal constituents of any body fluid that contains plasma proteins, their presence in skin perfusates or exudates does not necessarily implicate them in the condition under study. The presence of kinins themselves is more likely to be of pathogenetic significance. However, this has been firmly established only in few experimental or clinical instances, because it is extremely difficult to exclude generation and/or loss of kinins by artefact during the processing of samples. Numerous attempts have been made to confirm claims for pathological roles of kinins by observing whether a given phenomenon was intensified under the influence of drugs which stabilise kinins; alternatively, possible depression of the phenomenon after depletion of kininogen, or after administration of agents inhibiting the formation or activities of kinins, was studied. Finally, the effects of established anti-inflammatory drugs on the appearance and actions of kinins were examined. All these attempts face formidable difficulties. Antagonists available so far are rarely sufficiently potent, stable, selective and free of agonist activity or side-effects to bear the promise of useful results in in vivo experiments, and even less in clinical studies. Inhibitors of forming enzymes, some of them designed on the basis of the specificity of kallikreins for arginine and lysine bonds, invariably inhibit other serine proteases as well. Some interesting results have been obtained with compounds that inhibit kininases I or II, in spite of the fact that both enzymes exert important functions not connected with kinins. As the effectiveness of kinins is modulated by prostaglandins and possibly other endogenous factors, the impact of kinins on diseased skin could be due not only to their increased formation but also to enhanced sensitivity of target structures caused by potentiating autacoids. This is particularly relevant in the numerous skin conditions in which increased levels of prostaglandin-like substances have been found (see p. 313 and Chap. 29).

Evidence of altered responsiveness of the skin to kinins has been reported by MICHAELSSON (1970a, b), who examined the reactions to crude hog pancreatic kallikrein (Padutin, 2 or 4 units i.d.). In comparison with healthy individuals, immediate and late reactions were more pronounced in 15 cases of chronic urticaria and less pronounced in atopic patients and in psoriasis. The influence on the results of other enzymes present in Padutin (such as chymotrypsin, carboxypeptidase B) was not assessed. A rare instance of grossly reduced sensitivity to bradykinin was described by Greaves et al. (1970). As a result of intense hypersensitivity to catecholamines in a naevus anaemicus, high doses of bradykinin (10 µg i.d.) produced no vasodilatation, although they were fully effective in the skin adjacent to the naevus.

II. Inflammation and Related Conditions

Greaves and Sondergaard (1970a) and Sondergaard and Greaves (1970a) studied the participation of kinins in the response to ultraviolet radiation. The radiation was applied for 5 min to an area of forearm skin. During the 24 h following exposure, the skin was perfused through subdermal cannulae with sterile Tyrode's solution for 1.5–8 h. Histamine was found mainly 8–24 h after exposure, and prostaglandin-like activity even later. Kinins (0.1–3.2 ng/ml) were present in some subjects, but the levels and frequency were not higher than in non-irradiated areas. Continuous subcutaneous perfusion (Søndergaard and Greaves 1970b) was also used to study the effects on the skin of the mildly inflammatory tetrahydrofurfuryl nicotinate ointment. The ointment induced redness, oedema and proneness to dermographism. During these responses, the levels of kinins and histamine were not significantly increased above levels in untreated skin. A different pattern was found in perfusates from patients with chronic urticaria, in whom the incidence of histamine but not of kinins was higher (Plummer et al. 1977).

When suction bullae (Black et al. 1976) were produced in normal and trafuryl-treated skin, exudates from both bullae contained similar concentrations of kinin-like activity (Plummer et al. 1977). Winkelmann et al. (1965) also induced dermographism by tetrahydrofurfuryl nicotinate. Dermal perfusates contained low levels of bradykinin which disappeared after vascular occlusion; the kinin was therefore thought to be derived from blood plasma. No histamine was found.

Various modifications of cutaneous perfusion have been used in attempts to demonstrate kinins in skin afflicted by inflammatory or other diseases. In two series of 30 and 22 patients with contact eczema, Greaves and Søndergaard (1970c) and Greaves et al. (1971a, b), found kinin in perfusates of involved skin (0.1–3.4 ng/ml perfusate) in 12 and 7 patients, respectively. Both the incidence and the levels of the kinins were similar to those in perfusates from control subjects. A prostaglandin-like factor was consistently present. Of 16 patients with urticaria pigmentosa and factitious urticaria, only five had detectable concentrations ($\geqq 1$ ng/ml) of kinins in the perfusates (Greaves and Søndergaard 1970b). Winkelman et al. (1966), however, reported that in nine patients with urticaria pigmentosa, perfusates contained both histamine and bradykinin. Moreover, the latter was increased during dermographic reactions. Delaus and Winkelmann (1968) studied 11 patients with cold urticaria. When the perfused skin was challenged by the application of ice for 10 min, low levels of bradykinin were detected in all 11 patients. After comparable procedures kinins were found only in two out of six normal subjects. Søndergaard and Greaves (1971) applied a specific antigen challenge to skin in nine patients with respiratory allergy and produced an immediate weal and flare reaction. Kinins (0.2–0.9 ng/ml skin perfusate) were detected in five patients. Neither the incidence nor the concentrations differed from findings in normal subjects.

Hereditary angio-oedema (HAO) is characterised by episodes of acute localised non-inflammatory oedema in mucosae, subcutaneous tissue and skin (WILLIAMS et al. 1975). It is caused by absence or functional inadequacy of C1 inhibitor, inherited as a dominant autosomal trait. HAO is fully discussed in Chap. 23, Vol. I and Chap. 31, Vol. II. C1 inhibitor, as mentioned on p. 320, also inhibits clotting factor XII and plasma kallikrein, and thereby contributes to the control of intrinsic plasma kinin formation.

It is therefore likely that kinins are a factor in the pathogenesis of HAO, although the dominant role of excessive complement activity is well established. Early studies showed that sera from HAO patients had a reduced capacity for inhibiting plasma kinin formation (LANDERMANN et al. 1962). CURD et al. (1980) induced in five HAO patients localised swelling on the forearm by mechanical vibration of the skin for 10 min. Blisters were then produced by applying suction (200 mmHg) to the swollen skin for 30–60 min at 48°–51 °C. Double diffusion immunoprecipitation against monospecific antisera revealed in HAO and in control blister fluids clotting factor XII, pre-kallikrein, plasminogen, complement component C1s, α_1-antitrypsin and α_1-antichymotrypsin. HAO fluids contained about 15%–50% of the C1 inhibitor levels found in fluids from normal subjects. HMWK concentrations were also lower than normal. Active plasma kallikrein which cleaved ^{125}I-HMWK and released kinin from HMWK was present in all HAO fluids and only in two out of eight normal blisters. The patient's levels (10–44 µg/ml) approached the kallikrein potential found in normal plasma (50 µg/ml). The kallikrein in the blisters was inhibited by purified C1 inhibitor and by antibody against human plasma kallikrein.

The results support the view that in HAO tissue trauma triggers not only complement activation but also kinin formation which remains uncontrolled due to lack of C1 inhibitor. The forming enzyme is in all probability plasma kallikrein, as glandular kallikreins are not inhibited by C1 inhibitor. However, skin kallikrein (TOKI and YAMURA 1978–1980) cannot be fully excluded because its response to C1 inhibitor has not yet been established.

G. Concluding Remarks

The role of kinins in physiological and pathological processes is – with few exceptions – still not fully delineated. This is also true of their role in healthy and diseased skin in spite of the fact that, when applied to skin, relevant concentrations of kinins produce some or all of the signs of inflammation and related defence reactions.

Developments in recent years have influenced views on the role of kinins in divergent ways. On the one hand, claims for the biological importance of kinins are scrutinised with more scepticism. Although assay methods have improved, we are also more aware how easily the results may be distorted by artefact. Moreover, numerous endogenous substances have been discovered which could plausibly play the roles previously attributed to kinins. On the other hand, a multitude of close links have been revealed between kinins and other potent endogenous agonists (Fig. 4). Both fate and functions of each of these agonists are sub-

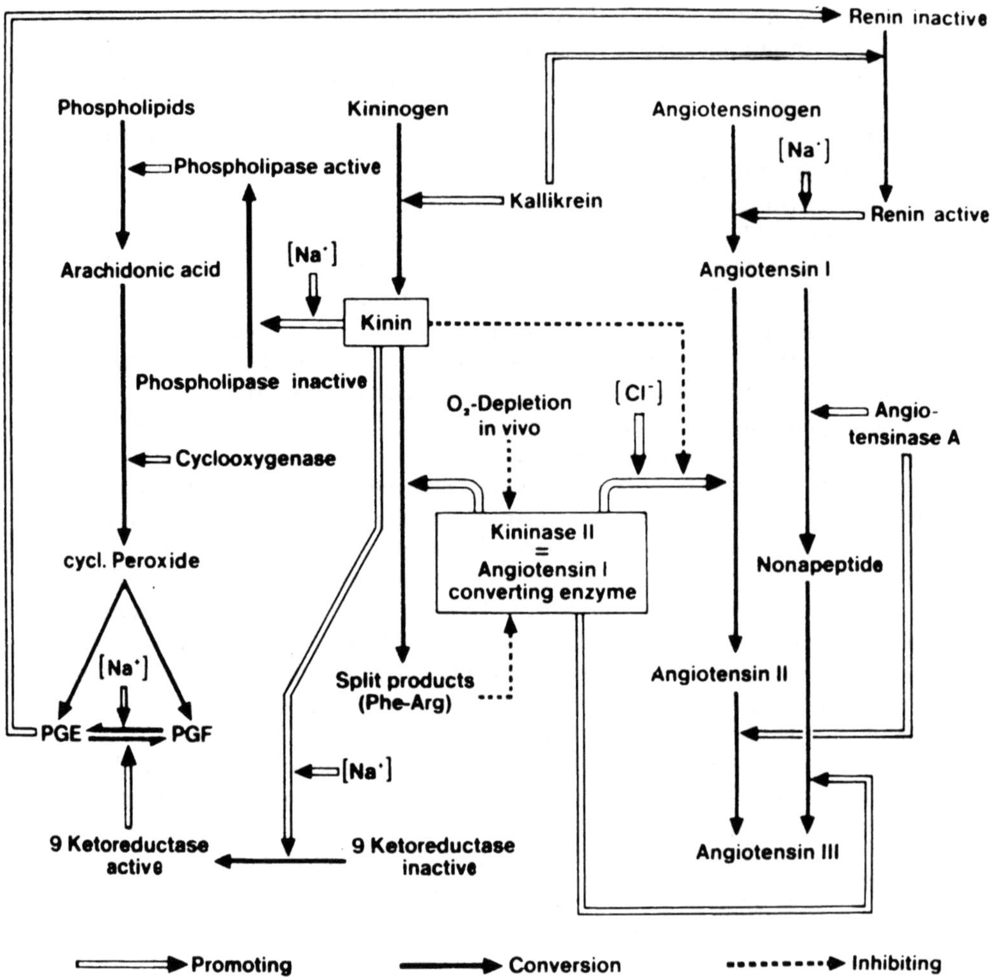

Fig. 4. Interactions between kallikreins and other autacoids. Kinins intervene in prostaglandin metabolism by activating the enzymes phospholipase A_2 and prostaglandin-9-keto reductase. Prostaglandins in turn potentiate kinin actions on blood vessels and afferent nerve endings. Kininase II (angiotensin I converting enzyme) has a high affinity for plasma kinins. These may therefore inhibit angiotensin I conversion. Some kallikreins may convert inactive to active renin. (HABERLAND and MCCONN 1979)

ject to modulating influences from most or all other participants. This implies that the impact of kinins on tissues may greatly increase even when their concentrations do not change. Seemingly ineffective, or even undetectable levels may produce symptoms because the concentration of a potentiating factor, such as a prostaglandin, has increased – an increase often induced by a kinin itself.

It is therefore not surprising that kinins have not been convincingly established as the sole begetters of any of the functions and conditions of the skin studied so far. Their contributory role as amplifying factors may, however, be more widespread than is suspected at present. The extent and importance of such contributions by kinins will become clearer only when potent selective inhibitors of their formation and antagonists of their actions are at our disposal.

Acknowledgements. I am very grateful to Miss C. A. BROUGH and Miss Q. JAYAWARDENA for the typing of the manuscript.

References

Barrett AJ, Starkey PM (1973) The interaction of α_2-macroglobulin with proteinases. Biochem J 133:709–724
Beck PW, Handwerker HO (1974) Bradykinin and serotonin effects on various types of cutaneous nerve fibres. Pflügers Arch 347:209–222
Black AK, Greaves MW, Hensby CN, Plummer NA (1976) A new method of obtaining human skin inflammatory exudate for pharmacological analysis. Br J Pharmacol 58:317
Carter RD, Joyner WL, Renkin EM (1974) Effect of histamine and some other substances on molecular selectivity of the capillary wall to plasma proteins and dextran. Microvasc Res 7:31–48
Chahl LA, Iggo A (1977) The effects of bradykinin and prostaglandin E_1 on rat cutaneous afferent nerve activity. Br J Pharmacol 59:343–347
Chung DW, Fujikawa K, McMullen BA, Davie EW (1986) Human plasma kallikrein, a zymogen to a serine protease that contains four tandem repeats. Biochemistry 25:2410
Cochrane CG, Griffin J (1979) Molecular assembly in the contact phase of the Hageman factor system. Am J Med 67:657–664
Cochrane CG, Revak SD (1980) Dissemination of contact activation in plasma by plasma kallikrein. J Exp Med 152:608–619
Curd JG, Prograis LJ, Cochrane CG (1980) Detection of active kallikrein in induced blister fluids of hereditary angioedema patients. J Exp Med 152:742–747
DeLaus FW, Winkelmann RK (1968) Kinins in cold urticaria. Arch Dermatol 98:67–74
Dorer FE, Kahn JR, Lentz KE, Levine M, Skeggs LT (1974) Hydrolysis of bradykinin by angiotensin-converting enzyme. Circ Res 34:824–827
Eisen V (1963) Observations on intrinsic kinin-forming factors in human plasma: the effect of acid, acetone, chloroform, heat and euglobulin separation on kinin formation. J Physiol 166:496–513
Eisen V (1964) Effect of hexadimethrine bromide on plasma kinin formation, hydrolysis of *p*-tosyl-*l*-arginine methyl ester and fibrinolysis. Br J Pharmacol 22:87–103
Eisen V (1970) Formation and function of kinins. In: Rotstein J (ed) The immunochemistry and biochemistry of connective tissue and its disease states. Rheumatology, vol 3. Karger, Basel, pp 103–168
Eisen V (1980) Kinins in physiology and pathology. Trends Pharmacol Sci 8:212–215
Eisen V, Glanville KLA (1969) Separation of kinin-forming factors in human plasma. Br J Exp Pathol 50:427–437
Erdös EG (ed) (1970) Bradykinin, kallidin and kallikrein. Handbook of experimental pharmacology, vol 25. Springer, Berlin Heidelberg New York
Erdös EG (1976) The kinins: a status report. Biochem Pharmacol 25:1663–1669
Erdös EG (ed) (1979a) Bradykinin, kallidin and kallikrein. Handbook of experimental pharmacology, vol 25, supplement. Springer, Berlin Heidelberg New York
Erdös EG (1979b) Inhibitors of kininases. Fed Proc 38:2774–2777
Erdös EG, Johnson AR, Boyden NT (1978) Hydrolysis of enkephalin by cultured human endothelial cells and by purified peptidyl dipeptidase. Biochem Pharmacol 27:843–848
Feldberg W, Lewis GP (1964) The action of peptides on the adrenal medulla: release of adrenaline by bradykinin and angiotensin. J Physiol 171:98–108
Ferreira SH (1972) Prostaglandins, aspirin-like drugs and analgesia. Nature New Biol 240:200–203

Ferreira SH (1980) Peripheral analgesia: mechanism of the analgesic action of aspirin-like drugs and opiate antagonists. Br J Clin Pharmacol 10:237–245
Fox RH, Hilton SM (1958) Bradykinin formation in human skin as a factor in heat vasodilatation. J Physiol 142:219–232
Frewin DB, McConnell DJ, Downey JA (1973) Is kininogenase necessary for human sweating? Lancet II:744
Frey EK, Kraut H, Werle E (1968) Das Kallikrein-Kinin-System und seine Inhibitoren. Enke, Stuttgart
Fritz H, Fink E, Truscheit E (1979) Kallikrein inhibitors. Fed Proc 38:2753–2759
Fritz H, Dietze G, Fiedler F, Haberland GL (eds) (1982) Recent progress on kinins. Agents and actions [Suppl], vol 9. Birkhäuser, Basel Boston Stuttgart
Fritz H, Back N, Dietze G, Haberland GL (eds) (1983) Kinins – III. Adv Exp Med Biol 156
Fujii S, Moriya H, Suzuki T (eds) (1979) Kinins – II. Adv Exp Med Biol 120: parts A and B. Plenum, New York
Gordon C, Schwarz V (1971) Cyclic AMP, bradykinin and sweat gland function. J Physiol 213:68–69
Greaves MW, Shuster S (1967) Responses of skin blood vessels to bradykinin, histamine and 5-hydroxytryptamine. J Physiol 193:255–267
Greaves MW, Søndergaard J (1970a) Pharmacologic agents released in ultraviolet inflammation studied by continuous skin perfusion. J Invest Dermatol 54:365–367
Greaves MW, Søndergaard J (1970b) Urticaria pigmentosa and factitious urticaria. Direct evidence for release of histamine and other smooth muscle-contracting agents in dermographic skin. Arch Dermatol 101:418–425
Greaves MW, Søndergaard J (1970c) A new pharmacological finding in human allergic contact eczema. Arch Dermatol 101:659–661
Greaves MW, Birkett D, Johnson C (1970) Nevus anaemicus: a unique catecholamine-dependent nevus. Arch Dermatol 102:172–176
Greaves MW, McDonald-Gibson W, Søndergaard JS (1971a) Recovery of prostaglandin-like fatty acids from human allergic contact eczema using a skin perfusion method. Br J Pharmacol 41:416 P
Greaves MW, Søndergaard JS, Holt G, McDonald-Gibson W (1971b) Recovery of prostaglandin-like fatty acids from human allergic contact eczema by the use of a skin perfusion method. J Pathol 103:8–9
Greenbaum LM (1979) Kininogenases of blood cells (alternate kinin generating systems). In: Erdös EG (ed) Bradykinin, kallidin and kallikrein. Springer, Berlin Heidelberg New York, pp 91–102 (Handbook of experimental pharmacology, vol 25, supplement)
Greenbaum LM, Margolius HS (eds) (1986) Kinins – IV. Adv Exp Med Biol 198:parts A and B
Griffin JH (1978) Role of surface in surface-dependent activation of Hageman factor (blood coagulation factor XII). Proc Natl Acad Sci USA 75:1998–2002
Haberland GL (1978) The role of kininogenases, kinin formation and kininogenase inhibition in post traumatic shock and related conditions. Klin Wochenschr 56:325–331
Haberland GL, McConn R (1979) A rationale for the therapeutic action of aprotinin. Fed Proc 38:2760–2767
Haberland GL, Rohen JW, Suzuki T, Blumel G, Schirren C, Huber P (1973–1977) Kininogenases, vols 1–4. Schattauer, Stuttgart
Hilton SM (1970) The physiological role of glandular kallikreins. In: Erdös EG (ed) Bradykinin, kallidin and kallikrein. Springer, Berlin Heidelberg New York, pp 389–399 (Handbook of experimental pharmacology, vol 25
Iggo A (1976) Peripheral and spinal "pain" mechanisms and their modulation. In: Bonica JJ, Albe-Fessard D (eds) Advances in pain research and therapy, vol 1. Raven, New York, pp 381–394
Ikeda K, Tanaka K, Katori M (1975) Potentiation of bradykinin-induced vascular permeability increase by prostaglandin E_2 and arachidonic acid in rabbit skin. Prostaglandins 10:747–758
Innis RB, Manning DC, Stewart JM, Snyder SH (1981) [^{3}H] Bradykinin receptor binding in mammalian tissue membranes. Proc Natl Acad Sci USA 78:2630–2634

Johnson AR (1979) Effects of kinins on organ systems. In: Erdös EG (ed) Bradykinin, kallidin and kallikrein. Springer, Berlin Heidelberg New York, pp 357–399 (Handbook of experimental pharmacology, vol 25, supplement)
Joyner WL, Carter RD, Raizes GS, Renkin EM (1974) Influence of histamine and some other substances on blood-lymph transport of plasma protein and dextran in the dog paw. Microvasc Res 7:19–30
Junnila SK, Jansen CT, Hopsu-Havu VK (1971) Elastase and trypsin inhibitors of human skin and serum. Partial purification and characterization. Acta Derm Venereol (Stockh) 51:251–256
Keele CA, Armstrong D (1964) Substances producing pain and itch. Arnold, London
Kunitz M, Northrop JH (1936) Isolation from beef pancreas of crystalline trypsinogen, trypsin, trypsin inhibitor and an inhibitor-trypsin compound. J Gen Physiol 19:991–1007
Landerman NS, Webster ME, Becker EL, Ratcliffe HE (1962) Hereditary angioneurotic edema. Deficiency of inhibitor for serum-globulin-permeability factor and/or plasma kallikrein. J Allergy 3:330–341
Leckie BJ (1981) Inactive renin: an attempt at a perspective. Clin Sci 60:119–130
Lewis GP, Reit E (1965) The action of angiotensin and bradykinin on the superior cervical ganglion of the cat. J Physiol 179:538–553
Majno E, Shea SM, Leventhal M (1969) Endothelial contraction induced by histamine type mediators: an electron microscopic study. J Cell Biol 42:647–672
McGiff JC (1980) Interactions of prostaglandins with the kallikrein-kinin and renin-angiotensin systems. Clin Sci 59:105–116
Michaelsson G (1970a) Effects of antihistamines, acetylsalicylic acid and prednisone on cutaneous reactions to kallikrein and prostaglandin E_1. Acta Derm Venereol (Stockh) 50:31–36
Michaelsson G (1970b) Decreased cutaneous reactions to kallikrein in patients with atopic dermatitis and psoriasis. Acta Derm Venereol (Stockh) 50:37–41
Moncada S, Ferreira SH, Vane JR (1973) Prostaglandins, aspirin-like drugs and the oedema of inflammation. Nature 246:217–219
Movat HZ (1979) The plasma kallikrein-kinin system and its interrelationship with other components of blood. In: Erdös EG (ed) Bradykinin, kallidin and kallikrein. Springer, Berlin Heidelberg New York, pp 1–89 (Handbook of experimental pharmacology, vol 25, supplement)
Newball HH, Berninger RW, Talamo RC, Lichtenstein M (1979) Anaphylactic release of a basophil kallikrein-like activity. I and II. J Clin Invest 64:457–465, 466–475
Nossel HL (1966) Activation of factors XII (Hageman) and XI (PTA) by skin contact. Proc Soc Exp Biol Med 122:16–17
Ogston D, Ogston CM, Ratnoff OD (1969) Studies on the clot-promoting effect of skin. J Lab Clin Med 73:70–77
Pisano JJ (1979) Kinins in nature. In: Erdös EG (ed) Bradykinin, kallidin and kallikrein. Springer, Berlin Heidelberg New York, pp 273–285 (Handbook of experimental pharmacology, vol 25, supplement)
Pisano JJ, Austen KF (eds) (1976) Chemistry and biology of the kallikrein-kinin system in health and disease. Fogarty Intern Center Proc, vol 27. US Government Printing Office, Washington DC
Plummer NA, Hensby CN, Black AK, Greaves MW (1977) Prostaglandin activity in sustained inflammation of human skin before and after aspirin. Clin Sci Mol Med 52:615–620
Regoli D, Barabé J (1980) Pharmacology of bradykinin and related kinins. Pharmacol Rev 32:1–46
Regoli D, Barabé J, Park WK (1977) Receptors for bradykinin in rabbit aortae. Can J Physiol Pharmacol 55:855–867
Rocha e Silva M, Beraldo WT, Rosenfeld G (1949) Bradykinin, a hypotensive and smooth muscle stimulating factor released from plasma globulin by snake venoms and by trypsin. Am J Physiol 156:261–273
Ryan JW, Ryan US (1976) Biochemical and morphological aspects of the actions and metabolism of kinins. In: Pisano JJ, Austen KF (eds) Chemistry and biology of the kal-

likrein-kinin system in health and disease. Fogarty Intern Center Proc 27:315–333, US Government Printing Office, Washington DC
Saameli K, Eskes TKAB (1962) Bradykinin and cardiovascular system: estimation of half-life. Am J Physiol 203:261–265
Schachter M (1969) Kallikreins and kinins. Physiol Rev 49:510–547
Schachter M (1980) Kallikreins (kininogenases) – a group of serine proteases with bioregulatory actions. Pharmacol Rev 31:1–17
Schwartz LB, Maier M, Spragg J (1986) Interaction of human LMWK with human mast cell tryptase. In: Greenbaum LM, Margolius HS (eds) Kinins – IV. Adv Exp Med Biol 198A:105–111
Shayman JA, Morrison AR (1985) Bradykinin-induced changes in phosphatidyl inositol turnover in cultured rabbit papillary collecting tubule cells. J Clin Invest 76:978
Søndergaard J, Greaves MW (1970 a) Pharmacological studies in inflammation due to exposure to ultraviolet radiation. J Pathol 101:93–97
Søndergaard J, Greaves MW (1970 b) The responses of the skin to tetrahydrofurfuryl nicotinate (Trafuril) studied by continuous skin perfusion. Br J Dermatol 82:14–18
Søndergaard J, Greaves MW (1971) Direct recovery of histamine from cutaneous anaphylaxis in man. Acta Derm Venereol (Stockh) 51:98–100
Stewart JM (1979) Chemistry and biologic activity of peptides related to bradykinin. In: Erdös EG (ed) Bradykinin, kallidin and kallikrein. Springer, Berlin Heidelberg New York, pp 227–272 (Handbook of experimental pharmacology, vol 25, supplement)
Toki N, Yamura T (1978) Plasma-kinin-forming enzyme in human skin: extraction and column chromatographic separation of plasma-kinin-forming enzyme and its inhibitor. Acta Derm Venereol (Stockh) 58:395–399
Toki N, Yamura T (1979) Kinin-forming enzyme in human skin: the purification and characterization of a kinin-forming enzyme. J Invest Dermatol 73:297–302
Toki N, Yamura T (1980) Further studies on a new kallikrein inhibitor in human skin –its purification and characterization. Arch Dermatol Res 267:301–311
Vane JR (1969) The release and fate of vasoactive hormones in the circulation. Br J Pharmacol 35:209–242
Vogt W, Dugal B (1976) Generation of an esterolytic and kinin-forming kallikrein–α_2-macroglobulin complex in human serum treated with acetone. Naunyn Schmiedebergs Arch Pharmacol 294:75–82
Wendt P, Blümel G (1979) Neutral proteases from granulocytes and their inhibitors. In: Haberland GL, Hamberg U (eds) Current concepts in kinin research. Pergamon Press, Oxford and New York, pp 73–81
Wiggins RC, Cochrane CG (1979) The autoactivation of rabbit Hageman factor. J Exp Med 150:1122–1133
Wiggins RC, Giclas PC, Henson PM (1981) Chemotactic activity generated from the fifth component of complement by plasma kallikrein of the rabbit. J Exp Med 153:1391–1404
Williams TJ, Morley J (1973) Prostaglandins as potentiators of increased vascular permeability in inflammation. Nature 246:215–217
Williams TJ, Peck MJ (1977) Role of prostaglandin-mediated vasodilatation in inflammation. Nature 270:530–532
Williams K, Rosen FS, Donaldson VH (1975) Observations on the ultrastructure of lesions induced in human and guinea pig skin by C1 esterase and polypeptide from hereditary angioneurotic edema (HANE) plasma. Clin Immunol Immunopathol 4:174–188
Wilner GD, Nossel HL, LeRoy EC (1968) Activation of Hageman factor by collagen. J Clin Invest 47:2608–2615
Winkelmann RK, Wilhelm CM, Horner FA (1965) Experimental studies on dermographism. Arch Dermatol 92:436–442
Winkelmann RK, Vesper LJ, Horner FA (1966) Histamine release from the skin of patients with cutaneous mast cell disease (urticaria pigmentosa). Acta Derm Venereol (Stockh) 46:279–284
Zacest R, Oparil S, Talamo RC (1974) Studies of plasma bradykininases using radiolabeled substrates. Aust J Exp Biol Med Sci 52:601–606

CHAPTER 20

Acetylcholine, Atropine and Related Cholinergics and Anticholinergics

M. A. Zar

A. Cholinergic Agents

I. Acetylcholine

There are sound reasons for believing that of all the neurotransmitters, acetylcholine (Fig. 1) was the earliest to appear during evolution (Michaelson 1974). It must be attributed to fortuitous coincidence that acetylcholine was also the first neurotransmitter to be isolated and identified in that role (Loewi 1921; Loewi and Navratil 1926). Brilliant though the first definite demonstration of neurohumoral transmission by Loewi was, it must be appreciated that firm theoretical background for his research had been laid down earlier by the meticulously accurate comparison of the response to acetylcholine and to stimulation of parasympathetic nerves (Dale 1914).

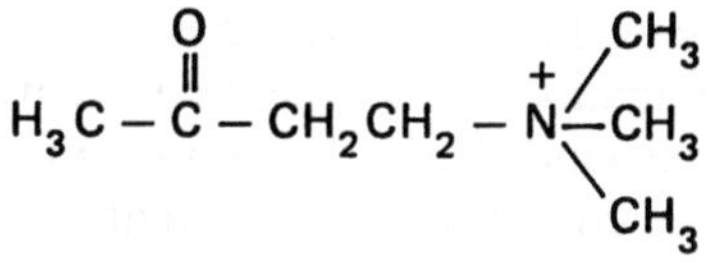

Fig. 1. Structure of acetylcholine

1. Distribution

a) Neuronal

Acetylcholine is present in the following classes of nerve and functions as their physiological neurotransmitter:

1. Preganglionic sympathetic and parasympathetic nerves
2. Postganglionic parasympathetic nerves
3. Postganglionic sympathetic nerves to eccrine sweat glands
4. Certain postganglionic sympathetic fibres to blood vessels
5. Motor nerves to the skeletal muscles
6. Certain central nervous system neurones.

b) Non-Neuronal

A number of mammalian tissues which are devoid of nerves have been shown either to contain acetylcholine or to have the ability to synthesise it. These include placenta, spermatozoa, red blood cells and platelets. The functional role of non-neuronal acetylcholine has remained uncertain and has been the subject of many speculative hypotheses (FELDBERG and LIN 1950; BURN and KOTTEGODA 1953; BRISCOE and BURN 1954; MCMAHON 1972; HARBISON et al. 1975). Many aspects of non-neuronal acetylcholine have been excellently reviewed by SASTRY and SADAVONGVIVAD (1978).

2. Biosynthesis and Storage

Acetylcholine is an ester of choline and it is synthesised in cholinergic neurones by acetylation of choline. The enzyme catalysing the reaction between choline and acetylcoenzymne A is choline acetyltransferase.

$$\text{choline} + \text{acetylcoenzyme A} \xrightarrow[\text{acetyltransferase}]{\text{choline}} \text{acetylcholine} + \text{coenzyme A}\,.$$

Choline is an essential constituent of human nutrition and is taken up by the cholinergic nerve axon through an active uptake process. Acetylcoenzyme A can be readily synthesised in the body and is present in most tissues, including the cholinergic neurones. Choline acetyltransferase is synthesised in the cell body of the neurones and then is transported to the axon terminal where the reaction takes place. Acetylcholine is formed in the cytoplasm of the axon terminal and is then stored within the synaptic vesicles. Not all acetylcholine is held within the vesicles; small quantities are also present in axonal cytoplasm.

3. Physiological Inactivation

Acetylcholine is virtually of no use as a therapeutic agent for two reasons. First, it acts on many organs and tissues, thus mitigating against its therapeutic use for an action at a specific site. Secondly, its action is extremely short lived owing to its rapid breakdown by tissue cholinesterases. Tissue cholinesterases are a group of enzymes which share the common property of combining with acetylcholine and then breaking it down to choline and acetic acid. The steps of the reaction are diagrammatically depicted below.

$$\begin{array}{c}\text{acetylcholine} + \text{cholinesterase} \\ \downarrow \\ \text{choline} + \text{acetylated cholinesterase} \\ \\ + H_2O \\ \downarrow \\ \text{choline} + \text{acetic acid} + \text{cholinesterase}\end{array}$$

4. Acetylcholine in Skin

There is ample experimental evidence for the presence of acetylcholine in skin. Although the efferent autonomic nerve supply to the cutaneous blood vessels and eccrine sweat glands anatomically belong to sympathetic division, the sympathetic nerve terminals to the eccrine sweat glands are atypical in being cholinergic. In their classic experiments on artificially perfused foot-pad of the cat, DALE and FELDBERG (1934) showed that electrical stimulation of sympathetic nerves led both to sweat secretion in the foot-pad and to the appearance of acetylcholine-like activity in the venous effluent. In one experiment, the authors tied ligatures in such a way that circulation to the sweat glands was stopped but circulation to the remainder of the foot remained intact. Stimulation of sympathetic nerves in this experiment led to intense vasoconstriction but caused neither sweat secretion nor the appearance of acetylcholine-like material in the venous effluent. Thus the experiment elegantly showed that the sympathetic nerve supply to the sweat glands alone was cholinergic.

Direct evidence for the presence of acetylcholine in human skin has come from the investigations of SCOTT (1962). His findings reveal that the acetylcholine content of normal human skin ranges from 0.1 to 3.2 μg/g with a mean value of 0.6 μg/g. These values for normal skin acetylcholine content, especially in the upper ranges, represent substantial amounts when compared with the acetylcholine contents of other tissues. For instance, the acetylcholine contents of cat superior cervical ganglion and guinea-pig ileum, tissues known to be richly endowed with cholinergic nerve terminals, are 17.3 μg/g and 6.5 μg/g respectively (BIRKS and MACINTOSH 1961; PATON and ZAR 1968). Two further components of cholinergic system which are invariably associated with cholinergically innervated tissues are the enzymes choline acetyltransferase, needed for acetylcholine synthesis, and the cholinesterases which catalyse the breakdown of released acetylcholine. No published data are available as yet regarding the presence of choline acetyltransferase in skin, but both direct and indirect evidence abounds for the presence of cholinesterases in human skin. Large quantities of cholinesterases have been biochemically demonstrated to be present in skin (HURLEY et al. 1953; MAGNUS and THOMPSON 1954; COTTON et al. 1973). The potentiation of dermal responses to acetylcholine by drugs known to prevent inactivation of acetylcholine provides indirect evidence for their presence in skin.

5. Pharmacological Actions

a) General

Acetylcholine initiates its pharmacological effect by activating two types of receptor in the body, the muscarinic and the nicotinic receptors. In general, muscarinic receptors are located in tissues innervated by postganglionic parasympathetic nerves and postganglionic sympathetic cholinergic nerves. Nicotinic receptors are situated on autonomic ganglia, skeletal muscle and adrenal glands. Both types of receptor are found in the central nervous system.

The muscarinic actions include:

1. Increased secretions from gastrointestinal, lacrimal, pancreatic, salivary and sweat glands
2. Contraction of
   a) the sphincter muscle of the iris
   b) ciliary muscle of the lens
   c) gastrointestinal smooth muscles, except the sphincters
   d) detrusor of the bladder
   e) bronchial smooth muscle

3. Relaxation of vascular smooth muscle
4. Slowing of heart rate

The nicotinic actions include excitation of autonomic ganglia, skeletal muscle and adrenal medulla.

b) Skin

Intradermal injection of acetylcholine causes short-lived but intense sweating and vasodilatation within the area of injection weal (Coon and Rothman 1939) as well as around the periphery of the weal (Rothman and Coon 1940). Intradermal injection of acetylcholine in low concentrations restricts the response to the area of the weal alone and higher concentrations of acetylcholine are needed for the diffuse effect. If acetylcholine is injected into skin which has been treated with a local anaesthetic or with a nicotinic receptor antagonist, hexamethonium, sweating and vasodilatation appear only within the area of the weal, and the diffuse effect is lost. Injection of nicotine into the skin evokes a response indistinguishable from the diffuse component of the acetylcholine response. The muscarinic receptor antagonist, atropine, blocks both local and diffuse effects of acetylcholine. These findings indicate that both sweating and vasodilator responses to injected acetylcholine have a dual origin: (a) a direct action on muscarinic receptors of sweat glands and blood vessels within the area of the weal, and (b) an indirect action, mediated through nicotinic receptors located on neuronal elements of the skin, which is responsible for the diffuse component of the acetylcholine response (Wada et al. 1952). The atropine susceptibility of all responses of acetylcholine implies that the neuronal elements involved in the indirect action of acetylcholine are themselves cholinergic.

More recently, Sato (1977) has investigated the effects of several pharmacological agents, including acetylcholine, on in vitro preparations of eccrine sweat glands. Acetylcholine evoked sweat secretion from the isolated glands, indicating a direct action on the sweat glands. Atropine readily antagonised the sudorific effect of acetylcholine, suggesting that the local receptors for acetylcholine are muscarinic in character.

A sex-related feature of the action of acetylcholine on eccrine sweat glands has been noted by several workers (Kahn and Rothman 1942; Gibson and Shelley 1948; Janowitz and Grossman 1950). They have found that intradermal administration of acetylcholine evokes much more intense sweating response in men than in women. There is also evidence for the view that sweating threshold to injected acetylcholine is higher in women than in men.

An odd finding associated with the action of acetylcholine in human eccrine sweat glands is the absence of supersensitivity to acetylcholine following postganglionic sympathetic denervation (Coon and Rothman 1941; Silver et al. 1963). It has been known for a long time that postganglionic denervation of an organ renders the cells of that organ supersensitive to the postganglionic neurotransmitter. The anomalous behaviour of human eccrine sweat glands has been explained by Lloyd (1965) on the assumption that these glands have a dual innervation, cholinergic and adrenergic (for which there is considerable experimental evidence), and that development of supersensitivity is limited to the adrenergic neurotransmitter.

SZABADI and his associates (MAPLE et al. 1982; VAN DEN BROCK et al. 1984; LONGMORE et al. 1985) have in recent years made an important contribution to understanding the factors influencing the sudorific action of cholinomimetics. They have found that eccrine sweat glands become substantially more sensitive to intradermally injected cholinomimetics in states of enhanced sympathetic activity (such as anxiety neurosis or in the presence of psychological or thermal stress). Furthermore, they have also found a greatly diminished sudorific response to intradermally injected cholinomimetics after systemic administration of ganglion-blocking agents. Their findings would appear to indicate that the pharmacological sensitivity of the eccrine sweat gland to cholinomimetic agents is governed by the ongoing activity in the innervating sympathetic fibres, and therefore provide an alternative but highly plausible explanation for the absence of hypersensitivity to acetylcholine in eccrine sweat glands following postganglionic sympathetic denervation.

Another type of sweat glands, the apocrine glands, are also found in man. In several mammalian species such as horses and cattle, the sweat glands are almost exclusively of apocrine type. Apocrine glands have a limited distribution in man, being principally located in axilla, scalp, face and external genitalia. Unlike eccrine secretion, acetylcholine does not stimulate apocrine secretion.

6. Functional Significance in Skin

a) Eccrine Sweat Secretion

It is generally agreed that the efferent autonomic nerve supply to the eccrine sweat glands, although anatomically sympathetic, is predominantly cholinergic, and that the function of acetylcholine originating in these nerves is to promote eccrine sweat secretion. Both of these conclusions are based upon sound scientific data, some of which have already been discussed. Briefly these are:

1. Acetylcholine is present in skin
2. Histochemically, the autonomic nerves to the eccrine sweat glands are cholinergic
3. Electrical stimulation of the nerves causes release of acetylcholine and enhancement of sweat secretion
4. Intradermal injection of acetylcholine causes sweat secretion
5. Drugs which antagonise acetylcholine diminish sweat secretion

b) Cutaneous Vasodilatation

There is experimental evidence for the presence of muscarinic receptors in cutaneous blood vessels and intradermal injection of acetylcholine leads to cutaneous vasodilatation. These findings might suggest an important role for acetylcholine in controlling the smooth muscle tone of skin blood vessels. However, there are no compelling reasons to accept this view. The cholinergic nerve supply to the skin blood vessels is sparse; stimulation of vasodilator nerves induces dilatation of blood vessels which is resistant to anticholinergic agents, suggesting the release of a neurotransmitter other than acetylcholine. Although as yet the identity of the non-cholinergic neurotransmitter of the vasodilator nerves remains unknown, it seems fair to conclude that acetylcholine does not play a significant physiological role in the induction of cutaneous vasodilatation.

c) Regeneration of Epithelial Layers of Skin

There is considerable experimental evidence that acetylcholine plays a vital role in the regeneration of amphibian limbs (SINGER 1959, 1974). Since the epithelial layers of mammalian skin have a high regenerating capacity, SASTRY and SADAVONGVIVAD (1978) have put forward the provocative idea of a possible link between skin acetylcholine metabolism and epithelial regeneration in mammals. In this context, it is worth recalling the findings of MEGAY (1935). He reported the presence of a substance in human sweat whose pharmacological profile (i.e. slowing of the frog heart, reduction of cat blood pressure, contraction of the frog rectus and leech muscles, potentiation of its actions by the anticholinesterase physostigmine and its antagonism by atropine) indicated it to be acetylcholine. Megay bioassayed the substance and in terms of acetylcholine he estimated its concentration in sweat as $0.1–0.2 \times 10^{-6}$ g/ml. Presence of such a high concentration of acetylcholine in sweat appears at first sight to be inconsistent with the normally highly efficient enzymatic acetylcholine breakdown mechanism in mammalian cholinergic synapses and might indicate a physiological role of acetylcholine in addition to sweat secretion and cutaneous vasodilatation.

d) Diseases Associated with Derangement of Cholinergic Mechanisms in Skin

α) *Atopic Dermatitis.* SCOTT (1962) has reported increased levels of acetylcholine in the skin of atopic subjects. Eccrine sweat glands of atopic subjects are also known to have greater sensitivity to acetylcholine (WARNDOFF 1970). The possibility that increased levels of acetylcholine in atopic skin might potentiate the effects of histamine locally and thus contribute to the disease has been raised by RYAN (1973). A neat theoretical framework for a role of acetylcholine in atopic dermatitis has been provided by MIER and COTTON (1976) on the following lines. Acetylcholine is known to raise intracellular levels of the nucleotide cyclic GMP. Cyclic GMP has a labilising influence on lysosomes. Raised levels of acetylcholine therefore, through the mediation of cyclic GMP, would increase the rate at which lysosomal enzymes are liberated, with consequent initiation of the chain of inflammatory events.

β) *Cholinergic Urticaria.* The characteristic feature of this form of urticaria is its association with the release of endogenous acetylcholine. That cholinergic neuronal activity is mandatory for cholinergic urticaria is shown by the observations that it can be inhibited by cervical sympathetic blockade or by the ganglion blocking agent hexamethonium bromide (HERXHEIMER 1956) as well as by local anaesthetics and by atropine.

II. Anticholinesterase Agents

A number of chemical substances, the so-called anticholinesterases, possess the property of combining with tissue cholinesterases and thereby preventing the enzymatic breakdown of acetylcholine. The net effect of the anticholinesterase action is to raise acetylcholine concentration in cholinergic synaptic clefts, resulting in a greatly potentiated and sustained effector tissue response.

According to the stability of anticholinesterase–cholinesterase complex, the anticholinesterases can be divided into two groups:

1. Reversible anticholinesterases. As the name indicates, these drugs combine with cholinesterases reversibly and the inhibition of the cholinesterases, therefore, is transient. Chemically, these drugs are carbamic esters and bear considerable resemblance to acetylcholine. Examples of this group are physostigmine, neostigmine and edrophonium.
2. Irreversible anticholinesterases. These anticholinesterases form irreversible complexes with the cholinesterases. Since the cholinesterases are permanently rendered non-available, the effects of the irreversible anticholinesterases are prolonged and only wear off when freshly synthesised cholinesterases become available at cholinergic synapses. Chemically the agents are organic phosphates. Examples of organic phosphate anticholinesterases are diisopropyl phosphorofluoridate, ecothiopate and certain insecticides (sarin, soman, parathion and malathion).

1. Pharmacological Actions

The effects of the anticholinesterases are almost entirely attributable to the preservation of released acetylcholine from cholinergic nerves and are therefore dependent upon the degree of cholinesterase inhibition, the functional integrity of cholinergic nerves and the intensity of neuronal activity at the cholinergic synapse.

In the skin, the visible effects of administering anticholinesterases are confined to the eccrine sweat glands. There have been many reports of enhanced sweating after the administration of these agents (Brun and Favre 1954; Randall and Kimura 1955). Organophosphorus anticholinesterases are used in agricultural farming as insecticides. Danger of accidental poisoning in man with these highly potent chemicals is ever-present due to their ability to gain entrance into the body through a variety of portals, including skin. Local sweating is an early and prominent feature of cutaneous absorption of these compounds.

III. Other Cholinomimetic Agents

A number of natural and synthetic compounds have the ability to stimulate acetylcholine receptors and are classified as directly acting cholinergics or cholinomimetics.

1. Stable Esters of Choline (Fig. 2)

a) Methacholine

Chemically methacholine closely resembles acetylcholine; the only difference from acetylcholine lies in the substitution of a hydrogen atom with a methyl group on the B-carbon atom. Methacholine possesses the usual muscarinic actions of acetylcholine but there is almost total lack of nicotinic actions. It is

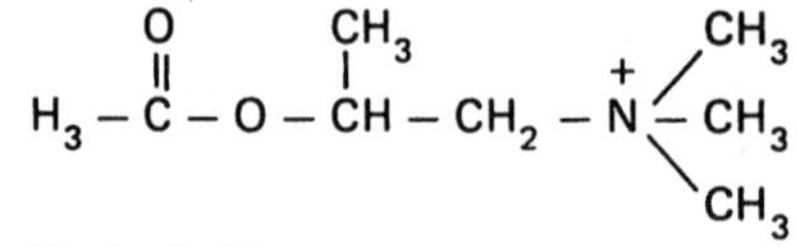

Methacholine

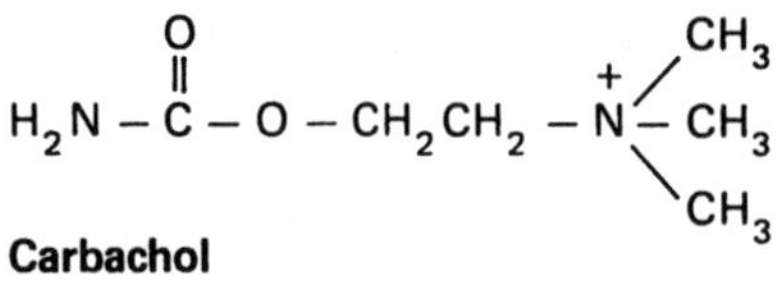

Carbachol

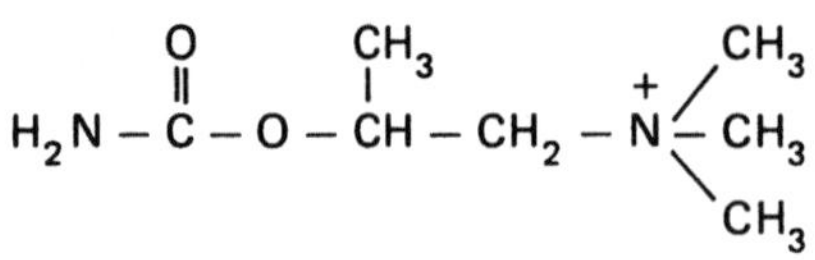

Bethanechol

Fig. 2. Structures of methacholine, carbachol and bethanechol

broken down by acetylcholinesterase but much more slowly than acetylcholine and therefore its effects are less fleeting than acetylcholine. Since acetylcholine receptors on eccrine sweat glands are muscarinic, administration of methacholine into skin, by injection or by iontophoresis, evokes profuse sweating. This effect has been reported to decline markedly on repeated administrations of methacholine (Thaysen et al. 1952).

b) Bethanechol

Chemically bethanechol differs from methacholine in the substitution of an amide moiety for the methyl group on the acyl carbon atom. In its pharmacological actions it resembles very closely methacholine, but its action is more prolonged because it is not broken down by cholinesterases.

c) Carbachol

Carbachol only differs from bethanechol in lacking the methyl group on the B-carbon atom. This slight alteration of chemical structure confers upon this compound the property of marked stimulant action on nicotinic receptors, while retaining resistance to hydrolysis by cholinesterases. In addition to its nicotinic actions, it also possesses very considerable muscarinic actions. When injected into skin, it produces a marked sweating response (Brun and Favre 1954).

2. Pilocarpine

Pilocarpine is a natural alkaloid from the leaves of a native South American plant, *Pilocarpus jaborandi*. Pilocarpine, like acetylcholine, possesses marked agonist action on muscarinic receptors but differs from acetylcholine in lacking any significant action on nicotinic receptors. It is a powerful stimulant of sweat secretion and acts on the sweat glands directly through the muscarinic receptors (RANDALL and KIMURA 1955). In the past, it had been used therapeutically as a diaphoretic in the belief that production of intense sweating would help in the removal of toxins from the body. JOHNSON and SPALDING (1974) have suggested the use of pilocarpine in abnormal sweating to determine the site of lesion. Since pilocarpine acts directly on the sweat glands, its injection will produce sweat if there is a lesion in the sympathetic nerve but not if the sweat glands themselves are nonfunctional. It should be borne in mind, however, that its potent actions at other muscarinic receptors and its long duration of action might produce many unpleasant side-effects. These include nausea, vomiting, abdominal cramps, marked salivation and lachrymation, miosis and ciliary spasm.

B. Anticholinergic Drugs

Drugs which act as antagonists against acetylcholine on muscarinic receptors have been called anticholinergic drugs. Other synonyms for this group of drugs are atropinic agents, muscarinic receptor antagonists, antimuscarinics, cholinergic blocking agents, spasmolytics, parasympatholytics etc. Anticholinergic drugs are extremely valuable therapeutic agents as well as immensely important tools for pharmacological and physiological research.

I. Chemistry

1. Belladonna Alkaloids and Their Derivatives

Belladonna alkaloids, atropine and hyoscine are present in several species of the Solanacea family and are formed by the combination of tropic acid with tropine,

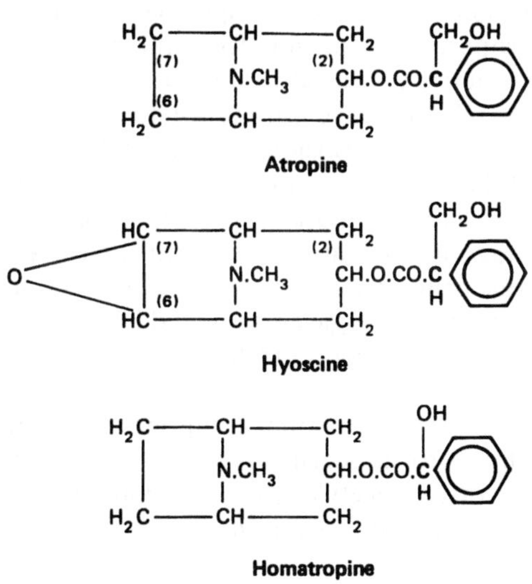

Fig. 3. Structures of atropine, hyoscine and homatropine

giving rise to atropine, and with oscine, forming hyoscine. Homatropine is formed by combining tropine with mandelic acid. Atropine methylnitrate, methylhyoscine hydrobromide and homatropine methylbromide are quaternary ammonium derivatives of atropine, hyoscine and homatropine respectively (Fig. 3).

2. Synthetic Anticholinergic Agents

a) Quaternary Ammonium Compounds

Several compounds possessing a quaternary ammonium moiety in their chemical structure are potent antagonists at muscarinic receptor sites. Examples of this class of drugs are mepenzolate bromide, methantheline bromide, oxyphenonium bromide, poldine methylsulphate, propantheline bromide and tridihexethyl chloride.

b) Other Synthetic Anticholinergic Agents

Other synthetic anticholinergic agents constitute a chemically diverse group. Examples of this group are provided by cyclopentolate hydrochloride, oxyphencyclimine hydrochloride, piperidolate hydrochloride and tropicamide.

II. Mechanism of Action

Anticholinergic drugs exert their pharmacological effect by competing with acetylcholine for attachment to muscarinic receptors. Whereas the combination of acetylcholine with the muscarinic receptor evokes the characteristic response of the host tissue cell, no such response is elicited by the anticholinergic drug–muscarinic receptor combination. Within the constraints of drug–receptor interaction, this phenomenon has been explained on the assumption that for successfully evoking a response, the drug must have affinity for binding with the receptor and additionally must possess another property, termed "intrinsic activity" by ARIENS (1954) or "efficacy" by STEPHENSON (1956). Accordingly acetylcholine, the agonist, is deemed to possess both properties, i.e. affinity for combining with the muscarinic receptor and the intrinsic activity, needed to initiate a response. Anticholinergic agents, according to this scheme, have affinity for the receptor but lack the intrinsic activity. Although the lack of intrinsic activity excludes them from functioning as agonists, by virtue of their affinity for the muscarinic receptor the anticholinergic agents can prevent acetylcholine–muscarinic receptor binding and can therefore successfully act as antagonists for acetylcholine at these sites. The reversible nature of the anticholinergic drug–muscarinic receptor interaction implies that the antagonism set up by the anticholinergic agent would be competitive and therefore can be overcome by raising the concentration of the agonist, acetylcholine.

Anticholinergic agents, having a quaternary ammonium component in their chemical configuration, possess, in addition to muscarinic receptor antagonist property, the ability to antagonise competitively acetylcholine at nicotinic sites. They can therefore interfere with the transmission at autonomic ganglia and neuromuscular junctions.

III. Pharmacological Actions

The pharmacological effects of anticholinergic agents are, in general, predictable from a knowledge of the distribution of muscarinic receptors in the body. However, the anticholinergic agents being pharmacological antagonists, the effects of their administration into any cholinergically innervated tissue or organ will depend upon the muscarinic receptor-mediated "cholinergic tone" of that tissue. For instance, administration of atropine induces tachycardia of variable magnitude in different individuals, by antagonising the action of neuronally released acetylcholine in the heart. The variability in the intensity of tachycardia in different individuals is a reflection of the individual variation in the degree of cholinergic vagal tone.

1. Cardiovascular System

The main effect of the anticholinergic agents is to increase the rate of the heart beat. There is little or no effect on arterial blood pressure, indicating the relative insignificance of the cholinergic control of the blood vessel tone. However, a fall in blood pressure can be induced by injecting cholinomimetics, and anticholinergic agents effectively antagonise this action of cholinomimetics.

2. Gastrointestinal Tract

The tone and motility of gut is reduced by anticholinergic agents.

3. Respiratory Tract

The smooth muscle of the respiratory tract is relaxed under the influence of anticholinergic drugs.

4. Eye

By blocking the muscarinic receptors located in the sphincter muscle of iris and ciliary muscle of the lens, the anticholinergic agents free these smooth muscles from their cholinergic nerve control and thus cause dilatation of the pupil and paralysis of accommodation.

5. Urinary Tract

Anticholinergic drugs, in general, relax the urinary tract smooth muscle and inhibit the contractions of ureter and urinary bladder.

6. Secretory Glands

The gastric, intestinal, bronchopharyngeal, salivary and lachrymal gland secretions are augmented by cholinergic agents. The anticholinergic drugs inhibit the secretions.

7. Central Nervous System

Muscarinic receptors for acetylcholine in the central nervous system are mostly located in the medulla and higher cerebral centres. Blockade of these receptors results in central excitation, the symptoms of which depend upon the severity of block and range from mild respiratory stimulation to hallucinations and delirium.

8. Skin

The studies on effects of anticholinergic agents on skin have been largely focussed on sweat glands. The inhibitory effect of anticholinergic agents on sweat secretion has been known for over 100 years. As early as 1878 OTT and WOOD FIELD had shown that atropine inhibited sweat secretion in cats. Shortly thereafter, TWEEDY and RINGER (1880) showed that injection of atropine and homatropine led to inhibition of sweating in man. Inhibition of sweating in man by iontophoretic application of atropine was reported by KUNO and KASHIWALSARA (1937). During the last 40 years several investigations have been carried out with the aim of finding an appropriate anticholinergic agent to treat the clinical condition of hyperhidrosis (excessive sweat secretion by eccrine glands). SHELLEY and HORWATH (1951) found the following order of effectiveness of anticholinergics in reducing sweating on iontophoretic application:

1. Hyoscine hydrobromide
2. Atropine methylnitrate
3. Atropine sulphate
4. Homatropine methylbromide
5. Hyoscine *N*-oxide hydrobromide

In a study carried out by MACMILLAN et al. (1964) on the antiperspirant action of topically applied anticholinergic agents, the authors concluded that the most effective compounds were esters of atropine and hyoscine. They attributed the effectiveness of these compounds to their high skin penetratability.

Propantheline spray to the sole of the foot was found to be an effective way of treating local hyperhidrosis by FRANKLAND and SEVILLE (1971). STOUGHTON et al. (1964) have found the topical use of hexapyronium bromide to produce satisfactory sweat suppression. GRICE and BETTLEY (1966) have used poldine methylsulphate as an antiperspirant and found that its topical application abolished sweating for 2–3 days. ABELL and MORGAN (1974) have produced effective local sweat inhibition by glycopyrronioum bromide iontophoresis.

Another type of sweat glands, the apocrine glands, have a limited distribution in man and are found chiefly in axilla. The secretion of apocrine glands is not under cholinergic control and anticholinergic agents have little or no effect on apocrine secretion (EVANS 1957).

Atropine, when administered in large doses, induces cutaneous vasodilatation. This is particularly noticeable in the cutaneous blood vessels of face (atropine flush). Since injection of acetylcholine dilates the blood vessels of the skin by its agonist action on muscarinic receptors, it is highly improbable that the anticholinergic action of atropine is the basis of "atropine flush". It is more likely that the anomalous vascular effect of atropine is due to a direct vasodilator action unrelated to its antagonism of acetylcholine.

IV. Adverse Effects

Undesirable effects even from topical use of anticholinergics in skin have frequently been reported, and usually appear as dryness of mouth, blurred vision, tachycardia and difficulty in micturition. The undesirable effects of anticholinergics are really an extension of their normal pharmacological actions and are attributable to their inability to distinguish between muscarinic receptors located in different sites. Lack of selectivity renders it very difficult to evoke the desired therapeutic effect of the anticholinergic therapy and yet avoid the accompanying adverse effects. By enhancing the concentration gradient of the anticholinergic drug between the local site and the systemic sites, topical application eases this difficulty but does not completely eliminate it.

V. Other Classes of Drug Possessing Pronounced Anticholinergic Activity

Several drugs well-known for their characteristic pharmacological properties and not normally listed as anticholinergic agents are remarkably potent as muscarinic receptor antagonists.

1. Histamine H_1-Receptor Antagonists

Most H_1-receptor antagonists possess anticholinergic activity. Ethanolamine derivatives (e.g. diphenhydramine) are probably most potent in this respect. Some of the side-effects of antihistamines, such as dryness of mouth, difficulty in micturition and reduced bronchopharyngeal secretions, are attributable to their anticholinergic activity.

2. Tricyclic Antidepressants

All clinically useful tricyclic antidepressants have a very marked anticholinergic effect, and many of their undesirable effects (dryness of mouth, blurred vision, constipation and urinary retention) are caused by blockade of muscarinic receptors in autonomically innervated organs.

3. Haloalkylamine α-Adrenoceptor Blocking Agents

Haloalkylamines possess anticholinergic activity, in addition to their other pharmacological properties such as α-adrenoceptor blockade, H_1-antagonism, 5-HT antagonism and blockade of neuronal re-uptake of noradrenaline. Phenoxybenzamine is probably the most commonly known prototype of this class of chemical compounds.

References

Abell E, Morgan K (1974) The treatment of idiopathic hyperhidrosis by glycopyrronium bromide and tap water iontophoresis. Br J Dermatol 91:87–91

Ariens EJ (1954) Affinity and intrinsic activity in the theory of competitive inhibition. Arch Int Pharmacodyn Ther 99:32–49

Birks R, MacIntosh FC (1961) Acetylcholine metabolism of a sympathetic ganglion. Can J Biochem Physiol 39:787–827
Briscoe S, Burn JH (1954) The formation of an acetylcholine-like substance by the isolated rabbit heart. J Physiol (Lond) 126:181–190
Brun R, Favre F (1954) Experiences sur la transpiration. 7. Examens par certaines substances. Dermatologia 108:257–270
Burn JH, Kottegoda SR (1953) Action of eserine on the auricles of the rabbit heart. J Physiol (Lond) 121:360–373
Coon JM, Rothman S (1939) Nature of the sweat response to drugs with nicotine-like action. Proc Soc Exp Biol Med 42:231–233
Coon JM, Rothman S (1941) The sweat response to drugs with nicotine-like action. J Pharmacol Exp Ther 73:1–11
Cotton DWK, Van den Hurk JJMA, Mier PD (1973) Hydrolysis of acetylcholine by soluble and particulate preparations of skin from normal and atopic subjects. Br J Dermatol 88:579–582
Dale HH (1914) The action of certain esters and ethers of choline and their relationship to muscarine. J Pharmacol Exp Ther 6:147–190
Dale HH, Feldberg W (1934) The chemical transmission of secretory impulses to the sweat glands of the cat. J Physiol (Lond) 82:121–128
Evans CL (1957) Sweating and sympathetic innervation. Br Med Bull 13:197–201
Feldberg W, Lin RCY (1950) Synthesis of acetylcholine in the wall of the digestive tract. J Physiol (Lond) 111:96–118
Frankland JC, Seville RH (1971) The treatment of hyperhidrosis with topical propantheline – a new technique. Br J Dermatol 85:577–581
Gibson TE, Shelley WB (1948) Sexual and racial difference in the response of sweat glands to acetylcholine and pilocarpine. J Invest Dermatol 11:137–142
Grice KA, Bettley FR (1966) Inhibition of sweating by poldine methosulphate (Nacton). Its use for measuring insensible perspiration. Br J Dermatol 78:458–464
Harbison RD, Olubadewo JO, Dwivedi C, Sastry BVR (1975) Proposed role of the placental cholinergic system in the regulation of fetal growth and development. In: Morselli PL, Garattini S, Sereni F (eds) Basic and therapeutic aspects of perinatal pharmacology. Raven, New York, pp 107–120
Herxheimer A (1956) The nervous pathway mediating cholinergic urticaria. Clin Sci 15:195–205
Hurley HJ, Shelley WB, Koelle GB (1953) The distribution of cholinesterases in human skin with special reference to eccrine and apocrine sweat glands. J Invest Dermatol 21:139–147
Janowitz HD, Grossman MI (1950) The response of sweat glands to some locally acting agents in human subjects. J Invest Dermatol 13:453–458
Johnson RH, Spalding JMK (1974) Disorders of autonomic nervous system. Blackwell, Oxford, pp 179–198
Kahn D, Rothman S (1942) Sweat response to acetylcholine. J Invest Dermatol 5:431–444
Kuno Y, Kashiwabara K (1937) An effective method for suppression of local sweating. Chin J Physiol 11:41–45
Lloyd DPC (1965) Cholinergy and adrenergy in the neural control of sweat glands. In: Curtis DR, McIntyre AK (eds) Studies in physiology. Springer, Berlin Heidelberg New York, pp 169–178
Loewi O (1921) Über humorale Übertragbarkeit der Herznervenwirkung. Pflügers Arch Ges Physiol 189:239–242
Loewi O, Navratil E (1926) Über humorale Übertragbarkeit der Herznervenwirkung. X. Mitteilung über das Schicksal des Vagusstoff. Pflügers Arch Ges Physiol 214:688
Longmore J, Bradshaw CM, Szabadi E (1985) Effects of locally and systemically administered cholinoceptor antagonists on the secretory response of human eccrine sweat glands to carbachol. Br J Clin Pharmacol 20:1–8
MacMillan FSK, Reller HH, Synder FH (1964) The antiperspirant action of topically applied anticholinergics. J Invest Dermatol 43:363–377

Magnus IA, Thompson RHS (1954) Cholinesterase activity of human skin. Br J Dermatol 66:163–174
Maple S, Bradshaw CM, Szabadi E (1982) Pharmacological responsiveness of sweat glands in anxious patients and healthy volunteers. Br J Psychiatry 141:154–161
McMahan D (1972) Chemical messengers in development: a hypothesis. Science 185:1012–1021
Megay KV (1935) Versuche an biologischen Testobjekten über die Natur des im Schweiß vorhandenen vagometischen Stoffes. Pflügers Arch Ges Physiol 236:159–165
Michaelson MJ (1974) Some aspects of evolutionary pharmacology. Biochem Pharmacol 23:2211–2224
Mier PD, Cotton DWK (1976) The molecular biology of skin, 1st edn. Blackwell, Oxford
Ott I, Wood Field GBW (1878) Sweat centres: the effect of muscarin and atropin on them. J Physiol (Lond) 1:193–195
Paton WDM, Zar MA (1968) The origin of acetylcholine released from guinea-pig intestine and longitudinal muscle strips. J Physiol (Lond) 194:13–33
Randall WC, Kimura KK (1955) The pharmacology of sweating. Pharmacol Rev 7:365–397
Rothman S, Coon JM (1940) Axon reflex response to acetylcholine in the skin. J Invest Dermatol 3:79–97
Ryan TJ (1973) Inflammation, fibrin and fibrinolysis. In: Jarret A (ed) The physiology and pathophysiology of the skin, vol 2. Academic, London, pp 745–777
Sastry BVR, Sadavongvivad C (1978) Cholinergic systems in non-nervous tissues. Pharmacol Rev 30:65–132
Sato K (1977) The physiology, pharmacology and biochemistry of the eccrine sweat glands. Rev Physiol Biochem Pharmacol 79:51–131
Scott A (1962) Acetylcholine in normal and diseased skin. Br J Dermatol 74:317–322
Shelley WB, Horvath PN (1951) Comparative study on the effect of anticholinergic compounds on sweating. J Invest Dermatol 16:267–274
Silver A, Versau A, Montagna W (1963) Studies of sweating and sensory function in cases of peripheral nerve injuries of the hand. J Invest Dermatol 40:243–258
Singer M (1959) The acetylcholine content of the normal forelimb regenerate of the adult newt, Triturus. Dev Biol 1:603–620
Singer M (1974) Neurotropic control of limb regeneration in the newt. Ann NY Acad Sci 228:308–321
Stephenson RP (1956) A modification of receptor theory. Br J Pharmacol Chemother 11:379–393
Stoughton RB, Chiu F, Fritsch W, Nurse D (1964) Topical suppression of eccrine sweat delivery with a new anticholinergic. J Invest Dermatol 42:151–155
Thayson JH, Schwartz IL, Dole VP (1952) Fatigue of sweat glands. Fed Proc 11:161–162
Tweedy J, Ringer S (1880) On the mydriatic properties of homatropin or oxytoluyltropein. With an account of its general physiological action. Lancet I:795–796
Van den Broek MD, Bradshaw CM, Szabadi E (1984) The effects of psychological "stressor" and raised ambient temperature on the pharmacological responsiveness of human eccrine sweat glands: implications for sweat gland hyper-responsiveness in anxiety states. Eur J Clin Pharmacol 26:209–213
Wada M, Arai T, Takagaki T, Nakagawa T (1952) Axon reflex mechanism in sweat responses to nicotine, acetylcholine and sodium chloride. J Appl Physiol 4:745–752
Warndorff JA (1970) The response of the sweat gland to acetylcholine in atopic subjects. Br J Dermatol 83:306–311

CHAPTER 21

Prostaglandins, Leukotrienes, Related Compounds and Their Inhibitors

S. D. BRAIN and T. J. WILLIAMS

A. Discovery of the Prostaglandins

KURZROK and LIEB (1930) found that fresh human semen induced either strong relaxation or strong contraction of the human uterus. Several years later, two independent workers, GOLDBLATT (1935) and VON EULER (1936), reported that seminal fluid contained smooth muscle-contracting and vasodepressor activity. VON EULER named the activity, which appeared to be lipid in nature, "prostaglandin". He discovered that it could be extracted from seminal fluid of several species and from ram vesicular gland (VON EULER 1936). The isolation of pure crystalline prostaglandins, from vesicular glands, was achieved in 1957 (BERGSTRÖM and SJÖVALL) and was followed by the first structure elucidations (BERSTRÖM et al. 1962). It is now known that the prostaglandins constitute a family of compounds of unique structure. These compounds are widely distributed in nature and have potent biological activities.

B. Biosynthesis of Prostaglandins

The structures of the prostaglandins suggested that they were derived from 20 carbon atom essential fatty acids, and it was discovered that incubation of these fatty acids with homogenised ram vesicular glands resulted in prostaglandin generation (VAN DORP et al. 1964; BERGSTRÖM et al. 1964). The enzyme system responsible was found to be present in the microsomal fraction of the homogenate. Arachidonic acid (eicosa-5,8,11,14-tetraenoic acid) gives rise to the 2-series prostaglandins. As arachidonic acid is the predominant essential fatty acid in animal tissues, the products of this acid have received more attention, although eicosa-8,11,14-trienoic and eicosa-5,8,11,14,17-pentaenoic acids also give rise to highly active prostaglandins of the 1- and 3-series respectively.

The naturally occurring prostaglandins are all 20 carbon atom fatty acids having a cyclo-pentane ring at C8–C12. Differences in the structure of this ring give rise to the sub-groups A, B, C, D, E, F, G, H, and I. The number of double bonds in the side chains determines the 1-, 2-, and 3-series prostaglandins denoted by the appropriate subscript.

Prostaglandin synthesis can be initiated by a wide range of stimuli which activate phospholipases. These enzymes liberate arachidonic acid from membrane phospholipids. Free arachidonic acid can then be acted on by an enzyme termed cyclo-oxygenase, as shown in Fig. 1. The first reaction in the cyclo-oxygenase

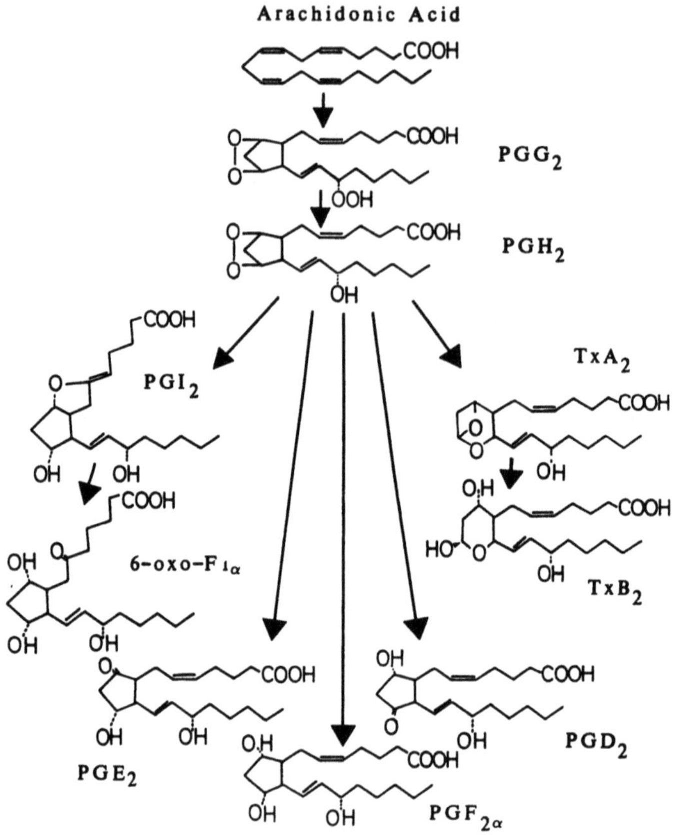

Fig. 1. Cyclo-oxygenase pathway

pathway is an oxygenation of C-11 of arachidonic acid to form the 15-hydroperoxy endoperoxide, PGG_2. PGG_2 contains the characteristic cyclopentane ring. The 15-hydroxy endoperoxide, PGH_2, is quickly generated from PGG_2 as the cyclo-oxygenase enzyme also possesses peroxidase activity (ANGGARD and SAMUELSSON 1964). The endoperoxides do have biological activities but many biological effects seen after arachidonic acid metabolism are better explained by further conversion of endoperoxides to other products. The endoperoxides are converted enzymatically, or in some cases spontaneously, to relatively stable products of which the most important are PGE_2, $PGF_{2\alpha}$, and PGD_2. More recently it has been discovered that two chemically unstable but highly biologically active products can also be synthesised from the endoperoxides. One of these products is thromboxane A_2 (TxA_2) (HAMBERG et al. 1975), which is not a prostaglandin by definition as it has an oxane ring at C8–C12. TxA_2 hydrolyses spontaneously to the relatively inert thromboxane B_2. The other unstable product is PGI_2 (MONCADA et al. 1976), which hydrolyses to the relatively inactive 6-oxo-$PGF_{1\alpha}$.

Although prostaglandins were discovered because of their presence and actions in animal reproductive organs, most tissues of the body, including the skin (JONSSON and ANGAARD 1972), are capable of synthesising prostaglandins. Further, prostaglandins have actions on many different tissues and there is evidence for their production and action in certain pathological conditions.

C. Action of Prostaglandins in Skin

I. Vasodilatation

PGE_1 and PGE_2 were found to be potent vasodepressor substances upon intravenous injection in animals owing to a lowering of peripheral resistance (for review see MALIK and MCGIFF 1976). In man, intravenous infusions of PGE_1 into the forearm were observed to induce erythema (BERGSTRÖM et al. 1959). Intradermal injection of PGE_1, PGE_2 and PGD_2 induces intense local erythema in man (SOLOMON et al. 1968; JUHLIN and MICHAELSSON 1969; SONDERGAARD and GREAVES 1971; FLOWER et al. 1976); however, $PGF_{1\alpha}$ and $PGF_{2\alpha}$ are relatively inactive (JUHLIN and MICHAELSSON 1969). At very high doses (1–5 μg) single intradermal injections of PGE_1 were reported to induce erythema persisting for 4–10 h. Lower doses, nearer to concentrations which might be expected in pathophysiological conditions, gave responses lasting approximately 1.5 h (CRUNKHORN and WILLIS 1971).

Prostacyclin, PGI_2 (JOHNSON et al. 1976), was discovered because of its relaxant effect on rabbit isolated coeliac and mesenteric arteries in vitro. It was subsequently found to be a potent inhibitor of platelet aggregation (MONCADA et al. 1976) and was also shown to dilate small vessels (arterioles) in the hamster cheek pouch (HIGGS et al. 1978b). In rabbit skin, using a radioactive xenon clearance technique to measure changes in blood flow, intradermally injected PGI_2 was found to induce vasodilatation with a potency comparable to PGE_2, whereas PGD_2, $PGF_{2\alpha}$ and 6-oxo-$PGF_{1\alpha}$ (a stable metabolite of PGI_2) exhibited very little activity (WILLIAMS 1979). Prostacyclin has also been observed to induce erythema when injected intradermally in man (HIGGS et al. 1979).

Because of its instability in aqueous solution, there is no direct evidence concerning the action of TxA_2 in skin. However, the actions of TxA_2 on vascular tissues in vitro (NEEDLEMAN et al. 1976) and in the mesenteric vascular bed in vivo (DUSTING et al. 1978) demonstrate strong vasoconstrictor activity.

The prostaglandin endoperoxides are also unstable, but in rabbit skin PGG_2 has been shown to induce transient vasoconstriction followed by vasodilatation, the second phase presumably being caused by a conversion to either PGE_2 or PGI_2 in the skin (LEWIS et al. 1977).

II. Oedema Formation

Quantitative measurement of oedema formation in animal skin is more easily achieved than measurement of changes in blood flow. On the other hand, measurement of oedema formation in human skin is, at best, semi-quantitative. For this reason several studies have been carried out on the actions of prostaglandins in animal skin, especially concerning the role of prostaglandins in oedema formation, but it is difficult to correlate these with changes in man.

In early studies, PGE_1 and PGE_2 were shown to cause protein leakage in rat skin (Kaley and Weiner 1968; Crunkhorn and Willis 1969, 1971) which could be suppressed by depletion of mast cell amines or by a mixture of antagonists to histamine and serotonin. Thus, these prostaglandins were thought to increase vascular permeability by releasing mast cell amines (Crunkhorn and Willis 1971). In guinea-pig skin, PGE_1 was demonstrated to cause protein leakage and was reported to be 10 times less potent than bradykinin as indicated by the diameters of extravasated protein-bound dye (Horton 1963). In a later study it was shown, by measuring the extravasation of ^{125}I-albumin, that prostaglandins were very weak in inducing protein leakage, PGE_1 being 1000–10 000 times less potent than bradykinin using this technique (Williams and Morley 1973).

The predominant effect of intradermal injections of E-type prostaglandins in man is erythema; however, whether these substances are also important in increasing microvascular permeability is open to question. In some studies weal formation in response to E-type prostaglandins has been reported whilst in others none was observed (Juhlin and Michaelsson 1969). Some workers have considered that prostaglandins are capable of increasing permeability indirectly by release of mast cell histamine in human skin (Sondergaard and Greaves 1971; Crunkhorn and Willis 1971). This indirect effect may vary between individuals (Crunkhorn and Willis 1971), which may explain differences in observations between different groups of workers.

The question of protein leakage induced by prostaglandins in human skin remains unresolved. However, a striking phenomenon which has come to light in animal skin experiments is synergism between prostaglandins and other mediators to induce oedema. In guinea-pig skin (Williams and Morley 1973; Thomas and West 1973) and rat paw (Moncada et al. 1973; Lewis et al. 1975), oedema responses to locally injected histamine and bradykinin were shown to be markedly enhanced by the addition of E-type prostaglandins. More recently, it has been reported that this phenomenon can be demonstrated in human skin (Basran et al. 1982). It has been suggested that oedema-potentiating activity is related to the vasodilator activity of prostaglandins (Williams and Peck 1977; Kopaniak et al. 1978). Using ^{133}Xe clearance to measure blood flow changes and ^{131}I-albumin extravasation to measure oedema formation in rabbit skin, it has been shown that the most powerful vasodilators, the E-type prostaglandins, are the most potent potentiators of bradykinin-induced oedema, whereas the weak dilators $PGF_{2\alpha}$ and PGD_2 are also weak oedema potentiators (Williams and Peck 1977; Williams 1979). The unstable potent vasodilator, PGI_2, has also been shown to be a potent oedema potentiator in rabbit skin (Peck and Williams 1978; Williams 1979) and rat paw (Komoriya et al. 1978), inducing detectable potentiation in doses as low as 1 ng in the rabbit (Williams 1979). Vasodilators other than prostaglandins can also potentiate oedema, e.g. adenosine (Williams and Peck 1977) and vasoactive intestinal polypeptide (Williams 1982).

These findings using exogenous agents have been extended by observations that vasodilator prostaglandins and the complement fragment C5a are generated together in rabbit skin in response to intradermally injected *Bordetella pertussis* organisms, boiled yeast cells or insoluble antibody–antigen complexes (Williams and Jose 1981). These two endogenous mediators co-operate to induce the observed oedema response. The mechanisms of the co-operation have been interpreted in the following manner. C5a induces increased venular permeability, which alone induces little oedema in a tissue with relatively low blood flow, such as skin. When a prostaglandin is present this causes arteriolar dilatation and consequently increased flow to venules, increased intravenular hydrostatic pressure and passive venular distension. Under these conditions leakage of plasma proteins is facilitated. It is possible that, in addition to the above mechanism where the mediators act remotely, synergism can also occur between prostaglandins and other mediators with both acting on the venule itself. Recent experiments demonstrating synergisms in perfused constant flow systems suggest that this can occur (AMELANG et al. 1981).

Human C5a has also been shown to induce oedema in rabbit skin when combined with a vasodilator prostaglandin (JOSE et al. 1981). Removal of carboxyl terminal arginine to produce C5a-des-Arg abolished histamine-releasing activity but had relatively little effect on oedema responses. This dissociated the permeability-increasing activity of C5a from effects on mast cells and it has been found that C5a (and leukotriene B_4, as discussed later in the text) increases permeability by triggering a rapid interaction between circulation polymorphonuclear leucocytes and venular endothelial cells (WEDMORE and WILLIAMS 1981).

III. Leucocyte Accumulation

It was reported that PGE_1 was chemotactic for rabbit polymorphonuclear leucocytes in an in vitro system, the Boyden chamber (KALEY and WEINER 1968). This was supported by subsequent observations on rabbit cells (HIGGS et al. 1975) but not by workers using guinea-pig, rat or human leucocytes (TURNER et al. 1975 a, b; FORD-HUTCHINSON et al. 1976). Alternatively, it has been proposed that thromboxane B_2 is chemotactic (BOOT et al. 1976) but the importance of this in vivo has not been ascertained. Other workers have produced evidence that prostaglandins can modulate the effects of chemotactic agents (O'FLAHERTY et al. 1979; CUNNINGHAM et al. 1981). Although the modulating effect may be of importance in vivo (FANTONE et al. 1983), it seems unlikely in retrospect that cyclo-oxygenase products of arachidonic acid are important chemotactic agents in their own right.

IV. Pain

In the early studies on the responses in human skin to injected PGE_1, it was observed that, while no overt pain was induced, there was a local "tenderness to pressure" (SOLOMON et al. 1968) or "hyperalgesia" (JUHLIN and MICHAELSSON 1969). The significance of this phenomenon was discovered subsequently (FERREIRA 1972) when it was found that perfusion of an exposed human skin blister base with PGE_1 was able to cause a striking enhancement of the pain response to histamine and bradykinin. More recent experiments using rats have shown that PGI_2 is also able to induce hyperalgesia (HIGGS et al. 1978 a; FERREIRA et al. 1978).

D. Generation of Prostaglandins in Skin

As prostaglandins are able to induce or modulate inflammatory changes in skin, many attempts have been made to evaluate the role of these substances in different inflammatory conditions. Based on investigations using a variety of other tissues, the proposal is that the inflammatory stimulus activates phospholipase in the skin, resulting in the liberation of arachidonic acid. Arachidonic acid is then converted to prostaglandin endoperoxides by cyclo-oxygenase and the endoperoxides are then converted to prostaglandins. In animal experiments prostaglandins and substances involved in their generation have been injected intradermally and changes in the microvasculature monitored using radioisotopic techniques.

Intradermal injection of phospholipase A_2 in the rabbit induces vasodilatation (VADAS et al. 1981) and these authors have suggested that inflammatory stimuli can cause the release of this enzyme from leucocytes. The enzyme can then release arachidonic acid from membranes of surrounding cells and arachidonate metabolites are responsible for the observed vasodilatation. In support of this idea, it has been reported that phospholipase A_2 can be detected in the lymph during tuberculin reactions in sheep (VADAS and HAY 1982). This is a mechanism additional to the more conventional idea of activation of phospholipase and release of arachidonic acid from the same cell. Intradermal injection of arachidonic acid into the rabbit (WILLIAMS 1979; VADAS et al. 1981) and guinea-pig (PECK et al. 1981) also induces vasodilatation, which can be abolished by inhibitors of prostaglandin synthesis (see Sect. J).

One of the unstable endoperoxides, PGG_2, was also shown to induce vasodilatation (LEWIS et al 1977). As stated previously, PGI_2 and PGE_2 are potent vasodilators: one or both of these are the likely products following intradermal injection of phospholipase A_2, arachidonic acid or prostaglandin endoperoxides. These results, taken together, suggest that the level of free arachidonic acid liberated by phospholipase is the limiting factor in prostaglandin-mediated vasodilatation and that cyclo-oxygenase is normally active within the skin, awaiting substrate. Some support for this is given by experiments in which homogenates of both normal and psoriatic human skin (HAMMERSTRÖM et al. 1979) or guinea-pig skin (RUZICKA and PRINTZ 1982) have been found to convert added arachidonic acid to prostaglandins. On the other hand, analysis of keratone slices taken from patients with psoriasis has revealed that although arachidonic acid levels are much higher in lesional skin than uninvolved skin, there is only a small increase in PGE_2 and $PGF_{2\alpha}$ levels (HAMMERSTRÖM et al. 1975).

What stimuli activate phospholipase in the skin? In the rabbit there is evidence that intradermal injection of killed *Bordetella pertussis* organisms or boiled yeast cells leads to indomethacin-sensitive vasodilatation which is due to prostaglandin generation (WILLIAMS and PECK 1977). In the Arthus reaction in the same species there is also evidence that prostaglandins mediate the observed vasodilatation (CRAWFORD et al. 1982).

Using perfusion through needles implanted in human skin, evidence for prostaglandin generation in allergic contact eczema and primary irritant dermatitis has been obtained (GREAVES et al. 1971; SONDERGAARD et al. 1974). Perfusate samples were separated by lipid extraction and thin layer chromatography and assayed for prostaglandins using isolated rat uterus and rat duodenum preparations. E- and F-type prostaglandins were detected by this technique but there was no evidence for increased prostaglandin production in two acute rections, urticaria factitia and cutaneous anaphylaxis (SONDERGAARD et al. 1974). Prostaglandin-like activity has also been detected in human burn blister fluid (ANGAARD and JONSSON et al. 1971).

In later experiments exudate was collected from suction bullae raised on human skin and assayed for prostaglandins using combined gas–liquid chromatography–mass spectrometry. Using this technique elevated levels of arachidonic acid and prostaglandins were detected following ultraviolet B and C irradiation (BLACK et al. 1978; KOBZA BLACK et al. 1978), in support of earlier experiments using skin perfusion and bioassay (SONDERGAARD and GREAVES 1970). Elevation of levels of arachidonic acid, PGE_2 and PGD_2 have been detected with a peak at 18–24 h after irradiation, whereas 6-oxo-$PGF_{1\alpha}$ (a stable metabolite of the unstable PGI_2) peaked earlier at 6 h (BLACK et al. 1978, 1980). However, prostaglandin levels were shown to be low at 48 h whilst erythema persisted (BLACK et al. 1980). This suggests that other vasodilator mediators are present at the later stages, or possibly that local arteriolar tone has been lowered by neural systems or local intrinsic effects on vascular smooth muscle.

X-ray irradiation has also been shown to induce the release of arachidonic acid, which is then converted to prostaglandins (ZIBOH et al. 1982). In these experiments skin samples were removed from pigs at different times following irradiation. Samples were found to have elevated levels of arachidonic acid and PGE_2. Interestingly, there was evidence of an increased conversion of PGE_2 to $PGF_{2\alpha}$ (which is less active as a vasodilator in most systems) with increasing time, owing to increased PGE_2-9-oxo reductase activity.

In addition to prostaglandin generation initiated by local stimulation of the skin, there is evidence that generalised vasodilatation can involve prostaglandins. Spontaneous facial and upper trunk flushing in postmenopausal women, flushing induced by alcohol consumption or by alcohol/propramide combination, and flushing induced experimentally by intravenous injection of nicotinic acid have all been shown to be sensitive to prostaglandin synthesis inhibitors (STRAKOSCH et al. 1980; BARNETT et al. 1980; PHILLIPS and LIGHTMAN 1981).

From these observations it would appear that prostaglandin generation in the skin is a component of many different types of inflammatory reaction. The relative importance of prostaglandins for their direct and co-operative effects compared with other mediators undoubtedly varies with the site and type of inflammatory reaction.

E. Discovery of the Lipoxygenase Pathway

A pathway, in addition to the cyclo-oxygenase pathway, for the metabolism of arachidonic acid was first described by HAMBERG and SAMUELSSON (1974) and NUGTEREN (1975). Their work revealed that platelets contain a soluble lipoxygenase which catalyses the hydroperoxidation of certain polyunsaturated acids. Arachidonic acid was found to be converted first into a hydroperoxy intermedi-

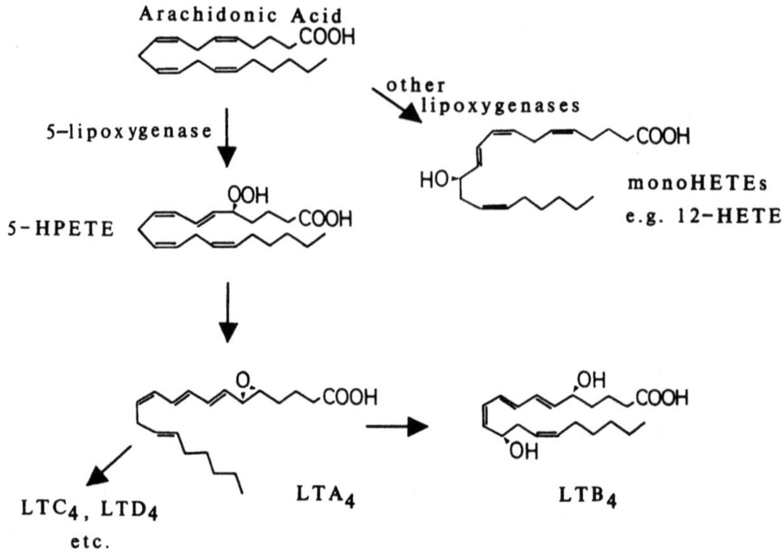

Fig. 2. Lipoxygenase pathway

ate, 12-L-hydroperoxy-5,8,10,14-eicosatetraenoic acid (12-HPETE), which was then rapidly reduced to the corresponding hydroxy acid, 12-L-hydroxy-5,8,10,14-eicosatetraenoic acid (12-HETE). It has since been discovered that other lipoxygenases can attack the arachidonic acid molecule at other positions, leading to the formation of several different monohydroxy acid products (5-HETE, 8-HETE, 11-HETE, and 15-HETE) (Fig. 2).

F. The Leukotrienes

Dr. Wasserman describes in Chap. 22 the 5-lipoxygenase pathway that leads to the production of LTC_4, LTD_4, and LTE_4, which are now considered to be the major active components of SRS-A. Additionally, as shown in Fig. 2, another leukotriene, LTB_4 [5(*S*),12(*R*)-dihydroxy-6,14-*cis*-8,10-*trans*-eicosatetraenoic acid], has been characterised (Borgeat and Samuelsson 1979a, bc; Sirois et al. 1981). LTB_4, like LTC_4, LTD_4, and LTE_4, is derived from arachidonic acid via oxygenation at C-5 to form initially 5-HPETE, which is rapidly metabolised to an unstable epoxide intermediate, LTA_4 (5,6-oxido-7,9,11,14-eicosatetraenoic acid). LTA_4 is then hydrolysed enzymically to form the dihydroxy acid, LTB_4. The polymorphonuclear leucocyte (PMN) is a major source of LTB_4. Borgeat and Samuelsson (1979a–c) have described the generation of several dihydroxy acids by PMNs which are either 5,12- or 5,6-dihydroxy acid metabolites of arachidonic acid. This is because non-enzymatic hydrolysis of the unstable LTA_4 leads to the formation of two 5,12 diastereoisomers of LTB_4, as well as two diastereoisomers of 5,6-dihydroxy-7,9,11,14-eicosatetraenoic acid. These compounds have been found to be of considerably less biological importance than LTB_4 (Ford-Hutchinson et al. 1981).

G. Action of Lipoxygenase Products in Skin

The more recently characterised leukotrienes have been found to be considerably more potent and to exhibit a much greater range of biological activities than the monohydroxy acid products. The biological properties of the leukotrienes are detailed more fully elsewhere (PIPER 1983; WILLIAMS 1983; BRAY 1983). The effects of lipoxygenase products in skin are described below.

I. Vasoactive Effects

Pure SRS-A (LTD_4 in the system used) obtained from sensitised guinea-pig lung perfused with antigen was found to have vasoconstrictor activity in guinea-pig skin, as measured using a ^{133}Xe clearance technique (WILLIAMS and PIPER 1980). Synthetic LTC_4 was also found to exhibit vasoconstrictor properties when injected into guinea-pig skin (DRAZEN et al. 1980) as observed using exclusion of intravenous Evans blue dye. A comparison of synthetic LTC_4 and LTD_4 was made by PECK et al. (1981) using ^{133}Xe clearance. They found that LTC_4 was not only more potent (1.6×10^{-11} mol/site caused a greater than 60% decrease in blood flow) but also had a steeper dose-response curve than LTD_4. This vasoconstrictor activity has now been observed in several different tissues. In man, BISGAARD and colleagues (1982), using a laser Doppler technique, have found that intradermal injection of LTC_4 and LTD_4 at higher doses (1 μg each) caused a rapid increase in cutaneous blood flow to levels 10 times higher than basal values. The increased blood flow was maintained for at least 20 min after injection and was similar to that due to histamine. In these experiments blood flow was measured at a spot at least 5 mm from the injection site to avoid injection artefacts. Other workers have suggested that a central pallor was seen at the site of injection of LTC_4 which could be due to a vasoconstriction (SOTER et al. 1983), but this view is not universally accepted (CAMP et al. 1983a). LTB_4 does not appear directly to affect blood flow in man, and an absence of effect on blood flow in rabbit skin has been observed (BRAY et al. 1981a).

II. Leucocyte Accumulation

12-HETE was found to be a mediator of neutrophil and eiosinophil chemotaxis soon after the discovery of the 12-lipoxygenase pathway in platelets (TURNER et al. 1975b; GOETZL et al. 1977). Since this finding it has been shown that other mono-HETE compounds have chemotactic and chemokinetic as well as other effects on leucocytes in vitro (GOETZL and SUN 1979; PALMER et al. 1980; O'FLAHERTY et al. 1981). The hydroperoxy compounds 5-HPETE, 11-HPETE, 12-HPETE and 15-HPETE were found to be more potent in affecting PMN movement than their respective mono-HETEs (PALMER et al. 1980). However, neither 5-HPETE, 5-HETE, 12-HPETE or 12-HETE significantly elevated leucocyte accumulation when injected intradermally into rabbit skin (HIGGS et al. 1981). The accumulation of leucocytes in human skin in response to a mono-HETE (12-HETE) has been observed, but only upon intradermal infusion of large quantities (DOWD et al. 1984).

More recently it has been found that LTB_4 is considerably more potent as a modulator of leucocyte function than the mono-HETE compounds. Indeed, it has been found to be one of the most potent chemotactic and chemokinetic compounds known for leucocytes in vitro (Ford-Hutchinson et al. 1980; Goetzl and Pickett 1980), with a potency similar to that of the fragment of the fifth component of complement, C5a (Bray et al. 1981 b).

A local infiltration of leucocytes, rich in neutrophils, has been observed after intradermal injection of LTB_4 into rabbit (Higgs et al. 1981), or human skin (Soter et al. 1983; Camp et al. 1983 a). LTB_4 has also been found to stimulate neutrophil accumulation into fluid-filled chambers placed over abraded skin in rabbit and man (Bray et al. 1981), and into human epidermis after topical application to human skin (Camp et al. 1984). The topical application of LTB_4 led to a response that was relatively delayed in onset but longer lasting. Histological analysis of biopsies taken 6–72 h after topical application of LTB_4 revealed a pronounced PMN infiltrate which was first observed 12 h after LTB_4 administration and was maximal at 24 h. At 24 h numerous vacuoles filled with neutrophils were apparent in the epidermis. LTC_4, LTD_4, and LTE_4 do not appear to affect leucocyte movement in vitro or accumulation in vivo.

III. Oedema Formation

Lipoxygenase products have been tested using rabbit, guinea-pig or rat and measuring the local accumulation of intravenously injected ^{125}I- or ^{131}I-albumin, as discussed previously. The monohydroxy and hydroperoxy acids tested have been found to be relatively inactive except upon injection of very high doses (10–20 μg), Higgs et al. (1981).

The effect of the leukotrienes on oedema formation is less straightforward. They have been found to be potent in increasing vascular permeability but the actions of LTC_4 and LTD_4 (as will be discussed first) are affected by their vasoconstrictor properties in certain species, whereas the action of LTB_4 (as will be discussed later) is related to its ability as a chemotactic agent.

Crude samples of SRS-A were found to cause extravasation of Evans blue dye in guinea-pig skin (Brocklehurst 1967). Pure SRS-A (LTD_4) was found to retain this activity as measured using ^{131}I-albumin (Williams and Piper 1980). Synthetic LTC_4 was not thought to affect vascular permeability in guinea-pig (Drazen et al. 1980) but it was found that the vasoconstrictor properties of LTC_4 were masking its ability to act as a mediator of increased vascular permeability. The injection of LTC_4 with a vasodilator prostaglandin does lead to protein exudation (Peck et al. 1981). However, synthetic LTD_4, which is a less potent vasoconstrictor agent, has been found to have more obvious activity in increasing vascular permeability in guinea-pig skin. The increased vascular permeability induced by LTD_4 when mixed with PGE_2 has been found to be inhibited by local injections of the SRS-antagonist FPL 55712, but not by the anti-histamine mepyramine (Peck et al. 1981). This suggests that the response in skin to LTD_4 is not dependent on a release of endogenous histamine. In human skin, LTC_4, LTD_4 as well as LTE_4 appear to cause weal and flare responses in an equipotent manner at low concentrations (SOTER et al. 1983; CAMP et al. 1983a). The responses are rapid in onset and can be observed up to 4 h after injection.

LTB_4 was found to induce increased vascular permeability in the rabbit skin (BRAY et al. 1981a; HIGGS et al. 1981; WEDMORE and WILLIAMS 1981). The oedema responses were found to be dependent not only on the presence of a vasodilator prostaglandin (as discussed previously) but also on the presence of PMNs (WEDMORE and WILLIAMS 1981). The dependence on the presence of PMNs was demonstrated by the failure of LTB_4, but not of histamine or bradykinin, to induce oedema in the skin of rabbits depleted of circulating PMNs. This increase in permeability is thought to be initiated by a rapid interaction of PMNs with the vascular endothelium and a similar mechanism of action is postulated for other chemotactic agents which also increase vascular permeability (WEDMORE and WILLIAMS 1981; WILLIAMS 1983). The injection of LTB_4 intradermally into human skin caused ill-defined areas of raised induration lasting more than 4 h and histological examination of biopsies showed perivascular infiltrates of neutrophils (SOTER et al. 1983; CAMP et al. 1983a). An enhancement of the inflammatory response was observed upon the concomitant administration of vasodilator prostaglandin. More recently it has been reported that a later inflammatory reaction observed 6–12 h after injection of LTB_4 is potentiated to a larger degree when PGE_2 is injected with LTB_4 (ARCHER et al. 1984).

The topical application of 5–500 ng LTB_4 to human skin caused well defined erythema and swelling which became visible 12 h after LTB_4 administration and persisted for several days (CAMP et al. 1984). The correlation of the time course of neutrophil accumulation with the visible response is further evidence for the dependence of the inflammatory response of LTB_4 on the presence of PMNs.

H. Generation of Lipoxygenase Products in Skin

The first results which suggested that a lipoxygenase pathway was active in skin were reported in 1975 by HAMMARSTRÖM and co-workers, at about the same time as the 12-lipoxygenase pathway was described. They found that free arachidonic acid and 12-HETE levels, as measured by gas chromatography–mass spectrometry, were much higher in keratome slices taken from lesional psoriatic skin than in those taken from the uninvolved skin. Prostaglandin levels (PGE_2 and $PGF_{2\alpha}$) were only slightly altered in lesional skin. A PMN infiltrate (mainly neutrophils) is a common characteristic of lesional psoriatic skin. 12-HETE is known to have chemotactic properties, as described previously. However, the more recent characterisation of LTB_4, which has been found to be at least 100 times more potent than 12-HETE as a chemotactic agent for PMNs in vitro, has prompted an investigation of the presence of LTB_4 in lesional psoriatic skin. Using a skin chamber technique, LTB_4 was found in samples obtained from abraded lesional skin but not in samples obtained from similar experiments carried out on either clinically normal psoriatic skin or from the skin of normal individuals. The LTB_4 was purified by reversed phase HPLC and detected in an in vitro assay of leucocyte motility (BRAIN et al. 1982a). In addition to LTB_4, biologically active amounts of 12-HETE were found, thus confirming the presence of considerable amounts in lesional skin. LTB_4 has also been detected in extracts from superficial psoriatic scale (Grabbe et al. 1982; Camp et al. 1983 b). Grabbe and co-workers have suggested that 5-HETE is also present but this is not supported by the findings of Camp and co-workers (1983). However, these workers do suggest the existence of several active lipoxygenases in lesional psoriatic skin. More recently, a LTC_4-like material (measured by use of a specific radioimmunoassay) has been detected in similar quantities to LTB_4 in chamber fluid from lesional psoriatic skin (Brain et al. 1984). Thus it appears that in psoriasis lipoxygenase products are formed that are capable of mediating not only the characteristic neutrophil accumulation but also the increased vascular permeability and erythema which feature in the pathology of this disease. The presence of LTB_4 has also been reported in samples of skin in allergic dermatitis (Barr et al. 1984) and in atopic dermatitis (Ruzicka et al. 1984). Interestingly, these skin diseases are not characterised by the continual presence of PMNs in lesional skin. Whether lipoxygenase products are produced by skin itself in sufficient quantities to initiate an inflammatory response is a question which has not been answered. Mono-HETEs have been suggested to be major metabolites of arachidonic acid in vitro in guinea-pig skin, in particular the epidermis, as well as in human and mouse epidermis (Ruzicka and Printz 1982; Hammarström et al. 1979; Ziboh et al. 1981). It has also been reported that cultured human keratinocytes stimulated with calcium ionophore produce trace amounts of LTB_4 (Brain et al. 1982). Thus, evidence is accumulating that lipoxygenase products are produced by skin under various conditions and could act to initiate, as well as to potentiate, an inflammatory response.

J. Inhibitors of Arachidonate Metabolism

The first suggestion that non-steroidal anti-inflammatory drugs (NSAIDs) affect the generation of arachidonic acid metabolites was provided by the findings of Piper and Vane (1969). They found that arachidonic acid increased the release of a rabbit aorta contracting factor (RCS), now known to be mainly TxA_2, from guinea-pig lungs in vitro during anaphylactic challenge, whilst aspirin inhibited this release. The association between arachidonic acid metabolites and aspirin became clear in 1971, when it was found that NSAIDs, such as aspirin, prevented the production of prostaglandins by human platelets (Smith and Willis 1971), guinea-pig lung (Vane 1971) and dog spleen (Ferreira et al. 1971).

Vane proposed that NSAIDs act by inhibiting the enzymes responsible for prostaglandin synthetase. This hypothesis has become well established and the inhibition of prostaglandin synthesis by NSAIDs has been demonstrated in many in vivo and in vitro systems. More importantly, the therapeutic efficacy of NSAIDs has been found to be closely related to their ability to inhibit prostaglandin synthesis. Additionally, a common side-effect of NSAIDs is gastrointestinal irritation. This is considered to be due to the inhibition of protective prostaglandins in the gastric mucosa.

NSAIDs have been found to inhibit peroxidation by cyclo-oxygenase at a specific point on the arachidonic acid molecule. Thus, NSAIDs are less potent at inhibiting other peroxidations, such as those catalysed by 5-, 8-, 12-, and 15-lipoxygenases, and cannot prevent the generation of thromboxanes or prostaglandins from endoperoxides. Several classes of compound are now known to inhibit prostaglandin synthesis (for reviews see FLOWER 1974; SHEN 1979) and these compounds inhibit in various ways (ROME and LANDS 1975).

The NSAIDs are important anti-inflammatory agents in man. This is thought to be because of the ability of prostaglandins to mediate vasodilatation and to potentiate the inflammatory properties of other mediators which produce oedema and pain (as previously discussed). Thus NSAIDs act to reduce oedema, redness and pain, three of the classic signs of inflammation. Intradermal injection of arachidonic acid into the skin of rabbit (WILLIAMS 1979) and guinea-pig (PECK et al. 1981) induces vasodilatation which, unlike the vasodilator actions of the prostaglandins, can be inhibited by NSAIDs. However, the NSAID indomethacin was found to have no effect on normal skin blood flow (WILLIAMS 1979). In man, indomethacin, either given orally or applied topically, completely prevented prostaglandin generation in response to ultraviolet irradiation by both B- and C-type wavelengths, but erythema was only partially suppressed (BLACK et al. 1978). This suggests that the inflammatory changes in response to irradiation may be due only in part to activation of cyclo-oxygenase. Interestingly, BLACK and coworkers (1978) found that there was an increase in free arachidonic acid in the non-irradiated skin that was treated with indomethacin. This result suggests that arachidonic acid is metabolised via the cyclo-oxygenase pathway in skin under normal conditions, but exposure of the skin to trauma during sample collection could be responsible for the findings.

The NSAIDs, although potent anti-inflammatory drugs, have not been found to be beneficial in the general treatment of chronic human skin disease (apart from lessening the relatively acute effects of sunburn). Moreover, the ability of aspirin and other NSAIDs to exacerbate urticaria and psoriasis is recognised. It has been suggested that inhibiting arachidonic acid metabolism via the cyclo-oxygenase pathway leads to increased metabolism via lipoxygenase enzymes. Free arachidonic acid and 12-HETE levels in lesional and uninvolved psoriatic epidermis has been shown to be increased by treating keratome slices with indomethacin before incubation in vitro (HAMMARSTRÖM et al. 1975). This finding is supported by the observation that indomethacin exacerbates pre-existing psoriatic lesions in vivo (KATAYAMA and KAWADA 1981), but a direct irritant effect of the topical indomethacin has not been excluded.

The characterisation of the leukotrienes and the possibility that these compounds are important inflammatory mediators in man, has led to the development of drugs which inhibit the lipoxygenase pathways as well as cyclo-oxygenase. It has been reported that benoxyprofen inhibits 5-lipoxygenase (DAWSON et al. 1982) as well as being an inhibitor of the cyclo-oxygenase pathway (CASHIN et al. 1977). Benoxyprofen has been found to be beneficial in the treatment of psoriasis vulgaris in both uncontrolled studies (ALLEN and LITTLEWOOD 1982; KINGSTON and MARKS 1982) and a double-blind placebo study (KRAGBALLE and HERLIN 1983). This drug has now been withdrawn from clinical use because of adverse toxic effects, but the results of the controlled trials showed that 75% of patients were considerably improved after benoxyprofen treatment. It is unknown whether the beneficial effects observed were directly due to inhibition of the generation of lipoxygenase products. However, it is considered that the beneficial effects were not directly related to the photosensitivity that occurs with benoxyprofen administration (KLIGMAN and KLAIDBEY 1982). This is because photosensitivity only occurred in regions exposed to sunlight whilst improvement in psoriasis was seen in both covered and exposed areas. These results suggest that the development of a drug which inhibits lipoxygenase enzymes could be beneficial for the treatment of psoriasis. Evidence is also accumulating for the presence of lipoxygenase products in certain types of dermatitis.

Drugs which inhibit both the cyclo-oxygenase and the lipoxygenase enzymes would act in a similar manner to anti-inflammatory steroids (AISs). However, AISs have been found to act via an alternate mechanism. This was originally implied by the finding that the addition of exogenous arachidonic acid overcame AIS- but not NSAID-induced inhibition of prostaglandin formation (GLYGLEWSKI et al. 1975). It was then discovered that AISs, unlike NSAIDs, do not act directly on prostaglandin synthetase, but suppress production of arachidonic acid metabolites in intact cells. This is caused by an indirect effect of AISs on phospholipase A_2 (BLACKWELL et al. 1978). AISs induce the synthesis of proteins that directly inhibit phospholipase A_2 (DANON and ASSOULINE 1978; FLOWER and BLACKWELL 1979). These proteins have been isolated from several sources. Material from guinea-pig lung and rat macrophages has been found to be similar, with a molecular weight of 15000, and has been named macrocortin (BLACKWELL et al. 1980). A higher molecular weight material (40000) has also been identified. This was released from rabbit neutrophils; it has been called lipomodulin (HIRATA et al. 1980).

Topically applied, AISs have been found to reduce the tissue concentration of arachidonic acid in psoriasis (HAMMARSTRÖM et al. 1977) and after systemic administration in a model of inflammatory skin disease (KOZBA BLACK et al. 1982). In the former experiments a reduction of both lipoxygenase and cyclo-oxygenase products was also observed.

It is well established that AISs are more efficient in suppressing inflammation than NSAIDs and AISs are widely used in the treatment of skin disease, but there are many side-effects associated with corticosteroid therapy. Thus a drug with similar inhibitory effects on arachidonic acid metabolism as AISs but without the associated side-effects could be beneficial in the treatment of certain skin diseases. Drugs specifically inhibiting 5-lipoxygenase and others inhibiting this enzyme and cyclo-oxygenase are under development in the pharmaceutical industry. These will be of considerable interest when available clinically.

K. Conclusion

We have endeavoured to establish the potential of arachidonic acid metabolites as inflammatory mediators in skin. The effects of cyclo-oxygenase products are potent and well characterised, but the use of prostaglandin synthetase inhibitors in skin disease is not of great benefit, unlike in other diseases such as rheumatoid arthritis, where these drugs given symptomatic relief. Certain lipoxygenase products, especially LTB_4, have potent effects on leucocytes and potent inflammatory effects in skin. The results obtained so far can only act to stimulate interest in the development of selective anti-lipoxygenase drugs. Whether such drugs, by interfering with the inflammatory process, will be of long-term benefit in skin diseases is a fundamental question which will no doubt be answered empirically.

Acknowledgement. S. D. Brain is supported by the Sir Halley Stewart Trust.

References

Allen BR, Littlewood SM (1982) Benoxaprofen: effect on cutaneous lesions in psoriasis. Br Med J Clin Res 285:1241

Amelang E, Prasad CM, Raymond RM, Grega GJ (1981) Interactions among inflammatory mediators on edema formation in the canine forelimb. Circ Res 49:298–306

Angaard E, Jonsson CE (1971) Effect of prostaglandins in lymph from scalded tissue. Acta Physiol Scand 81:440–447

Angaard E, Samuelsson B (1964) Prostaglandins and related factors 28. Metabolism of prostaglandin E_1 in guinea pig lung: the structures of two metabolites. J Biol Chem 239:4087–4102

Archer CB, Juhlin L, MacDonald DM, Morley J, Page CP (1984) Synergistic interaction between leukotriene B_4 (LTB_4) and prostaglandin E_2 (PGE_2) in human skin: a delayed-onset phenomenon. Br J Clin Pharmacol 17:610

Barnett AH, Spiliopoulos AJ, Keen H (1980) Blockade of chlorpropamide-alcohol flushing by indomethacin suggests an association between prostaglandins and diabetic vascular complications. Lancet II:164–166

Barr RM, Brain SD, Camp RDR, Cilliers J, Greaves MW, Mallett AL, Misch K (1984) Levels of arachidonic acid and its metabolites in the skin in human allergic and irritant contact dermatitis. Br J Dermatol 111:23–28

Basran GS, Morley J, Paul W, Turner-Warwick M (1982) Evidence in man of synergistic interaction between putative mediators of acute inflammation and asthma. Lancet I:935–937

Bergström S, Sjövall J (1957) The isolation of prostaglandin. Acta Chem Scand 11:1086

Bergström S, Duner H, von Euler US, Pernow B, Sjövall J (1959) Observations on the effects of infusion of prostaglandin E in man. Acta Physiol Scand 45:145–151

Bergström S, Ryhage R, Samuelsson B, Sjövall J (1962) The structure of prostaglandin E, F_1, and F_2. Acta Chem Scand 16:501–502

Bergström S, Danielsson H, Samuelsson B (1964) The enzymatic conversion of essential fatty acids into prostaglandins. J Biol Chem 239:4006–4008

Bisgaard H, Kristensen J, Sondergaard J (1982) The effect of leukotriene C_4 and D_4 on cutaneous blood flow in humans. Prostaglandins 23:797–801

Black AK, Greaves MW, Hensby CN, Plummer NA, Warin AP (1978) The effects of indomethacin on arachidonic acid and prostaglandins E_2 and F_2 levels in human skin 24 h after u.v.B. and u.v.C. irradiation. Br J Clin Pharmacol 6:261–266

Black AK, Fincham N, Greaves MW, Hensby CN (1980) Time course changes in levels of arachidonic acid and prostaglandins D_2, E_2, F_2 in human skin following ultraviolet B irradiation. Br J Clin Pharmacol 10:453–457

Blackwell GJ, Flower RJ, Nijkamp FP, Vane JR (1978) Phospholipase A_2 activity of guinea-pig isolated perfused lungs: stimulation, and inhibition by anti-inflammatory steroids. Br J Pharmacol 62:79–89

Blackwell GJ, Carnuccio R, Di Rosa M, Flower RJ, Parente L, Persico P (1980) Macrocortin: a polypeptide causing the anti-phospholipase effect of glucocorticoids. Nature 287:147–149

Boot JR, Dawson W, Kitchen EA (1976) The chemotactic activity of thromboxane B_2: a possible role in inflammation. J Physiol (Lond) 257:47

Borgeat P, Samuelsson B (1979 a) Transformation of arachidonic acid by rabbit polymorphonuclear leukocytes. J Biol Chem 254:2643–2646

Borgeat P, Samuelsson B (1979b) Metabolism of arachidonic acid in polymorphonuclear leukocytes. J Biol Chem 254:7865–7869

Borgeat P, Samuelsson B (1979c) Arachidonic acid metabolism in polymorphonuclear leukocytes: effect of calcium ionophore A23187. Proc Natl Acad Sci USA 76:2148–2152

Brain SD, Camp RDR, Dowd PM, Black AK, Woollard PM, Mallet AI, Greaves MW (1982a) Psoriasis and leukotriene B. Lancet II:762–763

Brain SD, Camp RDR, Leigh IM, Fort-Hutchinson AW (1982b) The synthesis of leukotriene B_4-like material by cultured human keratinocytes. J Invest Dermatol 78:328

Brain SD, Camp RDR, Charleson S, Dowd PM, Ford-Hutchinson AW, Greaves MW, Kobza Black A (1984) The release of LTC_4-like material from the involved lesional skin in psoriasis. Br J Clin Pharmacol 17:650

Bray MA (1983) The pharmacology and pathophysiology of leukotriene B_4. Br Med Bull 39:249–254

Bray MA, Cunningham FM, Ford-Hutchinson AW, Smith MJH (1981a) Leukotriene B_4: a mediator of vascular permeability. Br J Pharmacol 72:483–486

Bray MA, Ford-Hutchinson AW, Smith MJH (1981b) Leukotriene B_4: biosynthesis and biological activities. In: Piper PJ (ed) SRS-A and leukotrienes. Wiley, London

Bray MA, Ford-Hutchinson AW, Smith MJH (1981c) Leukotriene B_4: an inflammatory mediator in vivo. Prostaglandins 22:213–222

Brocklehurst WE (1967) The probable role of known mediators in hypersensibility reactions. In: Schild HO (ed) Proceedings of the 3rd International Pharmacological Meeting, vol II. Pergamon, Oxford, pp 67–75

Camp RDR, Coutts AA, Greaves MW, Kay AB, Walport MJ (1983a) Responses of human skin to intradermal injection of leukotrienes C_4, D_4, and B_4. Br J Pharmacol 80:497–502

Camp RDR, Mallet AI, Woollard PM, Brain SD, Kobza Black A, Greaves MW (1983b) The identification of hydroxy fatty acids in psoriatic skin. Prostaglandins 26:431–448

Camp RDR, Russell Jones R, Brain S, Woollard P, Greaves MW (1984) Production of intraepidermal microabscesses by topical application of leukotriene B_4. J Invest Dermatol 82:202–204

Cashin CH, Dawson W, Kitchen EA (1977) The pharmacology of benoxaprofen (2-[4-chlorophenyl]-α-methyl-5-benzoxazole acetic acid), LRCL 3794, a new compound with anti-inflammatory activity apparently unrelated to inhibition of prostaglandin synthesis. J Pharm Pharmacol 29:330–336

Crawford JP, Movat HZ, Ranadive NS, Hay JB (1982) Pathways to inflammation induced by immune complexes: development of the Arthus reaction. Fed Proc 41:2583–2587

Crunkhorn P, Willis AL (1969) Actions and interactions of prostaglandins administered intradermally in rat and man. Br J Pharmacol 36:216

Crunkhorn P, Willis AL (1971) Interaction between prostaglandins E and F given intradermally in the rat. Br J Pharmacol 41:507–512

Cunningham FM, Carter HR, Smith MJH, Ford-Hutchinson AW, Bray MA (1981) Aggregation of polymorphonuclear leucocytes (PMNs) by leukotriene B_4: effects of cyclooxygenase products and metabolic inhibitors. Agents Actions 11:583–584

Danon A, Assonline G (1978) Inhibition of prostaglandin biosynthesis by corticosteroids requires RNA and protein synthesis. Nature 273:552–554

Dawson W, Boot JR, Harvey J et al. (1982) The pharmacology of benoxaprofen with particular reference to effects on lipoxygenase product formation. Eur J Rheumatol Inflamm 5:61–68

Dowd PM, Kobza Black A, Woollard P, Camp RDR, Greaves MW (1984) The in vivo properties of 12-hydroxyeicosatetraenoic acid (12-HETE) in normal skin. J Invest Dermatol 82:413–414 (abstract)

Drazen JM, Austen KP, Lewis RA, Clark DA, Goto G, Marfat A, Corey EJ (1980) Comparative airway and vascular activities of leukotrienes C-1 and D in vivo and in vitro. Proc Natl Acad Sci USA 77:4354–4358

Dusting GJ, Moncada S, Vane JR (1978) Vascular actions of arachidonic acid and its metabolites in the perfused mesenteric and femoral beds of the dog. Eur J Pharmacol 49:65–72

Fantone JC, Marasco WA, Eglas LJ, Ward PA (1983) Anti-inflammatory effects of prostaglandin E_1: in vivo modulation of the formyl peptide chemotactic receptor on the rat neutrophil. J Immunol 130:1495–1497

Ferreira SH (1972) Prostaglandins, aspirin-like drugs and analgesia. Nature New Biol 240:200–203

Ferreira SH, Moncada S, Vane JR (1971) Indomethacin and aspirin abolish prostaglandin release from the spleen. Nature New Biol 231:237–239

Ferreira SH, Nakamura M, Abreu Castro MS (1978) The hyperalgesic effects of prostacyclin and PGE_2. Prostaglandins 16:31–37

Flower RJ (1974) Drugs which inhibit prostaglandin biosynthesis. Pharmacol Rev 26:33–67

Flower RJ (1978) Steroidal anti-inflammatory drugs as inhibitors of phospholipase A_2. Adv Prostaglandin and Thromboxane Leukotriene Res 3:105–112

Flower RJ, Blackwell GJ (1979) Anti-inflammatory steroids induce biosynthesis of a phospholipase A_2 inhibitor which prevents prostaglandin generation. Nature 278:456–459

Flower RJ, Harvey EA, Kingston WP (1976) Inflammatory effects of prostaglandin D_2 in rat and human skin. Br J Pharmacol 56:229–233

Ford-Hutchinson AW, Smith AW, Walker JR (1976) Chemotactic activity of solutions of prostaglandin E_1. Br J Pharmacol 56:345

Ford-Hutchinson AW, Bray MA, Doig MV, Shipley ME, Smith MJH (1980) Leukotriene B, a potent chemokinetic and aggregating substance released from polymorphonuclear leukocytes. Nature 286:264–265

Ford-Hutchinson AW, Bray MA, Cunningham FM, Davidson EM, Smith MJH (1981) Isomers of LTB_4 possess different biological potencies. Prostaglandins 21:143

Goetzl EJ, Pickett WC (1980) The human PMN leukocyte chemotactic activity of complex hydroxy-eicosatetraenoic acids (HETEs). J Immunol 125:1789–1791

Goetzl EJ, Sun FF (1979) Generation of unique monohydroxy-eicosatetraenoic acids from arachidonic acid by neutrophils. J Exp Med 150:406–411

Goetzl EJ, Woods JMW, Gorman RR (1977) Stimulation of human eosinophil and neutrophil PMN chemotaxis and random migration by 12-HETE. J Clin Invest 59:179–183

Goetzl EJ, Weller PE, Sun FF (1980) The regulation of human eosinophil function by endogenous mono-hydroxy-eicosatetraenoic acids (HETEs). J Immunol 124:926–933

Goldblatt MW (1935) Properties of humans seminal fluid. J Physiol (Lond) 84:208–218

Grabbe J, Czarnetzki BM, Mardin M (1982) Chemotactic leukotrienes in psoriasis. Lancet II:1464

Greaves MW, Sondergaard J, McDonald-Gibson W (1971) Recovery of prostaglandins in human cutaneous inflammation. Br Med J [Clin Res] 2:258–260

Gryglewski RJ, Panczenko B, Korbut R, Grodzinska L, Ocetkiewicz A (1975) Corticosteroids inhibit prostaglandin release from perfused mesenteric blood vessels of rabbit and from perfused lungs of sensitised guinea-pig. Prostaglandins 5:531–542

Hamberg M, Samuelsson B (1974) Prostaglandin endoperoxides. Novel transformation of arachidonic acid in human platelets. Proc Natl Acad Sci USA 71:3400–3404

Hamberg M, Svensson J, Samuelsson B (1975) Thromboxane: a new group of biologically active compounds derived from prostaglandin endoperoxides. Proc Natl Acad Sci USA 72:2994–2998

Hammerström S, Hamberg M, Samuelsson B, Duell EA, Stawiski M, Voorhees JJ (1975) Increased concentrations of nonesterified arachidonic acid, 12L-hydroxy-5,8,10,14-eicosatetraenoic acid, prostaglandin E_2, and prostaglandin F_2 in epidermis of psoriasis. Proc Natl Acad Sci USA 72:5130–5134

Hammerström S, Hamberg M, Duell EA et al. (1977) Glucocorticoid in inflammatory proliferative skin disease reduces arachidonic and hydroxyeicosatetraenoic acids. Science 197:994–995

Hammerström S, Lindgren JA, Marcelo C, Duell EA, Anderson TF, Voorhees JJ (1979) Arachidonic acid transformation in normal and psoriatic skin. J Invest Dermatol 73:180–183

Higgs EA, Moncada S, Vane JR (1978 a) Inflammatory effects of prostacyclin (PGI_2) and 6-oxo $PGF_{1\alpha}$ in the rat paw. Prostaglandins 16:153–162
Higgs EA, O'Grady J, Thrower PA, Moncada S (1979) Prostacyclin: inflammatory effects in human skin. Fourth international prostaglandin conference, May 1979, Washington DC, meeting abstracts, p 48
Higgs GA, McCall E, Youlten LJF (1975) A chemotactic role for prostaglandins released from polymorphonuclear leucocytes during phagocytosis. Br J Pharmacol 53:539–546
Higgs GA, Moncada S, Vane JR (1978 b) Prostacyclin (PGI_2) as a potent dilator of arterioles in the hamster cheek pouch. J Physiol (Lond) 526:43
Higgs GA, Salmon JA, Spayne JA (1981) The inflammatory effects of hydroperoxy and hydroxy acid products of arachidonate lipoxygenase in rabbit skin. Br J Pharmacol 74:429–433
Hirata F, Schiffmann E, Venkatasubramanian K, Salomon D, Axelrod J (1980) A phospholipase A_2 inhibitory protein in rabbit neutrophils induced by glucocorticoids. Proc Natl Acad Sci USA 77:2533–2536
Horton EW (1963) Action of prostaglandin E_1 on tissues which respond to bradykinin. Nature 200:892–893
Johnson RA, Morton DR, Kinner JH, Corman RR, McGuire JR, Sun FF, Whittaker N, Bunting S, Salmon J, Moncada S, Vane JR (1976) The chemical characterization of prostaglandin X (prostacyclin). Prostaglandins 12:915–928
Jonsson DC, Angaard E (1972) Biosynthesis and metabolism of prostaglandin E_2 in human skin. Scand J Clin Lab Invest 29:289–296
Jose PJ, Forrest MJ, Williams TJ (1981) Human C5a des Arg increases vascular permeability. J Immunol 127:2376–2380
Juhlin S, Michaelsson G (1969) Cutaneous vascular reactions to prostaglandins in healthy subjects and in patients with urticaria and atopic dermatitis. Acta Derm Venereol (Stockh) 49:251–261
Kaley G, Weiner R (1968) Microcirculatory studies with prostaglandin E_1. In: Ramwell PW, Shaw JE (eds) Prostaglandin symposium of the Worcester foundation for experimental biology. Interscience, New York, pp 321–328
Katayama H, Kawada A (1981) Exacerbation of psoriasis induced by indomethacin. J Dermatol 8:323–337
Kingston T, Marks R (1982) Benoxaprofen: effect on cutaneous lesions in psoriasis. Br Med J [Clin Res] 285:1741
Kligman AM, Kaidbey KH (1982) Phototoxicity to benoxaprofen. Eur J Rheumatol Inflamm 5:124–137
Kobza Black A, Greaves MW, Hensby CN, Plummer NA (1978) Increased prostaglandins E_2 and F_2 in human skin 6 and 24 h after ultraviolet B irradiation (290–320 nm). Br J Clin Pharmacol 5:431–436
Kobza Black A, Greaves MW, Hensby CN (1982) The effect of systemic prednisolone on arachidonic acid, and prostaglandin E_2 and F_2 levels in human cutaneous inflammation. Br J Clin Pharmacol 14:391–394
Komoriya K, Ohmori H, Azuma A, Kurozumi S, Hashimoto Y, Nicolaou KC, Barnette WF, Mangolda RL (1978) Prostaglandins 15:557–564
Kopaniak MM, Hay JB, Movat HZ (1978) The effect of hyperemia on vascular permeability. Microvasc Res 15:77–82
Kragballe K, Herlin T (1983) Benoxaprofen improves psoriasis: a double blind study. Arch Dermatol 119:548–552
Kurzrok R, Lieb CC (1930) Biochemical studies of human semen. II. The action of semen on the human uterus. Proc Soc Exp Biol Med 28:268–272
Lewis AJ, Nelson DJ, Sugrue MF (1975) On the ability of prostaglandin E_1 and arachidonic acid to modulate experimentally induced oedema in rat paw. Br J Pharmacol 55:51–56
Lewis GP, Westwick J, Williams TJ (1977) Microvascular responses produced by the prostaglandin endoperoxide PGG_2 in vivo. Br J Pharmacol 59:442
Malik KU, McGiff JC (1976) Cardiovascular actions of prostaglandins. In: Karim SMM (ed) Prostaglandins: physiological, pharmacological and pathological aspects. MTP Press, Lancaster, pp 103–200

Moncada S, Ferreira SH, Vane JR (1973) Prostaglandins, aspirin-like drugs and the oedema of inflammation. Nature 246:217–219

Moncada S, Gryglewski R, Bunting S, Vane JR (1976) An enzyme isolated from arteries transforms prostaglandin endoperoxides to an unstable substance that inhibits platelet aggregation. Nature 263:663–665

Needleman P, Moncada S, Bunting S, Vane JR, Hamberg M, Samuelsson B (1976) Identification of an enzyme in platelet microsomes which generates thromboxane A_2 from prostaglandin endoperoxides. Nature 261:558–560

Nugteren DH (1975) Arachidonate lipoxygenase in blood platelets. Biochim Biophys Acta 380:299–307

O'Flaherty JT, Kreutzer DL, Ward PA (1979) Effect of prostaglandins E_1, E_2, and F_2 on neutrophil aggregation. Prostaglandins 17(2):201–210

O'Flaherty JT, Thomas MH, Less CJ, McCall CE (1981) Neutrophil aggregating activity of monohydroxy-eicosatetraenoic acids. Am J Pathol 104:55–62

Palmer RMJ, Stepney RJ, Higgs GA, Eakins K (1980) Chemokinetic activity of arachidonic acid lipoxygenase products on leucocytes of different species. Prostaglandins 20:411–418

Peck MJ, Williams TJ (1978) Prostacyclin (PGI_2) potentiates bradykinin-induced plasma exudation in rabbit skin. Br J Pharmacol 62:464–465P

Peck MJ, Piper PJ, Williams TJ (1981) The effect of leukotrienes C_4 and D_4 on the microvasculature of guinea-pig skin. Prostaglandins 21:315–321

Phillips WS, Lightman SL (1981) Is cutaneous flushing prostaglandin mediated? Lancet I:754–756

Piper PJ (1983) Pharmacology of leukotrienes. Br Med Bull 39:255–259

Piper PJ, Vane JR (1969) Release of additional factors in anaphylaxis and its antagonism by anti-inflammatory drugs. Nature 223:29–35

Rome LH, Lands WFM (1975) Structural requirements for time-dependent inhibition of prostaglandin biosynthesis by anti-inflammatory drugs. Proc Natl Acad Sci USA 72:4863–4865

Ruzicka T, Printz MP (1982) Arachidonic acid metabolism in guinea-pig skin. Biochim Biophys Acta 711:391–397

Ruzicka T, Simmer T, Peskar BA, Braun Falco O (1984) Leukotrienes in skin of atopic dermatitis. Lancet I:222

Shen TY (1979) Prostaglandin synthetase inhibitors I. Handb Exp Pharmacol 50/11:305–347

Sirois P, Roy S, Borgeat P, Picard S, Corey EJ (1981) Structural requirement for the action of leukotriene B_4 on the guinea-pig lung: importance of double bond geometry in the 6,8,10-triene unit. Biochem Biophys Res Commun 99:385–390

Smith JB, Willis AL (1971) Aspirin selectively inhibits prostaglandin production in human platelets. Nature New Biol 321:235–237

Solomon LM, Juhlin L, Kirschbaum MM (1968) Prostaglandins on cutaneous vasculature. J Invest Dermatol 51:280–282

Sondergaard J, Greaves MW (1970) Pharmacological studies in inflammation due to ultraviolet radiation. J Pathol 101:93–97

Sondergaard J, Greaves MW (1971) Prostaglandin E_1: effect on human cutaneous vasculature and skin histamine. Br J Dermatol 84:424

Sondergaard J, Greaves MW, Jörgenson HP (1974) Recovery of prostaglandins in human primary irritant dermatitis. Arch Dermatol 110:556–558

Soter NA, Lewis RA, Corey EJ, Austen KF (1983) Local effects of synthetic leukotrienes (LTC_4, LTD_4, LTE_4 and LTB_4) in human skin. J Invest Dermatol 80:115–119

Strakosch CR, Jefferys DB, Keen H (1980) Blockade of chlorpropamide alcohol flush by aspirin. Lancet I:394–396

Thomas G, West GB (1973) Prostaglandins as modulators of bradykinin responses. J Pharm Pharmacol 25:747–748

Turner SR, Campbell JA, Lynn WS (1975a) Polymorphonuclear leukocyte chemotaxis toward oxidized lipid components of cell membranes. J Exp Med 141:1437–1441

Turner SR, Tainer JA, Lynn WS (1975b) Biogenesis of chemotactic molecules by the arachidonate lipoxygenase system of platelets. Nature 257:680–681

Vadas P, Hay JB (1982) The appearance and significance of phospholipase A_2 in lymph draining tuberculin reactions. Am J Pathol 107:285–291

Vadas P, Wasi S, Movat HZ, Hay JB (1981) Extracellular phospholipase A_2 mediates inflammatory hyperaemia. Nature 293:583–585

Van Dorp DA, Beerthuis RK, Nugteren DH, Vonkeman H (1964) The biosynthesis of prostaglandins. Biochim Biophys Acta 90:204–207

Vane JR (1971) Inhibition of prostaglandin synthesis as a mechanism of action for aspirin-like drugs. Nature 231:232–235

Von Euler US (1936) On the specific vasodilating and plain muscle stimulating substance from accessory genital glands in man and certain animals (prostaglandin and vesiglandin). J Physiol (Lond) 88:213–234

Wedmore CV, Williams TJ (1981) Control of vascular permeability by polymorphonuclear leucocytes in inflammation. Nature 289:646–650

Williams TJ (1979) Prostaglandin E_2, prostaglandin I_2 and the vascular changes of inflammation. Br J Pharmacol 65:517–524

Williams TJ (1982) Vasoactive intestinal polypeptide is more potent than prostaglandin E_2 as a vasodilator and oedema potentiator in rabbit skin. Br J Pharmacol 77:505–509

Williams TJ (1983) Interactions between prostaglandins, leukotrienes and other mediators of inflammation. Br Med Bull 39:239–242

Williams TJ, Jose PJ (1981) Mediation of increased vascular permeability after complement activation: histamine-independent action of rabbit C5a. J Exp Med 153:136–153

Williams TJ, Morley J (1973) Prostaglandins as potentiators of increased vascular permeability in inflammation. Nature 246:215–217

Williams TJ, Peck MJ (1977) Role of prostaglandin-mediated vasodilatation in inflammation. Nature 270:530–532

Williams TJ, Piper PJ (1980) The action of chemically pure SRS-A on the microcirculation in vivo. Prostaglandins 19:779–789

Ziboh VA, Marcelo CL, Voorhees JJ (1981) Induced lipoxygenation of arachidonic acid in mouse epidermal keratinocytes by calcium ionophore A23187, abstracted. J Invest Dermatol 76:307

Ziboh VA, Mallia C, Morhart E, Taylor JR (1982) Induced biosynthesis of cutaneous prostaglandins by ionizing irradiation (41362). Proc Soc Exp Biol Med 169:386–391

CHAPTER 22

Slow Reacting Substance of Anaphylaxis

S. I. WASSERMAN

A. Introduction

The term slow reacting substance was first used by FELDBERG and KELLAWAY over 40 years ago to refer to an activity which induced sustained contraction of smooth muscle following perfusion of lung tissue with cobra venom (FELDBERG and KELLAWAY 1938). Subsequently, sensitised guinea-pig (KELLAWAY and TRETHEWIE 1940) and human lung tissue (BROCKLEHURST 1962) was demonstrated to generate similar activity upon challenge with specific antigen. The contractile activity generated by this principle obtained from lung was shown to be unaffected by agents which totally inhibited histamine-induced contractile responses and, therefore, it was felt to be a new principle of anaphylactic reactions and termed "slow reacting substance of anaphylaxis" (BROCKLEHURST 1953). Until recently SRS-A was defined by its chromatographic behaviour, sensitivity to sulphatases and lipoxygenase, resistance to protease and inhibitibility by a semisynthetic blocking agent FPL 55712 (AUGSTEIN et al. 1973). Based upon these or similar criteria, SRS-A has been identified in anaphylactic diffusates of perfused guinea-pig heart–lung preparations (KELLAWAY and TRETHEWIE 1940), human nasal polyp (KALINER et al. 1973) or lung fragments (ORANGE et al. 1971), dispersed lung cells (LEWIS et al. 1974), enriched populations of lung mast cells (PATERSON et al. 1976), the blood of guinea-pigs dying from anaphylaxis (STECHSHULTE et al. 1973), and guinea-pig skin after anaphylactic challenge (JONES and KAY 1974), and from human neutrophils (CONROY et al. 1976), rat mast cells (YECIES et al. 1979 a), rat basophil leukaemia tumour and cells (YECIES et al. 1979 b), and rat macrophages (BACH et al. 1979) after interaction with the calcium ionophore A 23187.

The precise site and mechanism of the *immunological* generation of SRS-A have been best assessed in vivo in the rat peritoneal cavity and in vitro in human lung. In the rat SRS-A can be generated following activation by either IgE- or IgG_a-dependent reactions (ORANGE and AUSTEN 1969). In the former, short (2 h) but not long (24 h) term passive sensitisation with antibody prepares the animal for antigen-induced SRS-A release, whereas either regimen permits histamine release. IgG_a-dependent reactions occur only after short periods (1–3 h) of passive sensitisation. The generation of SRS-A by IgG_a-dependent reactions requires neutrophils and complement and is enhanced in the presence of neutrophilic exudates and inhibited in the presence of eosinophilic or mononuclear leucocyte infiltrates (ORANGE et al. 1969). In human lung tissue SRS-A generation during IgE-dependent reactions decreases as mast cells are enriched (PATERSON et al.

1976), suggesting that even in IgE-dependent reactions non-mast cell sources of SRS-A exist. In fact, if sensitised human lung cells are challenged with small amounts of antigen or if cells are sensitised passively with limited quantities of IgE-containing serum and then challenged, SRS-A generation occurs primarily as an intracellular event without appreciable SRS-A release (LEWIS et al. 1974).

The regulation of SRS-A generation and release resembles, superficially, that of histamine but differs in several important ways. Thus, agents which elevate intracellular levels of cyclic 3′5′-adenosine monophosphate (cAMP) inhibit generation and release of SRS-A, whereas agents which lower cAMP or enhance intracellular levels of cyclic 3′5′-guanosine monophosphate (cGMP) enhance SRS-A generation and release (ORANGE et al. 1971; LEWIS et al. 1974; KALINER et al. 1972). It is of interest, however, that SRS-A generation is more sensitive to cGMP or cholinergic augmentation and to cAMP and β-adrenergic inhibition than is the release of histamine (ORANGE et al. reviewed in AUSTEN et al. 1974). Thus, if lung fragments are incubated with isoproterenol and then with cGMP prior to antigen challenge, histamine release occurs but SRS-A generation is inhibited. Moreover, SRS-A release is irreversibly inhibited by DFP, whereas the release of histamine is restored by removing the DFP by washing. Similar inhibitibility of SRS-A generation and release despite histamine release has been noted in the presence of both cytochalasin A and B (ORANGE 1975). These data all suggest that while the processes which result in mast cell activation and the release of the preformed mediator histamine and in the generation and release of SRS-A share some important early biochemical steps, they also diverge at some critical point.

B. Physico-chemical Characterisation

The high biological potency and small amount of SRS-A generated by immunological reactions created an obstacle to the isolation and identification of the mediator. Early work established that SRS-A was acidic, resistant to proteolysis, soluble in 80% ethanol but not in chloroform or acetone, was unstable in acid, and was inactivated by peroxides of ether and butanol (BROCKLEHURST 1953). Numerous groups endeavouring to purify SRS-A developed chromatographic procedures which took advantage of the known solubility and stability properties of SRS-A (ORANGE et al. 1973; MORRIS et al. 1978; TAKAHASHI et al. 1976; DRAZEN et al. 1979). These procedures included extraction of crude samples into 80% ethanol, base hydrolysis in 0.1 N NaOH for 30 min, sequential chromatography on Amberlite XAD-2 or 8, silicic acid, and DEAE-cellulose resins, and reversed phase high pressure liquid chromatography. These procedures were sufficient to permit spark source spectroscopy which gave evidence of sulphur in active SRS-A preparations, and to demonstrate the mediator's acidic nature, its protein-binding properties, its ultraviolet spectrum with peak absorption at 280 nm, its low molecular weight, and its sensitivity to arylsulphatase inactivation and to oxidation, but its resistance to phospholipases A_2 and C, to proteases and to 15-hydroxy prostaglandin dehydrogenase (ORANGE et al. 1973, 1974; MORRIS et al. 1978; TAKAHASHI et al. 1976; DRAZEN et al. 1979).

Unequivocal proof of the chemical structure and of purification of SRS-A, however, required two nearly simultaneous, but independent, advances which occurred between 1976 and 1979. The first group of advances regarded generation of SRS-A itself. Work by BACH and PARKER and their collaborators demonstrated that radioactive arachidonic acid could be incorporated into partially purified SRS-A (JAKSCHIK et al. 1977a; BACH et al. 1977). Following the finding by ORANGE et al. that SRS-A production could be enhanced by cysteine and other thiols (ORANGE and CHANG 1975; ORANGE and MOORE 1976), confirmed and extended by BACH (BACH and BRASHLER 1978), it was also demonstrated that ^{35}S could be incorporated into immunologically generated SRS-A (DAWSON et al. 1975) as a sulphur-containing metabolite of arachidonic acid. The acceptance of this information was hampered, however, by the unique ultraviolet spectrum of SRS-A (MORRIS et al. 1978), which was unlike any known prostaglandin metabolite of arachidonate, and the fact that insufficient amounts of radiolabelled material were available to perform structure proof studies.

The second set of advances which finally permitted structural identification of SRS-A concerned the growth in understanding of metabolism of arachidonic acid, most particularly the definition and elucidation of its lipoxygenase-dependent metabolic products. The key to unravelling the structure of SRS-A was the finding of a lipoxygenase-dependent family of metabolites termed leukotrienes (SAMUELSON et al. 1979, 1980) which possessed three conjugated double bonds and which, therefore, demonstrated an ultraviolet absorption spectrum with maximum at 280 nm. This understanding, taken together with the knowledge that arachidonic acid enhanced SRS-A production (JAKSCHIK et al. 1977b) whereas inhibition of arachidonate metabolism by both lipoxygenase and cyclo-oxygenase pathways (BACH et al. 1979; JAKSCHIK et al. 1977c), but not by cyclo-oxygenase alone (WALKER 1973), blocked SRS-A production, led to the surmise that SRS-A was a product of lipoxygenation of arachidonic acid. Previous chemical information suggested that thiols can form thioether adducts with lipid hydroperoxides (GARDNER et al. 1977), and led to the suggestion that SRS-A was generated by a similar reaction. Amino acid analysis of SRS-A then defined three lipid moieties with biological activity and substituted with amino acids at the 6 position; the moieties were composed of intact glutathione (PARKER et al. 1979; HAMMARSTROM et al. 1979), or its cleavage products cysteinylglycine (ORNING et al. 1980; MORRIS et al. 1980) or cysteine (SAMUELSON and HAMMARSTROM 1980), with the most active moiety being the cysteinylglycine fraction (DRAZEN et al. 1980; SOK et al. 1980).

Thus, the interest in SRS-A and in arachidonic acid metabolism coincided and now permits a clear understanding of the biochemical pathways which generate SRS-A (Fig. 1). Following cell activation by immunological means (guinea-pig or human lung) or by calcium ionophore (basophil leukaemia cells, mononuclear leucocytes) or phagocytic stimuli (ROUZER et al. 1980) (mononuclear leucocytes), arachidonic acid is liberated from intact membrane phospholipids by the action of phospholipase A_2 or from diacylglycerol by diacylglycerol lipase (KENNERLY et al. 1979). The action of lipoxygenase upon liberated arachidonic acid then generates 5-hydroperoxy eicosatetraenoic acid (5-HPETE), which is metabolised to the unstable leukotriene A. Glutathione (Cys-Gly-Glu) is then added to leukotriene A by formation of a thioether bond at the 6 position to form leukotriene C, one of the three forms of SRS-A. Transglutaminase, a ubiquitous enzyme, then

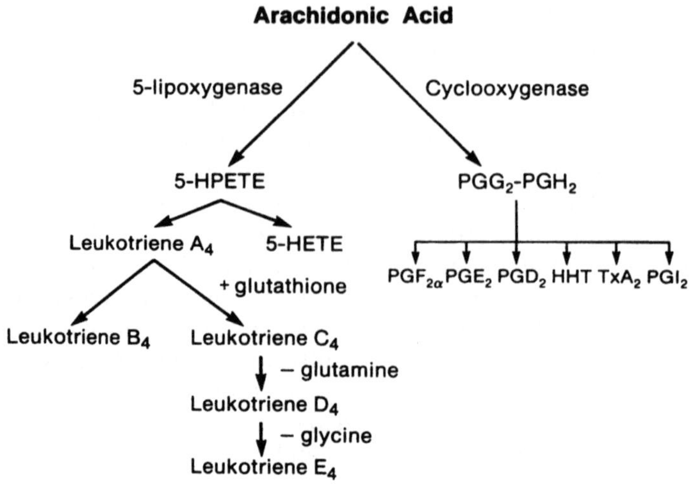

Fig. 1. Proposed biosynthetic pathway for leukotrienes (SRS-A)

cleaves the terminal glutamine from leukotriene C to form the most active SRS-A, leukotriene D. The final cleavage of glycine by an amino peptidase yields a molecule possessing cysteine as its sole amino acid, and termed leukotriene E. The stereochemistry of SRS-A is crucial and has been shown to require 7,9-*trans*-11,14-*cis* configuration for biological activity (SOK et al. 1980; PARKER et al. 1980).

C. Functional Characterisation

Initial biological studies employed crude or only partially purified material; however the findings regarding SRS-A developed with such preparations have been confirmed employing pure and/or synthetic molecules. Partially purified SRS-A induces sustained contraction of many smooth muscle preparations, including guinea-pig ileum and airway and human airway tissue, with particular potency on human peripheral airway tissue (DRAZEN et al. 1979). SRS-A is relatively ineffective in contracting the oestrous rat uterus or the ascending gerbil colon (ORANGE et al. 1979), preparations sensitive to bradykinin and prostaglandins E_1 and $F_{2\alpha}$, respectively. Inhalation of aerosols of partially purified preparations of SRS-A at high concentrations also induces bronchospasm in primates.

Utilising pure or synthetic leukotrienes, similar smooth muscle contractile profiles have been identified. Thus the specific activity of leukotriene C, D and E upon the guinea-pig ileum is approximately 1000, 4000 and 75 units/nmol, respectively, where 1 unit represents a contractile activity equivalent to that of 5 ng histamine base (SOK et al. 1980). All three leukotrienes also cause contraction of airways smooth muscle both in vitro and in vivo. Thus, leukotriene C, D and E contract guinea-pig peripheral airways at molar concentrations approximately 1/1000, 1/100000 and 1/10000 those of histamine required to induce similar contractions (DRAZEN et al. 1980). Contraction of guinea-pig central airways by leukotriene C and D is accomplished at concentrations 1/100 those of histamine, further indicating not only the potency of the leukotrienes but their relative selectivity for the peripheral airways (DRAZEN et al. 1980). In addition, leukotriene C and D are pulmonary agonists in vivo in guinea-pigs, inducing falls in pulmonary conductance and dynamic compliance after intravenous injection of 0.5 µg/kg. Such alterations in pulmonary mechanics occur 30–45 s after infusion of the mediator and persist for 5 min (DRAZEN et al. 1980). The bronchoconstrictive effects of synthetic leukotriene C_4 are enhanced in the presence of propranolol, suggesting the leukotriene is capable of inducing the release of adrenal catecholamines (WELTON et al. 1981). The leukotrienes are also potent constrictors of human bronchial smooth muscle (DAHLENN et al. 1980). Leukotriene C and D at concentrations greater than 10 ng/kg are systemic vasodepressors in anaesthetised guinea-pigs; however, in unanaesthetised animals hypertension is initially noted, followed within minutes by hypotension. In the skin of guinea-pigs, leukotriene C in amounts in excess of 10 ng is a vasoconstrictor whereas leukotriene D and E are both vasodilators active in amounts above 5 ng (DRAZEN et al. 1980; LEWIS et al. 1980). These effects occur at 1/100th the molar concentrations of histamine required to induce similar changes in vasopermeability and are potentiated by PGE_1 (WILLIAMS and PIPER 1980). Leukotriene D (0.5 µg) is also an active vasodilator in human skin (SOTER et al. 1983).

Less well characterised SRS-A moieties have been shown to act synergistically with histamine to contract guinea-pig ileum and to augment cardiac rate and induce dysrhythmias (LEVI and BURKE 1980). Recent observations also suggest that SRS-A or similar compounds could mediate smooth muscle contraction induced by other agonists, as indicated by the fact that C5a-induced tracheal contraction can be inhibited by FPL 55712 (REGAL and PICKERING 1981). The leukotrienes, in turn, may regulate inflammatory events indirectly through the induction of other biologically active materials. Thus, thromboxane A_2, prostacyclin and PGE_2 may be generated following the interaction of a variety of animal tissues with leukotrienes C_4 and D_4 (FOLCO et al. 1981; FEUERSTEIN et al. 1981; TERISHITA et al. 1981; ENGINEER et al. 1978). Moreover, the bronchoconstrictive and cardiovascular effects of intravenous leukotriene C_4 in guinea-pigs may be inhibited not only by the SRS-A inhibitor FPL 55712 but also by indomethacin (SCHIANTARELLI et al. 1981). These findings suggest a complex interrelationship of the various pathways of arachidonic acid metabolism in the induction of smooth muscle contraction, and add another level of complexity to the problem of non-steroidal anti-inflammatory drug-induced disorders.

D. Inactivation of SRS-A

Several enzymatic and non-enzymatic pathways for SRS-A inactivation have been defined. As sequential conversion of leukotriene C to E would be expected to alter the biological potency of the "SRS-A", it is important to differentiate inactivation from structural modification of the amino acid portion of the molecule.

Oxidation of SRS-A was the first defined mechanism for its destruction. Thus, peroxides of ether (BROCKLEHURST 1953), hydrogen peroxide (ORANGE et al.

1973), soybean lipoxygenase (SIROIS 1979) or mast cell peroxidase (HENDERSON and KALINER 1979) can degrade SRS-A. In addition, the intact eosinophil (WASSERMAN et al. 1975a) obtained from patients with hypereosinophilia can inactivate SRS-A in a time- and concentration-dependent manner. As such eosinophils are known to be "activated" (BASS et al. 1980) and to spontaneously generate superoxide and other metabolic products of oxygen, it is conceivable that eosinophils act to degrade SRS-A by oxidative pathways.

E. Summary

SRS-A, initially described more than 40 years ago, is now known to be a family of metabolites of arachidonic acid termed leukotriene C, D, and E. These molecules may be generated by immunological and non-immunological stimuli from a wide variety of cells and tissues. Such molecules are potent constrictors of airway smooth muscle, alter vascular permeability and affect vascular contractility. The discovery of their structure now permits identification of the cells which generate these principles, definition of the pathways of their inactivation and elucidation of the amounts and location of these mediators in defined disease states.

References

Augstein J, Farmer JB, Lee TB, Sheard P, Tattersall ML (1973) Selective inhibition of slow reacting substance of anaphylaxis. Nature (New Biol) 245:215–217

Austen KF, Lewis RA, Stechschulte DJ, Wasserman SI, Leid RW, Goetzl EJ (1974) Generation and release of chemical mediators of immediate hypersensitivity. In: Brent L, Holoborow J (eds) Progress in immunology II, vol 2. North-Holland, Amsterdam, pp 61–71

Bach MK, Brashler JR (1978) Stimulated production of slow reacting substance by mercaptans from ionophore A 23187-induced mononuclear cells: mercaptan structure-activity studies. Life Sci 23:2119–2126

Bach MK, Brashler JR, Gorman RR (1977) On the structure of slow reacting substance of anaphylaxis: evidence of biosynthesis from arachidonic acid. Prostaglandins 14:21–38

Bach JK, Brashler JR, Brooks CD, Neerken AJ (1979) Slow reacting substances: comparison of some properties of human lung SRS-A and two distinct fractions from ionophore-induced rat mononuclear cell SRS. J Immunol 122:160–165

Bass DA, Grover WH, Lewis JC, Szeida P, DeChatelet LR, McCall CE (1980) Comparison of human eosinophils from normals and patients with eosinophilia. J Clin Invest 66:1265–1273

Brocklehurst WE (1953) Occurrence of an unidentified substance during anaphylactic shock in cavy lung. J Physiol (Lond) 120:16

Brocklehurst WE (1962) Slow reacting substance and related compounds. In: Kallos P, Waksman BH (eds) Progress in allergy, vol 6. Karger, Basel, pp 539–558

Conroy MA, Orange RP, Lichtenstein LM (1976) Release of slow reacting substance of anaphylaxis (SRS-A) from human leukocytes by the calcium ionophore A 23187. J Immunol 116:1677–1681

Dahlenn SE, Hedquist P, Hammerstrom S, Samuelson B (1980) Leukotrienes are potent constrictors of human bronchi. Nature 288:484–485

Dawson W, Lewis RA, Tomlinson R (1975) The release of ^{35}S-labelled material and SRS-A from immunologically challenged guinea pig lungs. J Physiol (Lond) 247:37–38

Drazen JM, Lewis RA, Wasserman SI, Orange RP, Austen KF (1979) Differential effects of a partially purified preparation of slow reacting substance of anaphylaxis (SRS-A) on guinea pig tracheal spirals and parenchymal strips. J Clin Invest 63:1–5

Drazen JM, Austen KF, Lewis RA, Clark DA, Goto G, Marfat A, Corey EJ (1980) Comparative airway and vascular activation of leukotrienes C-1 and D in vivo and in vitro. Proc Natl Acad Sci USA 77:4354–4358

Engineer DM, Morris MR, Piper PJ, Sirois P (1978) The release of prostaglandins and thromboxanes from guinea pig lung by slow reacting substance of anaphylaxis, and its inhibition. Br J Pharmacol 64:211–216

Feldberg W, Kellaway CH (1938) Liberation of histamine and formation of lysolecithin-like substance of cobra venom. J Physiol (Lond) 94:187–226

Feuerstein N, Foegh M, Ramwell PW (1981) Leukotrienes C_4 and D_4 induce prostaglandin and thromboxane release from rat peritoneal macrophages. Br J Pharmacol 72:389–391

Folco G, Hansson G, Grastrom E (1981) Leukotriene C_4 stimulates TXA_2 formation in isolated sensitised guinea pig lungs. Biochem Pharmacol 30:2491–2493

Gardner HW, Kleiman R, Weisleder D, Inglett GE (1977) Cysteine adds to lipid hydroperoxide. Lipids 12:655–660

Hammarstrom S, Murphy RC, Samuelson B, Clark DA, Mioskowski C, Corey EJ (1979) Stereochemistry of leukotriene C-1. Biochem Biophys Res Commun 91:1266–1272

Henderson WR, Kaliner MA (1979) Mast cell granule peroxidase: location, secretion, and SRS-A inactivation. J Immunol 122:1322–1328

Jakschik BA, Falkenhein SF, Parker CW (1977a) Precursor role of arachidonic acid in slow reacting substance from rat basophil leukemia cells. Proc Natl Acad Sci USA 74:4577–4581

Jakschik BA, Kulczycki A, MacDonald HH, Parker CW (1977b) Release of slow reacting substance (SRS) from rat basophilic leukemia (RBL-1) cells. J Immunol 119:618–622

Jakschik BA, Falkenhein S, Parker CW (1977c) Precursor role of arachidonic acid in release of slow reacting substance from rat basophilic leukemia cells. Proc Natl Acad Sci USA 74:4577–4581

Jones DG, Kay AB (1974) Passive sensitization of guinea pig skin in vitro for the antigen-induced release of anaphylactic mediators. Clin Exp Immunol 16:213–221

Kaliner MA, Orange RP, Austen KF (1972) Immunological release of histamine and slow reacting substance of anaphylaxis from human lung. IV. Enhancement by cholinergic and alpha adrenergic stimulation. J Exp Med 136:556–567

Kaliner MA, Wasserman SI, Austen KF (1973) The immunological release of chemical mediators from human nasal polyps. N Engl J Med 289:277–281

Kellaway CH, Trethewie ER (1940) The liberation of a slow-reacting smooth muscle-substance in anaphylaxis. Q J Exp Physiol 30:121–145

Kennerly DA, Sullivan TJ, Sylvester P, Parker CW (1979) Diacylglycerol-arachidonic acid release J Exp Med 150:1039–1044

Levi R, Burke JA (1980) Cardiac anaphylaxis: SRS-A potentiates and extends the effects of released histamine. Eur J Pharmacol 62:41–49

Lewis RA, Wasserman SI, Goetzl EJ, Austen KF (1974) Formation of SRS-A in human lung tissue and cells before release. J Exp Med 140:1133–1146

Lewis RA, Drazen JM, Austen KF, Clark DA, Corey EJ (1980) Identification of the C (6)-S-conjugation of leukotriene A with cysteine as a naturally occurring slow reacting substance of anaphylaxis. (SRS-A). Biochem Biophys Res Commun 96:271–277

Morris HR, Taylor GW, Piper PJ, Sirois P, Tippins JR (1978) Slow reacting substance of anaphylaxis. Purification and characterization. FEBS Lett 87:203–206

Morris HR, Taylor GW, Piper PJ, Samhoun MN, Tippins JR (1980) Slow reactin substances (SRS-S): the structure identification of SRS-S from rat basophilic leukemia (PRL-1) cells. Prostaglandins 19:185–201

Ohnishi H, Kosuzume H, Kitamura Y, Yamaguchi K, Nobuhara M, Suzuki Y (1980) Structure of slow reacting substance in anaphylaxis (SRS-A). Prostaglandins 20:655–663

Orange RP (1975) Dissociation of the immunologic release of histamine and slow reacting substance of anaphylaxis from human lung using cytochalasins A and B. J Immunol 114:182–186
Orange RP, Austen KF (1969) Slow reacting substance of anaphylaxis. Adv Immunol 10:105–144
Orange RP, Chang PL (1975) The effect of thiols on immunologic release of slow reacting substance of anaphylaxis. J Immunol 115:1072–1077
Orange RP, Moore EG (1976) The effect of thiols on the immunologic release of slow reacting substance of anaphylaxis. II. Other vitro and in vivo models. J Immunol 116:392–397
Orange RP, Stechschulte DJ, Austen KF (1969) Cellular mechanisms involved in the release of slow reacting substance of anaphylaxis. Fed Proc 28:1710–1715
Orange RP, Austen WG, Austen KF (1971) Immunologic release of histamine and slow-reacting substance of anaphylaxis from human lung. I. Modulation by agents influencing cellular levels of cyclic 3′5′-adenosine monophosphate. J Exp Med 134:136–148
Orange RP, Murphy RC, Karnovsky ML, Austen KF (1973) The physiochemical characteristics and purification of slow-reacting substance of anaphylaxis. J Immunol 110:760–770
Orning L, Hammarstrom S, Samuelson B (1980) Leukotriene D: a slow reacting substance from rat basophilic leukemia cells. Proc Natl Acad Sci USA 77:2014–2017
Parker CW, Huber MM, Hoffman MK, Falkenhein SF (1979) Characterization of the two major species of slow reacting substance from rat basophilic leukemia cells as glutathionyl thioethers of eicosatetraenoic acids oxygenated at the 5 position. Evidence that peroxy groups are present and important for spasmogenic activity. Prostaglandins 18:673–686
Parker CW, Falkenhein S, Huber MM (1980) Sequential conversion of the glutathionyl side chain of slow reacting substance (SRS) to cysteinyl-glycine and cysteine in rat basophilic leukemia cells stimulated with A 23187. Prostaglandins 20:863–884
Parker CW, Koch DA, Huber MM, Falkenhein S (1980) Arylsulfatase inactivation as a major mechanism when ordinary commercial preparations of the enzyme are used. Prostaglandins 20:887–905
Paterson NAM, Wasserman SI, Said JW, Austen KF (1976) Release of chemical mediators from partially purified human lung mast cells. J Immunol 117:1356–1362
Regal VF, Pickering PJ (1981) $C5_a$ induced tracheal contraction: effect of an SRS-A antagonist and inhibition of arachidonate metabolism. J Immunol 126:313–316
Rouzer CA, Scott WA, Cohn ZA, Blackburn P, Manning JM (1980) Mouse peritoneal macrophages release leukotriene C in response to a phagocytic stimulus. Proc Natl Acad Sci USA 77:2928–2932
Samuelson B, Hammarstrom S (1980) Nomenclature for leukotrienes. Prostaglandins 19:645–647
Samuelson B, Borgeat P, Hammarstrom S, Murphy RC (1979) Introduction of a nomenclature: leukotrienes. Prostaglandins 17:785–787
Samuelson B, Borgeat P, Hammarstrom S, Murphy RC (1980) Leukotrienes: a new group of biologically active compounds. Adv Prostaglandin Thromboxane Leukotriene Res 6:1–18
Schiantarelli P, Bongrani S, Folco G (1981) Bronchospasm and pressor effects induced in the guinea pig by leukotriene C_4 are probably due to release of cyclooxygenase products. Eur J Pharmacol 73:363–366
Sirois P (1979) Inactivation of slow reacting substance of anaphylaxis (SRS-A) by lipoxidase. Prostaglandins 17:395–404
Sok DE, Pai JK, Atarache V, Sim CJ (1980) Characterization of slow reacting substances (SRS's) of rat basophilic leukemia (RBL-1) cells. Effect of cysteine on SRS-profile. Proc Natl Acad Sci USA 77:6481–6485
Soter NA, Lewis RA, Corey EJ, Austen KF (1983) Local effects of synthetic leukotrienes in human skin. J Invest Dermatol 80:115–119
Stechschulte DJ, Orange RP, Austen KF (1973) Detection of slow reacting substance of anaphylaxis (SRS-A) in plasma of guinea pigs during anaphylaxis. J Immunol 111:1585–1589

Takahashi H, Webster ME, Newball HH (1976) Separation of slow reacting substance of anaphylaxis (SRS-A) from human lung into four biologically active fractions. J Immunol 117:1039–1044

Terishita ZI, Fukui H, Hirata M, Terao S, Ohkawa S, Nishikawa K, Kikuchi A (1981) Coronary vasoconstriction and PGI release by leukotrienes in isolated guinea pig hearts. Eur J Pharmacol 73:357–361

Walker JL (1973) The regulatory function of prostaglandins in the release of histamine and SRS-A from passively sensitised human lung tissue. In: Bergstrom S, Bernhard S (eds) Advances in the biosciences, vol 9. Pergamon, Braunschweig, pp 235–239

Wasserman SI, Goetzl EJ, Austen KF (1975a) Inactivation of human SRS-A by intact human eosinophils and by eosinophil arylsulfatase. J Allergy Clin Immunol 55:72 (Abstract)

Welton AF, Crowley HJ, Miller DA, Yaremko B (1981) Biological activities of a chemically synthesized form of leukotriene E_4. Prostaglandins 21:287–296

Williams TJ, Piper PJ (1980) The action of chemically pure SRS-A on the microcirculation in vivo. Prostaglandins 19:779–789

Yecies LD, Wedner HJ, Johnson SM, Jakschik BA, Parker CW (1979a) Slow reacting substance (SRS) from ionophore A 23187 stimulated peritoneal mast cells of the normal rat. I. Conditions of generation and initial characterization. J Immunol 122:2083–2089

Yecies LD, Watanabe A, Parker CW (1979b) Slow reacting substance from ionophore A 23187-stimulated basophilic leukemia cells and peritoneal mast cells in the rat. I. Purification and comparison during sequential Sephadex LH-20 and thin layer chromatography. Life Sci 25:1909–1916

CHAPTER 23

Complement

A. G. BIRD

A. Introduction

Complement is a system of factors occurring in blood and tissue fluids that are activated characteristically by antibody–antigen interactions. The consequences of the activation of complement are the release of products which mediate a number of biologically significant events. In common with other major humoral effector systems, including the clotting, kinin and fibrinolytic systems, complement is composed of a series of discrete proteins present as precursor molecules. These, when activated by a variety of stimuli, result in a reaction sequence which proceeds in a predefined direction. Complement, like the other triggered enzyme systems, displays the inherent amplification kinetics of a chain reaction and each protein product becomes a catalyst for the next in sequence. The net result is a cascade phenomenon during which intermediate proteins and breakdown products possessing potent biological activity are produced. The whole system is under stringent internal control and regulation, but initial activation produces a rapid local inflammatory response at the activation site. The products of complement activation comprise all the major functions necessary for acute inflammation and their release in vasodilatation, increased capillary permeability, neutrophil chemotaxis and subsequent particle phagocytosis.

A key feature of the complement cascade is its integration into the humoral immune system as its key effector mechanism. Thus following antibody–antigen reactions during which antibody provides the recognition signal, the complement sequence and its resulting products result in the inflammatory responses which are such a major feature of antibody interactions. Without a fully functioning and integral complement system the ability of antibody to effectively localise and remove particulate antigen is severely impaired and can result in persistent or progressive infection or immune complex mediated tissue damage.

The protein chemistry of the complement reaction sequences and their important control influences will be discussed briefly. More details of the biochemistry can be found in the reviews by MÜLLER-EBERHARD (1975) and PORTER (1977). The relationship of complement to inflammatory and cellular interactions will be discussed and finally their importance in the initiation of tissue inflammation and damage of particular relevance to the skin will be presented.

B. The Classical Pathway

The activation of the complement system involves two discrete initial cascade systems comprising the classical and alternative pathways of activation (Table 1), both of which result in the bulk reaction sequence which involves the cleavage of C3. C3 cleavage should be regarded as the pivotal reaction of the system, and the resulting cleavage products are responsible for the majority of the biological consequences of complement activation. The route of classical complement activation is initiated by immune complexes of antigen and immunoglobulins of IgM class or IgG sub-classes 1, 2 or 3 in humans. In addition, the sequence can be triggered by a number of non-immunoglobulin-derived interactions, including polyanions, C-reactive protein and phosphoryl choline containing derivatives, and it can also be directly activated by certain micro-organisms, especially Oncorna viruses.

Initiation of the sequence begins with C1 fixation. C1 is a molecular complex of three subunit proteins C1q, C1r, and C1s, and interaction with the activating

Table 1. Characteristics of individual complement components

	Previous terminology	Cleavage products	Molecular weight	Mean serum concentration (μg/ml)
Classical pathway				
C1q	–	–	400000	70
C1r	–	$C\overline{1r}$	170000	48
C1s	–	$C\overline{1s}$	85000	35
C4	–	C4a, C4b	200000	450
C2	–	C2a, C2b	115000	25
C3	–	C3a, C3b	190000	1250
Alternative pathway				
C3b	–	C3b, C3c, C3d, C3e	170000	–
B	C3 pro-activator: glycine-rich β-glycoprotein	Ba, Bb	95000	200
D	C3 pro-activator convertase	–	25000	2
P	Properdin	–	200000	25
Membrane attack complex				
C5	–	C5a, C5b	206000	80
C6	–	–	128000	75
C7	–	–	121000	55
C8	–	–	154000	54
C9	–	–	79000	60
Control proteins				
$C\overline{1}INH$	–	–	105000	180
C4BP	–	–	570000	250
H	βIH	–	150000	475
I	C3B inactivator, KAF	–	100000	35

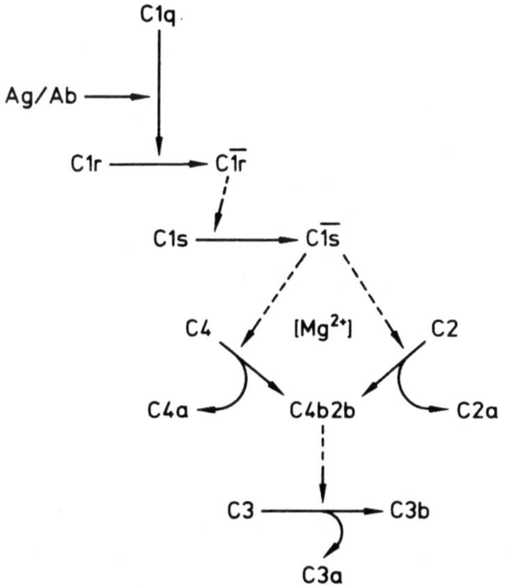

Fig. 1. Events leading to the generation of the classical pathway convertase C4b2b. A bar (e.g. $C\overline{1r}$) denotes a component in its active form exerting enzymic action on the next sequential component

moiety occurs via the C1q recognition subunit. C1q has a complex structure which has been visualised by electron microscopy and comprises a central protein core surrounded by six collagen-like arms which terminate in pod-like structures that are responsible for the initial recognition. Binding of a minimum of two arms results in steric change in the molecule and activates C1r, a serine esterase which in turn cleaves C1s in the first enzymic reaction of the sequence to form $C\overline{1s}$, which carries the major enzymic site of the $C\overline{1}$ molecule. $C\overline{1}$ molecule or C1 esterase has two natural substrates, the next two components of the sequence, C4 and C2. The fluid phase protein C4 is cleaved into two major fragments, C4a and C4b. C4b has the capacity to covalently bind to any activating surface in the vicinity by a short-lived binding site. However, even in optimal circumstances only a small proportion of this split product binds before the binding site decays, the residuum remaining as soluble (and inactive) C4b. The C4a fragment is released immediately into the fluid phase and plays no further role in pathway activation.

The next component in the sequence, C2, is similarly cleaved by $C\overline{1s}$ and the efficiency of breakdown is greatly enhanced by the binding by C4b of the native C2 in the presence of magnesium ions. Once again two reaction products result: C2a (formerly C2b) is released into the fluid phase, but C2b (formerly C2a) remains fixed to the bound C4b to form the C4b,2b complex, which is the next sequential enzyme. Each molecule of $C\overline{1s}$ is capable of cleaving many C4 and C2 molecules and, therefore, this step represents one of considerable amplification in the classical pathway.

C4b,2b is the C3 convertase generated by the classical pathway. The cleavage of the native C3 molecule is performed by C2b and the role of C4b in the complex is to bind the C3 in a satisfactory orientation for the reaction. Again, the action of C3 convertase represents a further amplification step but the magnitude of this is partially offset by the rapid decay of the convertase which results from dissociation of C2b (Fig. 1).

Cleavage of the two chain C3 is the bulk reaction of the sequence. The surface-bound C3 is split initially into two fragments: C3a, a 9000-dalton peptide which is released as a free fragment and is a reaction product with potent anaphylatoxin activity, and C3b, which has an available but short-lived binding site for the surface on which C4b2b (or alternative pathway) activation has taken place. Binding of the unstable C3b is inefficient and even in ideal conditions only a proportion of available molecules bind. Bound C3b is the active form of C3, playing a role in the generation of the next sequential enzyme, C5 convertase, but also itself having the effector property of interaction with phagocytic cells bearing C3b receptors.

C. The Alternative Pathway

The alternative pathway of complement activation represents the activation limb of this effector system available to provide a spontaneous defence against bacterial and fungal infection not necessarily dependent upon initial antibody recognition of the infection. The alternative pathway comprises two further serum components B and D. Factor D, unlike other complement components, circulates in the active state as a serine esterase. It is capable of splitting factor B once this latter component has undergone a conformational change by binding to the C3 conversion product C3b in the presence of magnesium ions. The complex of C3b and split factor B, C3bBb, becomes a convertase for further native C3, cleaving it in an analogous way to the classical convertase, C4b2b, and resulting in production of further C3b. A third factor, properdin (P), further stabilises the activity of C3bBb by binding C3b.

Thus, the alternative pathway, once primed with C3b, provides a positive amplification loop (Fig. 2). The positive reinforcement is limited by two control proteins, factors H and I, which together convert C3b to its inactive form, C3bi. Factor H has a major inhibiting effect on C3bBb. It effectively competes with B and Bb for binding to C3b and, therefore, will compete both for formation and for dissociation of the C3bBb complex. In addition it increases the susceptibility of C3b to cleavage by factor I. Factor I is a circulating active enzyme which acts on C3b, cleaving the alpha chain of this molecule, destroying much of its biological activity and resulting in production of C3bi. C3bi is then further degraded to C3c and C3d fragments by serine esterases (e.g. trypsin) which are not formally part of the complement system. Therefore, factors H and I play the major role in controlling the degree of activation of the alternative pathway, which is constantly "ticking over" as a result of continuous spontaneous generation of small amounts of C3b (RUDDY 1980). An increased rate of formation of C3b results from classical pathway activation or other enzymic degradation of C3.

The competition between factors B and H for binding to surface C3b is greatly influenced by the nature of the activating surface. The proportions of carbohydrate and sialic acid influence events, carbohydrate decreasing and sialic acid increasing the natural affinity of C3b for factor H. Some cell membranes or microbial surfaces with very low sialic acid content are potent activators of the alternative pathway as a result of the "protected surface" that they offer C3bBb from

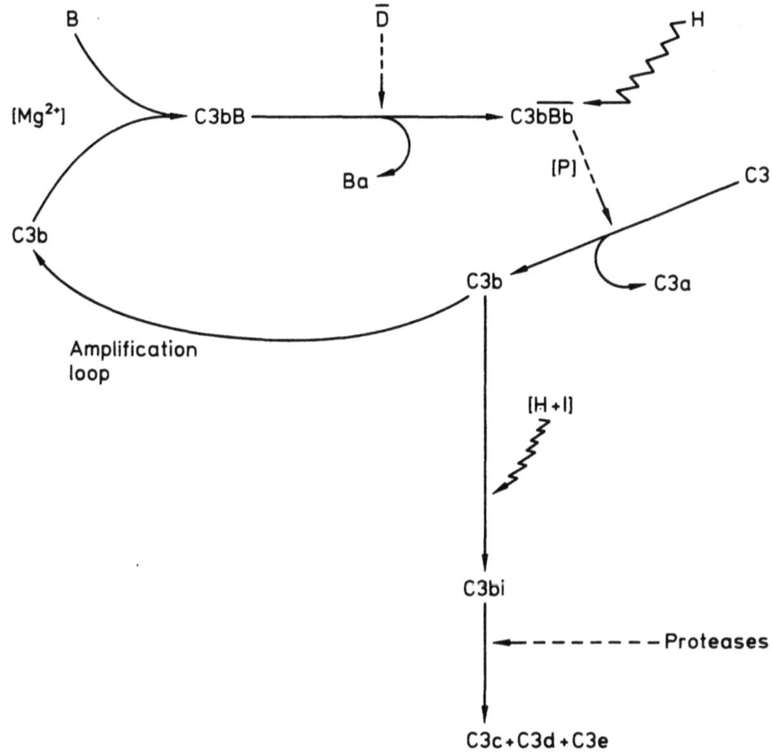

Fig. 2. Generation of the alternative pathway C3 convertase C3bBb and the subsequent cleavage of C3. Points of inactivation of the amplification loop by factors H and I are shown

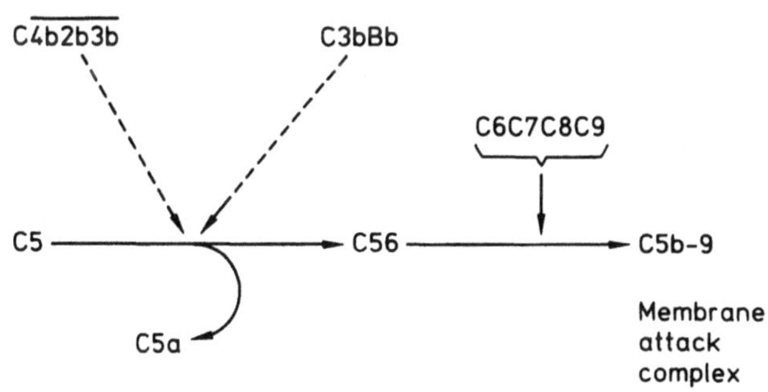

Fig. 3. Generation of the membrane attack complex following the cleavage of C5

the action of factors H and I. Most bacteria, yeast cell walls and all plants lack sialic acid. They support alternative pathway activation and, therefore, promote expression of activated C3 on their surface, expediting their removal by opsonisation or by activation of the membrane attack complex of complement (see later) (Fig. 3). Conversely, genetic deficiency or in vitro depletion of factors H and I results in the uncontrolled consumption of C3 by the C3bBb convertase. Fulminant activation of the alternative pathway is seen in two other situations. The serum of some patients with mesangio-capillary glomerulonephritis and most patients with partial lipodystrophy contains an IgG auto-antibody which stabilises the C3bBb complex from the action of factors H and I. The resulting hypocomplementaemia affects levels of factors B and C3 but early classical pathway component levels are unaffected. Also, experimental addition of cobra venom factor (CVF) to mammalian serum results in profound hypocomplementaemia affecting alternative pathway components. The active principle in CVF has recently been shown to be cobra C3b, which is resistant to displacement and degradation by factors H and I. Again the result is rapid and total C3 depletion is induced by unhindered firing of the amplification loop.

D. Terminal Events in the Complement Cascade

Following C3 cleavage, complement activation, whether initiated by classical or alternative channels, proceeds via a final common pathway culminating in the self-assembly of the membrane attack complex. C5 is split by the same convertases responsible for C3 breakdown (i.e. C3bBb or C4b2b) but this reaction requires appropriate presentation of C5. C3b, when either surface bound or in the fluid phase, can bind C5 and render the molecule susceptible to the C3/5 convertase. C3b itself is required and if combined with Bb or degraded to C3bi it is ineffective in this function. C5 conversion results in the generation of the split fragment C5a, which has important biological activity (see later), and its release is the result of the last enzymic reaction of the pathway. As with C2, C4 and C3 cleavage, the larger fragment C5b remains attached loosely to the activation site to continue the reaction. C5b contains a binding site for C6 and is eventually released from C3b. Following their association, C56 further incorporates C7, which creates a trimolecular complex with exposed hydrophobic binding sites. If the complex is not rapidly bound, it will decay in the fluid phase. The release and reattachment at this stage allows the final attack complex to associate at membrane sites at some distance from the initial site of fixation and is responsible for the phenomenon of lysis of bystander cells. Once C567 is bound, the site of the subsequent complement lesion is localised and lysis is induced by the incorporation of one molecule of C8 followed by two or three molecules of C9. Although some membrane damage can result from insertion of C8 alone, the efficiency of the process is markedly increased by the presence of C9. The final mechanism of membrane damage induced by the attack complex is still unclear.

E. Biological Activities Resulting from Complement Activation

The significance of complement activation results from the expression of split products and recognition molecules with profound biological activity in addition to the final common pathway of membrane lysis. The major activities are summarised in Table 2. The key split products are either released from the cleaved fragment and then serve as soluble mediators or are retained on the activation complex as bound intermediates and initiate recognition. Thus both released and

Table 2. Cell populations bearing complement receptors and biological effects of complement products

Receptor type	Cell	Biological function
C4a	Mast cell/basophil	Exocytosis
C3a	Mast cell/basophil Neutrophil Eosinophil	Exocytosis
Cell-bound C3b (CR1)	Erythrocyte Neutrophil Monocyte/macrophage B-lymphocyte Glomerular podocyte	Immune complex clearance Enhanced phagocytosis and exocytosis As above ? Antigen focussing Immune complex localisation
Cell-bound C3di (CR2)	B-lymphocyte	?
Cell-bound C3bi (CR3)	Neutrophil Monocyte/macrophage NK cell	Exocytosis Enhanced phagocytosis ? Enhance killing
C3e	Neutrophil	Blood recruitment from bone marrow
C5a	Mast cell/basophil Neutrophil Monocyte/macrophage	Exocytosis and leukotriene synthesis Chemotaxis, exocytosis, increased adhesion and CR1 expression Chemotaxis, exocytosis ? Leukotriene synthesis

retained cleavage fragments act as ligands for specific receptors on a variety of cells. The net result is the initiation of an inflammatory reaction at the site of complement fixation and subsequent phagocytosis and removal of the complexed antigen. The cleaved and retained intermediate products operate synergistically to co-ordinate the local inflammation, cell recruitment and ultimate clearance of the infective agent against which the humoral immune system is directed.

I. Diffusible Products

Membranes of mast cells and basophils have receptors for the cleavage products of C3 and C5, C3a and C5a respectively. Both molecules have anaphylatoxin activity and will degranulate local tissue mast cells in an analogous way to that produced by cross-linking of IgE molecules by antigen, although the interaction occurs via different mast cell receptors. Although both fragments express activity, C5a is considerably more potent in its action on mast cells and also exerts powerful effects on the neutrophil, causing directional chemotaxis and increased levels of metabolism. Moreover, C5a release induces neutrophil vessel margination as a result of increased endothelial adhesiveness. The cells are sequestered in capillary beds, particularly in the lung, where lysosomal release can result in local tissue damage associated with reciprocal blood neutropenia. C5a-induced neutrophil deposition in the pulmonary circulation has been incriminated in the neutropenia following haemodialysis (Craddock et al. 1977) and in the more severe pulmonary abnormalities associated with the adult respiratory distress syndrome (Hammarschmidt et al. 1980). The combined effects of C3a and C5a are to induce vasodilatation, neutrophil margination, chemotaxis and activation within tissues adjacent to sites of complement consumption.

C3e is a further diffusible fragment of biological importance. It is a 10000-dalton peptide derived from C3c by as yet undefined proteolytic enzymes. The fragment induces neutrophil recruitment from bone marrow reserves and experimental challenge induces leucocytosis. C3e is, therefore, of importance in increasing the availability of neutrophil at sites of complement-induced inflammation and thus in potentiating the effectiveness of C3a and C5a.

II. Retained Complement Products

The next major group of properties involve the ability of complement intermediate molecules to serve as cell recognition moieties. This recognition is mediated largely by C3b and its breakdown products and to a lesser extent by C4b. C3 intermediates present on particulate or cell surfaces activating complement can interact with complement receptors present on a wide range of cells (Table 2). The result of these interactions may be adherence of the particle only (red cells), opsonisation or enzyme exocytosis from neutrophils and activation of intracellular polymorph-killing enzymes. The range of activities expressed is determined by the nature of the cell or the relevant receptors involved.

Three different receptors for fragments of C3 have now been characterised and reclassified as complement receptors (CR) 1–3 (Kazatchkine and Nydegger 1982). CR1 is the C3b receptor also known as the immune adherence receptor and has been characterised as a 250000-dalton glycoprotein which is found on a wide variety of cell types, including erythrocytes, polymorphs, macrophages and mast cells. Erythrocyte CR1 receptors may facilitate the removal of circulating immune complexes of antigen/antibody expressing C3b whereas the receptors on macrophages and polymorphs aid phagocytosis and also induce the exocytosis of lysosomal enzymes from these cells. The CR1 receptor can, therefore, act as an opsonising receptor for C3b-coated particles in the same way as Fcγ receptor interactions with neutrophil and macrophage receptors. Fc interactions are capable of enhancing intracellular killing and thus co-operation between C3b and Fc binding induces both particle ingestion and effective degradation. There is also evidence that the intermediate C4b can bind to the CR1 receptor and is responsible for similar biological effects. CR1 expressed on B- and T-lymphocytes is thought to be of importance in initiation and regulation of the immune response by antigen (Klaus and Humphrey 1977).

The receptor for C3d is now reclassified as CR2 and is a 72000-dalton glycoprotein thought to be only expressed on a major subpopulation of B-lymphocytes; its function is, therefore, not clearly linked directly with inflammation or particle phagocytosis. The functional role of the CR2 receptor is still unclear, but it is known to be very closely associated with the route used by Epstein-Barr virus to infect human B lymphocytes (YEFENOF and KLEIN 1977; FINGEROTH et al. 1984).

The CR3 receptor has specificity for C3bi and was initially described on neutrophil and macrophage/monocyte membranes. The realisation that the monoclonal antibody MAC-1 recognises CR3 has now led to the localisation of CR3 on the natural killer cell subset also. CR3 is involved in the phagocytosis of C3bi particles by neutrophils and monocytes and also enhances the antibody-dependent cytotoxicity of killer cells (FEARON 1984).

III. Membrane Lysis

The cytolytic properties of complement were the first activities attributed to the system in studies on the phenomenon of immune haemolysis and the requirements of a heat labile factor for the lysis of certain bacterial strains in the presence of antibody. Complement lesions can be formed in many types of biological membrane, including those of nucleated cells, bacteria and enveloped viruses and the artificial membranes of liposomes. In general, although nucleated cells are susceptible to complement lesion damage, the scope of membrane repair mechanisms in these cells results in their relative resistance to complete cell lysis. Amongst bacteria it is believed that membrane lysis is particularly important in defence against *Neisseria* species. In humans with specific defects of terminal components of the membrane attack complex, severe recurrent infection with these organisms usually forms their only clinical susceptibility to bacterial infection. Disorganisation of the lipid bilayer in the membranes of enveloped viruses also makes these agents particularly susceptible to damage by the membrane attack complex (BERRY and ALMEIDA 1968).

IV. Solubilisation of Immune Complexes

Immune complexes of antigen/antibody exert the majority of their biological effects through their interaction with the complement system. The consequences are determined by the nature of the complex itself and its site of deposition or localisation. However, in addition, the interaction of complement with complexes also alters their biological properties. MILLER and NUSSENZWEIG (1975) provided the first evidence that complement could play a role in reducing the size of large insoluble immune complexes to smaller "soluble" moieties of less than 19S. It was demonstrated that solubilisation occurred rapidly and involved the fixation of C3b on the surface of the complex almost exclusively by activation of the alternative pathway. This surprising dependence on the alternative complement limb, which is usually regarded as antigen independent, has recently been confirmed by workers who have demonstrated the phenomenon in a medium where only purified alternative pathway components have been introduced (FUJITA et al. 1981). C3bBb is deposited on the surface of the complex, where it appears to be partially protected from the action of factor I. Large quantities of C3b remain on the complex and result in dissociation of large insoluble complexes to smaller soluble units bearing surface C3b. The C3b is apparently inserted by covalent bonding and probably sterically inhibits re-association, perhaps by inhibiting Fc–Fc interactions. The biological significance of this phenomenon is that it may protect the individual from the phlogistic effects of large complexes such as those found in the circulation in systemic lupus erythematosus. The presence of C3b on the soluble aggregates also allows their interaction with C3b receptor bearing cells, including monocytes/macrophages of the reticulo-endothelial system and neutrophil polymorphs, and plays a key role in their physiological removal. The interaction of C3b on immune complexes with the dendritic reticular cells of lymph node germinal centres also appears to play a major role in the generation of immunological memory (KLAUS and HUMPHREY 1977).

F. Regulation of Complement Activation

The self-perpetuating and accelerating nature of the complement cascade and the superimposition of the continuously cycling alternative pathway loop necessitate that the system is maintained under tight regulatory control. The importance of such control measures is manifest since in clinical medicine it is deficiency of the control proteins rather than deficiencies of the integral components of the complement cascade which gives rise to the most florid clinical manifestation that result from unbridled complement activation.

Modulation of the system occurs at two major levels. The first level of control is the result of the inherent inefficiency and instability of the early components. Following their generation, many complement intermediate fragments such as C4b and C3b are only transiently active. If they fail to locate their binding sites, they rapidly decay. Experimental evidence suggests considerable inefficiency of binding of these intermediates (COOPER and MÜLLER-EBERHARD 1967). In addition, the enzyme systems generated by these bound intermediates are subject to rapid internal decay. The classical and alternative pathway C3 convertases have short half-lives and the active site of the complex disintegrates, leaving the residual intermediates free to regenerate again if further native C2 or factor B is available (MÜLLER-EBERHARD 1975). This internal redundancy plays a major role in limiting the capacity of complement activation to become uncontrolled.

Superimposed on these innate mechanisms are a series of serum proteins which form natural inhibitors of the sequence and limit the cascade at major points of amplification.

I. $C\bar{1}$ Inhibitor

$C\bar{1}$ inhibitor ($C\bar{1}$INH), a plasma α_2-globulin, is a glycoprotein with an approximate molecular weight of 90000. It inhibits $C\overline{1r}$, $C\overline{1s}$ and a number of other serine esterases by irreversibly blocking their active combining sites. $C\bar{1}$INH will only interact and inhibit active enzymes and is ineffective upon native C1r or C1s. It is probable that $C\bar{1}$INH functions as a false substrate for serine esterases but then fails to redissociate and is, therefore, removed as it inhibits. $C\bar{1}$INH also inhibits other plasma proteases, including plasmin-activated Hageman factor and kallikrein of the other triggered enzyme systems (SCHREIBER et al. 1973).

Inherited deficiency of C$\bar{1}$INH produces the clinical syndrome of hereditary angio-oedema. In this condition uncontrolled activity of C$\bar{1}$ results in consumption of the substrates C2 and C4 and marked reduction in their serum levels. C2 and C4 consumption is associated with intermediate fragment release and in particular to appearance of poorly characterised C2 kinin which results in clinical manifestations. The defect occurs as a familial trait and is inherited as an autosomal dominant since the quality of inhibitor synthesised by heterozygotes is insufficient to control the activity of the classical pathway. Clinical features include painless, non-irritant swellings of the extremities and face, including mucous membranes and tongue. Attacks are sometimes precipitated by local trauma. In children, in particular, the swelling may be preceded by a transient skin rash and in some kindreds the disease may take a purely abdominal form in which mucosal and mesenteric oedema induces abdominal pain, diarrhoea and vomiting. Involvement of the larynx or epiglottis during an attack can result in acute respiratory obstruction and is a cause of fatality in untreated patients.

There are two inherited forms; one is associated with a total reduction of inhibitor levels measured both antigenically and functionally, while in the other ($<15\%$ of cases) a non-functional variant molecule is present in normal levels but there is a marked reduction in functional activity. In all active cases there is marked reduction of the early components C2 and C4 in serum and also of C$\bar{1}$INH in the common form of the disease. C1q levels are characteristically normal, however. Detection of affected families is reliably achieved by quantitation of serum C4 levels as a screening test in suspected cases followed by antigenic and functional quantitation of C$\bar{1}$INH levels in those with low C4 levels.

The condition can be satisfactorily treated by prophylactic administration of the plasmin inhibitors ε-aminocaproic and tranexamic acid, or preferably by the use of non-virilising anabolic steroids danazol or stanazolol. The use of these latter agents results in the enhanced synthesis of C$\bar{1}$INH by the normal heterozygote gene to levels adequate for normal homeostasis of classical pathway activation in both forms of the inherited condition. This leads to clinical, but not total immunochemical, correction of the defects (GELFAND et al. 1976). Activation of the classical complement pathway results in consumption of C$\bar{1}$INH but this is usually balanced by increased synthesis. Rarely, in association with lymphoma or benign paraproteinaemia, an acquired form of C$\bar{1}$INH deficiency is observed in which clinical features are associated with reduction of C2, C4 and C$\bar{1}$INH levels but also of C1q in patients without a relevant family history. Such patients also respond well to the same prophylaxis as is administered for the hereditary forms of the disease.

II. C4 Binding Protein

C4 binding protein (C4BP), a β-globulin, binds only to activated C4b via multiple binding sites. Its association with C4b renders the latter susceptible to the action of factor I, which cleaves C4b to further fragments C4c and C4d. Moreover, C4BP displaces C4b from C42 in an analogous way to factor H action on bound C3b, again exposing C4b to the action of factor I. C4BP, therefore, has a key role in limiting classical pathway amplification.

III. Factors H and I

The major importance of the two control proteins factors H and I is in controlling the availability of C3b for the generation of C5 convertase and the feedback loop of the alternative pathway, as has already been described. Congenital deficiency or experimental serum removal of either protein results in rapid depletion of factor B and C3 as a result of alternative pathway deregulation. Persistence of undegraded C3b enables factor D to cleave factor B and results in $C\overline{3bBb}$, which remains active due to the failure of control normally imposed by the synergistic action of factors H and I. Genetic deficiencies of both control proteins have now been described. In each deficiency there is marked reduction in serum C3 levels, and predisposition to recurrent bacterial infections is seen in a similar way to that described in C3 component deficiency (see later).

G. Inherited Deficiencies of Complement Components

Genetic deficiencies of individual complement components are rare. Nevertheless, these experiments of nature yield the same insight into the physiological function as other inborn errors of metabolism have into other biochemical and metabolic pathways. Many of the single component defects described have been associated with skin disease manifestations and, therefore, are worthy of detailed consideration.

Deficiencies of almost every individual component have now been described, but clinical presentation can be grouped under three major categories depending on the position of the component in the reaction sequence (AGNELLO 1978; LACHMANN and ROSEN 1978).

Table 3. Disease associations reported in patients with genetic deficiencies of individual complement components or inhibitors

Component	Disease associated
C1q	SLE syndrome
C1r	SLE syndrome, chronic GN
C1s	SLE
C4	SLE, GN
C2	SLE, GN, Henoch-Schönlein purpura, DLE, chronic vasculitis. Also compatible with apparent good health
C3	Recurrent severe bacterial infections, arthritis
C5	SLE, recurrent Gram-negative infections
C6	Recurrent neisserial infections, SLE. Healthy individuals reported
C7	Neisserial infections, GN and SLE
C8	Neisserial infections and SLE
H, I	Recurrent severe bacterial infections, haemolytic anaemia, GN
$C\bar{1}INH$	Hereditary angio-oedema, SLE, DLE, GN

DLE, discoid lupus erythematosus; GN, glomerulonephritis; SLE, systemic lupus erythematosus.
References: AGNELLO (1978), LACHMANN and ROSEN (1978), DONALDSON et al. (1977)

I. Bacterial Infections

The key role of complement in initiation of recognition and removal of bacteria in association with antibody has already been stressed. Single component deficiency might, therefore, be expected to predispose to bacterial infections in a similar way to antibody deficiency in hypogammaglobulinaemic states. In practice, it appears that the alternative and clinical pathways can cross-compensate for deficiency in either activation limb and the terminal membrane attack complex is not of key importance in destruction of many bacterial species. In fact, only absolute absence of C3, whether the result of homozygous C3 deficiency or secondarily as the result of continuous alternative pathway C3 consumption consequent upon factor H or I deficiency, is associated with widespread susceptibility to bacterial infection. The crucial role of C3 in neutrophil recruitment (C3e), chemotaxis (C5a) and opsonisation (C3b) is exemplified by the severity of pyogenic infections in the affected cases.

Deficiencies of individual components of the membrane attack complex also predispose to bacterial infections but with a much more limited range of individual species. Neisserial infections, usually systemic and recurrent, are characteristic of late component deficiencies and suggest that complement-induced membrane damage may be an important factor in the host resistance to these organisms, which are usually handled normally in immunoglobulin deficiency.

II. Auto-immune or Immune Complex Diseases

Deficiencies of individual early components of the classical pathway do not usually present with bacterial susceptibility. Compensation by alternative pathway fixation of C3 is presumably adequate for bacterial defence. However, such individuals have a high incidence of immune complex disease and in particular of discoid and systemic lupus erythematosus (SLE). In addition other dermatological manifestations, including Henoch-Schönlein purpura, and other evidence of vasculitis have been described.

There are two potential explanations for this strong association. The first invokes the finding that two key complement components of the classical pathway, C4 and C2, are encoded by genes lying within the major histocompatibility complex (MHC) of mice and man. It is, therefore, postulated that disease susceptibility is not the direct result of the factor deficiency itself but is determined by other MHC disease susceptibility genes which are in linkage disequilibrium with the complement alleles. The high incidence of SLE in patients totally deficient in C2 (Agnello 1978) and carrying one null gene for C4 (Fielder et al. 1983) gives some support for this hypothesis. However, the increasing recognition that inherited deficiencies of early components (e.g. C1q, C1r) not linked to MHC genes are also strongly associated with SLE-like states argues strongly that it is the deficiencies themselves rather than linked disease susceptibility genes which are responsible. Further support for this concept derives from the observation that patients with hereditary angio-oedema who are chemically depleted of early classical component also have a markedly increased incidence of SLE-like disease (Donaldson et al. 1977).

These pooled clinical observations strongly suggest that it is the early component deficiencies themselves which predispose to immune complex mediated disease and attention has been shifted to the role of the classical pathway in solubilisation and removal of formed immune complexes. The importance of complement in reducing the inflammatory potential of immune complexes by dissolution has been discussed earlier. PORTER (1983) has speculated in detail how polymorphic variants of complement components may result in impaired or inhibited activation of complement. The efficiency of activation of the total sequence may determine the extent of susceptibility to immune complex damage resulting from failure of immune aggregate dissolution. Inefficiency may, therefore, result from the absence of compatible polymorphic molecules in the sequence as much as from the total deficiency resulting from possession of a null allele.

Finally, isolated complement component deficiencies may result in the failure to successfully eliminate organisms of relatively low pathogenicity. The importance of complement in removal of certain viral strains has already been emphasised, and failure of the complement sequence may then favour viral persistence and continued production of whole viral particles or subunit antigens. Production of such antigens in the presence of an intact humoral response would result in a continuous source of immune complex formation, with the potential for precipitation of SLE-like syndromes.

H. Complement in Skin Disease

The epidermal basement membrane of the skin and the vessels of the dermis are frequent sites for the localisation of antibody–antigen reactions. Deposition of immune complexes, whether found locally or at a distance, results in secondary activation of complement by both classical and alternative pathways. Similar involvement of the complement cascade results from the localisation of auto-antibodies directed against integral skin antigens which characterise the bullous skin diseases.

I. Immune Complex Disease

Systemic lupus erythematosus is associated with the tissue deposition of immune complexes found particularly between nuclear antigens, especially double-stranded DNA, and their respective antibodies. In the skin, these complexes are typically found in granular deposition at the dermo-epidermal junction, where they can be detected by direct immunofluorescence of biopsied skin. These complexes contain early classical pathway complement components, indicating that activation is predominantly by the classical pathway. Factors responsible for the anatomical distribution of complexes in the skin in this condition still remain undefined.

Small vessels of the dermis represent the most important site for localisation of immune complexes in other connective tissue disorders, particularly rheumatoid arthritis and Sjögren's syndrome. Mixed essential cryoglobulinaemia, chronic hepatitis B carriage and post-streptococcal antigenaemia are also associated with skin immune complex deposition in a similar distribution. Whilst circulating immune complexes and evidence of complement activation can often be detected in the serum of such patients, the prominent involvement of complement in these lesions is most clearly demonstrated by direct immunofluorescence examination of affected skin. All experimental evidence in these conditions suggests that the pathological damage occurs as a direct result of immune complex activation of complement and the consequent inflammation and neutrophil recruitment. Mast cell degranulation and neutrophil enzyme exocytosis cause severe damage to vessel integrity and eventual disruption of local tissues. Histologically, the appearances are those of a leucocytoclastic vasculitis which is again the result of the complement activation. In many of these conditions, the skin deposition of immune complexes is associated with the presence of serum cryoprecipitates. These complexes have physical characteristics which result in their precipitation at temperatures below 37 °C and this may be a major factor in their preferential localisation within the skin, particularly in the cooler peripheries.

The uncontrolled fluid phase activation of early complement components which occurs in $C\bar{1}INH$ deficiency induces a marked increase in the dermal vessel permeability and results in angio-oedema. However, C5 activation is minimal and, therefore, neutrophil recruitment is not prominent in the tissues in this hereditary disease.

II. Bullous Skin Diseases

Dermatological diseases in which epidermal blisters are prominent form a heterogeneous group of disorders but in all there is direct evidence for the presence of activated complement in the sites of the disease. However, deposition of complement components and immunoglobulins may be insufficient to explain the full clinical picture since the reactants also often appear prominently in apparently normal skin adjacent or distant to sites of active disease. Therefore, additional factors may be important in the initiation or full expression of the clinical disease.

Pemphigus vulgaris and bullous pemphigoid are chronic blistering skin diseases affecting predominantly middle-aged and elderly patients respectively. In each, an IgG auto-antibody appears against defined skin antigens and is usually present in excess in the serum of affected individuals. In pemphigus, the auto-antibody is deposited in the intercellular spaces between epidermal cells, and C1q, C4, and C3 are also present at this site in perilesional skin. However, a recent report (Hashimoto et al. 1983) has questioned the role of complement in the production of these lesions and has provided evidence that the activity of auto-antibody lies in its specificity for local plasminogen activator. Nevertheless, it is probable that at least the complement activation present at sites of active disease will exacerbate the pathological damage and favour the localisation of further tissue reactive auto-antibody.

In bullous pemphigoid, auto-reactive IgG is directed sub-epidermally at the level of the lamina lucida in a linear distribution along the epidermal basement membrane. Complement is often, although not invariably, seen in the same distribution. The variation in complement deposition may be explained by the recent observation that the IgG present is often predominantly and sometimes exclusively of the IgG4 sub-class (BIRD et al. 1985). Human IgG4 fails to fix complement by the classical pathway, although when aggregated it will activate the alternative pathway under appropriate conditions. Unlike pemphigus, pemphigoid is not reproducibly induced in higher primate skin by transfer of serum containing the auto-antibody, and this finding suggests that other factors may also be involved in the initiation of the blister lesions.

A related condition, herpes gestationis, appears exclusively in pregnancy and in the immediate post-partum period. It manifests as a self-limiting disease sometimes transmitted transplacentally to the neonate. Immunofluorescence of the skin reveals prominent linear C3 deposition either with or without early classical pathway components, but immunoglobulin deposition is often sparse or absent. Recent work suggests that failure to visualise immunoglobulin is the result of insensitivity of immunofluorescence and that small amounts of IgG are consistently present in the basement membrane zone (JORDON et al. 1976). The relationship between the antibody found in herpes gestationis and that found in pemphigoid remains unclear but the absolute temporal association of the former condition with pregnancy or the early post-partum period suggests a fundamental difference in aetiology.

Dermatitis herpetiformis is the final major skin blistering condition in which complement may play a role in pathogenesis. The eruption in this disease is characterised by much smaller vesicles which are intensely irritant. Histologically, the early lesions show micro-abscesses comprising mainly polymorphs which appear prominently over the dermal papillae. In more advanced lesions the papillary apices separate from the overlying epidermis. The immunopathology of the lesions is characteristic, with IgA deposition, usually in granular form, localised with C3 along the dermo-epidermal junction, maximally over dermal papillary tips. Immunological deposits are as obvious in biopsies from unaffected skin. This finding indicates that further factors in addition to the reactants seen are necessary to produce overt lesions. Use of antisera against early classical complement components and alternative pathway factors indicates that complement activation is the result of predominant alternative pathway activity (PROVOST and TOMASI 1974). This finding is in keeping with the known inability of IgA to efficiently activate the classical pathway after complexing with antigen. The source of the IgA and its antigenic specificity are still uncertain in this disease and no antibody with reactivity for skin antigens is present in the serum, unlike in pemphigus and pemphigoid.

A clue to the source of the IgA may well lie in the strong association between dermatitis herpetiformis and bowel disease; in some patients small intestinal disease indistinguishable from coeliac disease is present and in a much larger proportion macroscopic or microscopic mucosal lesions are present without overt clinical bowel disease. Removal of gluten from the diet leads to clinical improvement in the skin disease. Recent evidence suggests that the IgA in the skin is dimeric although it does not contain secretory component (UNSWORTH et al. 1982). This finding suggests that the IgA may arise from a mucosal origin but since it does not contain secretory component it is not in the full secreted form.

References

Agnello V (1978) Complement deficiency states. Medicine 57:1–23
Berry DM, Almeida JD (1968) The morphological and biological effects of various antisera on avian infectious bronchitis virus. J Gen Virol 3:97–105
Bird P, Friedmann PS, Ling N, Bird AG, Thompson RA (1986) Subclass distribution of IgG autoantibodies in bullous pemphigoid. J Invest Dermatol 86:21–25
Cooper NR, Müller-Eberhard HJ (1967) Molecular measurement of the fourth component of human complement. In: Peeters H (ed) 15th Colloquium of Protides of the Biological Fluids. Elsevier, Amsterdam, pp 453–457
Craddock PR, Hammerschmidt D, White JG, Dalmass AP, Jacob HS (1977) Complement (C5a) induced granulocyte aggregation in vitro. J Clin Invest 60:260–264
Donaldson VH, Hess EV, McAdams AJ (1977) Lupus erythematosus-like disease in three unrelated women with hereditary angio-neurotic oedema. Ann Intern Med 86:312–313
Fearon DT (1984) Cellular receptors for fragments of the third component of complement. Immunology Today 5:105–110
Fielder AHL, Walport MJ, Batchelor JR, Rynes RJ, Black CM, Dodi IA, Hughes GRV (1983) A family study of the MHC of patients with SLE. Null alleles of C4A and C4B may determine disease susceptibility. Br Med J [Clin Res] 286:425–428
Fingeroth JD, Weiss JJ, Tedder TF, Strominger JL, Biro PA, Fearon DT (1984) Epstein-Barr virus receptor of human B lymphocytes is the C3d receptor CR2. Proc Natl Acad Sci USA 81:4510–4514
Fujita T, Takata Y, Tamura N (1981) Solubilisation of immune precipitates by six isolated alternative pathway proteins. J Exp Med 154:1743–1751
Gelfand JA, Sherin RJ, Alling DW, Frank MM (1976) Treatment of hereditary angio-oedema with danazol. N Engl J Med 295:1444–1448
Hammarschmidt DE, Weaver LJ, Hudson LD, Craddock PR, Jacob HS (1980) Association of complement activation and elevated plasma C5a with adult respiratory distress syndrome: pathophysiological relevance and possible prognostic value. Lancet I:947–949
Hashimoto K, Shafran KM, Webber PS, Lazarus GS, Singer KH (1983) Anti-cell surface pemphigus autoantibody stimulates plasminogen activator activity of human epidermal cells. J Exp Med 157:259–272
Jordon RE, Heine KG, Tappeiner G, Bushkell LL, Provost TT (1976) The immunopathology of herpes gestationis: immunofluorescence studies and chracterisation of "HG factor". J Clin Invest 57:1426–1431
Kazatchkine MD, Nydegger UE (1982) The human alternative pathway: biology and immunopathology of activation and regulation. Prog Allergy 30:193–234
Klaus GGB, Humphrey JH (1977) The generation of memory cells. I. The role of C3 in the generation of B memory cells. Immunology 33:31–45
Lachmann PJ, Rosen FS (1978) Genetic defects of complement in man. Springer Semin Immunopathol 1:107–119
Miller GW, Nussenzweig V (1975) A new complement function: solubilisation of antigen-antibody aggregates. Proc Natl Acad Sci USA 72:418–422
Müller-Eberhard HJ (1975) Complement. Annu Rev Biochem 44:697–715
Porter RR (1977) Structure and activation of the early components of complement. Fed Proc 36:2191–2196
Porter RR (1983) Complement polymorphism, the major histocompatibility complex and associated diseases: a speculation. Mol Biol Med 1:161–168
Provost TT, Tomasi TB (1974) Evidence for activation of complement via the alternative pathway in skin diseases. II. Dermatitis herpetiformis. Clin Immunol Immunopathol 3:178–186
Ruddy S (1980) Function of the control proteins of the classical and alternative complement activation pathways. In: Thompson RA (ed) Recent advances in clinical immunology, vol 2. Churchill Livingstone, Edinburgh, pp 91–111

Schreiber AD, Kaplan AP, Austen KF (1973) Inhibition of C$\bar{1}$ INH of Hageman factor fragment activation of coagulation, fibrinolysis and kinin-generation. J Clin Invest 52:1402–1409

Unsworth DJ, Payne AW, Leonard JN, Fry L, Holborow EJ (1982) The IgA in dermatitis herpetiformis is dimeric. Lancet I:478–480

Yefenof E, Klein G (1977) Membrane receptor stripping confirms the association between EBV receptors and complement receptors on the surface of human B lymphoma lines. Int J Cancer 20:347–352

CHAPTER 24

Neutrophil and Eosinophil Chemotaxis and Cutaneous Inflammatory Reactions

A. J. WARDLAW and A. B. KAY

A. Introduction

Neutrophil and eosinophil leucocytes are derived principally from the bone marrow although colony forming cells circulate in the peripheral blood. Together with basophils they form the granulocyte series of white cells. Mature granulocytes are incapable of division ("end cells") and have a life span of days once they have entered the circulation. Granulocytes exhibit random and directional locomotion (chemokinesis and chemotaxis), possess varying degrees of phagocytic activity, and actively secrete lysosomal enzymes and other biological agents following contact with an appropriate stimulus. Many of these properties are shared by mononuclear phagocytes but the neutrophil, and to a lesser degree the eosinophil, is more rapidly mobilised to the sites of inflammation.

Neutrophils are traditionally considered as primary phagocytic cells with the ability to rapidly ingest and kill common invading micro-organisms such as staphylococci and Gram-negative rods. However, the cell has the potential for extracellular digestion of basement membranes and other connective tissue and might play an additional role in facilitating the movement of other inflammatory cells to the sites of injury (WRIGHT 1982).

Eosinophils are also capable of phagocytosis but it seems unlikely that this is a major function. There is considerable evidence to support the view that eosinophils act principally in adaptive immunity against helminthic larvae and that they adhere to, and secrete their enzymes on to, the surface of worms which have been appropriately opsonised by antibody and/or complement (KAY 1980, 1985). Large numbers of eosinophils also appear in various hypersensitivity states, and it has been suggested that, in reactions associated with the release of mast cell constituents, they might also play a role in the inactivation of pharmacological mediators.

Neutrophils and eosinophils, like other phagocytic cells, display directional (chemotaxis) and increased random (chemokinesis) locomotion. There has been considerable debate as to whether cell accumulation in vivo is due wholly or in part to chemotaxis (chemotaxis being essentially an in vitro observation). Nevertheless, a number of diseases have been recognised in which there is a primary defect in neutrophil locomotion. Many of these conditions are associated with cutaneous manifestations. Isolated defects of eosinophil locomotion have not been described.

In this short review neutrophil and eosinophil chemotaxis and the regulation of the chemotactic response in the context of skin disorders will be discussed.

Mention will also be made of certain diseases characterised by cutaneous neutrophilia and/or eosinophilia and whether these conditions involve the participation of specific chemotactic factors for these cell types.

B. Methods of Measuring Chemotaxis and Cell Accumulation

Granulocytes undergo random locomotion even in a resting state, involving movement in a straight line for short periods with frequent random turning events. Stimulation results in an increased rate of random locomotion (orthokinesis), which under the influence of a chemical stimulus is termed chemokinesis, and an increased rate of turning (klinokinesis). Chemotaxis involves directional locomotion down a concentration gradient and is the most effective and likely mechanism for cell accumulation in vivo. Locomotion is also affected by the physical environment (contact guidance), e.g. neutrophils will preferentially migrate down the long axis of collagen fibres (for a fuller discussion, see WILKINSON 1985). Standard locomotion assays are ideal for assessing whether factors possess chemotactic activity but do not readily distinguish between the various types of locomotion. Detailed study of the nature of the locomotive response requires expensive equipment and cannot be performed on large populations of cells (WILKINSON 1986).

Investigation of granulocyte locomotion may be performed in vitro and in vivo. The most commonly used in vitro method is that of Boyden (BOYDEN 1962). The Boyden chamber consists essentially of two compartments divided by a micropore filter. In the upper (cell) compartment a suspension of granulocytes (neutrophils or eosinophils) is placed and in the bottom (test) compartment the chemotaxin is introduced. Following an incubation period, the filter is removed, fixed and stained. Cell locomotion is expressed either as the number of cells which have traversed the entire thickness of the filter (lower surface count) or by measuring the distance which the "leading front" of cells has travelled from the upper surface into the filter. A recent modification of this technique has involved the use of multiwell microchemotaxis chambers (FALK et al. 1980). These are simpler to use, require less cells and chemoattractant and make possible more reproducible results than individual chambers.

The Boyden chamber method does not distinguish between chemokinesis and chemotaxis. Such distinction can be achieved by placing varying concentrations of the chemoattractant above and below the filter to see the extent to which abolition of the gradient diminishes the directional locomotive response. Another technique, which is rapid and simple, involves placing the cells in agar in a microtitre well together with the chemoattractant. This measures only chemokinesis. Alternatively, cells are placed in a well cut in agar separated from the chemoattractant placed in another well. Identification of different types of cell is more difficult using the agar method, so that pure populations of cells are necessary.

Pure populations of neutrophils are obtained by dextran sedimentation of heparinised peripheral blood (to remove erythrocytes) followed by a Ficoll-Paque density gradient to remove mononuclear cells. Eosinophils require a slightly more complex separation using blood from eosinophilic donors to obtain pure populations (VADAS 1979), although Ficoll-Paque separation of blood from donors with a high percentage of eosinophils may be acceptable if neutrophil contamination is not important. In vitro investigation of cell locomotion is relatively straightforward using the above methods. A detailed appraisal of these methods is given in an excellent monograph by WILKINSON (1982).

In vivo in experimental animals cell accumulation following the application of a "chemotactic" stimulus (either a chemoattractant or a procedure likely to provoke an inflammatory response) can be performed in several body cavities, i.e. peritoneum, joints and the pleural space, as well as the skin and foot pad. Radiolabelling of the granulocytes followed by reinjection to see if they accumulate at sites of inflammation can also give useful information. The skin in humans is a convenient site for the study of in vivo responses to chemoattractants. These can either be injected into the skin and biopsies taken or the surface of the skin abraded and the chemoattractant, followed by a cover slip or filter, applied and removed at time intervals before fixing and staining. A dental burr can be used to abrade the skin; however, this gives a high background of neutrophils. Suction blisters raised by applying approximately 140 mmHg of negative pressure offers an alternative technique. This causes division of the skin at the dermo-epidermal junction. Fluid collecting in the blister can be analysed or the blister deroofed and a chamber placed on the abraded area. Low background counts can be obtained by debriding the skin with sellotape, which is repeatedly applied and removed over a defined area until capillary points are observed (NORRIS et al. 1979). A chamber containing saline may be placed on the abraded skin and used to analyse mediators generated either spontaneously by diseased skin or after stimulation such as allergen challenge (BRYANT and KAY 1977).

C. Chemotactic Factors for Neutrophils and Eosinophils

Many factors with in vitro chemotactic activity for granulocytes have been described but the biological significance of many of them is unclear. To establish the physiological relevance of a chemotactic mediator, several criteria need to be fulfilled as suggested by DALE and FOREMAN (1984) for neurotransmitters. These include:

1. Establishment of in vitro and in vivo activity
2. Recovery of elevated levels of the mediator in the affected tissue during a pathological event
3. Demonstration of mechanisms for the generation and metabolism of the mediator in the tissue
4. Modification of the pathological event by a specific inhibitor

Few, if any, putative chemotactic mediators have fulfilled all these criteria in any disease process. One of the difficulties with the in vitro description of chemotactic activity is that the factors are rarely compared with mediators of known activity, so their relative effectiveness is often difficult to determine. There are numerous potent neutrophil chemotactic factors but few mediators, either fully or partially characterised, which are effective eosinophilotactic mediators. Most of

Table 1. Known chemotactic factors and their activities

Class	Factor	Cell type	Activity (in vitro)	Structure	Origin
Complement	C5a	N E	S M	11 200-dalton basic protein	From C5 following activation of complement by classical or alternative pathway
	C567	N E		435 000-dalton protein	Activation of terminal complement components
Cell associated					
(a) *Lipids*	(1) Lipoxygenase (LTB_4, 5-HETE, 12-HETE, 15-HPETE)	N E	S M	5S, 12R-dihydroxy-6-14-CRS-8-10-*trans*-eicosatetraenoic acid (LTB_4) 5-Hydroxy-eicosatetraenoic acid (5-HETE)	Membrane derived primarily from neutrophils and macrophages
	(2) PAF-acether	N E	S S	1-*O*-Alkyl-2-acetyl-sn-glycero-3-phosphocholine	Many cell types
(b) *Amines*	Histamine	E	W	β-Imidazole ethylamine	Mast cell and basophil granule
(c) *Peptides*	High molecular weight NCA (HMW-NCA)	N	Not known	600 000-dalton neutral protein	Mast cell associated but definite cell of origin unknown
	ECF-A				
	(1) Tetrapeptides	E	W	Val-Gly-Ser-Glu Ala-Gly-Ser-Glu	Mast cell granule
	(2) Higher MWT peptides	E	Not known	600- to 1000-dalton acidic peptide	Mast cell granule
(d) *Lymphokines and monokines*	Eosinophil chemotactic factor	E	Not known	25 000- to 40 000-dalton glycoprotein	T-lymphocytes (mitogen and antigen stimulated)
	Lymphocyte-derived chemotactic factor (LDCF)	N	S	12 000-dalton glycoprotein	T-lymphocytes (mitogen stimulated)
	Tumour necrosis factor (TNF)	N		17 000-dalton glycoprotein	Monocytes T-lymphocytes
	Granulocyte–macrophage colony stimulating factor (GM-CSF)	N		18 000-dalton glycoprotein	T-lymphocytes

Abbreviations: N, neutrophils; E, eosinophils; S, strong; M, moderate; W, weak.

these are not specific for eosinophils. Chemotactic factors can be divided into those derived from plasma (the complement and coagulation systems), those that are cell associated and those derived from micro-organisms (bacteria and parasites) (Table 1).

I. Complement-Derived Factors

The component of complement most extensively studied is a fragment derived from the fifth component of complement (C5a). Human C5a is an 11200-dalton single polypeptide chain, cationic glycoprotein which also has potent spasmogenic (anaphylatoxic) activity (GERARD and HUGLI 1981). It has an ED_{50} in chemotaxis of 1.2×10^{-9} *M* and is rapidly converted to C5a-des-Arg by a serum carboxypeptidase which cleaves the C-terminal arginyl residue. Purified C5a-des-Arg is about 10–15 times less potent for neutrophil chemotaxis than C5a and retains only 0.1% of the spasmogenic activity of C5a. However, in plasma C5a-des-Arg is complexed with an anionic polypeptide which enhances its chemotactic activity to equal that of C5a. This complex probably accounts for most of the chemotactic activity in zymosan-activated serum (PEREZ et al. 1986). C5a and C5a-des-Arg have been shown to be very effective chemoattractants in vitro and in vivo for neutrophils. They are less active for eosinophils, at least in vitro. C5a-des-Arg has been isolated from inflammatory sites in animal models and from psoriatic skin scales. C5a-des-Arg is probably degraded by phagocytic cells following interaction and internalisation by specific membrane receptors.

II. Cell-Derived Factors

Lipid mediators derived from membrane phospholipid after cell stimulation have been found to be potent chemoattractants. In particular these include lipids derived by the 5-lipoxygenase pathway from arachidonic acid and the phospholipid PAF-acether, 1-*O*-alkyl-2-acetyl-sn-glycero-3-phosphocholine (platelet activating factor) (HANAHAN et al. 1980). The most well studied arachidonic acid derivative is leukotriene B_4 (LTB_4) (5,12,dihydroxy-eicosatetraenoic acid). This is a potent and effective chemoattractant for neutrophils (optimum dose 10^{-7} *M*) which is active in vitro and in vivo (FORD-HUTCHINSON et al. 1980; GOETZL 1980) with a potency equal to C5a and which produces a prolonged weal and flare and neutrophil infiltration when injected (CAMP et al. 1983; MOVA et al. 1984). The major LTB_4 metabolite 20-hydroxy-LTB_4 has equal activity to LTB_4 for neutrophils but the related lipids, 5-HETE, 15-HPETE and 5,12-diHETE, are less active (DAHINDEN et al. 1984). LTB_4 has also been detected in extracts of scales from psoriatic skin (CAMP et al. 1985). It is inactivated by oxidation to 20-hydroxy- and carboxy-LTB_4, which occurs intracellularly in neutrophils (LEWIS and AUSTEN 1984). It appears, however, to be relatively stable in both plasma and perfused lung tissue. PAF-acether is one of the most effective eosinophil chemoattractants so far described and has equal activity for neutrophils with an optimum dose of 10^{-6} *M* (WARDLAW et al. 1986). PAF-acether has potent inflammatory activity when injected into skin (ARCHER et al. 1985) and has been detected in plasma in patients with cold urticaria (GRANDEL et al. 1986). PAF-acether is derived from the inactive lyso-PAF by acetylation and is rapidly metabolised to lyso-PAF by the action of a specific acetyl hydrolase found in the cytoplasm of neutrophils, eosinophils and platelets as well as plasma. It is intriguing that both these mediators can activate granulocytes which also generate and metabolise them. Another interesting feature is that much of the lipid generated appears to remain cell associated and is not released into the extracellular medium, suggesting they may have autocrine activity modulating intracellular events.

Both these mediators are generated by a wide variety of cell types and this has tended to diminish the role of the mast cell as a source of chemotactic mediators. Mast cell derived and associated chemotactic factors are diverse in both chemical structure and potency and in some cases have proved elusive to characterise fully. Histamine has been reported as having chemotactic activity for eosinophils (CLARKE et al. 1975), but like the two ECF-A tetrapeptides, this activity is very weak.

Eosinophil chemotactic factor of anaphylaxis (ECF-A) is an activity which was first described in supernatants from sensitised human lung challenged with allergen which eluted on Sephadex G25 with a molecular weight of 300–1000 daltons (KAY and AUSTEN 1971). Purification of eosinophil chemotactic activity derived from human lung extracted with Butanol revealed three peaks: 10000, 2000–4000 and 300–1000 daltons. The latter peak was due to a family of acidic peptides two of which, the ECF-A tetrapeptides, Val-Gly-Glu and Ala-Gly-Ser-Glu, were fully characterised (GOETZL and AUSTEN 1975). However, the relative inactivity of these peptides compared to the unfractionated supernatant suggests that the higher molecular weight activity also justifies further investigation. A chemotactic activity which can be reliably detected in the plasma of man is high molecular weight neutrophil chemotactic activity (HMW-NCA). This 600000-dalton neutral peptide has been identified in serum in clinical models of asthma (e.g. antigen and exercise challenge) (LEE et al. 1982) and several types of urticaria (solar, cold, cholinergic (SOTER et al. 1979, 1980; WASSERMAN et al. 1977). It is released in association with histamine and has been found in nasal washing after antigen challenge in sensitised individuals and in supernatants from chopped lung after challenge with anti-IgE (O'DRISCOLL et al. 1983). As it is found in clinical models and in vitro studies associated with mast cell activation and its appearance is inhibited by DSCG, it is thought to be mast cell derived although there is no direct evidence that this is the cell source. Increased neutrophil chemotactic activity in the serum of acute asthmatics has been demonstrated. This resolved as several high molecular weight peaks on gel filtration. One of these peaks had the biochemical characteristics of HMW-NCA. Little activity was observed in mild asthmatics or non-asthmatic controls (BUCHANAN et al. 1986). The exact chemical nature of HMW-NCA is still unknown, so its potency, physiological relevance and cell source are difficult to determine.

III. Lymphokines

There are several reports of eosinophil chemotactic activity in T-lymphocyte culture supernatants from both human and animals after mitogen or antigen stimulation which remains poorly characterised. These factors appear to have a molecular weight in the range of 40000 daltons and have been reported as specific for eosinophils (WADEE and SHER 1980). Their effectiveness is unknown as they have not been compared with potent chemically characterised mediators such as PAF-acether. Tumour necrosis factor (TNF) and granulocyte/macrophage colony stimulating factor (GMCSF) have been reported as chemotactic for neutrophils (MING et al. 1987), although the effectiveness of these monokines is still not clear. Mitogen-stimulated mononuclear cell supernatants have considerable neutrophil chemotactic activity which has a molecular weight of 12000 daltons on gel filtration and a variable isoelectric point (MAESTRELLI et al. 1987). This may be related to the lymphocyte-derived monocyte chemotactic factor (LDCF) described by ALTMAN and KIRCHNER (1972).

IV. Coagulation Products

Products of the coagulation cascade which attract neutrophils (and to a lesser extent, eosinophils) in chemotaxis include fibrin and fibrinogen degradation products, fibrinopeptide B and kallikrein (KAY and KAPLAN 1975). These observations might have significance in fibrinolysis and wound healing.

V. Micro-organisms

Bacterial products and material derived from metazoan parasites have neutrophil and eosinophil chemotactic properties. A number of synthetic bacterial analogues, the formyl-methionyl peptides, have a wide range of potency which is dependent on this primary structure. The most widely studied is formyl-methionyl-leucinyl-phenylalanine (f-MLP) (SCHIFFMAN et al. 1975), which is very active for human neutrophils (optimum 10^{-8}–10^{-7} *M*) and has some activity for human eosinophils.

D. Other Properties of Chemotactic Factors

One feature of granulocyte biology which has been of increasing interest is the activation of these cells in vivo and in vitro. This is most clearly seen with eosinophils where two populations, normal density and light density cells, can be identified by density gradient centrifugation (SPRY 1985). The normal density cells are present in eosinophilic and non-eosinophilic subjects and the light density eosinophils are present in subjects with an eosinophilia irrespective of the cause. The light density cells have several features which suggest they are in an activated state, including increased membrane receptor expression, increased cytotoxic capacity and increased mediator generation both spontaneously and after stimulation (SHAW et al. 1985). It is likely that they have an increased chemotactic response. Tissue eosinophils such as those recovered from bronchoalveolar lavage fluid appear to be of the light density type. Many mediators, including most chemotactic factors, can stimulate normal density eosinophils and neutrophils to mimic many of the changes observed when the cells are in an activated state. Thus receptor expression, cytotoxicity, mediator generation and formation of oxygen radicals as well as density changes can all be stimulated in vitro by various mediators. This suggests that light density cells may be derived from normal density cells by the action of inflammatory mediators. Thus granulocyte infiltration alone may not be sufficient to cause tissue damage. The cells may also need to be in an activated state.

E. Control of Neutrophil and Eosinophil Locomotion

The control of neutrophil and eosinophil locomotion is still poorly understood. Levels of intracellular cyclic nucleotides appear to influence leucocyte locomotion. Cyclic 3′,5′-adenosine monophosphate (cAMP), its analogues and agents such as isoprenaline and the theophyllines which increase cAMP levels inhibit chemotaxis. An increase in cyclic 3′,5′-guanosine monophosphate (cGMP) appears to enhance cell locomotion (KLEBANOFF and CLARK 1978) although in a more recent report, enhancement of chemotaxis was associated with a decrease in the intracellular concentration of cGMP (LANE and LAMBKIN 1984). Chemotaxis is thought to be energy dependent; however, inhibition of energy metabolism and depletion of ATP by 2-deoxyglucose actually increased neutrophil chemotaxis to fMLP despite a decrease in random migration and other cellular functions such as phagocytosis and superoxide anion generation. The enhancement of chemotaxis was magnesium but not calcium dependent. Chemotaxis is assumed to be mediated by binding of the chemotactic agent to a specific membrane receptor. This causes activation of the cell, possibly through the phosphatidyl inositol pathway, which leads to a locomotory response probably via cross-linking of actin and myosin, which are plentiful in the granulocyte cytoplasm (STOSSEL et al. 1984). Further evidence that chemotaxis is dissociated from other cellular functions such as phagocytosis, possibly by way of receptors of differing affinity, was obtained by comparison of the responses of rat and human neutrophils to LTB_4. Rat neutrophils which have a high affinity receptor aggregate but do not chemotax to LTB_4, whereas human PMNs with a similar high affinity but also a low affinity receptor undergo both aggregation and chemotaxis (KREISLE et al. 1985).

In vivo to get to the site of the inflammatory response, cells need to migrate through the vascular endothelium. Adhesion to the vascular endothelium and the cellular matrix is therefore a key component of cell locomotion. Adhesion appears to be a complicated phenomenon mediated by several surface glycoproteins (SPRINGER et al. 1987). Chemotactic agents enhance expression of these receptors and deficiency of them leads to a severe immunocompromised state, leucocyte adhesion deficiency. Enhancement of expression is also magnesium but not calcium dependent. Other factors which directly influence movement of leucocytes include plasma membrane charge and the liberation and oxygenation of arachidonic acid (SNYDERMAN and GOETZL 1981).

In vitro phenomena which are of interest but poorly understood are high dose inhibition and deactivation. Above the optimum concentration of a mediator there is a decline in the responsiveness of cells to a chemotactic stimulus without any evidence of cytotoxicity. Deactivation is observed if the cells are preincubated with a chemoattractant often at suboptimal doses. This leads to non-responsiveness to the same agent (but not other chemoattractants) when tested in a Boyden chamber. This may be a mechanism for inhibiting further cell locomotion to maximise localisation at the site of inflammation.

F. Defects in Chemotaxis

I. Impaired Generation of Chemoattractants

In certain situations chemotactic factor generation is either partially or totally suppressed. In man, the best described situations relate to disorders of the complement system in which there is an inability to produce C5a. Patients have been described with deficiency in the C3b inactivator (with consequent decrease in C3 concentrations) (ALPER et al. 1970) and of C5. C5 deficiency can be hereditary, and associated with lupus erythematosus (ROSENFELD et al. 1976), or C5 is functionally impaired, as in patients with Leiner's disease (JACOBS and MILLER 1972). In these patients cutaneous lesions have been described as well as more severe generalised infections.

II. Inhibition of Chemotaxis

Inhibitors of chemotaxis have been observed in the serum and plasma in a variety of disorders. These include recurrent pyodermas, subcutaneous abscesses and cutaneous candidiasis. The chemical properties of these factors, and whether their elaboration is primary or secondary, are unknown.

III. Intrinsic Disorders of Granulocyte Locomotion

Virtually all the information on intrinsic disorders of granulocytes is on neutrophils. Primary dysfunction of eosinophils has not been reported. In intrinsic defects the abnormality is with the cells themselves rather than a serum factor. Defects have been noted in a wide variety of conditions but the mechanism is rarely understood. They are often associated with severe immunocompromised state and manifest in childhood with recurrent infections particularly of the skin, buccal cavity, and gums. Defects in chemotaxis are often accompanied by other defects in neutrophil function such as cytotoxicity but this is not always the case. As mentioned earlier, leucocyte adhesion deficiency, a rare syndrome with about 30 reported cases, has recently been described; in this syndrome there is abnormal neutrophil locomotion both in vivo and in vitro due to a genetic defect in cell adherence. The condition is characterised by recurrent severe infections, death occurring at a young age in severe cases (ANDERSON et al. 1985). It is possible that other conditions such as the "lazy leucocyte syndrome" described in 1971 may also have similar defects (MILLER et al. 1971). In this condition, patients have recurrent infection associated with depressed neutrophil chemotaxis and random migration. Accumulation of leucocytes to "skin window" sites is also impaired.

A number of patients have been described with very high levels of IgE, eczema and recurrent staphylococcal abscesses. The condition is referred to as the "hyperimmunoglobulinaemia E syndrome" and is associated with depressed neutrophil chemotaxis (Buckley 1979). It is not entirely clear whether this syndrome is part of the spectrum of atopic dermatitis. A relationship between atopic dermatitis, severe infections and depressed neutrophil chemotaxis has been reported but in uncomplicated atopic dermatitis neutrophil locomotion in vitro was normal.

Depressed neutrophil chemotaxis has been reported in acrodermatitis enteropathica and can be reversed by the administration of zinc. Impaired neutrophil chemotaxis has also been reported in congenital ichthyosis, the Wiskott-Aldrich syndrome, measles and mycosis fungoides. A defect in neutrophil chemotaxis, as well as alterations in cellular immunity, were reported in chronic mucocutaneous candidiasis. The abnormalities were reversed following treatment with dialysable transfer factor (Lawton et al. 1977).

G. Skin Diseases Associated with Neutrophil Infiltration

Neutrophils are present, to a greater or lesser extent, at some time in the natural history of all infective skin lesions. In addition to the conditions described above, in which there are defects in either cell locomotion or in the capacity to generate chemotactic agents in vivo, cutaneous manifestations are typical of disorders of impaired phagocytic microbial killing. These include chronic granulomatous disease, Chédiak-Higashi syndrome, myeloperoxidase deficiency and glucose-6-phosphate dehydrogenase deficiency (reviewed in Dahl et al. 1981). Psoriasis is a common disease which is characterised by neutrophil infiltration and has been extensively studied as a human model of inflammation. Several lipid mediators, including PAF acether, LTB_4 and 12-HETE, have been identified in psoriatic scales and blister fluid from affected skin.

H. Skin Disease Associated with Eosinophil Infiltration

Probably the commonest cause of a cutaneous eosinophilia is reinfection with helminithic larvae. Examples include schistosomiasis and *Strongyloides* and filarial infections. In these situations there is intense itching, mast cell involvement and a raised total and specific IgE concentration. A similar but much less marked skin reaction is observed 4–6 h following a "skin prick test" in sensitised individuals, using common inhaled allergens such as pollen extract or the house dust mite. It is assumed that in these situations eosinophils accumulate as the result of the release of mast cell derived chemotactic factors such as the ECF-A peptides and histamine, although the realisation that potent eosinophilotactic factors such as PAF-acether and LTB_4 are not primarily mast cell derived together with the awareness that other cell types such as platelets, macrophages and eosinophils can be triggered by an IgE-mediated stimulus has called this view into question.

An interesting syndrome, Wells' disease (Wells and Smith 1979), has been described, in which an eosinophilic cellulitis occurs characterised by acute cutaneous swelling which slowly resolves. Histologically eosinophils, eosinophilic granules and flame figures are present in the dermis. Eosinophilic skin lesions are present in 50% of the subjects with hypereosinophilic syndrome and eosinophil-associated vascular lesions are common in allergic granulomatous disease, Chédiak-Higashi syndrome and polyarteritis nodosa.

Eosinophil infiltration into the skin is occasionally observed in pemphigus, herpes gestationis and dermatitis herpetiformis.

J. Conclusions

Infiltration of the tissues with neutrophils and eosinophils is a key feature of the host response to infection and tissue insult and is a distinctive feature of many skin diseases. Many agents have been described in vitro which are effective chemoattractants for neutrophils, with the lipid derivative of arachidonic acid (LTB_4 and its metabolites), C5a and PAF-acether possibly having the greatest physiological relevance although the role of lymphocytes and monokines, whose chemotactic activity remains poorly characterised, may be central to the regulation of tissue infiltration by granulocytes. PAF-acether is one of the most active eosinophilotactic agents but is not specific for eosinophils. Specificity may be induced by interaction with other mediators or specific eosinophilic chemoattractants may be derived from mast cells or lymphocytes. Cell locomotion is a complex phenomenon at a molecular level and its regulation in vivo and in vitro is still relatively poorly understood. In general, skin accumulation of neutrophils is associated with bacterial infections, and conditions in which chemotaxis is impaired often result in susceptibility to severe infections. Eosinophil accumulation in the skin is most often seen in helminthic diseases where the migrating larvae may attract the eosinophils by parasite-derived chemotaxins, or possibly through mast cell degranulation or a T cell mediated response.

References

Alper CA, Abramson N, Johnston RB, Jandl JH, Rosen FS (1970) Increased susceptibility to infection associated with abnormalities of complement-mediated functions and of the third component of complement (C3). Engl J Med 282:349–358

Altman GC, Kirchner H (1972) The production of a monocyte chemotactic factor by agammaglobulinaemic chicken spleen cells. J Immunol 109:1149–1151

Anderson DC, Schmalsteig FC, Finegold MJ et al. (1985) The severe and moderate phenotypes of heritable Mac-1, LFA-1 deficiency: their quantitative definition and relation to leukocyte dysfunction and clinical features. J Infect Dis 152:668–689

Archer CB, Page CP, Morley J, MacDonald DM (1985) Accumulation of inflammatory cells in response to intracutaneous platelet activating factor PAF-acether in man. Br J Dermatol 112:285–290

Boyden S (1962) The chemotactic effect of mixtures of antibody and antigen on polymorphonuclear leucocytes. J Exp Med 115:453–466

Bryant DH, Kay AB (1977) Cutaneous accumulation in atopic and non-atopic individuals. The effect of an ECF-A tetrapeptide and histamine. Clin Allergy 7:211–217

Buchanan DR, Cromwell O, Kay AB (1986) Neutrophil chemotactic activity (NCA) in acute severe asthma. J Allergy Clin Immunol 77:183 (abstract)

Buckley RH (1979) Hyperimmunoglobulin E, undue susceptibility to infection, and depressed immunologic function. In: Hodes H, Kagen BM (eds) Pediatric immunology. Science Medicine Publishing, New York, pp 219–247

Camp RDR, Coutts AA, Greaves MW, Kay AB, Walport MJ (1983) Responses of human skin to intradermal injection of leukotrienes C_4, D_4, and B_4. Br J Pharmacol 80:497–502
Camp RD, Mallet AI, Cunningham FM et al. (1985) The role of chemoattractant lipoxygenase production in the pathogenesis of psoriasis. Br J Dermatol 113 [Suppl 28]:98–103
Clarke RAF, Collins JI, Kaplan AP (1975) The selective eosinophil chemotactic activity of histamine. J Exp Med 142:1462–1476
Dahinden CA, Clancy RM, Hugli TE (1984) Specificity of leukotriene B_4 and structure-function relationships for chemotaxis of human neutrophils. J Immunol 133:1477–1482
Dahl MV, Cates KL, Quie PG (1981) Deficiency of phagocyte function and related disorders. In: Safai B, Good RA (eds) Immunodermatology. Plenum, New York, p 425
Dale MM, Foreman JC (1984) Textbook of immunopharmacology. Blackwell Scientific, Oxford
Falk W, Goodwin RH Jr, Leonard EJ (1980) A 48-well microchemotaxis assembly for rapid and accurate measurement of leucocyte migration. J Immunol Methods 33:239–247
Ford-Hutchinson AW, Bray MA, Doig MV, Shipley ME, Smith MJH (1980) Leukotriene B_4, a potent chemotactic and aggregating substance released from polymorphonuclear leukocytes. Nature 286:264–265
Gerard C, Hugli TE (1981) C5a: a mediator of chemotaxis and cellular release reactions. In: Becker EL, Simon AS, Austen KF (eds) Biochemistry of the acute allergic reactions. Alan R. Liss, New York, pp 147–160
Goetzl EJ (1980) Mediators of immediate hyper-sensitivity derived from arachidonic acid. N Engl J Med 303:822–825
Goetzl EJ, Austen KF (1975) Purification and synthesis of eosinophilotactic tetrapeptides of human lung tissue: identification as eosinophil chemotactic factor of anaphylaxis. Proc Natl Acad Sci USA 72:4123–4127
Grandel KE, Farr RS, Wanderer AA, Eisenstadt TC, Wasserman SI (1986) Association of platelet-activating factor with primary acquired cold urticaria. N Engl J Med 313:405
Hanahan DJ, Demopoulos CA, Leihr J, Pinckard RN (1980) Identification of platelet-activating factor isolated from rabbit basophils of acetyl, glyceryl ether phosphorylcholine. J Biol Chem 255:5514–5516
Jacobs JC, Miller ME (1972) Fatal familial Leiner's disease: a deficiency of the opsonic activity of serum complement. Pediatrics 49:225–232
Kay AB (1980) The role of the eosinophil in physiological and pathological processes. In: Thompson RA (ed) Recent advances in clinical immunology. Churchill Livingstone, Edinburgh London, pp 113–143
Kay AB (1985) Eosinophils as effector cells in immunity and hypersensitivity disorders. Clin Exp Immunol 62:1–12
Kay AB, Austen KF (1971) The IgE-mediated release of an eosinophil leucocyte chemotactic factor from the human lung. J Immunol 107:899–902
Kay AB, Kaplan AP (1975) Chemotaxis and haemostasis. Br J Haematol 31:417–422
Klebanoff SJ, Clark RA (1978) Chemotaxis. In: The neutrophil: function and clinical disorders. North-Holland, Amsterdam, Chap 3
Kreisle RA, Parker CW, Griffin GL, Senior RM, Stenson UF (1985) Studies of leukotriene B_4-specific binding and function in rat polymorphonuclear leucocytes. Absence of a chemotactic response. J Immunol 134:3356–3363
Lane TA, Lambkin GE (1984) A reassessment of the energy requirements for neutrophil migration, adenosine triphosphate depletion extraction chemotaxis. Blood 64:986–993
Lawton JWM, Darg C, Pepper D, Kay AB (1977) Human transfer factor prepared by dialysis, ultrafiltration and gel chromatography: biological activity in local transfer of skin sensitivity. J Immunol Methods 16:119–129

Lee TH, Nagy L, Nagakura T, Walport MJ, Kay AB (1982) Identification and partial characterization of an exercise-induced neutrophil chemotactic factor in bronchial asthma. J Clin Invest 69:889–899

Lewis RA, Austen KF (1984) The biologically active leukotrienes. Biosynthesis, metabolism, receptors, functions and pharmacology. J Clin Invest 73:889–897

Maestrelli P, Tsai J-J, Kay AB (1987) Human lymphocyte-derived neutrophil chemotactic activities. J Allergy Clin Immunol 79:160 (abstract)

Miller ME, Oski FA, Harris MB (1971) Lazy leucocyte syndrome. A new disorder of neutrophil function. Lancet I:665–669

Ming WJI, Bersani L, Mantovani A (1987) Tumour necrosis factor is chemotactic for monocytes and polymorphonuclear leukocytes. J Immunol 138:1469–1474

Movat HZ, Rett LC, Burrowes CE, Johnston MG (1984) The in vivo effect of leukotriene B_4 on polymorphonuclear leucocytes and the microcirculation: comparison with activated complement (C5a-des-Arg) and enhancement with prostaglandin E_2. Am J Pathol 115:233–244

Norris DA, Lipman SH, Wertin WL (1979) Human monocyte chemotaxis: a quantitative in vivo technique. J Invest Dermatol 72:81–84

O'Driscoll BRC, Lee TH, Cromwell O, Kay AB (1983) Immunologic release of neutrophil chemotactic activity from human lung tissue. J Allergy Clin Immunol 72:695–701

Perez HD, Chenoweth DE, Goldstein IM (1986) Attachment of human C5a-des-Arg to its cochemotaxin is required for maximum expression of chemotactic activity. J Clin Invest 78:1589–1595

Rosenfeld SI, Kelly MF, Leddy JP (1976) Hereditary deficiency of the fifth component of complement in man. I. Clinical, immunochemical, and family studies. J Clin Invest 57:1626–1634

Schiffman E, Corcoran FA, Wahl SM (1975) *N*-Formylmethionine peptides as chemoattractants for leucocytes. Proc Natl Acad Sci USA 72:1059–1062

Shaw RJ, Walsh GM, Cromwell O, Moqbel R, Spry CJF, Kay AB (1985) Activated human eosinophils generate SRS-A leukotrienes following IgG-dependent stimulation. Nature 316:150–152

Snyderman R, Goetzl EJ (1981) Molecular and cellular mechanisms of leukocyte chemotaxis. Science 213:830–837

Soter NA, Wasserman SI, Pathak MA, Parrish JA, Austen KF (1979) Solar urticaria: release of mast cell mediators into the circulation after experimental challenge. J Invest Dermatol 72:282 (abstract)

Soter NA, Wasserman SI, Austen KF, McFadden ER (1980) Release of mast cell mediators and alterations in lung function in patients with cholinergic urticaria. N Engl J Med 302:604–608

Springer TA, Dustin ML, Kishimoto TK, Marlin SD (1987) The lymphocyte function-associated LFA-1, CD2 and LFA-3 molecules: cell adhesion receptors of the immune system. Ann Rev Immunol 5

Spry CJF (1985) Synthesis and secretion of eosinophil granule substances. Immunol Today 6:332–335

Stossel TP, Hartwig JH, Yin HL, Southwick FS, Zaner KS (1984) The motor of leucocytes. Fed Proc 42:2760–2763

Vadas MA, David JR, Butterworth A, Pisani NT, Singok TA (1979) A new method for the purification of human eosinophils and neutrophils and a comparison of the ability of these cells to damage schistosomula of *Schistosoma mansoni*. J Immunol 122:1228–1238

Wadee AA, Sher R (1980) The effects of a soluble factor released by sensitized mononuclear cells incubated with *S. haematobium* ova on eosinophil migration. Immunology 41:989–995

Wardlaw AJ, Moqbel R, Cromwell O, Kay AB (1986) Platelet-activating factor. A potent chemotactic and chemokinetic factor for human eosinophils. J Clin Invest 78:1701–1706

Wasserman SI, Soter NA, Center DM, Austen KF (1977) Cold urticaria. Recognition and characterisation of a neutrophil chemotactic factor which appears in serum during experimental challenge. J Clin Invest 60:189–196

Wells GC, Smith NP (1979) Eosinophilic cellulitis. Br J Dermatol 100:101–109
Wilkinson PC (1982) Chemotaxis and inflammation, 2nd edn. Churchill Livingstone, Edinburgh
Wilkinson PC (1985) Random locomotion, chemotaxis and chemokinesis. Immunol Today 6:273–278
Wilkinson PC (1986) Locomotion and chemotaxis of leucocytes. In: Weir (ed) Handbook of experimental immunology, vol 2. Cellular immunology. Blackwell, Oxford, Chap 51
Wright DG (1982) The neutrophil as a secretory organ of host defense. In: Gallin JI, Fauci AS (eds) Advances in host defense mechanisms, vol 1. Raven, New York, pp 75–110

CHAPTER 25

Neuropeptides and the Skin

S. D. BRAIN and J. A. EDWARDSON

A. Introduction

Recent studies on the distribution and physiological actions of regulatory peptides in skin indicate the growing relevance of this subject to clinical dermatology. Peptides form the largest group of extracellular messengers in the body, functioning in at least three modes as hormones, neurotransmitters and also as locally acting paracrine regulators. The endocrine role of peptides has, until recently, received most attention, and the actions of melanocyte-stimulating hormone (MSH) and related hormones on the integument have been the major focus for research in this area. The isolation and characterisation of several hypothalamic

Table 1. Neuropeptides in skin

Peptide	Amino acids	Location	Possible functions
Substance P	11	Primary sensory neurones; C-fibre nerves supplying cutaneous blood vessels; possible intra-epidermal endings	Sensory – possibly nociceptive; stimulates histamine release from mast cells; increases microvascular permeability
Vasoactive intestinal polypeptide	28	Nerves supply cutaneous blood vessels; eccrine sweat glands, sebaceous glands and some hair follicles; Merkel cells	Vasodilatation; stimulates sweat gland production; potentiates some effects of acetylcholine; possibly regulates Merkel cell–axon complex
Neuropeptide Y	36	Sympathetic noradrenergic neurones	Vasoconstriction; potentiates some effects of noradrenaline
Calcitonin gene-related peptide	37	Primary sensory neurones; C-fibre nerves supplying cutaneous blood vessels	Vasodilatation
Somatostatin	14 (or 28)	Cellular location unknown – levels too low for immunocytochemical detection	Releases histamine from mast cells but generally a powerful inhibitor of endocrine and paracrine secretion
Neurotensin	13	Cellular location unknown	Releases histamine from mast cells

releasing hormones in the early 1970s led to the discovery that peptides such as thyrotrophin-releasing hormone (TRH), gonadotrophin-releasing hormone and somatostatin are widely distributed throughout the nervous system. Most of the regulatory peptides of the gastrointestinal tract are found in the nervous system and their distribution embraces sensory, motor and inter-neurones and includes both sympathetic and parasympathetic divisions of the autonomic nervous system (see EMSON 1979; BLOOM and POLAK 1983; for review). Peptides such as substance P, vasoactive intestinal peptide (VIP), neuropeptide Y (NPY) calcitonin-gene related peptide (CGRP) and others (Table 1) are widely distributed in the nerve supply to peripheral tissues, including the skin. This distribution almost certainly explains why many of the peripheral effects of autonomic nerve stimulation cannot be prevented by cholinergic or adrenergic blocking agents. Neuropeptides such as substance P and CGRP are mainly contained in small unmyelinated C-fibre nerves. These sensory fibres not only transmit signals to the central nervous system but also have efferent roles which consist in the release of biologically active peptides. These nerves have long been thought to play an essential role in the local skin reaction to irritant chemicals or tissue damage (LEWIS 1927). In other circumstances they can also act as a neurogenic component to modulate inflammatory conditions. The concept of peptides released from nerves having an inflammatory role in conditions such as skin diseases is becoming increasingly popular (FARBER et al. 1986).

In addition to this neural localisation there is evidence for the presence of immunoreactive enkephalin (HARTSCHUH et al. 1979) and VIP (HARTSCHUH et al. 1983) in Merkel cells in the skin of several mammalian species, suggesting a local paracrine function for these substances.

B. Biochemistry of Peptide-Mediated Signalling

Most regulatory peptides are synthesised as part of an inactive macromolecular precursor which is subsequently cleaved post-ribosomally by specific processing endopeptidases to yield the much smaller biologically active peptide. The use of recombinant nucleic acid technology to identify precursor sequences for oligopeptides represents a major advance of recent years. Precursors of several peptides known to act on skin have been characterised in this way; these include the common precursor for adrenocorticotrophic hormone (ACTH), MSH and β-endorphin (NAKANISHI et al. 1979), the human enkephalin precursor (COMB et al. 1982), the substance P precursor (NAWA et al. 1983), the CGRP precursor (AMARA et al. 1982) and a macromolecule which gives rise to both VIP and PHI (peptide, histidine, isoleucine), so-called because the porcine molecule consists of a 28 amino acid sequence with histidine and isoleucine at the N- and C-terminals respectively (ITOH et al. 1983). Thus, in some cases the precursor gives rise to a family of hormonal peptides with diverse structures and biological activities (e.g. ACTH- and β-endorphin), while other precursors contain multiple sequences of the same peptide (e.g. TRH, RICHTER et al. 1984) or closely related peptides (e.g. both met- and leu-enkephalin sequences occur in pro-enkephalin).

Recently, the primary structures of two types of substance P precursor (protachykinins) from bovine brain have been determined (NAWA et al. 1983) and one of these contains a sequence homologous to kassinin, a peptide found in amphibian skin that has been shown in some mammalian systems to be more potent than substance P. The novel sequence is flanked by paired Lys-Arg bonds which are sites for proteolytic cleavage during processing. The presumed secretory product is neurokinin A (or substance K). As a result of this and similar studies substance P is now considered to be just one member of the tachykinin family. Evidence, to date, indicates that there are two other major tachykininins, neurokinin A (as mentioned above), also known as substance K or neuromedin L (KIMURA et al. 1983, NAWA et al. 1983; MINAMINO et al. 1984), and neurokinin B, also known as neuromedin K (KIMURA et al. 1983; KANGAWA et al. 1983). All the peptides have similar C-terminal amino acid sequences. Three different types of receptor have been described so far. Each has a preferential affinity for one of the tachykinins: SP-P receptors for substance P, SP-E receptors for neurokinin A and SP-K receptors for neurokinin B (see BUCK and BURCHER 1986).

The neuropeptide calcitonin gene-related peptide (CGRP) was discovered by the elegant studies of ROSENFELD and colleagues (AMARA et al. 1982; ROSENFELD et al. 1983). Their studies revealed that the calcitonin gene in the rat encodes for two different mRNAs: the macromolecular precursor of the hormone calcitonin (which is important for protecting the body against calcium loss) and the macromolecular precursor of CGRP. Alternative RNA splicing of the transcripts of the calcitonin gene ultimately results in the production of mRNA for CGRP. The mechanism which allows the precursor for either calcitonin or CGRP to be generated is not yet understood. The calcitonin gene can also encode for CGRP in man (EDBROOKE et al. 1985). Calcitonin, as expected, is the predominant expression product in the thyroid whilst CGRP is the predominant product in certain C-fibres nerves, as mentioned previously. It has been found to be localised in certain nerves with substance P (LUNDBERG et al. 1985).

Biochemical mechanisms for the disposal of peptides in peripheral tissues are poorly understood but are likely to resemble mechanisms in the central nervous system, where a number of membrane-bound neutral endopeptidases have been characterised (EDWARDSON and MCDERMOTT 1985). There is, for example, an enzyme present on synaptic membranes which has a molecular mass of about 500000 and which cleaves substance P between the Gln^6-Phe^7, Phe^7-Phe^8, and Phe^8-Gly^9 bonds with equal facility (LEE 1982). Investigations on the substrate requirements of some of these peptidases have led to the development of selective inhibitors and this represents one of the most likely points for therapeutic intervention in peptide-mediated processes. Although uptake mechanisms, such as those that exist for catecholamines, have not been described for intact peptides, there is some evidence that peptide fragments produced by proteolytic cleavage may be taken up into cells. Active uptake of the C-terminal heptapeptide, substance P (5-11), into rat brain and rabbit spinal cord slices has been described by NAKATA et al. (1981).

C. Distribution and Functions of Peptides in Mammalian Skin

The number of peptides which have been suggested to be neuropeptides is continually increasing. These peptides have a bewildering array of biological actions suggesting an extremely complex system. Discussion in this chapter will be mainly limited to the possible roles of the tachykinins, CGRP and VIP in physiological and pathological conditions. The presence of these peptides in both animal and human skin has been demonstrated and extensive research into the effects of each of the peptides on the cutaneous microvasculature has been carried out.

The contribution of substances released from nerves to the inflammatory response has been studied for many years. The term "neurogenic inflammation" was originally coined by BRUCE (1913). He found that application of mustard oil to the skin of cats caused an immediate inflammatory response consisting of erythema and oedema, which was abolished if the nerve was sectioned distal to the dorsal root ganglion. Lewis, who defined the triple response to describe injury in human skin, suggested that neurogenic inflammatory reactions are based on axon reflexes taking place in the end branches of sensory nerves. As mentioned earlier, the nerve fibres that are thought to play an important role in neurogenic inflammation are the unmyelinated sensory neurones of the C-fibre group. The mustard oil, capsaicin, selectively depletes these fibres of their contents with the result that subsequent application of irritants or mustard oil fails to induce an inflammatory response (JANCSO et al. 1967). Experiments using capsaicin have led to a greater understanding of the role of the C-fibres in inflammation, especially as the C-fibres of rats treated with capsaicin at a neonatal stage are irreversibly damaged (JANCSO et al. 1977).

I. Tachykinins and CGRP

"Substance P" was the name given by VON EULER and GADDUM in 1931 to standard powdered extracts of brain or intestine which had powerful acetylcholine-like effects on vascular smooth muscle and blood pressure but which could not be blocked by atropine. The active principle was first isolated and characterised by CHANG and co-workers in 1971 and shown to have the structure: Arg^1-Pro^2-Lys^3-Pro^4-Gln^5-Gln^6-Phe^7-Phe^8-Gly^9-Leu^{10}-Met^{11}-NH_2. It was the first neuropeptide to be isolated, and partly for this reason it has been extensively studied. Immunocytochemical studies using antibodies raised against synthetic substance P have shown that this peptide is widely distributed in the nervous system. A subpopulation (25%) of small bipolar cells located in the spinal dorsal root ganglia are rich in substance P. These primary afferent sensory neurones are the source of small diameter C-fibres which project centrally to the substantia gelatinosa of the dorsal horn of the spinal cord, and there is good evidence that these fibres convey nociceptive stimuli (JESSEL 1982).

Studies with radioactively labelled amino acids have shown that substance P synthesised in the perikarya of dorsal root ganglion cells is transported peripherally by fast axoplasmic flow, indicating a peripheral role for this peptide. Substance P in cutaneous nerves is found around blood vessels and also in fibres

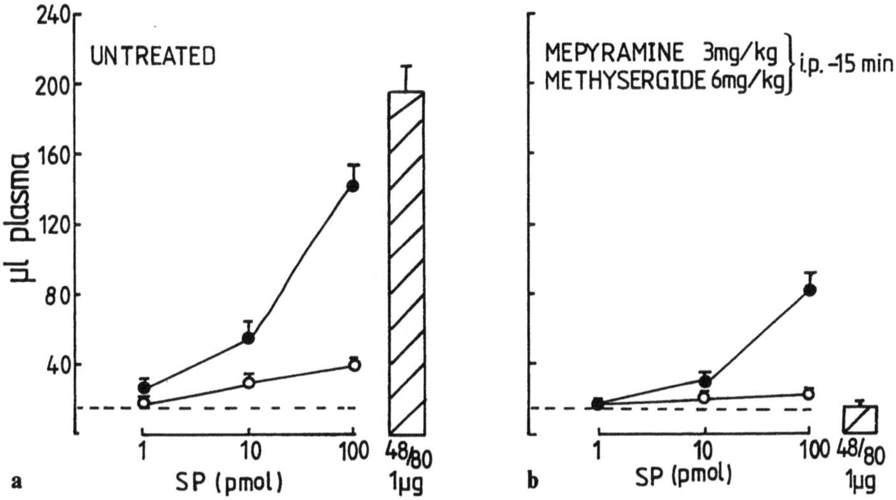

Fig. 1 a, b. Effect of substance P on oedema formation in the rat skin. Oedema was measured in the rat skin by the extravascular accumulation of intravenously injected ^{125}I-albumin (see BRAIN and WILLIAMS 1985 for methods). **a** Substance P (*SP* 1–100 pmol,-○-) induced oedema in a dose-dependent manner in untreated rats. The oedema was potentiated by the presence of the vasodilator CGRP (10 pmol,-●-); see text for explanation. **b** Rats were treated with the histamine antagonist mepyramine (3 mg/kg) and the 5-HT antagonist (methysergide 6 mg/kg), both given intraperitoneally 15 min before intradermal injections. Oedema induced by compound 48/80, which degranulates mast cells, was completely abolished in these rats. In contrast, oedema was still observed at substance P + CGRP-injected sites. Each point is mean ± SEM of experiments carried out in four rats

which penetrate the epidermis (BLOOM and POLAK 1983). The bioactive effects of substance P on non-neural tissues suggest an important role in efferent pathways, and studies using animals have done much to clarify a role for substance P as an inflammatory mediator.

Substance P can increase vascular permeability and induce vasodilatation in certain inflammatory models (LEMBECK and HOLZER 1979; GAMSE et al. 1980; LEMBECK and GAMSE 1982). It is potent at releasing histamine from isolated mast cells and tissues. Thus, the actions of released histamine (and 5-hydroxytryptamine, 5-HT, which is contained in rodent mast cells) are partially responsible for the actions of substance P. However, as shown in Fig. 1, where oedema induced by an agent that degranulates mast cells (compound 48/80) is totally inhibited by a combination of the histamine antagonist mepyramine and the 5-HT antagonist methysergide in rat skin, substance P can still induce oedema in a dose-dependent manner. This suggests that substance P can have a direct action in increasing microvascular permeability in the cutaneous microvasculature that is independent of mast cell activation.

More recently, however, using molecular biological techniques, the neurokinins and also CGRP have been discovered. In the periphery CGRP can be found in similar nerve fibres as the tachykinins (SENAPATI et al. 1986; MULDERRY et al. 1985). It is now thought that these novel peptides could be responsible for some of the neurogenically mediated responses that were previously ascribed to substance P in both animals and man. The analysis of experiments carried out with the tachykinins and CGRP requires some care. All tachykinins induce hypotension when injected intravenously into the guinea-pig (HUA et al. 1984). Similar results have been found with CGRP. This could suggest that these neuropeptides have similar actions as vasodilators. However, they probably only reach high intravenous levels after intensive systemic nervous stimulation. It seems more important to analyse responses to the neuropeptides in organs when given locally, at an extravascular site. This situation is more similar to when the neuropeptide is released from a nerve in vivo.

The intradermal injection of each of the three tachykinins into rat skin leads to local oedema formation (GAMSE and SARIA 1985). Oedema was measured by the extravascular accumulation of intravenously injected Evan's Blue. Although each tachykinin induced increased microvascular permeability there was a potency difference (neurokinin B > neurokinin A > substance P). In contrast to the tachykinins, CGRP does not cause increased microvascular permeability but is a potent vasodilator when injected intradermally (BRAIN et al. 1985). Blood flow was assessed using a multi-site ^{133}xenon clearance technique. In previous studies it was found that the ability of vasodilator prostaglandins to potentiate inflammatory oedema correlated with their potency as vasodilators in animal skin (WILLIAMS and PECK 1977; WILLIAMS 1979; see Chap. 21, this book). These findings led to the hypothesis that increased arteriolar blood flow causes increased intravenular pressure and in turn increased plasma leakage at sites where mediators of increased microvascular permeability are present. As CGRP is a vasodilator, a similar potentiation of inflammatory oedema to that observed with the prostaglandins should be observed with CGRP. The ability of CGRP to potentiate oedema induced by mediators of increased microvascular permeability has been examined in two species. In the rabbit, where the experiments using vasodilator prostaglandins were performed, CGRP potentiated oedema formation induced by two types of mediator of increased microvascular permeability (BRAIN and WILLIAMS 1985). Histamine, bradykinin and platelet activating factor (PAF) increase microvascular permeability by a direct effect on endothelial cells whilst C5a, fmlp and leukotriene B_4 increase microvascular permeability by a mechanism dependent on the presence of polymorphonuclear leucocytes (WEDMORE and WILLIAMS 1981). The ability of CGRP to potentiate oedema induced by these two types of mediator of increased microvascular permeability is indicative that the potentiation is a consequence of the vasodilator activity of CGRP. Substance P, although a potent vasodilator when given intravenously in the rabbit, does not have any discernible effects on microvascular permeability or blood flow when injected intradermally in the rabbit. This could be partly due to a lack of mast cells in the skin of the rabbit. In contrast, in the rat, where substance P and the tachykinins are potent mediators of increased microvascular permeability, CGRP potentiated inflammatory oedema induced by all the tachykinins (GAMSE and SARIA 1985; BRAIN and WILLIAMS 1985; see Fig. 1).

These studies suggest that, if released from nerves, tachykinins can induce inflammatory oedema. Additionally CGRP, as a consequence of its potent vasodilator activity, can act as a neurogenic component of inflammation to potentiate

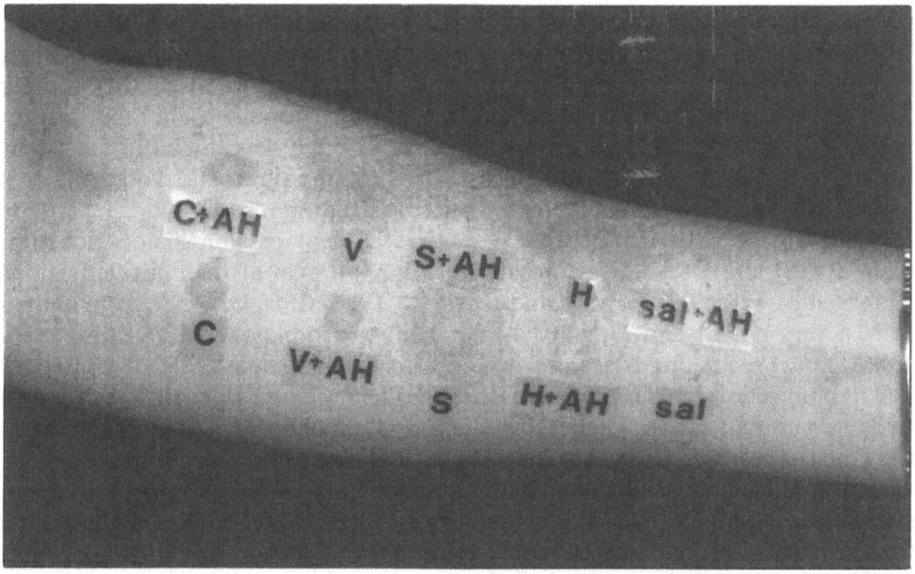

Fig. 2. Effect of an anti-histamine on neuropeptide-induced responses in human skin. CGRP (*C*, 10 pmol), vasoactive intestinal peptide (*V*, 10 pmol), substance P (*S*, 10 pmol), histamine (*H*, 500 pmol) and saline (*sal*, 50 µl) were injected intradermally into the human forearm with and without the histamine H_1 antagonist chlorpheniramine (*AH*, 1 µg). The picture was taken 10 min after injections. VIP, substance P and histamine all induced the classical triple response consisting of weal, local reddening and flare. Chlorpheniramine inhibited the flare in each case, suggesting that substance P and VIP both act to release histamine in human skin. CGRP induced only local reddening and this was unaffected by chlorpheniramine

oedema induced not only by the tachykinins but also by any other mediator of increased microvascular permeability. The important question is: how relevant are these findings to human skin?

The ability of substance P to stimulate the triple response in a similar manner to histamine (as shown in Fig. 2) was initially described by HAGERMARK and coworkers in 1978. Several lines of evidence suggest that substance P induces the triple response partially through histamine release from mast cells. Firstly, substance P releases histamine from isolated human skin mast cells (BENYON et al. 1987). Secondly, anti-histamines inhibit the substance P-induced flare in a similar manner to the histamine-induced flare (HAGEMARK et al. 1978; FOREMAN et al. 1983; see Fig. 2). The experiments of FOREMAN et al. (1983) revealed that the N-terminal basic amino acids are necessary for substance P to release histamine from mast cells and thus to induce a flare in human skin. Interestingly, although neurokinins A and B have a similar carboxy terminal sequence to substance P, neurokinin B has no basic amino peptides in its N-terminal whilst neurokinin A has only one (Lys). Therefore one would not expect neurokinin A and B to be as potent as substance P in inducing a flare. The work of DEVILLIER et al. (1986) has revealed exactly these findings. FOREMAN showed that the substance P-induced weal response was mediated in a different manner from the flare response, i.e. not through histamine release and not through basic N-terminal amino acids. The tachykinins have similar potency in inducing a weal in human skin. This is probably related to their similarity in amino acid structure at the C-terminal. Thus the studies carried out so far suggest that tachykinins can act in a pro-inflammatory manner via two mechanisms: (a) through degranulating mast cells leading to histamine and protease release, (b) through a direct effect on the microvasculature as a mediator of increased microvascular permeability.

Calcitonin-gene related peptide, in contrast to the tachykinins, does not stimulate a weal and flare unless injected into human skin at high concentrations (Brain et al. 1985; Piotrowski and Foreman 1986). It is, however, as in animal skin, a very potent and long-lasting cutaneous vasodilator (Brain et al. 1985, 1986). Local reddening is seen in response to sub-picomole doses of CGRP. A 100 fmol injection of CGRP induces local reddening that can be observed for up to 20 min after injection. Higher doses (e.g. a 10 pmol dose) induce local reddening that can be observed in skin for up to several hours after injection (see Fig. 2). The local reddening is associated with increased blood flow in human skin (measured using a laser Doppler blood flow meter). The local blood flow remains high for 1–2 h after injection of a 10 pmol dose and then decreases back towards basal levels as the local erythema subsides (Brain et al. 1986). It has been suggested that CGRP could have an important role in the control of blood flow in physiological and also pathological conditions. If CGRP is released at inflammatory sites it could contribute to the reddening which is of course one of the cardinal signs of inflammation and is normally observed in inflammatory diseases of the skin. Additionally, as CGRP is a vasodilator, the results in animal skin suggest that it should be able to potentiate weal formation induced by an agent that increases microvascular permeability. This has not yet been demonstrated in human skin, possibly because it is difficult to measure local weal formation accurately.

Lewis carried out a detailed study of the skin and suggested the triple response to explain the acute response of skin to injury. The flare is one component; it can spread for up to several centimetres around the injection site. We have described how substance P can initiate the flare by a mechanism dependent on mast cell-derived histamine. We would now like to analyse what actually mediates the vasodilatation to cause the flare. The flare is due to an axon reflex and thus requires an intact neuronal pathway between the site of nociceptor stimulation, i.e. injury, and/or substance P release/injection. The importance of histamine in stimulating the axon reflex-mediated flare at the afferent site is obvious as antihistamines inhibit the flare when given with the agonist (Hagermark et al. 1978; Foreman et al. 1983, see Fig. 2). Histamine and substance P have also been suggested to be released at the efferent end of the axon at a site in close association with arterioles to mediate the increased blood flow of the flare (Lembeck 1983; Foreman et al. 1983). We agree that substance P can initiate the flare through histamine release, but we consider that CGRP is a candidate for the agent which is released close to arterioles to mediate the flare. This is because when CGRP is injected intradermally into human skin it induces a local reddening with very low doses (15 fmol) that has a similar time course to that of the flare. In contrast, sufficient substance P has to be injected to induce a triple response (1 pmol) before a cutaneous effect is observed (Brain unpublished observations, Brain et al.

1986). These experiments suggest that the release of extremely low amounts of CGRP from nerves in close association with arterioles could cause the necessary local increased blood flow that gives the appearance of the flare, whereas released substance P would not be able to induce a similar effect at such low concentrations.

II. Vasoactive Intestinal Peptide

Vasoactive intestinal peptide (VIP) is a 28 amino acid hormone isolated in 1980 from porcine duodenum (see SAID 1984, for a review). The human precursor has a molecular weight of 200 000 and contains a 27 amino acid peptide known as PHM which is co-secreted. (PHM is almost identical to porcine PHI but has a C-terminal *m*ethionine instead of an *i*soleucine residue.) PHM has marked structural homology with VIP and both peptides are related to the glucagon secretion family. VIP has a wide distribution in both the central and the peripheral nervous system. It coexists with acetylcholine in the cholinergic nerve supply to various endocrine glands and is also co-localised with enkephalins and catecholamines in adrenomedullary chromaffin granules. It has modulatory effects on the release of a large constellation of transmitters, hormones and paracrine regulators.

Immunocytochemical studies have shown the distribution of VIP in mammalian skin (BLOOM and POLAK 1983; HARTSCHUH et al. 1983; HARTSCHUH et al. 1984). Similar patterns of immunoreactive VIP fibres are found in the skin of pig, dog, cat and man. The major cutaneous supply of VIP fibres is associated with larger arteries of the deep and superficial vascular plexus of the reticular layer and the arterial sections of arteriovenous anastomoses. A much less dense innervation of small arteries and veins of all sizes is observed and, in contrast to the reticular layer, papillary loop vessels do not appear to contain VIP terminals. A profuse VIP-containing nerve supply is found in eccrine sweat glands and endings are associated with the microvasculature, gland cells and myoepithelial cells but only rarely with duct cells. There is also evidence that at least a few primary sensory neurones contain VIP (SCHULTZBERG et al. 1980).

Vasoactive intestinal peptide, like the tachykinins and CGRP, has become known as a vasodilator because of its marked hypotensive effect when given systemically. However, detailed studies of its effect after intradermal injection into human skin suggest that it has its own unique profile of activity. ANAND et al. (1983) showed it stimulates a weal and flare. Figure 2 shows that like substance P, the flare is inhibited by co-injection with anti-histamines, suggesting that VIP stimulates the flare by a similar mechanism to substance P, through histamine release. A closer study of Fig. 2 reveals that even in the presence of the anti-histamine an intense local reddening is observed at the VIP injection site. This is because VIP, like CGRP, also has the ability to stimulate a local protracted increased blood flow. This is more easily observed after the weal and flare has faded. In a study in which the cutaneous responses to intradermal injections of CGRP and VIP were compared, VIP induced a protracted local reddening in each of the individuals studied. In each case, however, it was less intense and shorter acting than that induced by CGRP (BRAIN et al. 1986). Because VIP-containing fibres innervate the sweat glands it has been suggested that VIP plays a role in thermoregulation. The blood supply to both male and female human genitalia appears to be under extensive control by VIP and this peptide is considered as a major transmitter responsible for the vascular engorgement which occurs during sexual excitation (see POLAK and BLOOM 1984 for a review). Since some Merkel cells are found without adjacent neuronal contacts, HARTSCHUH et al. (1984) have speculated that peptides such as VIP and enkephalin which they contain may subserve a paracrine function, acting locally on adjacent cells such as keratinocytes, mast cells or endothelial cells.

III. Other Neuropeptides

Several other peptides are known to be localised in neurones which innervate the skin. Somatostatin (SRIF, somatotrophin-release inhibiting factor) exists in two forms containing 14 and 28 amino acids and is found in the CNS, sympathetic nerves, D cells of the endocrine pancreas and cells of the gastrointestinal mucosa. It is a powerful inhibitor of the secretion of most peptide hormones and also blocks many hormone actions by a direct effect on target cells. In man SRIF is present in skin in concentrations much lower than substance P or VIP and cannot be reliably demonstrated by immunocytochemical techniques. In the pig, concentrations of SRIF in skin range from 5 to 13 pmol/g (BLOOM and POLAK 1983). Other peptides secreted by the gastrointestinal tract and which are also present in skin at low concentrations are bombesin, enkephalins, and neurotensin. The latter hormone is normally present in blood in low picomolar concentrations. Levels rise rapidly after a fat meal to reach concentrations sufficient to inhibit gastric secretion, alter intestinal mobility and induce vasoconstriction in adipose tisses. Neurotensin has been shown to trigger histamine release from mast cells in vitro (SYDBOM 1982) and was reported to be more potent than substance P at low doses (10^{-8} *M*). However, the maximum response to neurotensin released only 10% of the total histamine, whereas much greater responses were obtained with substance P. Circulating levels of neurotensin after a meal do not appear to be sufficiently high to alter mast cell function but it is likely that much higher local concentrations will occur in the vicinity of nerve terminals.

The autonomic nerve supply to the cardiovascular system also contains several other neuropeptides with potent actions on blood vessels. Neuropeptide Y (NPY) is a 36 amino acid peptide widely distributed in nerve fibres, where it frequently coexists with noradrenaline (LUNDBERG et al. 1982). It potentiates the effects of noradrenaline and has powerful vasoconstrictor activity.

IV. Neuropeptides as Growth Factors?

Another important insight into the potential of these agents is gained from the study of the effect of neuropeptides on cultured cells. Substance P is known to have several activities in vitro that suggest a potential role as an immunomodulator (see MCGILLIS et al. 1987). It stimulates proliferation of certain T-lymphocytes (PAYAN et al. 1983). More importantly, substance P and neurokinin A have recently been found to stimulate the growth of cultured arterial smooth muscle cells and cultured human skin fibroblasts in a potent manner (NILSSON et al.

1985). This is an important finding as it is the first suggestion that substances released from nerves could play a role in the control of cell proliferation. Proliferation of specific cell types at appropriate times is an essential step in the healing of wounds. Several animal models have been used to investigate wound healing. KISHIMOTO (1984) has reported that substance P-containing nerve fibres (i.e. C-fibres) are regenerated during the healing of burn wounds in guinea-pig skin. In contrast, SENAPATI et al. (1986) have shown a depletion of neuropeptides (substance P and CGRP) during a 2-week period after wounding the rat skin. Although these two studies appear contradictory, this is not necessarily so. The decrease observed in the latter study could be due to an increased release from terminals. The significance of the neuropeptide changes is uncertain in the absence of turnover data. However, these findings are not only of interest to wound healing, but also to situations where there is increased cell proliferation, such as psoriasis, where normal control is lost and cell proliferation, especially of the keratinocytes, proceeds at a greatly increased rate. We anticipate that in the future the ability of different neuropeptides to affect the growth of different cell types will be extensively examined.

D. Conclusions and Clinical Implications

The clinical relevance of the neuropeptides is as yet unknown. An important step forward in the understanding of their importance in skin conditions will come from studies into how and when these peptides are released. Such studies are at an early stage. One study has been carried out by WALLENGREN and co-workers (1986) where substance P and VIP were measured by radioimmunoassay in samples taken from patients with inflammatory skin diseases. Spontaneous blisters contained high amounts of substance P immunoreactivity, particularly in bullous pemphigoid, whilst VIP levels were low. Additionally, suction blisters from inflamed skin often contained substance P, but not VIP. These results do implicate substances released from C-fibres in inflammatory skin disease.

A conclusive role of the neuropeptides will only be assumed after the demonstration that specific inhibitors of neuropeptide release or action have anti-inflammatory effects. From the data presented here, it is obvious that future research into the role of the neuropeptides in inflammatory conditions and the development of specific antagonists is of considerable importance.

Acknowledgement. S. D. Brain is supported by the Sir Halley Stewart Trust.

References

Amara SG, Jonas V, Rosenfeld MG, Ong ES, Evans RM (1982) Alternative RNA processing in calcitonin gene expression generates mRNAs encoding different polypeptide products. Nature 298:240–244

Anand A, Bloom SR, McGregor GP (1983) Topical capsaicin pretreatment inhibits axon reflex vasodilatation caused by somatostatin and vasocactive intestinal polypeptide in human skin. Br J Pharmacol 78:665–669

Benyon RC, Church MK, Lowman MA (1987) Histamine release from human dispersed skin mast cells induced by substance P. Br J Pharmacol 90:102

Bloom SR, Polak JM (1983) Regulatory peptides and the skin. Clin Exp Dermatol 8:3–18
Bradbury AF, Smyth DG (1983) Substrate specificity of an amidating enzyme in porcine pituitary. Biochem Biophys Res Commun 112:372–376
Brain SD, Williams TJ (1985) Inflammatory oedema induced by synergism between calcitonin gene-related peptide (CGRP) and mediators of increased vascular permeability. Br J Pharmacol 86:855–860
Brain SD, Williams TJ, Tippins JR, Morris HR, MacIntyre I (1985) Calcitonin gene-related peptide is a potent vasodilator. Nature 313:54–56
Brain SD, Tippins JR, Morvis HR, MacIntyre I, Williams TJ (1986) Potent vasodilator activity of calcitonin gene-related peptide in human skin. J Invest Dermatol 87:533–536
Bruce AN (1913) Vasodilator axon reflexes. Q J Exp Physiol 6:339–354
Buck SH, Burcher E (1986) The tachykinins: a family of peptides with a brood of "receptors". Trends in Pharmacological Sciences 7:65–68
Chang MM, Leeman SE, Nrall HD (1971) Amino acid sequence of substance P. Nature 232:86–87
Comb M, Seeburg PH, Adelman J, Eiden L, Herbert E (1982) Primary structure of the human met- and leu-enkephalin precursor and its mRNA. Nature 295:663–666
Devillier P, Regoli D, Asseraf A, Descurs B, Marsac J, Renoux M (1986) Histamine release and local responses of rat and human skin to substance P and other mammalian tachykinins. Pharmacology 32:340–347
Edbrooke MR, Parker D, McVey JH, Riley JH, Sorenson GD, Pettengell OS, Graig RK (1985) Expression of the human calcitonin/CGRP gene in lung and thyroid carcinoma. EMBO J 4:715–724
Edwardson JA, McDermott JR (1985) Metabolism of neuropeptides at brain and pituitary sites. Biochem Soc Trans 13:50–53
Emson PC (1979) Peptides as neurotransmitter candidates in the mammalian CNS. Prog Neurobiol 13:61–116
Farber EM, Nickoloff BJ, Recht B, Fraki JE (1986) Stress, symmetry and psoriasis: possible role of neuropeptides. J Am Acad Dermatol 14:305–311
Foreman JC, Jordan CC, Oehme P, Renner H (1983) Structure-activity relationships for some substance P-related peptides that cause wheal and flare reactions in human skin. J Physiol 335:449–465
Gamse R, Saria A (1985) Potentiation of tachykinin-induced plasma protein extravasation by calcitonin gene-related peptide. Eur J Pharmacol 114:61–66
Gamse R, Holzer P, Lembeck F (1980) Decrease of substance P in primary afferent neurones and impairment of neurogenic plasma extravasion by capsaicin. Br J Pharmacol 68:207–213
Hagermark O, Hokfelt T, Pernow B (1978) Flare and itch induced by substance P in human skin. J Invest Dermatol 71:233–235
Hartschuh W, Weihe E, Büchler M, Helmsteadter V, Feurle GE, Forssmann WG (1979) Met-enkephalin-like immunoreactivity in Merkel cells. Cell Tissue Res 201:343–348
Hartschuh W, Weihe E, Reinecke M (1983) Peptidergic (neurotensin, VIP, substance P) nerve fibres in the skin. Immunohistochemical evidence of an involvement of neuropeptides in nociception, pruritus and inflammation. Br J Dermatol 109 (Suppl 25):14–17
Hartschuh W, Reinecke M, Weihe E, Yanaihara N (1984) VIP-immunoreactivity in the skin of various mammals: immunohistochemical, radioimmunological and experimental evidence for a dual localization in cutaneous nerves and Merkel cells. Peptides 5:239–245
Hua XY, Lundberg JIM, Theodorsson-Norheium E, Brodin E (1984) Comparison of cardiovascular and bronchoconstrictor effect of substance P, substance K and other tachykinins. Naunyn Schmiedebergs Arch Pharmacol 328:196–201
Itoh N, Obata K, Yanaihara N, Okamoto H (1983) Human preprovasoactive intestinal polypeptide contains a novel PHI-27-like peptide, PHM-27. Nature 304:547–549
Jancso G, Kiraly E, Jancso-Gabor A (1977) Pharmacologically induced selective degeneration of chemosensitive primary sensory neurons. Nature 270:741–743

Jancso N, Jancso-Gabor A, Szolcsanyi J (1967) Direct evidence of neurogenic inflammation and its prevention by denervation and by treatment with capsaicin. Br J Pharmacol 31:138–151
Jessel TM (1982) Substance P in nociceptive sensory neurones. Ciba Found Symp 91:225–248
Kangawa K, Minamino N, Fukuda A, Matsuo H (1983) Neuromedin K: a novel mammalina tachykinin identified in porcine spinal cord. Biochem Biophys Res Commun 114:533–540
Kimura S, Okada M, Sugita Y, Kanzawa I, Munekata E (1983) Novel neuropeptides, neurokinin α and β, isolated from porcine spinal cord. Proc Jpn Acad Ser B 59:101–104
Kishimoto S (1984) The regeneration of substance P containing nerve fibres in the process of burn wound healing in the guinea pig skin. J Invest Dermatol 83:219–233
Lee CM (1982) Enzymic inactivation of substance P in the central nervous system. Ciba Found Symp 91:165–179
Lembeck F (1983) Mediators of vasodilatation in skin. Br J Dermatol 109 (Suppl 25):1–9
Lembeck F, Gamse R (1982) Substance P in peripheral sensory processes. In: Porter R, O'Connor M (eds) Substance P in the nervous system. Pitman, London, pp 35–54
Lembeck F, Holzer P (1979) Substance P as a mediator of antidromic vasodilatation and neurogenic plasma extravasation. Naunyn Schmiedebergs Arch Pharmacol 310:175–183
Lewis T (1927) The blood vessels of the human skin and their responses. Shaw, London
Lundberg JM, Terenius L, Hokfelt T, Martling CR, Tatemoto K, Mutt V, Polak JM, Bloom SR, Goldstein M (1982) Neuropeptide Y (NPY)-like immunoreactivity in peripheral noradrenergic neurones and effects of NPY on sympathetic function. Acta Physiol Scand 116:447–480
Lundberg JM, Franco-Cereceda A, Hua X, Hokfelt T, Fischer JA (1985) Co-existence of substance P and calcitonin gene-related peptide like-immunoreactivities in sensory nerves in relation to cardiovascular and bronchoconstrictor effects of capsaicin. Eur J Pharmacol 108:315–319
McGillis JP, Organist ML, Payan DG (1987) Substance P and immunoregulation. Fed Proc 46:196–199
Minamino N, Kangawa K, Fukuda A, Matsuo H (1984) Neuromedin L: a novel mammalian tachykinin identified in porcine spinal cord. Neuropeptides 4:157–166
Mulderry PK, Ghatei MA, Rodrigo J, Allen JM, Rosenfeld MG, Polak JM, Bloom SR (1985) Calcitonin gene-related peptide in cardiovascular tissues of the rat. Neuroscience 14:947–954
Nakanishi S, Inoue A, Kita T, Nakamura M, Chang ACY, Cohen SN, Numa S (1979) Nucleotide sequences of cloned cDNA for bovine β-lipotrophin precursor. Nature 278:423–427
Nakata Y, Kusaka Y, Yajima H, Segawa T (1981) Active uptake of substance P carboxy-terminal heptapeptide (5-11) into rat brain and rabbit spinal cord slices. J Neurochem 37:1529–1534
Nawa H, Hirose T, Takashima H, Inayama S, Nakamishi S (1983) Nucleotide sequences of cloned cDNAs for two types of bovine brain substance P precursor. Nature 306:32–36
Nilsson J, von Euler AM, Dalsgaard CJ (1985) Stimulation of connective tissue cell growth by substance P and substance K. Nature 315:61–63
Payan DG, Brewster DR, Goetzl EJ (1983) A specific stimulation of human T-lymphocytes by substance P. J Immunol 131:1613–1615
Piotrowski W, Foreman JC (1986) Some effects of calcitonin gene-related peptide in human skin and on histamine release. Br J Dermatol 114:37–46
Polak JM, Bloom SR (1983) Distribution of bombesin, somatostatin, substance P and vasoactive intestinal polypeptide in feline and procine skin. Life Sci 32:2827–2936
Polak JM, Bloom SR (1984) Localization and measurement of VIP in the genitourinary system of man and animals. Peptides 5:225–230

Richter K, Kawashima E, Egger R, Kreil G (1984) Biosynthesis of thyroptrophin releasing hormone in the skin of *Xenopus laevis:* partial sequence of the precursor deduced from cloned cDNA. EMBO J 3:617–621
Rosenfeld MG, Mermod JJ, Amara SG, Swanson LW, Sawchenko PE, Rivier J, Vale WW, Evans RM (1983) Production of a novel neuropeptide encoded by the calcitonin gene via tissue specific RNA processing. Nature 304:129–135
Said SI (1984) Vasoactive intestinal polypeptide (VIP): current status. Peptides 5:143–150
Schultzberg M, Hokfelt T, Lundberg JM, Fuxe K, Mutt V, Said S (1980) Distribution of VIP-neurones in the peripheral and central nervous system. Endocrinol Jpn 1:23–30
Senapati A, Anand P, McGregor GP, Ghatei MA, Thompson RPH, Bloom SR (1986) Depletion of neuropeptides during wound healing in rat skin. Neurosci Lett 71:101–105
Sydbom A (1982) Histamine release from isolated rat mast cells by neurotensin and other peptides. Agents Actions 12:91–93
von Euler VS, Gaddum JH (1931) An unidentified depressor substance in certain tissue extracts. J Physiol (Lond) 72:74–87
Wallengren J, Ekman R, Moller H (1986) Substance P and vasoactive intestinal peptide in bullous and inflammatory skin disease. Acta Derm Venereol (Stockh) 66:23–28
Wedmore CV, Williams TJ (1981) Control of vascular permeability by polymorphonuclear leukocytes in inflammation. Nature 289:646–650
Williams TJ (1979) Prostaglandin E_2, prostaglandin I_2 and the vascular changes of inflammation. Br J Pharmacol 65:517–524
Williams TJ, Peck MJ (1977) Role of prostaglandin-mediated vasodilation in inflammation. Nature 270:530–532

CHAPTER 26

Polyamines

J.C. ALLEN

A. Occurrence and Metabolism of Polyamines

I. Introduction

It seems strange that the ubiquitous bio-organic cations the polyamines have, until recently, been consigned to the boxroom of biochemistry. Nevertheless, it is now apparent that they play a variety of subtle and sometimes essential roles. This chapter emphasises their participation in normal and abnormal epidermis; restrictions on space make it quite impossible to do justice to many general recent developments, and the reader is referred to more comprehensive treatments for further information (COHEN 1971; TABOR and TABOR 1976; JÄNNE et al. 1978; GAUGAS 1980).

II. Structure and Occurrence

The diamine *putrescine* (buta-1,4-diamine), $H_2N(CH_2)_4NH_2$, is usually included amongst the biological polyamines because of its structural and biosynthetic relationships. The others are its *mono-* and *bis*-aminopropyl derivatives *spermidine* (*N*-(3-aminopropyl)buta-1,4-diamine), $H_2N(CH_2)_3NH(CH_2)_4NH_2$, and spermine (*N,N'-bis*(3-aminopropyl)buta-1,4-diamine), $H_2N(CH_2)_3 \cdot NH(CH_2)_4NH(CH_2)_3NH_2$. All three are ubiquitous in eukaryotes, but prokaryotes lack spermine. A significant proportion of polyamines exist in vivo as *N*-acetyl derivatives (ABDEL-MONEM and OHNO 1977; SEILER et al. 1980a; BACHRACH and SEILER 1981), and a further fraction is covalently bound to the γ-glutamyl side chains of certain proteins (FOLK et al. 1980; WILLIAMS-ASHMAN and CANNELLAKIS 1980). SEILER (1980) has reviewed methods of polyamine estimation, some of which are appropriate to a level of a few picomoles.

Levels of putrescine, spermidine and spermine in normal human epidermis have been determined as 12, 125, and 200 nmol/g wet weight respectively (BÖHLEN et al. 1978), 29, 100, and 132 nmol/g wet weight (LAUHARANTA et al. 1981) or 3.6, 47, and 52 nmol/µg DNA (PROCTOR et al. 1979). All three sets of data were obtained using full-thickness biopsy punches. In mouse epidermis, frequently used as an experimental model, the levels in nmol/g wet weight are: putrescine, 19; spermidine, 1400; spermine, 610 (SEILER and KNÖDGEN 1979).

The polyamines are toxic. In mice, spermine has an LD_{50} of 0.128 mmol/kg, but spermidine is only about 2% as effective (TABOR and ROSENTHAL 1956).

III. Polyamine Metabolism

The pathway of mammalian polyamine biosynthesis is well established, and is depicted in Fig. 1. Putrescine is formed by decarboxylation of L-ornithine, the reaction being catalysed by L-ornithine decarboxylase (EC 4.1.1.17). *S*-Adenosyl-L-methionine decarboxylase (EC 4.1.1.50) converts *S*-adenosyl-L-methionine to *S*-(5'-adenosyl)-3-methylthiopropylamine, and the propylamino group of this compound is transferred to putrescine by spermidine synthase (EC 2.5.1.16). Spermine is formed by a similar transfer to the aminobutyl group of spermidine, catalysed by a separate aminopropyl transferase, spermine synthase (EC 2.5.1.-). Ornithine decarboxylase appears to be a cytoplasmic enzyme, whereas *S*-adenosylmethionine decarboxylase occurs in both the cytoplasm and the nucleus (McCORMICK 1977).

Polyamine catabolism is by no means so clear. Acetylation and oxidation appear to be the transformations involved (SEILER et al. 1980a, 1981a). Enzymes have been identified which acetylate spermidine and spermine to form acetylpolyamines, which are then excreted in the urine together with free polyamines. However, acetylation may not be merely a prelude to excretion, since the acetylpolyamines are better substrates than the free polyamines for polyamine oxidase (HÖLLTÄ 1977; SEILER et al. 1980b), which converts acetylspermine to spermidine and acetylspermidine to putrescine, releasing aminopropionaldehyde and acetic acid in each case (Fig. 1). There is also a distinct polyamine oxidase found in ru-

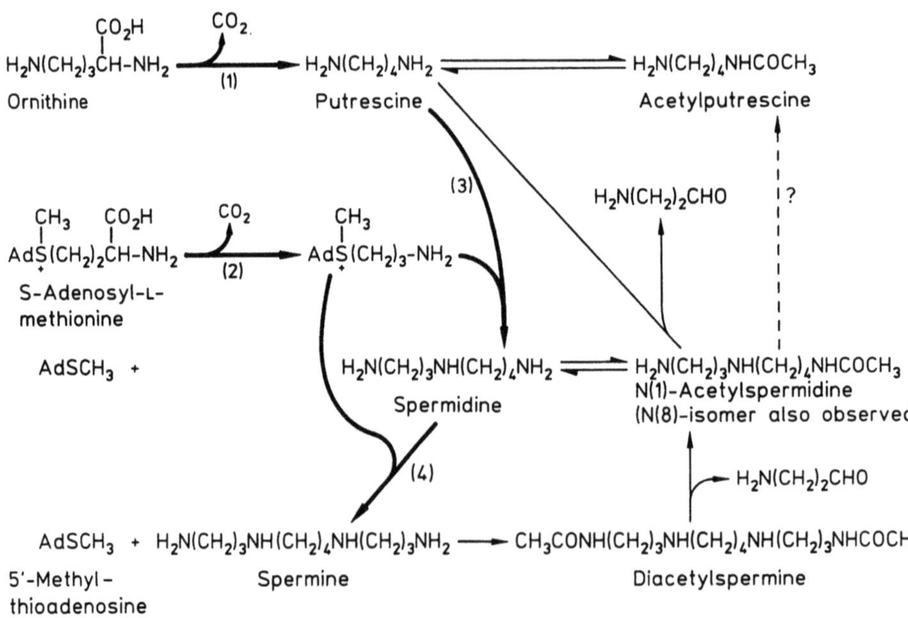

Fig. 1. Biosynthesis (*bold arrows*) and catabolism of polyamines in mammalian tissues. The biosynthetic enzymes are: (*1*) L-ornithine decarboxylase, (*2*) *S*-adenosyl-L-methionine decarboxylase, (*3*) spermidine synthetase, (*4*) spermine synthetase. For present understanding of enzymes involved in catabolism, see Sect. A.III

minant serum, which oxidatively deaminates spermine and spermidine (but not putrescine) to aminoaldehydes, NH_3 and H_2O_2 (TABOR et al. 1964); diamine oxidases which convert putrescine to aminobutyraldehyde have been found in cells (QUASH et al. 1979; SUNKARA et al. 1981) and in human pregnancy serum and retroplacental fluid (ILLEI and MORGAN 1979).

Much remains to be elucidated in this area. The occurrence of the oxidases, the function of the degradative oxidative pathway spermine → spermidine → putrescine (MORGAN 1980) and the role of the aminoaldehydes, which are potent cell growth inhibitors (BACHRACH 1970; ALLEN et al. 1979), are all the subject of intensive research.

It has now also been shown that polyamines may be covalently conjugated to proteins (AIGNER-HELD and DAVES 1980), and in particular may be bound to the γ-glutamyl residues in those proteins that serve as substrates for Ca^{2+}-dependent transglutaminases (FOLK et al. 1980; WILLIAMS-ASHMAN and CANNELLAKIS 1980). This reaction, which has been shown to increase during rapid growth (HADDOX and RUSSELL 1981), is likely to have considerable physiological relevance (WILLIAMS-ASHMAN and CANNELLAKIS 1980; RENNERT et al. 1980).

IV. Regulation of Polyamine Biosynthesis

The exquisite modulation of polyamine biosynthesis is normally effected by regulation of ornithine decarboxylase, which is rapidly inducible, has a short half-life and is subject to specific modulators of activity.

A rapid and large increase in ornithine decarboxylase activity is a universal and apparently obligatory biological response to stimulation of growth (MAUDSLEY 1979; CANNELLAKIS et al. 1979). In epidermis, appropriate stimulants of ornithine decarboxylase activity include hair plucking, wounding, ultraviolet irradiation, epidermal growth factor, cAMP and tumour-promoting phorbol esters. These are discussed in more detail in Sect. D.I. The increase in enzymic activity stems from de novo synthesis, but the necessary sequence which precedes induction is not clear: recent work (NEBERT et al. 1980; LANZ and BRUNE 1981) has not supported earlier hypotheses invoking cAMP and protein kinases (BACHRACH 1980).

The in vivo half-life of ornithine decarboxylase is generally less than 1 h (CANNELLAKIS et al. 1979), which implies its rapid inactivation, and it is not surprising that the enzyme should be subject to inhibition in vivo by the polyamines themselves. Micromolar levels are effective (HELLER et al. 1978). However, the inhibition is not a direct effect, and appears to involve the induction of a specific macromolecular inhibitor (FONG et al. 1976). Although this so-called antizyme has been apparently well-characterised (CANNELLAKIS 1981), a report (RUSSELL 1981) has suggested that the post-translational inhibition of ornithine decarboxylase by transglutaminase-mediated putrescine incorporation may also play a significant and perhaps alternative role in the regulation of ornithine decarboxylase. Certainly, the antizyme has never been found in epidermal tissue.

Once the initial increase in ornithine decarboxylase activity has caused a sufficient rise in the putrescine level, the rates of synthesis of spermidine and spermine are determined by the activity of the other polyamine biosynthetic enzymes.

S-Adenosylmethionine decarboxylase is stimulated by putrescine and inhibited by spermidine (ALHONEN-HONGISTO 1980; MAMONT and DANZIN 1981), and there is some evidence of hormonally induced changes in the aminopropyl transferases (KAPYAHÖ et al. 1980).

V. Inhibitors of Polyamine Biosynthesis

Reviews of the more widely used inhibitors of polyamine biosynthesis have appeared (GAUGAS 1980; CALDERARA et al. 1981).

The most useful inhibitors of ornithine decarboxylase in vivo are (a) α-difluoromethylornithine (DFMO), an active-site-directed irreversible inhibitor ("suicide" inhibitor), (b) α-methylornithine, a competitive inhibitor, and (c) 1,3-diaminopropane, which seems to operate through induction of antizyme. The most widely used inhibitor of *S*-adenosylmethionine decarboxylase is MGBG (or methyl-GAG), $X{=}C(CH_3)CH{=}X$, where X is $=NNHC(=NH)NH_2$; this is almost universally (and incorrectly) referred to as methylglyoxal *bis*(guanylhydrazone). Results from its use must be interpreted with caution, since it can cause in vivo effects which are not due solely to polyamine depletion (PLESHKEWYCH et al. 1980; SEPPÄNEN et al. 1981).

B. Polyamines and Growth

I. Polyamines and Cell Growth

Quiescent cells which have been stimulated to divide invariably exhibit a rapid and large increase in ornithine decarboxylase activity in G_1-phase, and HEBY and ANDERSSON (1980) consider this to be the trigger for the whole ensuing programme of polyamine metabolism. Continuously dividing cells which have been synchronised by mitotic selection show two peaks of ornithine decarboxylase activity and polyamine accumulation, which occur in late G_1-S and G_2-M (HEBY et al. 1976; HEBY and ANDERSSON 1980).

Although an increase in cellular polyamine levels is not necessary for the initiation of DNA synthesis (RUPNIAK and PAUL 1978a), optimal rates of DNA synthesis appear to demand a certain minimal level of polyamines (KNUTSON and MORRIS 1978), and if non-transformed cells are substantially deprived of polyamines they are arrested in G_1-phase (HEBY et al. 1976; BOYNTON et al. 1976). Polyamine-limited transformed cells, however, are able to progress to S- or G_2-phase before they cease growth (HEBY et al. 1978; RUPNIAK and PAUL 1978b; SUNKARA et al. 1979a), and RUPNIAK and PAUL (1980) were the first to exploit this difference, to selectively kill SV40-transformed mouse 3T3 fibroblasts in the presence of normal 3T3 cells.

Although most work has centred on the involvement of polyamines in DNA synthesis and metabolism, their influence on other aspects of cell proliferation is likely to be great. Depletion of polyamines by treatment of cells with MGBG or α-methylornithine (see Sect. A.V.) produces changes in chromosome condensation (SUNKARA et al. 1979a) and a reduction in cytokinesis (SUNKARA et al.

1979 b), and polyamine starvation of polyamine-auxotrophic CHO cells results in disappearance of actin filaments and microtubules (POHJANPELTO et al. 1981). Their roles in differentiating cells are still obscure, but ornithine decarboxylase activity has been shown to decline progressively with age of fibroblasts (DUFFY and KREMZENER 1977), and there is evidence that inhibition of polyamine synthesis in keratinocytes causes the onset of differentiation (PROCTOR et al. 1980).

In sharp contradistinction to the growth-promoting effects of the polyamines themselves, the aminoaldehydes formed by their oxidation or oxidative deamination are potent mitotic inhibitors (HIGGINS et al. 1969; BACHRACH 1970; ISRAEL et al. 1973; ALLEN et al. 1977; GAHL and PITOT 1978). Although oxidised polyamines are reported to eventually degrade to yield the cytotoxic product acrolein (ALARCON 1970), the reversibility (BYRD et al. 1977) and cycle phase specificity (ALLEN et al. 1979) of mitotic inhibition by oxidised polyamines suggest that acrolein is not responsible. There is increasing evidence of a physiological role for oxidised polyamines (MAURER and MASCHLER 1979; GAUGAS and DEWEY 1979; MORGAN and ILLEI 1980).

II. Polyamines and Tissue Growth

All rapid growth is accompanied by increased polyamine synthesis and metabolism (for a review, see JÄNNE et al. 1978), including mammalian embryogenesis, compensatory growth and tissue repair, a whole variety of stressful inducements to growth, and growth triggered by hormones or growth factors. The reader is referred to comprehensive reviews for further details, except for stimulants of epidermal growth, which are discussed in Sect. D.I.

III. The Biochemical Role of Polyamines

Although the polyamines are vital participants in cell proliferation and a range of other biological events, they clearly lack the kind of molecular complexity which is so often associated with biochemical specificity. This paradox between structural simplicity and functional significance should not, however, cause any intellectual difficulty. Ca^{2+} and Mg^{2+} are structurally simple yet biologically important, and they exhibit specificity in the sense that they are not usually interchangeable. Likewise, although polyamines may often be supplanted by simple metallic cations in in vitro biological systems, this rarely leads to optimal activity.

It is possible to identify three biochemical distinctions between polyamines and metal cations. Polyamines can be made intracellularly, and supply can be rapidly increased or diminished. The cationic groups on the polyamines are *spaced,* making it possible for them to bind to a number of separate anionic or electronegative groups to form a kind of molecular scaffold (QUIGLEY et al. 1978; BOLTON and KEARNS 1978), and these cationic groups may be eliminated by oxidation or acylation, allowing the scaffold to be converted into a more permanent, covalently linked structure.

Exemplification of these characteristics may be found in the involvement of polyamines in DNA structure (HUSE et al. 1978; BLOOMFIELD et al. 1980; BURTON et al. 1981) and synthesis (KNUTSON and MORRIS 1978; WALLACE et al. 1981), RNA structure (KARPETSKY et al. 1977; QUIGLEY et al. 1978; BOLTON and KEARNS 1978; NÖTHIG-LASLO et al. 1981) and synthesis (WALLACE and KEIR 1979), synthesis of poly(ADP-ribose) (TANIGAWA et al. 1980; KAWAMURA et al. 1981) and synthesis of proteins (IGARASHI et al. 1978; IGARASHI et al. 1980). Polyamines also influence the activity of a wide variety of enzymes: some of the changes are probably largely due to ionic strength effects [it is usually forgotten that, for example, addition of 1 m*M* spermine to 100 m*M* univalent buffer at pH = pKa (<7.5) gives a 10% increase in ionic strength] and may have little physiological significance, but many important enzymes are markedly affected (BACHRACH et al. 1978; WRIGHT et al. 1978; NORDLIE et al. 1979; TASHIMA et al. 1981; LUTAYA and GRIFFITHS 1981).

C. Polyamines and Hyperproliferative Diseases

This subject has been extensively reviewed by JÄNNE et al. (1978). There is a general pattern of increased putrescine and spermidine synthesis and accumulation in a whole variety of animal tumours and other types of hyperproliferative condition, which is accompanied by elevated levels of polyamines and their acetyl derivatives in urine and tissue fluids. The original observation by RUSSELL (1971) aroused considerable interest in the diagnostic potential of assay of excreted polyamines, but initial expectations (RUSSELL 1977) have not been completely fulfilled (COHEN 1977; CANNELLAKIS et al. 1979). Nevertheless, a general doubling of polyamine excretion occurs in hyperproliferative diseases (CANNELLAKIS et al. 1979), and increasing use of polyamine assays in monitoring progess of chemotherapy seems likely. It has been hypothesised (RUSSELL et al. 1978a) that, in cancer patients receiving chemotherapy, spermidine excretion indicates tumour-cell "kill", and putrescine excretion reflects the size of the "growth fraction" of the tumour (i.e. the recruitment of cells into the proliferative compartment). There is some evidence for this (RUSSELL et al. 1978; NISHIOKA et al. 1980), but the question is by no means settled (HEBY and ANDERSSON 1978; ROMANO et al. 1980).

D. Polyamines in Skin

I. Induction of Polyamine Biosynthesis in Skin

The close link between growth stimulation and induction of polyamine biosynthesis is qualitatively the same in skin as in any other tissue. STÜTTGEN (1968) was the first to undertake any detailed study of polyamine levels in skin. It is noteworthy that topical applications of putrescine or spermine themselves are able to induce DNA synthesis in mouse epidermis (GANGE and DEQUOY 1980).

1. Wound Healing

Elevation of ornithine decarboxylase activity is one of the earliest detectable events in healing of rat skin (MITZUTANI et al. 1974; MORRISON and GOLDSMITH

1978), reaching a maximum at 12 h after wounding and returning to normal by about 48 h. The spermidine concentration is increased, with a maximum after 24 h; spermine levels show a smaller elevation. The synthetic corticosteroid betamethasone, which is able to inhibit wound healing, delays the increase in ornithine decarboxylase activity by about 12 h.

PROBST and KREBS (1975) showed that in mouse epidermis, stimulation of DNA synthesis by hair plucking produced a rapid and large rise in ornithine decarboxylase activity, both in the matrix cells of the hair follicle and in the basal cells of the interfollicular epidermis. The plucking of *growing* hair did not induce any change in ornithine decarboxylase activity, which well illustrates the relation of ornithine decarboxylase induction to *stimulation* of growth: the induction is not simply a stress effect, since neighbouring unstimulated skin showed no significant increase. Likewise, skin massage gives no detectable rise in ornithine decarboxylase activity, but abrasion does (CLARK-LEWIS and MURRAY 1978).

It has been shown that the effects of certain drugs on the increases in ornithine decarboxylase activity brought about by hair plucking and phorbol ester tumour promoters (see Sect. D.I.4) are different, as are the magnitude and timing of these increases (LESIEWICZ et al. 1980). In particular, the response to the phorbol ester was mainly epidermal, whereas hair plucking elicited a response in both epidermis and dermis. In addition, the effects of the combined stimuli were additive, thus providing convincing evidence that the stimuli induce ornithine decarboxylase by different mechanisms.

2. Epidermal Growth Factor

Epidermal growth factor has been shown to induce increases in ornithine decarboxylase activity in mouse skin and in cultures of chick embryo epidermis (STASTNY and COHEN 1970), and in cultures of human fibroblasts (DIPASQUALE et al. 1978). In both in vitro cases there was an increase of transport of polyamines into the cells, which relates to the observations that very low concentrations of putrescine (10^{-8}–10^{-7} *M*) can act as a growth factor for fibroblasts (HAM 1964; POHJANPELTO and RAINA 1972; POHJANPELTO 1973), and that this putrescine is actively transported and concentrated by proliferating cells (POHJANPELTO 1976). WALLACE and KEIR (1981) have recently emphasised the close relationship of the growth status of cells to selective and active import and export of polyamines.

3. Ultraviolet Irradiation

Extensive irradiation of cells can so damage their DNA as to inhibit the induction of ornithine decarboxylase. The effects of various forms of irradiation, including psoralen plus UV-A light (360 nm), on the induction of ornithine decarboxylase in Chinese hamster fibroblasts have been compared (BEN-HUR and RIKLIS 1981).

Lower doses of radiation can greatly enhance ornithine decarboxylase activity. In hairless mouse epidermis irradiated with UV-B light (290–320 nm), the maximum value, which occurred 28 h after irradiation, was 80–350 times the basal level (LOWE et al. 1978; LOWE 1980). Induction of ornithine decarboxylase requires de novo synthesis (LICHTI et al. 1980), occurs only when a radiation dose sufficient to produce oedema or erythema is given (VERMA et al. 1979), and can be prevented by corticosteroid treatment (GANGE and DEQUOY 1980). SEILER and KNÖDGEN (1979) used a similar model, and showed that UV irradiation produced a rapid rise in epidermal putrescine and a slower but steady increase in spermidine. Oral or intraperitoneal administration of DFMO (see Sect. A.V) eliminated these changes. The same group (SEILER et al. 1981 b) have demonstrated increased urinary excretion of polyamines in UV-irradiated mice. Certain anti-oxidants, probably acting as free-radical scavengers, are able to reduce radiation induction of ornithine decarboxylase (PETERSON et al. 1980).

A comparison of the effects of UV-A radiation plus either topical 8-methoxypsoralen or anthracene in the hairless mouse (GANGE 1981) showed that PUVA treatment gave a longer depression of DNA synthesis and a greater delay before expression of ornithine decarboxylase than did UV-A plus anthracene.

4. Tumour Promoters

A single application of a few nanomoles of 12-*O*-tetradecanoylphorbol-13-acetate (TPA) or other tumour-promoting phorbol esters is able to produce a dramatic rise (200- to 400-fold) in ornithine decarboxylase activity in epidermal cells (O'BRIEN et al. 1975a); it was later shown that the effect is not confined to the epidermis (KISHORE and BOUTWELL 1981). A subsequent but lesser rise in epidermal *S*-adenosylmethionine decarboxylase also occurs, and leads to an accumulation of the polyamines. These increases can be prevented by pre-treatment with cycloheximide or azacytidine. Non-phorbol tumour promoters produce similar qualitative increases in activity to those caused by TPA, but non-tumour-promoting phorbols and other hyperplastic agents produce no increase in ornithine decarboxylase activity (O'BRIEN et al. 1975b; O'BRIEN 1976; O'BRIEN and DIAMOND 1978). Since the 4 α-epimers of tumour-promoting phorbol esters are neither tumour promoting nor ornithine decarboxylase inducing (FUJIKI et al. 1980), the mechanism of the enzyme's induction by tumour-promoting phorbols appears to be receptor mediated, and is thus probably quite distinct from the mechanism of induction by ultraviolet light. Indeed, the effect of a combination of the two stimuli is synergic (LICHTI et al. 1981).

It was originally suggested (O'BRIEN 1976) that the induction of ornithine decarboxylase by TPA represents an obligatory stage of its tumour-promoting property. Closer investigation, however, has by no means fully supported this hypothesis, and present evidence is confusing. LANZ and BRUNE (1981) have shown that addition of TPA to confluent 3T3 fibroblasts precipitates a sequence of prostaglandin release, ornithine decarboxylase induction, DNA synthesis and cell proliferation. However, the proliferative effects of TPA were independent of both prostaglandin release and polyamine synthesis. They attribute previous findings of inhibition of TPA-induced ornithine decarboxylase in mouse skin by indomethacin (VERMA et al. 1977) to non-specific effects of the prostaglandin synthetase inhibitor. In contrast to this apparent lack of a causal relationship is the observation that topical application of putrescine itself to mouse skin, after treatment with TPA, can prevent both ornithine decarboxylase induction and papilloma formation (WEEKES et al. 1980); however, the explanation given of a high level of putrescine being necessary for tumour promotion seems untenable in the light of the reported facile and active transport of polyamines by proliferating cells (WALLACE and KEIR 1981).

Although treatment with TPA gives rise to an increase in epidermal cAMP (GRIMM and MARKS 1974), this is apparently dissociated from the effect of TPA on polyamine biosynthesis (MUFSON et al. 1977). Despite the fact that injection of cholera toxin produces localised increases in both cAMP and ornithine decarboxylase activity in mouse epidermis (MURRAY et al. 1980), and although there is a parallelism between induction of ornithine decarboxylase and cAMP-dependent protein kinases in the induction of drug-metabolising enzymes in the liver (NEBERT et al. 1980; KANO and NEBERT 1981), no causal relationship has yet been shown to exist.

There have been reports on the influence of a number of other agents on ornithine decarboxylase induction by TPA, including retinoids (VERMA and BOUTWELL 1977), local anaesthetics (YUSPA et al. 1980) and anthralin (DE YOUNG et al. 1981). The latter, although itself a weak tumour promoter, under certain dose regimes actually inhibited TPA-induced ornithine decarboxylase activity.

Clarification of this whole area is likely to result from further investigation of exactly which epidermal cell population is influenced by TPA. It has now been reported (LICHTI et al. 1981) that it is the proliferating basal cells and not the differentiating suprabasal cells which are affected. Furthermore, the ornithine decarboxylase activity of cells with a high *natural* proliferation rate was low, and so it would appear that the induction of the enzyme is not a prerequisite for high continual rates of cell proliferation.

II. Polyamines and Hyperproliferative Diseases of the Skin

1. Psoriasis

Since elevated polyamine synthesis and accumulation are so closely associated with rapid growth, it is not surprising to find that the greatly increased epidermopoiesis in psoriasis leads to elevated polyamine levels in the skin, blood and urine of psoriatic patients. Early findings using only a few subjects actually suggested a reduction of skin polyamines in psoriasis (STÜTTGEN 1968; PROCTOR et al. 1975), but all other studies have shown a statistically significant increase in putrescine, spermidine and spermine over normal controls in involved psoriatic epidermis (RUSSELL et al. 1978b; BÖHLEN et al. 1978; GROSSHANS et al. 1978; PROCTOR et al. 1979; LAUHARANTA et al. 1981). Activities or ornithine decarboxylase and *S*-adenosylmethionine decarboxylase were elevated sixfold (RUSSELL et al. 1977; RUSSELL et al. 1978b). Russell's group (RUSSELL et al. 1978b) also found that levels of all three polyamines were increased in the uninvolved skin of psoriatic patients. However, the situation is not entirely clear, since BÖHLEN et al. (1978) found no elevation of spermidine or spermine in uninvolved skin whereas PROCTOR et al. (1979) found no elevation of putrescine. Both groups used full-thickness skin samples and apparently similar methodology. A degree of increase of polyamine levels in uninvolved skin might be expected, because of the considerably higher DNA synthesis compared to that in normal skin of healthy subjects (MARKS 1978).

The concentration of spermine in the whole blood of psoriatic patients is increased two- to threefold above normal, and there is also an elevation of spermidine (PROCTOR et al. 1975; COOPER et al. 1978). More detailed investigation (COOPER et al. 1978) reveals increased amounts associated with the erythrocytes but not with the leucocytes. There is also a rise in urinary excretion of polyamines in psoriasis (RUSSELL et al. 1978b; SAKAKIBARA and YOSHIKAWA 1979), but further work on a larger group of patients will be required to resolve conflicting results on specific polyamines.

MARCELO and VOORHEES (1980) have attempted to link the expression of psoriasis to observed changes in cyclic nucleotides, prostaglandins and polyamines in the disease, but the evidence for any causal relationship involving the polyamines is so far poor. Undoubtedly, changes in polyamine levels and synthesis occur, and treatment of psoriasis with the glucocortioid diflorasone (RUSSELL et al. 1978b), dithranol (BÖHLEN et al. 1978), a topical regimen consisting of anthralin, salicylic acid, tar baths, corticosteroids and ultraviolet irradiation (PROCTOR et al. 1979), or oral retinoids and PUVA treatment (LAUHARANTA et al. 1981) all markedly reduce these increases. But it has yet to be shown that treatment with an inhibitor of polyamine biosynthesis such as DFMO (see Sect. A.V) is successful in producing a clinical improvement.

2. Skin Cancers

It is worth noting that the purportedly tissue-specific in vitro inhibitor or "chalone" (ALLEN and SMITH 1979; IVERSON 1981) of melanoma cell proliferation has been identified as spermidine (DEWEY 1978). There is no evidence so far to suggest that its effect is anything but artefactual.

Measurements of polyamines in skin cancer have all so far shown increases above control levels. Increased concentrations in blood, in the erythrocytes in particular, were observed in a study which included seven patients with melanoma (TAKAMI and NISHIOKA 1980). In a much more detailed investigation, TOWNSEND et al. (1976) measured urinary polyamines from 79 melanoma patients who were classified according to clinical stage and activity of disease. There was a good correlation between activity and excretion. Although increase in excretion of any single polyamine was an unreliable guide to prognosis, it was considered that serial determinations of all three polyamines would be useful in predicting response to therapy and prognosis. There has recently been a preliminary report on excretion of acetylpolyamines by patients with melanomas (SEILER et al. 1981a, b).

Urinary and hepatic polyamine levels and hepatic ornithine decarboxylase activity were found to be increased in methylcholanthrene-induced mouse fibrosarcomas (SUKUMAR and NAGARAJAN 1978), and a direct relation between therapy and polyamine accumulation and excretion was shown.

In a clinical study, SCALABRINO et al. (1980) found that the levels of both ornithine decarboxylase and *S*-adenosylmethionine decarboxylase were increased in basal cell epitheliomas but not so much as in squamous cell carcinomas. This correlates well with the relative growth rates of the tumours: squamous cell carcinoma is a rapidly growing invasive tumour, but basal cell epitheliomas proliferate at a much lower rate.

Acknowledgements. I am grateful to several workers for allowing me to have details of their work before publication, to Dr. C.J. SMITH for a critical reading of the manuscript, and to JANET BUTLER for typing it.

References

Abdel-Monem MM, Ohno K (1977) Polyamine metabolism. II. *N*-(Monoaminoalkyl) and *N*-(polyaminoalkyl) acetamides in human urine. J Pharm Sci 66:1195–1197

Aigner-Held R, Daves GD (1980) Polyamine metabolites and conjugates in man and higher animals: a review of the literature. Physiol Chem Phys Med NMR 12:389–400

Alarcon RA (1970) Acrolein. IV. Evidence for the formation of the cytotoxic aldehyde from enzymatically oxidised spermine or spermidine. Arch Biochem Biophys 137:365–375

Alhonen-Hongisto L (1980) Regulation of *S*-adenosylmethionine decarboxylase in Ehrlich ascites-carcinoma cells grown in culture. Biochem J 190:747–754

Allen JC, Smith CJ (1979) Chalones. A reappraisal. Biochem Soc Trans 7:584–592

Allen JC, Smith CJ, Curry MC, Gaugas JM (1977) Identification of a thymic inhibitor ("chalone") of lymphocyte transformation as a spermine complex. Nature 267:623–625

Allen JC, Smith CJ, Hussain JI, Thomas JM, Gaugas JM (1979) Inhibition of lymphocyte proliferation by polyamines requires ruminant plasma polyamine oxidase. Eur J Biochem 102:153–158

Bachrach U (1970) Oxidised polyamines. Ann NY Acad Sci 171:939–956

Bachrach U (1980) The induction of ornithine decarboxylase in normal and neoplastic cells. In: Gaugas JM (ed) Polyamines in biomedical research. Wiley, Chichester, pp 81–107

Bachrach U, Seiler N (1981) Formation of acetylpolyamines and putrescine from spermidine by normal and transformed chick embryo fibroblasts. Cancer Res 41:1205–1208

Bachrach U, Katz A, Hochman J (1978) Polyamines and protein kinase I. Induction of ornithine decarboxylase and activation of protein kinase in rat glioma cells. Life Sci 22:817–822

Ben-Hur E, Riklis E (1981) Inhibition of induced ornithine decarboxylase activity in Chinese hamster cells by gamma irradiation, far ultraviolet light and psoralen plus near ultraviolet light: a comparative study. Int J Radiat Biol 39:527–535

Bloomfield VA, Wilson RW, Rau DC (1980) Polyelectrolyte effects in DNA condensation by polyamines. Biophys Chem 11:339–343

Böhlen P, Grove J, Beya MF, Koch-Weser J, Henry MH, Grosshans E (1978) Skin polyamine levels in psoriasis: the effect of dithranol therapy. Eur J Clin Invest 8:215–218

Bolton PH, Kearns DR (1978) H-bonding interactions of polyamines with the 2′-OH of RNA. Nucleic Acids Res 5:1315–1324

Boynton AL, Whitfield JF, Isaacs RJ (1976) A possible involvement of polyamines in the initiation of DNA synthesis by human WI-38 and mouse BALB/3T3 cells. J Cell Physiol 89:481–488

Burton DR, Forsén S, Ramansson P (1981) The interaction of polyamines with DNA. A ^{23}Na NMR study. Nucleic Acids Res 9:1219–1228

Byrd WJ, Jacobs DM, Amoss MS (1977) Synthetic polyamines added to cultures containing bovine sera reversibly inhibit in vitro parameters of immunity. Nature 267:621–623

Calderara CM, Zappia V, Bachrach D (eds) (1981) Advances in polyamine research, vol 3. Raven, New York

Cannellakis ES, Viceps-Madore D, Kyriakidis DA, Heller JS (1979) The regulation and function of ornithine decarboxylase and of the polyamines. In: Horecker BL, Stadtman ER (eds) Current topics in cellular regulation, vol 15. Academic, New York, pp 155–202
Cannellakis ES, Heller JS, Kyriakidis DA (1981) The interaction of ornithine decarboxylase with its antizyme. In: Calderara CM, Zappia V, Bachrach U (eds) Advances in polyamine research, vol 3. Raven, New York, pp 1–13
Clark-Lewis I, Murray AW (1978) Tumor promotion and the induction of ornithine decarboxylase activity in mechanically stimulated mouse skin. Cancer Res 38:494–497
Cohen SS (1971) Introduction to the polyamines. Prentice-Hall, New Jersey
Cohen SS (1977) Meeting report: conference on polyamines and cancer. Cancer Res 37:939–942
Cooper KD, Shukla JB, Rennert OM (1978) Polyamine compartmentalization in various disease states. Clin Chim Acta 82:1–7
Dewey DL (1978) The identification of a cell culture inhibitor in a tumour extract. Cancer Lett 4:77–84
De Young LM, Helmes CT, Chao WR, Young JM, Miller V (1981) Paradoxical effect of anthralin on 12-*O*-tetradecanoylphorbol-13-acetate-induced mouse epidermal ornithine decarboxylase activity, proliferation, and tumour promotion. Cancer Res 41:204–208
Di Pasquale A, White D, McGuire J (1978) Epidermal growth factor stimulates putrescine transport and ornithine decarboxylase activity in cultivated human fibroblasts. Exp Cell Res 116:317–323
Duffy PE, Kremzner LT (1977) Ornithine decarboxylase activity and polyamines in relation to aging of fibroblasts. Exp Cell Res 108:435–440
Folk JE, Park MH, Chung SI, Schrode J, Lester EP, Cooper HL (1980) Polyamines as physiological substrates for transglutaminases. J Biol Chem 225:3695–3700
Fong WF, Heller JS, Cannellakis ES (1976) The appearance of an ornithine decarboxylase inhibitory protein upon the addition of putrescine to cell cultures. Biochim Biophys Acta 428:456–465
Fujiki H Mori M, Sugimura T, Hirota M, Ohigashi H, Koshimizu K (1980) Relationship between ornithine decarboxylase-inducing activity and configuration at C-4 in phorbol ester derivatives. J Cancer Res Clin Oncol 98:9–13
Gahl WA, Pitot HC (1978) Reversal by aminoguanidine of the inhibition of proliferation of human fibroblasts by spermidine and spermine. Chem Biol Interact 22:91–98
Gange RW (1981) Epidermal ornithine decarboxylase activity and thymidine incorporation following treatments with ultraviolet A combined with topical 8-methoxypsoralen or anthracene in the hairless mouse. Br J Dermatol 105:247–255
Gange RW, Dequoy PR (1980) Topical spermine and putrescine stimulated DNA synthesis in the hairless mouse epidermis. Br J Dermatol 103:27–32
Gaugas JM (ed) (1980) Polyamines in biomedical research. Wiley, Chichester
Gaugas JM, Dewey DL (1979) Evidence for serum binding of oxidised spermine and its potent G_1-phase inhibition of cell proliferation. Br J Cancer 39:548–557
Grimm W, Marks F (1974) Effect of tumor-promoting phorbol esters on the normal and isoproterenol-elevated level of adenosine-3′,5′-cyclic monophosphate in mouse epidermis in vivo. Cancer Res 34:3128–3134
Grosshans E, Henry M, Bohlen P, Grove J, Beya MF, Koch-Weser J (1978) Skin polyamine levels in psoriasis – the effect of therapy. J Invest Dermatol 70:227
Haddox MK, Russell DH (1981) Increased nuclear conjugated polyamines and transglutaminase during liver regeneration. Proc Natl Acad Sci USA 78:1712–1716
Ham RG (1964) Putrescine and related amines as growth factors for a mammalian cell line. Biochem Biophys Res Commun 14:34–38
Heby O, Andersson G (1978) Tumour cell death: the probable cause of increased polyamide levels in physiological fluids. Acta Pathol Microbiol Immunol Scand [A] 86:17–20
Heby O, Andersson G (1980) Polyamines and the cell cycle. In: Gaugas JM (ed) Polyamines in biomedical research. Wiley, Chichester, pp 17–34
Heby O, Gray JW, Lindl PA, Marton LJ, Wilson CB (1976) Changes in L-ornithine decarboxylase activity during the cell cycle. Biochem Biophys Res Commun 71:99–105

Heby O, Andersson G, Gray JW (1978) Interference with S and G_2 phase progression by polyamine synthesis inhibitors. Exp Cell Res 111:461–464
Heller JS, Chen KY, Kyriakidis DA, Fong WF, Cannellakis ES (1978) The modulation of the induction of ornithine decarboxylase by spermine, spermidine and diamines. J Cell Physiol 96:225–234
Higgins ML, Tillman MC, Rupp JP, Leach FL (1969) The effect of polyamines on cell culture cells. J Cell Physiol 74:149–154
Hölltä E (1977) Oxidation of spermidine and spermine in rat liver: purification and properties of polyamine oxidase. Biochemistry 61:91–100
Huse Y, Mitsui Y, Iitaka Y, Miyaki K (1978) Preliminary X-ray studies on the interaction of salmon sperm DNA with spermine. J Mol Biol 122:43–53
Igarashi K, Watanabe Y, Nakamura K, Kojima M, Fujiki Y, Hirose S (1978) Effect of spermidine on *N*-formylmethionyl-ᵗRNA binding to 30S ribosomal subunits and on *N*-formylmethionyl-ᵗRNA-dependent polypeptide synthesis. Biochem Biophys Res Commun 83:806–813
Igarashi K, Kojima M, Watanabe Y, Maeda K, Hirose S (1980) Stimulation of polypeptide synthesis by spermidine at the level of initiation in rabbit reticulocyte and wheat germ cell-free systems. Biochem Biophys Res Commun 97:480–486
Illei G, Morgan DML (1979) The distribution of polyamine oxidase activity in the fetomaternal compartments. Br J Obstet Gynaecol 86:873–877
Israel M, Zoll EC, Muhammed N, Modest EJ (1973) Synthesis and antitumor evaluation of the presumed cytotoxic metabolites of spermine and *N,N'*-bis(3-aminopropyl)nonane-1,9-diamine. J Med Chem 16:1–5
Iversen OH (1981) The chalones. In: Baserga R (ed) Handbook of experimental pharmacology, vol 57. Springer, Berlin Heidelberg New York, pp 491–550
Jänne J, Pösö H, Raina A (1978) Polyamines in rapid growth and cancer. Biochim Biophys Acta 473:241–293
Käpyaho K, Pösö H, Jänne J (1980) Role of propylamine transferases in hormone-induced stimulation of polyamine biosynthesis. Biochem J 192:59–63
Kano I, Nebert DW (1981) Ornithine decarboxylase induction in liver and hepatoma-derived cell cultures. No detectable differences between control and 3-methylcholanthrene-treated cells. Mol Pharmacol 20:172–178
Karpetsky TP, Hieter PA, Frank JJ, Lery CC (1977) Polyamines, ribonucleases and the stability of RNA. Mol Cell Biochem 17:89–99
Kawamura M, Tanigawa Y, Kitamura A, Miyake Y, Shimoyama M (1981) Effect of polyamines on purified poly(ADP-ribose) synthetase from rat liver nuclei. Biochim Biophys Acta 652:121–128
Kishore GS, Boutwell RK (1981) Induction of mouse hepatic ornithine decarboxylase by skin application of 12-*O*-tetradecanoylphorbol-13-acetate. Experientia 37:179–180
Knutson JC, Morris DR (1978) Cellular polyamine depletion reduces DNA synthesis in isolated lymphocyte nuclei. Biochim Biophys Acta 520:291–301
Lanz R, Brune K (1981) Dissociation of tumour-promoter-induced effects on prostaglandin release, polyamine synthesis and cell proliferation of 3T3 cells. Biochem J 194:975–982
Lauharanta J, Kousa M, Kapyaho K, Linnamaa K, Mustakallio K (1981) Reduction of increased polyamine levels in psoriatic lesions by retinoid and PUVA treatments. Br J Dermatol 105:267–272
Lesiewicz J, Morrison DM, Goldsmith LA (1980) Ornithine decarboxylase in rat skin. 2. Differential response to hair plucking and a tumor promoter. J Invest Dermatol 75:411–416
Lichti U, Bowden GT, Patterson E, Ben T, Yuspa SH (1980) Germicidal ultraviolet light induces ornithine decarboxylase in mouse epidermal cells and modifies the induction caused by phorbol ester tumor promoters. Photochem Photobiol 32:177–182
Lichti U, Patterson E, Hennings H, Yuspa SH (1981) The tumor promoter 12-*O*-tetradecanoylphorbol-13-acetate induces ornithine decarboxylase in proliferating basal cells but not in differentiating cells from mouse epidermis. J Cell Physiol 107:261–270
Lowe NJ (1980) Epidermal ornithine decarboxylase, polyamines, cell proliferation and tumor promotion. Arch Dermatol 116:822–825

Lowe NJ, Verma AK, Boutwell RK (1978) Ultraviolet light induces epidermal ornithine decarboxylase. J Invest Dermatol 71:417–418
Lutaya G, Griffiths JR (1981) Rapid formation of spermine in skeletal muscle during tetanic stimulation. FEBS Lett 123:186–188
Mamont PS, Danzin C (1981) In vitro and in vivo regulation of *S*-adenosyl-L-methionine decarboxylase by polyamines. In: Calderara CM, Zappia V, Bachrach U (eds) Advances in polyamine research, vol 3. Raven, New York, pp 123–135
Marcelo CL, Voorhees JJ (1980) Cyclic nucleotides, prostaglandins and polyamines in psoriasis. Pharmacol Ther 9:297–310
Marks R (1978) Epidermal activity in the involved and uninvolved skin of patients with psoriasis. Br J Dermatol 98:399–404
Maudsley DB (1979) Regulation of polyamine biosynthesis. Biochem Pharmacol 28:153–161
Maurer HR, Maschler R (1979) The influence of spermine, spermidine and various sera on T-lymphocyte and granulocyte colony growth in vitro. Z Naturforsch [C] 34:452–459
McCormick F (1977) Polyamine metabolism in enucleated mouse L-cells. J Cell Physiol 93:285–292
Mizutani A, Inoue H, Takeda Y (1974) Changes in polyamine metabolism during wound healing in rat skin. Biochim Biophys Acta 338:183–190
Morgan DML (1980) Polyamine oxidases. In: Gaugas JM (ed) Polyamines in biomedical research. Wiley, Chichester, pp 285–302
Morgan DML, Illei G (1980) Polyamine – polyamine oxidase interaction: part of maternal protective mechanism against fetal rejection. Br Med J [Clin Res] 280:1295–1297
Morrison DM, Goldsmith LA (1978) Ornithine decarboxylase in rat skin. J Invest Dermatol 70:309–313
Mufson RA, Astrup EG, Simsiman RC, Boutwell RK (1977) Dissociation of increases in levels of 3′,5′-cyclic AMP and 3′,5′-cyclic GMP from induction of ornithine decarboxylase by the tumour promoter 12-*O*-tetradecanoylphorbol-13-acetate in mouse epidermis in vivo. Proc Natl Acad Sci USA 74:657–661
Murray AW, Solanki V, Froscio M, Rogers A (1980) Effects of cholera toxin on ornithine decarboxylase activity in mouse skin. J Invest Dermatol 75:508–511
Nebert DW, Jensen NM, Perry JW, Oka T (1980) Association between ornithine decarboxylase induction and the A_h locus in mice treated with polycyclic aromatic hydrocarbons. J Biol Chem 255:6836–6842
Nishioka K, Ezaki K, Hart JS (1980) A preliminary study of polyamines in the bone marrow plasma of adult patients with leukaemia. Clin Chim Acta 107:59–66
Nöthig-Laslo V, Weygand-Durasevic I, Zivkovic T, Kucan Z (1981) Binding of spermine to tRNA stabilizes the conformation of the anticodon loop and creates strong binding sites for divalent cations. Eur J Biochem 117:263–267
Nordlie RC, Johnson WT, Comatzer WE, Twedell GW (1979) Stimulation by polyamines of carbamyl phosphate: glucose phosphotransferase and glucose-6-phosphate phosphohydrolase activities of multifunctional glucose-6-phosphatase. Biochim Biophys Acta 585:12–23
O'Brien TG (1976) The induction of ornithine decarboxylase as an early, possibly obligatory, event in mouse skin carcinogenesis. Cancer Res 36:2644–2653
O'Brien TG, Diamond L (1978) Ornithine decarboxylase, polyamines and tumor promoters. In: Slaga TJ, Sivak A, Boutwell RK (eds) Carcinogenesis, vol 2. Raven, New York, pp 273–287
O'Brien TG, Simsiman RC, Boutwell RK (1975 a) Induction of the polyamine-biosynthetic enzymes in mouse epidermis by tumor-promoting agents. Cancer Res 35:1662–1670
O'Brien TG, Simsiman RC, Boutwell RK (1975 b) Induction of the polyamine-biosynthetic enzymes in mouse epidermis and their specificity for tumor production. Cancer Res 35:2426–2433
Peterson AO, McCann V, Black HS (1980) Dietary modification of UV-induced epidermal ornithine decarboxylase. J Invest Dermatol 75:408–410

Pleshkewych A, Kramer DL, Kelly E, Porter CW (1980) Independence of drug action on mitochondria and polyamines in L1210 leukemia cells treated with methylglyoxal-*bis* (guanylhydrazone). Cancer Res 40:4533–4540
Pohjanpelto P (1973) Relationship between putrescine and the proliferation of human fibroblasts in vitro. Exp Cell Res 80:137–142
Pohjanpelto P (1976) Putrescine transport is greatly increased in human fibroblasts initiated to proliferate. J Cell Biol 68:512–520
Pohjanpelto P, Raina A (1972) Identification of a growth factor produced by human fibroblasts as putrescine. Nature 235:247–249
Pohjanpelto P, Virtanen I, Hölltä E (1981) Polyamine starvation causes disappearance of actin filaments and microtubules in polyamine-auxotrophic CHO cells. Nature 293:475–477
Probst E, Krebs A (1975) Ornithine decarboxylase activity in relation to DNA synthesis in mouse interfollicular epidermis and hair follicles. Biochim Biophys Acta 407:147–157
Proctor MS, Fletcher HV, Shukla JB, Rennert OM (1975) Elevated spermidine and spermine levels in the blood of psoriasis patients. J Invest Dermatol 65:409–411
Proctor MS, Wilkinson DI, Orenberg EK, Farber EM (1979) Lowered cutaneous and urinary levels of polyamines with clinical improvement in treated psoriasis. Arch Dermatol 115:945–949
Proctor MS, Liu SCC, Wilkinson DI (1980) Effect of methylglyoxal *bis*-guanylhydrazone on polyamine biosynthesis, growth and differentiation of cultured keratinocytes. Arch Dermatol Res 269:61–68
Quash G, Keolouangkhot T, Gazzolo L, Ripoli H, Saez S (1979) Diamine oxidase and polyamine oxidase activities in normal and transformed cells. Biochem J 177:275–282
Quigley GJ, Teeter MM, Rich A (1978) Structural analysis of spermine and magnesium ion binding to yeast Phe-tRNA. Proc Natl Acad Sci USA 75:64–68
Rennert OM, Chan WY, Griesmann G (1980) Polyamine-peptide conjugates: proposed functions. Physiol Chem Phys Med NMR 12:441–450
Romano M, Cecco L, Cerra M, Montuori R, De Rosa C (1980) Polyamines as biological markers of the effectiveness of therapy in acute leukemia. Tumori 66:677–687
Rupniak HT, Paul D (1978 a) Lack of a correlation between polyamine synthesis and DNA synthesis by cultured rat liver cells and fibroblasts. J Cell Physiol 96:261–263
Rupniak HT, Paul D (1978 b) Regulation of the cell cycle by polyamines in normal and transformed fibroblasts. In: Campbell RA, Morris DR, Bartos D, Daves GD, Bartos F (eds) Advances in polyamine research, vol 1. Raven, New York, pp 117–126
Rupniak HT, Paul D (1980) Selective killing of transformed cells by exploitation of their defective cell cycle control by polyamines. Cancer Res 40:293–297
Russell DH (1971) Increased polyamine concentrations in the urine of human cancer patients. Nature 233:144
Russell DH (1977) Clinical relevance of polyamines as biochemical markers of tumor kinetics. Clin Chem 23:22–27
Russell DH (1981) Posttranslational modification of ornithine decarboxylase by its product putrescine. Biochem Biophys Res Commun 99:1167–1172
Russell DH, Combest WL, Duell EA, Stawiski MA, Anderson T, Voorhees JJ (1977) Increased ornithine decarboxylase and *S*-adenosylmethionine decarboxylase activities in involved and uninvolved skin samples from patients with psoriasis. Fed Proc 36:970
Russell DH, Durie BGM, Salmon SE (1978 a) Polyamines as predictors of success and failure in chemotherapy. Lancet II:797–799
Russell DH, Combest WJ, Duell EA, Stawiski MA, Anderson TF, Voorhees JJ (1978 b) Glucocorticoid inhibits elevated polyamine biosynthesis in psoriasis. J Invest Dermatol 71:177–181
Sakakibara S, Yoshikawa K (1979) Urinary polyamine levels in patients with psoriasis. Arch Dermatol 265:133–137
Scalabrino G, Pigatto P, Ferioli ME, Modena D, Puevari M, Carú A (1980) Levels of activity of the polyamine biosynthetic decarboxylases as indicators of degree of malignancy of human cutaneous epitheliomas. J Invest Dermatol 74:122–124

Seiler N (1980) Assay of polyamines in tissues and body fluids. In: Gaugas JM (ed) Polyamines in biomedical research. Wiley, Chichester, pp 435–461
Seiler N, Knödgen B (1979) Effects of ultraviolet light on epidermal polyamine metabolism. Biochem Med 21:168–181
Seiler N, Bolkenius FN, Knödgen B (1980a) Acetylation of spermidine in polyamine catabolism. Biochim Biophys Acta 633: 181–190
Seiler N, Bolkenius FN, Knödgen B, Mamont P (1980b) Polyamine oxidase in rat tissues. Biochim Biophys Acta 615:480–488
Seiler N, Koch-Weser J, Knödgen B, Richards W, Tardif C, Bolkenius FN, Schecter P, Tell G, Mamont P, Fozard J, Bachrach U, Grosshans E (1981a) The significance of acetylation in the urinary excretion of polyamines. In: Calderara CM, Zappia V, Bachrach U (eds) Advances in polyamine research, vol 3. Raven, New York, pp 197–211
Seiler N, Richards W, Knödgen B (1981b) UV-induced changes in urinary polyamine excretion in the mouse. Arch Dermatol Res 270:25–32
Seppänen P, Alhonen-Hongisto L, Jänne J (1981) Death of tumor cells in response to the use of a system of stimulated polyamine uptake for the transport of methyl-glyoxal *bis*-(guanylhydrazone). Eur J Biochem 118:571–576
Stastny M, Cohen S (1970) Epidermal growth factor IV. The induction of ornithine decarboxylase. Biochim Biophys Acta 204:578–589
Stüttgen G (1968) Basic low molecular weight amine content of the skin. Fette Seifen Anstrichsm 70:667–669
Sukumar S, Nagarajan B (1978) Effect of various therapeutic treatments on polyamine contents in experimental fibrosarcoma. Indian J Biochem Biophys 15:169–172
Sunkara PS, Pargac MB, Nishioka K, Rao PN (1979a) Differential effects of inhibition of polyamine biosynthesis on cell cycle traverse and structure of the prematurely condensed chromosomes of normal and transformed cells. J Cell Physiol 98:451–457
Sunkara PS, Rao PN, Nishioka K, Brinkley BR (1979b) Role of polyamines in cytokinesis of mammalian cells. Exp Cell Res 119:63–68
Sunkara PS, Ramakrishna S, Nishioka K, Rao PN (1981) The relationship between the levels and rates of synthesis of polyamines during mammalian cell cycle. Life Sci 28:1497–1506
Tabor CW, Rosenthal SM (1956) Pharmacology of spermine and spermidine. Some effects on animals and bacteria. J Pharmacol Exp Ther 116:139–155
Tabor CW, Tabor H (1976) 1,4-Diaminobutane (putrescine), spermidine and spermine. Annu Rev Biochem 45:285–306
Tabor CW, Tabor H, Bachrach U (1964) Identification of the aminoaldehydes produced by the oxidation of spermine and spermidine with purified plasma amine oxidase. J Biol Chem 239:2194–2203
Takami H, Nishioka K (1980) Raised polyamines in erythrocytes from melanoma-bearing mice and patients with solid tumours. Br J Cancer 41:751–756
Tanigawa Y, Kawasaki K, Imai Y, Shimoyama M (1980) Effect of polyamines on ADP-ribosylation by chick-embryo-liver nuclei. Biochim Biophys Acta 608:82–95
Tashima Y, Hasegawa M, Lane LK, Schwartz A (1981) Specific effects of spermine on Na^+, K^+-adenosine triphosphatase. J Biochem (Tokyo) 89:249–255
Townsend RM, Banda PW, Marton LJ (1976) Polyamines in malignant melanoma. Cancer 38:2088–2092
Verma AK, Boutwell RK (1977) Vitamin A acid (retinoic acid), a potent inhibitor of 12-*O*-tetradecanoylphorbol-13-acetate-induced ornithine decarboxylase activity in mouse epidermis. Cancer Res 37:2196–2201
Verma AK, Rice HM, Boutwell RK (1977) Prostaglandins and tumor promotion: inhibition of tumor promoter-elicited ornithine decarboxylase activity in epidermis by inhibitors of prostaglandin synthesis. Biochem Biophys Res Commun 79:1160–1166
Verma AK, Lowe NJ, Boutwell RK (1979) Induction of mouse epidermal ornithine decarboxylase activity and DNA synthesis by ultraviolet light. Cancer Res 39:1035–1040
Wallace HM, Keir HM (1979) The effect of spermidine on RNA polymerase II activity in isolated nuclei from baby-hamster kidney cells (BHK-21/C13). Biochem Soc Trans 7:1086–1087

Wallace HM, Keir HM (1981) Uptake and excretion of polyamines from baby hamster kidney cells (BHK-21/C13). The effect of serum on confluent cultures. Biochim Biophys Acta 676:25–30
Wallace HM, Duff PM, Pearson CK, Keir HM (1981) The effect of polyamines on DNA synthesis in vitro. Biochim Biophys Acta 652:354–357
Weekes RG, Verma AK, Boutwell RK (1980) Inhibition by putrescine of the induction of epidermal ornithine decarboxylase and tumor promotion caused by 12-*O*-tetradecanoylphorbol-13-acetate. Cancer Res 40:4013–4018
Williams-Ashman HG, Cannellakis ZN (1980) Transglutaminase-mediated covalent attachment of polyamines to proteins: mechanisms and potential physiological significance. Physiol Chem Phys Med NMR 12:457–472
Wright RK, Buehler BA, Schott SN, Rennert OM (1978) Spermine and spermidine, modulators of the cell surface enzyme adenyl cyclase. Pediatr Res 12:830–833
Yuspa SH, Lichti U, Ben T (1980) Local anaesthetics inhibit induction of ornithine decarboxylase by the tumor promoter 12-*O*-tetradecanoylphorbol-13-acetate. Proc Natl Acad Sci USA 77:5312–5316

CHAPTER 27

Proteolytic Enzymes in Relation to Skin Inflammation

G. VOLDEN and V. K. HOPSU-HAVU

A. Introduction

Inflammation is a series of events which in part follow each other and in part are contemporary. The acute phase is followed by resolution or the healing phase. A variety of enzymes and their natural inhibitors are known to operate in the inflammatory process.

In conjunction with their inhibitors and enhancers, proteases regulate the inflammatory process by producing vaso-active and chemotactic substances, and in addition maintain a critical balance between protein synthesis and degradation. Proteolytic enzymes involved in inflammation may stem from local tissue and blood elements.

In this review we intend to take a general look at proteases and proteolytic processes present in skin, in the cells which migrate to the site of inflammation and in those which are activated from different exudating serum components.

B. Proteases of the Skin

I. Classification

1. Proteinases

Proteolytic enzymes or proteases are proteins which are capable of catalysing the hydrolysis of peptide bonds (Fig. 1). Those proteases which preferentially attack large peptides or proteins and split bonds located somewhere in the middle of the peptide sequence are called proteinases or endopeptidases. Further subclassification of proteases is based on the catalytic mechanism which can be revealed by using different kinds of modifying substance, i.e. inhibitors and enhancers. The four subclasses recognised in the Enzyme Nomenclature (1978) are the serine, cysteine or thiol, aspartic or carboxyl and metallo proteinases. The main proteinolytic enzymes of human skin have been the subject of intense study. Historical review, data from species other than humans and analytical details have been dealt with in a recent review (HOPSU-HAVU et al. 1977).

Serine proteases are the most common of mammalian proteinases and are active at the neutral or slightly alkaline pH range. A trypsin-like alkaline serine protease of human skin is well characterised (FRÄKI and HOPSU-HAVU 1975). This proteinase causes a prolonged increase in vascular permeability as well as leucocyte accumulation at the injection site in skin. The chymotrypsin-like protease of

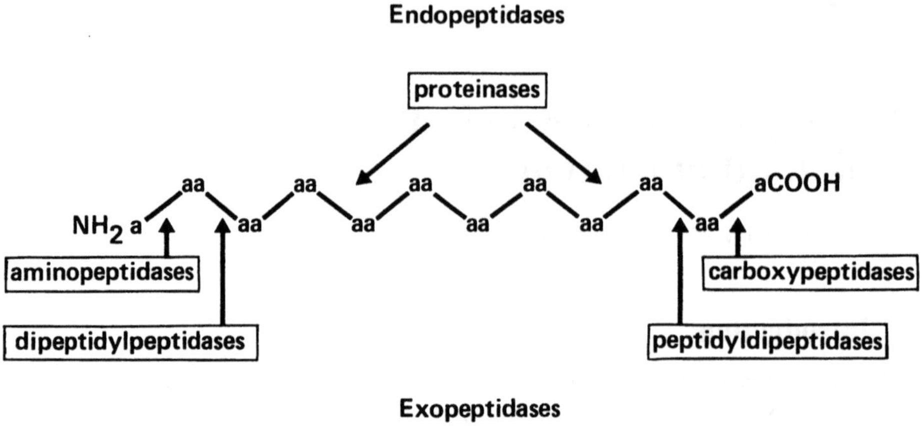

Fig. 1. Schematic presentation of the sites of attack by various proteolytic enzymes

human skin resembles cathepsin G obtained from human leucocytes (SCHMIDT and HAVEMANN 1974) and spleen (STARKEY and BARRETT 1976 b). It causes a very intensive infiltration of leucocytes in mice after intraperitoneal injection as well as a cytotoxic effect in vitro.

Kallikreins are enzymes which liberate vaso-active peptides from precursor kininogens of plasma. Kallikreins have been separated from human skin (TOKI and JAMURA 1978).

The main cysteine proteinase of human skin is closely similar to cathepsin B_1 of other tissues (FRÄKI 1976). This proteinase may also hydrolyse collagen at non-helical cross-linking portions as well as other proteins in skin.

The main acid or carboxyl proteinases of human skin include a proteinase closely similar to cathepsin D. This proteinase is lysosomal and may be active in cellular protein catabolism and responsible for hydrolysis of amorphous structures of connective tissue matrix (glycoproteins).

A collagenolytic metallo enzyme is known to be released by human skin fibroblast cultures. The enzyme may be secreted as a pro-enzyme which is then activated by a proteolytic process (STRICKLIN et al. 1977).

2. Exopeptidases

A large group of proteolytic enzyme do not attack large proteins or, if they do, they release amino acids or dipeptides from the ends of the protein without attacking the peptide bonds in the middle. These enzymes are called exopeptidases or simply peptidases.

Aminopeptidase in skin extracts of different species was demonstrated by FRUTON as early as 1946. The same author presented evidence about the presence of tripeptide aminopeptidase or aminotripeptidase in skin structures. The presence of aminopeptidases A and B as well as of carboxypeptidases is also known.

Dipeptide hydrolases or dipeptidases have not been thoroughly characterised in skin structures. Dipeptidylpeptidases I, II, III, and IV are all known to be present in skin structures.

The joint hydrolytic capacity of all skin proteases cna be expected to be sufficient for breakdown of any skin proteins to smaller peptides. Some of the proteolytic enzymes may have more specific functions in physiological or pathological skin conditions.

II. Cellular Localisation and Functions

1. Localisation

The major proportion of all intracellular proteases is located in lysosomes. Cathepsin B_1 and D are the main lysosomal proteinases of skin (HOPSU-HAVU and FRÄKI 1980). MIER and VAN DEN HURK (1976b) found that the activities of cathepsin B, C and D and of arylamidase are low in both epidermis and dermis compared with most other tissues.

Following damage or a stimulus to the skin it is reasonable to expect that all enzymes appearing free within the lysosomal membrane will have been liberated simultaneously. Some of the lysosomal enzymes are, however, bound to the lysosomal membrane and are liberated only after more pronounced cell damage. The release of lysosomal proteases is regulated by both labilising and stabilising factors found in normal serum (MIYAMOTO and TERAYAMA 1973). Such stabilising factors are important in limiting the lysosomal involvement in inflammatory processes.

The exact location of the neutral skin proteases is not known. It is possible that skin mast cell granules are an important storage site. Most of the exopeptidases are easily extracted with dilute buffer solutions and they are thus soluble in the cytosol in skin, as in other tissues. This is in agreement with the notion that small peptides, split from ingested proteins by lysosomal proteases, can escape to the cytosol and are here split further to amino acids. A number of exopeptidases as well as proteinases in different tissues are known to be located in cell surface membranes and to be activated or released during active growth or immunological stimulation. No exact data of this kind are available on skin enzymes.

2. Functions

The proteolytic enzymes play an essential role in the catabolism of the cellular proteins as part of the normal turnover of the cell organelles. Similarly, they break down phagocytosed material within the lysosomes. The enzymes may also be excreted and thus function extracellularly both in physiological and pathological conditions. The total proteolysis may depend on the proteases of local tissue structures, e.g. keratinocytes or fibroblasts, on those carried to the site by inflammatory cells or on activation of several proteinolytic cascade systems of plasma, e.g. complement, kallikrein, clotting and fibrinolysis. The essential plasma components can diffuse to the diseased tissue site as a consequence of the pathological process, e.g. inflammation.

Since the actual proteolytic activity is a consequence of the amount of proteases and the amount of protease inhibitors at a particular tissue site, both of these need to be discussed in order to understand the whole mechanism. Protease inhibitors can be derived from local cellular sources as well as from plasma, which in fact is rich in various protease inhibitors. Practically all proteolytic activity can be inhibited by these inhibitors and thus the ordinary effects of proteolysis are eliminated. Similarly, proteolytic activity can be affected by certain drugs used to modify disease processes.

C. Proteases in Inflammatory Cells

I. Granulocytes

The neutrophil granules are known to store a number of proteolytic enzymes (Table 1). The enzymes are synthesised during the early stages of maturation in bone marrow and are located mainly in the azurophil and specific granules (lysosome-like). Either the enzymes of the azurophil granules or of the specific granules can be preferentially released to the surrounding of the cell during phagocytosis or chemical stimulation.

Cathepsin G attacks several natural substrates like fibrinogen, casein, proteoglycans and complement components. It is readily inhibited by human plasma inhibitors α_2-macroglobulin, α_1-trypsin inhibitor and α_1-chymotrypsin inhibitor.

Elastase of human neutrophils has also been purified and at least three major isoenzymes are known (STARKEY 1977). The enzyme hydrolyses very rapidly insoluble elastin as well as a broad variety of other natural substrates, e.g. fibrinogen, collagen, haemoglobin, histone, proteoglycans and complement components. Inhibitors are α_2-macroglobulin and α_1-antitrypsin as well as low concentrations of gold thiomalate (JANOFF 1970).

Non-specific collagenase or neutral proteinase activity in the azurophil granules of neutrophils is also a well defined entity (LAZARUS et al. 1968; OHLSSON and OHLSSON 1973). Soluble collagen is split to one- and three-quarter fragments as by other mammalian collagenases. In addition to collagen, some other proteins like proteoglycans, fibrinogen and complement fractions are hydrolysed. Specific collagenase of neutrophils is a metallo enzyme and different from the non-specific one. It is located in the specific granules of human neutrophils (MURPHY et al. 1977).

Neutrophil-secreted plasminogen activator is a trypsin-like serine proteinase with very restricted specificity (CHRISTMAN et al. 1977). Its only known macromolecular substrate is plasminogen, the concentrations of which are very high in plasma and extracellular fluids. Plasminogen activator activity in resting neutrophils is very low, but when stimulated, intense secretion takes place.

The powerful proteases of neutrophils may serve the cell by digesting the phagocytosed material, e.g. bacteria and tissue debris. They may, however, cause damage to the host, if released. A number of mechanisms can cause liberation of granular enzymes, for example cell death, regurgitation during phagocytosis, release during migration, secretion and reversed endocytosis (SMOLEN and WEISSMANN 1978). Death and lysis of neutrophils is a common occurrence in inflammatory areas, being caused by bacterial toxins and similar agents. A massive release of neutrophil granules takes place as regurgitation of enzymes during the attendant phagocytosis. Granular discharge already occurs before vacuoles are closed

Table 1. Proteinase of phagocytes

Neutrophils	Macrophages
Elastase	Cathepsin B
Cathepsin G	Cathepsin D
Neutral proteinase	Elastase
Cathepsin B	Collagenase
Cathepsin D	Plasminogen activator
Collagenase, specific	Azocaseinase
Plasminogen activator	

and thus marked release of enzymes into the extracellular space may take place. An extension of the same phenomenon is so-called frustrated phagocytosis or reversed endocytosis, which occurs when there is phagocytosis of too large particles which cannot be fully internalised. Similarly, continued activation and secretion can be demonstrated on IgA-coated surfaces. The surface is considered as a large particle and the attendant phagocytosis leads to continued granular release. This may result in very high local protease concentrations. A similar phenomenon may take place on IgG-coated biological membranes, e.g. the basement membranes or cartilage surfaces, and lead to considerable destruction, at least as long as the protease inhibitors can be excluded from this microcompartment.

The enzyme release may also occur due to soluble compounds. C5a generated during complement system activation stimulates fusion of granules and plasma membranes, thus causing the release of granular material (GOLDSTEIN and WEISSMANN 1974). The extracellular proteases may, in this case, be more easily inactivated by protease inhibitors and thus the damage may be less. However, there may be local compartments where considerable tissue damage ensues.

II. Macrophages

The intracellular levels of proteinases in macrophages are low compared to those found in neutrophils. However, the amounts delivered extracellularly may become quite large with time, especially in the case of inflammatory macrophages. The main neutral proteinase of activated macrophages is plasminogen activator (UNKELESS et al. 1974). The other neutral proteinases are collagenase (WAHL et al. 1974; WERB and GORDON 1975a), elastase (WERB and GORDON 1975b) and metallo proteinase degrading azocasein (GORDON 1976). Overall, the proteinase activities of the macrophages and neutrophils resemble each other.

Proteinase release from macrophages, as from neutrophils, may take place both during phagocytosis and as an active secretion. However, the lack of active neutral proteinase containing granules in macrophages is a distinguishing feature. Macrophages become activated after phagocytosis of certain kinds of particle, for example zymosan and erythrocytes, and the metabolic burst appears to be an initial event in the activation process (BAGGIOLINI et al. 1978).

III. Other Inflammatory Cells

Eosinophils have an important function in allergic reactions, where they phagocytize antigen–antibody precipitates. They contain a number of hydrolytic enzymes. Their proteolytic activity is, however, fairly low, even though at least cathepsin D, collagenase and an aminopeptidase are known to be present. Most of the proteolytic hydrolysis is certainly due to cathepsin D, as in lysosomal compartments of many other cell types.

The presence of lymphocytes in inflammation is considered an indication of disease chronicity. Their protease activity is low. Significant high local protease concentrations may, however, be formed during activation of lymphocytes, e.g. immune reactions.

Mast cell granules are very rich in proteolytic enzymes. The best known is the rat peritoneal mast cell "chymase". It was recently shown by SEPPÄ (1978) that mast cell chymase is the main protease of rat skin homogenate. Human mast cells are also known to contain a trypsin-like serine protease (GLENNER and COHEN 1960; GLENNER et al. 1962). The other proteases of mast cells are present in very low concentrations and have not been thoroughly characterised.

D. Plasma-Derived Proteases

As soon as the intravascular proteins, which basically represent substrates, and the extravascular compartment proteases come into contact with each other, a number of plasma-derived systems are activated. They include the kinin system, the complement system and clotting and fibrinolytic systems, which all are series of proteolytic events [see comprehensive reviews by RYAN and MAJNO (1977), MOVAT (1979a, b), and MINTA and MOVAT (1979)].

E. Protease Inhibitors and Enhancers

The chemical activity of proteases at any tissue site is modified by enzyme inhibitors and enhancers. They may be derived from local tissue sources or from plasma. The amount depends on the local rate of synthesis and flow of plasma proteins outside the circulation, which may be greatly increased in pathological conditions.

I. Plasma-Derived Inhibitors

Human plasma is a rich source of protease inhibitors and altogether nine inhibitors are known and have been characterised (Table 2). There is at least one inhibitor for every possible protease which could be activated or generated at the site of tissue processes. The main inhibitor of serum is α_1-proteinase inhibitor (α_1-antitrypsin), which reacts with a wide range of serine proteinases. The proteinases are secondarily transferred to α_2-macroglobulin, which is in practice an irreversible inhibitor. It mediates the elimination via phagocytosis of the enzymes from circulation and tissues.

Table 2. Proteinase inhibitors in human plasma and skin

Inhibitor	Mol.wt.	Conc. (μmol/l)	Examples of proteases inhibited
Plasma			
1. α_1-Proteinase inhibitor (α_1-A)	52000	5.2	Elastase, nonspecific collagenase, cathepsin G, plasmin
2. α_1-Antichymotrypsin (α_1-AC)	69000	6.4	Chymotrypsin, cathepsin G
3. Antithrombin III (AT III) + heparin	65000	4.0	Thrombin, factor Xa plasmin kallikrein
4. α_2-Macroglobulin (α_2-M)	720000	3.6	Plasmin, kallikrein, cathepsin G, collagenase, elastase etc.
5. Inter-α-trypsin inhibitor (ITI)	160000	2.8	Trypsin, chymotrypsin
6. C_1-esterase inhibitor (C1 INA)	70000	2.4	C_1-esterase, plasmin, kallikrein
7. α_2-Plasmin inhibitor (α_2-PI)	70000	0.9	Plasmin
8. β_1-Collagenase inhibitor (β_1-CI)	40000	n. 0.4	Collagenases (metallo enzymes)
9. α_1-Cysteine proteinase inhibitor	90000	?	Cysteine proteinases, cathepsin B and C
Skin			
Cysteine proteinase inhibitor (epidermal)	13000	?	Cysteine proteinases, cathepsin B and C etc.
Skin kinin-forming enzyme inhibitor	57000	?	Kinin-forming enzyme of skin

Adapted from FRITZ (1980).

II. Inhibitors in Skin

Human skin extracts and thus normal human skin itself are known to contain fairly large amounts of α_1-proteinase inhibitor, while α_2-macroglobulin is present only in minimal concentrations (FRÄKI and HOPSU-HAVU 1972). A less well characterised high molecular weight proteinase inhibitor of trypsin and another lower molecular weight inhibitor of trypsin and elastase are also known (JUNNILA et al. 1971). An inhibitor of human skin kinin-forming enzyme which does not affect plasmin, thrombin and plasma kallikrein or C_3-esterase, trypsin and chymotrypsin has also been reported (TOKI and YAMURA 1978). An inhibitor located histochemically principally in epidermis and other keratinising epithelia (JÄRVINEN et al. 1978) and originally reported in rat skin by JÄRVINEN and HOPSU-HAVU (1975) and JÄRVINEN (1976) was separated in human skin by FRÄKI (1976). This is a low molecular weight, 13000-dalton inhibitor of cysteine proteinases. It does not affect serine, metallo or carboxyl proteinases. HIBINO et al. (1980) suggested that the epidermal SH-protease inhibitor can be excreted from the epidermal cells into the dermis, where SH-protease activity is involved in the inflammatory process. The inhibitor resembles the SH-proteinase inhibitor found in the healing phase of the Arthus reaction and it has been suggested that it is secreted by mononuclear cells (HAYASHI 1975). The other cysteine proteinase inhibitors, with higher molecular weights of around 74000 (JÄRVINEN 1976) or 90000 (SASAKI et al. 1977), which have been found in skin extracts are probably derived from serum.

III. Protease Enhancers

A collagenase activity enhancing factor has been derived from human serum (SELTZER et al. 1978). Human skin extracts have also been shown to contain a proteinase binding, stabilising and enhancing factor which affects numerous proteases (FRÄKI et al. 1979). These, as well as other possible enhancing substances, may have important regulatory roles in affecting local protease activities at limited tissue compartments.

F. Proteases in Different Phases of Inflammation

I. Vasodilatation and Exudation

The early phase of inflammation characterised by increased vasodilatation and vascular leakage can be produced by tissue injury and liberation of several factors with vaso-active potency, e.g. vaso-active amines and peptides, slow reacting substance (SRS), prostaglandins and non-enzymatic lysosomal components. These will not be dealt with further in this review, since they are not parts of lytic systems.

II. Chemotaxis

Many of the lymphokines released by the lymphocytes are chemotactic (WILKINSON 1974). The most potent chemo-attractants are formed, however, when the complement components C_3 and C_5 are activated by proteases in the course of inflammatory reactions. This may be a consequence of antigen–antibody reactions but also a direct effect of proteases, e.g. granulocyte proteases, since both elastase and cathepsin G can generate a chemotactic effect from C_3 and C_5 (VENGE and OLSSON 1975; WARD et al. 1978). WARD and HILL (1970) have also shown that a neutral protease from heart tissue could generate chemotactic activity from C_3.

Human skin proteinases may also induce chemotaxis by cleaving fibrinogen and activating Hageman factor or kallikrein (RUDDY et al. 1972), and by activating complement (LAZARUS et al. 1975). Injection into the peritoneal cavity of normal mice of a serine proteinase purified from human skin induces accumulation of polymorphonuclear leucocytes. This response is not seen after intraperitoneal injection of C_5-deficient mice (THOMAS et al. 1977). Inhibition by proteinase inhibitors provides evidence that proteolysis is necessary for the chemotactic effect of this serine proteinase, which generated more chemotactic activity than other common proteinases.

Non-complement-derived chemotactic factors can be produced by proteolysis from collagen by collagenase, from fibrin by plasmin, from various tissue proteins by neutral and acid proteinases and from IgE and IgM (HAYASHI 1975; GREENBAUM 1979). Peptides derived from degraded collagen are also chemotactic for fibroblasts (POSTLETHWAITE et al. 1976). Exogenous chemo-attractants and proteases capable of producing chemotaxins from tissue proteins are excreted by many micro-organisms (WARD et al. 1972).

III. Fibrin Deposition and Fibrinolysis

The deposition of fibrin in tissue or intravascularly plays an important role in inflammatory processes. Several irritants, antigen–antibody complexes and factors involved in cell-mediated hypersensitivity may activate the clotting system via different pathways. Chemotactic fibrinopeptides are produced and the fibrin network may serve to support cellular infiltration and capillary propagation (ASTRUP 1968). The fibrinolytic process participates in the inflammatory process by degrading fibrin. Many cell types seem to participate in the fibrinolytic process. Stimulated granulocytes and macrophages secrete plasminogen activators which convert plasminogen to plasmin, resulting in fibrinolytic activity (UNKELESS et al. 1973).

RYAN et al. (1971) investigated a number of inflammatory skin conditions and concluded that there is normal or increased fibrinolysis in the acute oedematous response which follows injury to the dermis. Consequently there is little fibrin in such skin. By contrast there is loss of fibrinolysis in the inflammatory response which follows minor injury to epidermis. This is often associated with heavy deposition of fibrin. The epidermis was suggested to contribute inhibitors of fibrinolysis and thus cause fibrin deposition and consequently white cell infiltration.

The findings of IZAKI et al. (1979) suggest that fibrinolysis is regulated in the lesions by proteinases and inhibitors. The role of T cells was also evaluated, and it was found that fibrin deposition is an unspecific tissue reaction, in contrast to fibrinolysis, which is T cell dependent.

IV. Repair and Chronic Inflammation

1. Repair

The increase in the activities of proteolytic enzymes may be of importance in protein digestion in the repair of the damaged area, as has been shown with rabbit macrophages (DINGLE et al. 1973). The monocyte–macrophage system of cells participates actively in phagocytosis and intracellular digestion, and the cells thus serve as the major scavengers of the body at the site of inflammation. They hydrolyse cell debris and foreign materials both intracellularly after ingestion and extracellularly after release of stored enzymes. Macrophages perform this function in all types of inflammation, including infection (PAGE et al. 1974), during the cleaning of wounds.

Cathepsin D and B_1 are the main proteases in lysosomes and are quite important in connection with intracellular as well as extracellular protein catabolism. The activity of cathepsin D appears to be increased in the degradative phases of skin inflammation and may thus be involved in dermal connective tissue catabolism (LAZARUS 1974).

RYAN and CARDIN (1967) demonstrated that lysates of purified lysosomes induce mitoses in cultured cells. It is known that silica induces selectively the release of lysosomal enzymes, and ALLISON et al. (1968) proposed that the overgrowth of fibroblasts is due to such a release. Thus the activation of connective tissue cells during inflammation due to release of lysosomal enzymes should be responsible for tissue repair.

2. Chronic Inflammation

If the inflammatory stimulus is not eliminated, chronic inflammation will develop. Granulocytes which predominate at the inflammatory site during the first 10–15 h are replaced by mononuclear cells (DAVIS and ALLISON 1976), first by lymphocytes and later mainly by macrophages. Macrophages secrete neutral proteinases (see Sect. C.II) and complement components (BENTLEY et al. 1976; MCCLELLAND and VAN FURTH 1976).

Chemotactic activity is induced by cleavage products of C_3 and C_5 by the proteases. Purified C_{3b} added to macrophage cultures increases secretion of hydrolytic enzymes into the culture medium (SCHORLEMMER et al. 1976). Some of the secreted enzymes are proteinases which are able to cleave C_3 (SCHORLEMMER and ALLISON 1976), whereby more C_{3b} is formed. Hence more macrophages migrate to the lesion and the process continues. The other cleavage product, C_{3a}, is highly lytic (SCHORLEMMER et al. 1977). Thus proteases are released from lytic cells as well as from macrophages induced by C_{3b} secretion. The serial stimulations that occur should be regarded as essential to the pathogenesis of chronic inflammation.

G. Inflammatory Skin Conditions

I. Immune Reactions

Neutral SH-dependent proteases have been purified from the *Arthus reaction* in rabbit skin (KOONO et al. 1968) and in guinea-pig skin (OHYAMA 1974). HAYASHI et al. (1975) demonstrated the release of vaso-active compounds from immunoglobulins. Intradermal injection of the partially purified enzyme induced a marked inflammatory reaction similar to the Arthus reaction. Leucocyte cathepsin G and elastase may also initiate the degradation of the vascular basement membrane that occurs in the Arthus reaction in vivo (JANOFF 1972; STARKEY 1977). Probably cathepsin D and E are also active. Collagenase has been shown to destroy basal membranes in in vitro experiments (LAZARUS et al. 1968; JANOFF and ZELIGS 1968).

Proteolytic enzymes may activate complement and the cleavage products of C_3 and C_5 have chemotactic activity. No leucocytes are deposited at the site of immune complex deposition in decomplemented animals (WARD and COCHRANE 1965) and the Arthus reaction does not occur in neutropenic animals (STETSON 1951; COCHRANE et al. 1959), indicating that the presence of neutrophils is essential for eliciting the lesion. The essential components in the Arthus reaction are thus neutrophil leucocytes, complement and immune complexes.

A low molecular weight inhibitor of thiol proteinases was found in the Arthus skin reaction site at a later stage. This thermostabile inhibitor may be derived from monocytes and may modulate the Arthus skin reaction (HAYASHI 1975). A second Arthus skin proteinase inhibitor was found in rabbit serum (TOKAJI 1971). Several of the normal serum inhibitors may also be involved (see Sect. E.I).

Proteolytic enzymes can stimulate B- and T-lymphocytes, as has been shown with granulocyte elastase and cathepsin G (VISCHER et al. 1976). This is best explained by a proteolytic attack on the cell membrane. A direct effect of both elastase and cathepsin G on the cell membrane enhances the neutrophil ingestion rate of immune complexes (HÄLLGREN and VENGE 1976).

Reversed endocytosis (frustrated phagocytosis) can be expected to result in high local concentrations of proteases. The proteases are released into the microcompartment between the cell membrane and the Ig-coated surface. It is likely that reversed endocytosis with release of proteolytic enzymes may be important in diseases in which immune complexes are deposited on fixed structures, e.g. the basement membrane.

Crude homologous lysosomal extract or semipurified cathepsin D induces proteolysis of IgG, resulting in IgG fragments which are auto-antigens capable of inducing inflammation (FEHR et al. 1974). In some instances proteases may alter or damage the tissue in such a way that it becomes foreign to the body and in this way form the basis for an immunological reaction.

In *photocontact reactions* the photosensitising agent is possibly altered by light in such a way that it becomes a contact allergen (BURCKHARDT 1941). The reaction is clinically and histologically indistinguishable from contact allergic reaction (WILLIS and KLIGMAN 1968). It was shown in cell cultures that some photosensitising substances are concentrated in lysosomes (ALLISON and PATON 1965; ALLISON et al. 1966), and cell death was observed after exposure of the cultures to the appropriate wavelength. Increased enzyme activities were demonstrated in suction blister fluid after latent periods of 10–24 h; the length of the latent period depended on the intensity of the reaction, being shortest for strong reactions (VOLDEN and THUNE 1979b). The increased activities were always observed some hours after the first visible sign of erythema. Despite the fact that the photosensitising agents are probably concentrated in lysosomes, the lysosomal membrane does not rupture more easily after irradiation with the appropriate wavelength than in common sunburn reactions.

II. Psoriasis

Psoriasis is characterised by an increased rate of proliferation of epidermal cells associated with inflammatory reactions in the dermis. Psoriatic lesions are known to have an increased proteolytic activity (STÜTTGEN and WÜRDEMANN 1959; STEIGLEDER and BLOMAYER 1964; SCHWARTZE and HANSON 1971; NEUFAHRT and HENCKEL 1974). Acid hydrolases, including cathepsin B, C, and D and arylamidase, are increased in psoriatic lesions compared with uninvolved psoriatic epidermis and controls (MIER and VAN DEN HURK 1976a). FRÄKI and HOPSU-HAVU (1976) found that the pattern of several proteolytic enzymes partly differed from those of the normal human skin. High activity of plasminogen activator was found in psoriatic epidermis and evidence exists that more than one component of plasminogen activator is present in psoriatic scales (FRÄKI et al. 1978).

Proteases have been shown to induce cell division and an escape from contact inhibition (BURGER 1970). The growth of transformed cells can be selectively inhibited by protease inhibitors (SCHNEBLI and BURGER 1972). Changes in cell surfaces are believed to be involved in the regulation of growth brought about by the action of proteases or by viral transformation. Similarities exist between psoriatic lesions and other tissues characterised by fast cell proliferation. This is supported by the demonstration of accumulation of cancer-related glycoprotein, a split product of a protease inhibitor of trypsin and chymotrypsin, in plasma and urine of psoriatic patients (FINE et al. 1979).

A typical feature of psoriatic lesions is the migration of polymorphonuclear leucocytes into the epidermis. This could be caused by several mechanisms. The presence of activated components of complement in the psoriatic lesion has been shown by BRAUN-FALCO et al. (1977). Recently, TAGAMI and OFUJI (1977) have extracted leukotactic complement activation products from psoriatic scales. Activation of complement may be due to release of epidermal proteases that are capable of activating complement by non-immunological mechanisms (LAZARUS et al. 1977). The high efficacy of a serine protease from human epidermis in activating complement and thus inducing polymorphonuclear leucocyte accumulation in vivo as well as in vitro has been demonstrated (THOMAS et al. 1977). Activation of complement could also be due to (a) the highly active plasminogen activator of psoriatic epidermis, (b) antigen–antibody reactions between stratum corneum antigen and the circulating auto-antibodies described by KROGH (1970) or (c) antigen–antibody reactions between basal cell nuclear antigen and the antibodies which can be extracted from the circulating lymphocytes (CORMANE et al. 1979). The initial events might be caused by certain external trauma (JABLONSKA et al. 1979). Once at the site, the leucocytes can liberate a number of proteases, which again can activate complement and aggravate the lesion. Similarly, proteolytic enzymes may play a part in the activation of phospholipases which in turn liberate arachidonic acid and several of its metabolites, which act as chemo-attractants (HAMMARSTRÖM et al. 1975).

III. Infections

Inflammation induced by micro-organisms should be regarded as a prime defence mechanism on the part of the organism. Granulocytes and macrophages move towards the micro-organisms and ingest them (phagocytosis). Proteolytic enzyme leakage occurs in connection with the phagocytosis. Proteases, especially cathepsin G, have both bactericidal and fungicidal effects (STARKEY 1977; ODEBERG et al. 1975). Human cathepsin D also has a bactericidal effect at pH 6.4 which can be blocked by pepstatin (THORNE et al. 1976). This enzyme is probably involved in different kinds of inflammatory process (BARRETT 1977). Elastase is capable of hydrolysing peptidoglycans of the bacterial cell wall. The granulocyte proteases thus together digest the bacteria.

Elastase and cathepsin G contribute to the inflammatory process by hydrolysing the major proteins of connective tissue (STARKEY and BARRETT 1976a, b). Human neutrophil elastase has been shown in pus in free form, but also in the form of complexes with the inhibitors α_1-antitrypsin and α_2-macroglobulin, indicating that a critical balance exists between proteases and inhibitors in bacterial inflammations.

The Shwartzman haemorrhagic reaction is induced by injection of bacterial endotoxin. The first injection is intracutaneous and this is followed by intravenous injections on consecutive days. It has been shown that cathepsin D is involved in this reaction. The Shwartzman reaction is inhibited by soya bean trypsin inhibitor and partially by pepstatin (LIN and WILLIAMS 1975). The release of the bacterial endotoxins, lipopolysaccharides, from bacteria invading the tissue is responsible for the initiation of damage to the tissue (KASS and WOLFF 1973). The Swartzman haemorrhagic reaction is probably the result of direct destruction of basal membranes by collagenase, leading to extravasation of blood cells since basal membranes are destroyed by collagenase in vitro (LAZARUS et al. 1968; JANOFF and ZELIGS 1968). It has been shown that *Salmonella typhosa* incubated in the presence of leucocyte extracts, trypsin, lysozyme and inflammatory exudates release considerable amounts of bacterial lipopolysaccharides in contrast to the modest release from controls incubated only in phosphate buffered saline. Pre-treatment with the known protease inhibitor Pms-F completely abolished the release of bacterial lipopolysaccharides (GINSBURG et al. 1977). This indicates that the factors capable of releasing bacterial lipopolysaccharides are probably proteolytic enzymes. Lysozyme, trypsin and proteases released from leucocytes were thus all able to induce release of bacterial lipopolysaccharides from the surface of the bacteria. Bacterial lipopolysaccharides play a role in activation of complement as well as triggering the Shwartzman haemorrhagic reaction (GINSBURG et al. 1977). Regarding the physiological role of bacterial lipopolysaccharides, it is known that they may trigger lymphocyte transformation and induce release of lymphokines (PEAVY et al. 1972), which are potent chemotactic factors (WILLOUGHBY 1978).

Increased concentrations of proteases have been found in many virally transformed cells as compared with uninfected controls (BOSMAN 1969, 1972; BOSMAN et al. 1974). Virally transformed cells also exhibit and secrete fibrinolytic activity which was not found in corresponding control cultures (UNKELESS et al. 1973; OSSOWSKI et al. 1973). They are known to activate complement components and to produce chemotactins (WARD et al. 1972).

Among the numerous protease inhibitors of human plasma, α_2-macroglobulin is the only one which contributes to the defence against invasive micro-organisms. α_2-macroglobulin inhibits the neutral proteinases of *Pseudomonas aeruginosa* and of *Staphylococcus aureus* and *Proteus vulgaris* (KUEPPERS and BEARN 1966). It further inhibits the keratinase of *Trichophyton mentagrophytes* (YU et al. 1972). Thus α_2-macroglobulin reduces the inflammatory process induced by these micro-organisms. Since α_2-macroglobulin is a high molecular weight plasma glycoprotein, it mainly escapes from the capillaries under conditions of increased vascular permeability such as inflammatory reactions. In severe infection, endotoxins from Gram-negative bacteria can cause consumption of plasma protease inhibitors via the classical routes (by activation of the clotting, fibrinolytic and complement cascades by system-specific proteinases such as thrombokinase or plasminogen activator). The consumption may also be due to unspecific degradation of plasma factors by leucocyte proteinases and proteases released from platelets, macrophages, mast cells and endothelial cells. The inhibitory capacity may be overstressed, leading to circulatory complications such as (a) disseminated intravascular coagulation caused by activation of the clotting and fibrinolytic systems, and (b) anaphylactic responses induced by activation of the complement system (FRITZ 1980).

IV. Ionising Irradiation and Ultraviolet Light

1. Ionising Irradiation

The proteolysis of casein has been shown to be significantly increased in X-ray irradiated rat skin (STÜTTGEN and SCHILLING 1959). The activity of rat liver cathepsin B is increased 24 h after whole-body X-ray irradiation with 800 rads (KOCMIERSKA-GRODZA and GOUTIER 1971). The late post-irradiation increase in lysosomal enzyme activities may be due to stimulated synthesis of new enzymes or activation of pre-existing latent enzymes (RENE and EVANS 1970; RENE et al. 1971; PARIS et al. 1969), or may be caused by invading inflammatory cells. The finding that following irradiation the activity of enzymes in the Golgi vesicles is increased before their incorporation into lysosomes speaks in favour of increased synthesis (HUGON and BORGERS 1965).

2. Ultraviolet Light

WEISSMANN and DINGLE (1961) demonstrated that 98% of cathepsin was released after a high dose of ultraviolet irradiation of rat liver lysosomal suspension, and that hydrocortisone effectively reduced the release. This result was subsequently confirmed by others (WEISSMANN and FELL 1962; DESAI et al. 1964). The latter in vitro study additionally showed that the release was dose dependent.

Irradiation with UV-B in healthy volunteers and patients with chronic polymorphic light dermatosis caused dose-related increases in the levels of acid hydrolases, including arylamidase, in suction blisters after a latent period of 11–18 h, which is some hours after the appearance of the erythema (VOLDEN 1978; VOLDEN and THUNE 1979a). The pattern of enzyme release in these studies, together with additional studies of the activities of cathepsin B and C and arylamidase in epidermis, blister fluid and dermis shortly after irradiation and 18 h later (VOLEN unpublished data), indicates that the epidermal lysosomes are the major source, although a certain diffusion of enzymes from invading leucocytes should not be excluded. These studies indicate that the lysosomal enzymes, including arylamidase, are released as a consequence of the inflammatory reaction. These results are comparable with those of OLSON and NORDQUIST (1966) and HÖNIGSMANN et al. (1974), who demonstrated electron microscopically lysis of lysosomes in the later stages of the ultraviolet reaction.

Free marker enzymes for different subcellular fractions released into suction blister fluid have been used to reveal the extent of damage to membranes and structures in UV-B induced erythematous reactions (VOLDEN et al. 1980). Increased release of enzymes from cytosol, lysosomes and the plasma membrane indicated cell damage. The failure to detect marker enzymes for mitochondria, peroxisomes and the microsomal fraction, however, suggests that cell lysis did not occur to any significant extent.

As to the mechanism, a direct vascular effect of the sunburn rays (UV-B) is unlikely since many hours elapse before the clinical symptoms are seen. Further research is needed to clarify the initial event of the inflammatory reaction caused by ultraviolet light.

3. Porphyria

The possible role of proteolytic enzymes in photolysis of protoporphyrin-treated human fibroblasts has been ruled out (WAKULCHIK et al. 1980). It was found that collagenous and non-collagenous proteins were lost at the same rate following irradiation, indicating the action of proteolytic enzymes.

In vivo experiments in the active phases of porphyria cutanea tarda demonstrated increased free lysosomal enzymes, including arylamidase, in suction blister fluid following irradiation with violet light (VOLDEN and THUNE 1979c). Increased enzyme activities were first seen some hours after the first visible sign of erythema, indicating lysosomal damage as a consequence of the inflammatory reaction.

H. Deficiencies of Proteases and Their Inhibitors

I. Protease Deficiencies

In the conditions discussed so far there is increased activation or levels of proteases. In theory, one would expect that inborn errors, for example, might lead to a lack of key enzymes instrumental in destroying active mediators, and thus exaggerate the inflammatory reactions. Such an enzyme defect is actually known. MATHEWS et al. (1980) have described a family in which low levels of carboxypeptidase N were connected with episodic angio-oedema. This enzyme is also referred to as serum carboxypeptidase B, kininase I or anaphylatoxin inactivator; it inactivates C3a, C4a, C5a, bradykinin, kallikrein and fibrinopeptides. Inactivation of C5a and lysyl-bradykinin by the serum was markedly prolonged. Family study pointed to autosomal recessive inheritance.

II. Inhibitor Deficiencies

Protease inhibitors are proteins capable of inhibiting proteolytic enzymes (see Sect. H). In subjects with decreased levels of protease inhibitors in blood or locally in tissues, an increase in the protease activity may occur. A low inhibitor activity may be due to a defect in the rate of synthesis, a defect in the quality of the inhibitor for genetic reasons, or increased consumption because of disease processes. A well-known example is deficiency of C1-esterase inhibitor in patients with hereditary angio-oedema (DONALDSON and EVANS 1973). Local injury or stress is suggested to activate plasminogen, which activates C1-esterase. In the absence of C1-esterase inhibitor this causes the liberation of vascular permeability increasing fractions from the second component of complement (C2 kinins). Administration of purified C1-esterase inhibitor successfully aborts attacks of oedemá (SCHULZ 1974; GADEK et al. 1980). BLEUMINK and DOEGLAS (1977) and DOEGLAS and BLEUMINK (1974) demonstrated markedly decreased amounts of protease inhibitors in patients with cold contact urticaria and angio-oedema. A markedly decreased α_1-antitrypsin level was found in 14 out of 21 family members of a patient with recurrent angio-oedema in whom deficiency of α_1-antitrypsin and antichymotrypsin was found (BLEUMINK and DOEGLAS 1980). Decreased antichymotrypsin was found in eight patients and chronic urticaria in three patients in whom protease deficiency was also demonstrated. All these studies indicate that those with decreased protease inhibitor levels are predisposed to develop angio-oedema and/or chronic urticaria. Low levels of inter-α-trypsin inhibitor or α_1-antichymotrypsin with normal or increased levels of C1-esterase were also found in some forms of urticaria by EFTEKHARI et al. (1980).

J. Proteases and Pharmaceutical Agents

A number of chemical compounds used as anti-inflammatory agents in clinical medicine or as modifiers of inflammation in laboratory research are known to affect proteolytic enzyme systems. Following administration of cortisol, prednisolone, indomethacin or oxyphenylbutazone but not salicylate, HOUCK and PATEL (1965) and HOUCK et al. (1967, 1968) found a collagenolytic proteinase with a pH optimum at 5.5 and a neutral proteinase capable of hydrolysing chymotrypsin substrates in rat skin and in fibroblast cultures. This proteinase preparation caused an increase of vascular permeability by injection into the skin (LYCKE et al. 1967). Anti-inflammatory agents were suggested to induce the synthesis of collagenolytic proteinase via depression of an operon in skin fibroblasts. Corticosteroids at physiological concentrations have also been shown to inhibit the synthesis of neutral collagenase by fibroblasts in vitro (KOOB et al. 1974). Glycocorticoid derivatives are also potent inhibitors of plasminogen activator production by macrophages (VASSALLI et al. 1976). This occurs at physiological hormone concentrations and does not affect a variety of other macrophage functions. The potency of anti-inflammatory steroids to inhibit plasminogen activator production correlates with their general anti-inflammatory potency, suggesting that their anti-inflammatory potency may partly be due to their inhibition of plasminogen activator production by inflammatory cells. Degranulation of leucocytes and thus the release of proteases is inhibited by a number of factors which increase the intracellular cAMP, e.g. theophylline, prostaglandin E, β-adrenergic agonists, colchicine, vinblastine and adrenal glycocorticoids. An enhanced release is caused by cGMP, cholinergic agonists and phorbol myristate acetate (GOLDSTEIN 1975). All these agents are thus bound to affect the course of the inflammatory reactions. Of special interest is the fact that many of the agents that inhibit release of lysosomal enzymes are also effective against leucocyte motility or phagocytosis per se.

The lysosomal mucopolysaccharide degradation in cultured fibroblasts is strongly inhibited by $1–2 \times 10^{-5}$ *M* chloroquine (LIE and SCHOFIELD 1973). Chloroquine 50–100 m*M* inhibited cathepsin B_1 but not cathepsin D (WIBO and POOLE

1974). Dapsone (0.1 mg/ml) has been shown to inhibit guinea-pig skin glycosidases and arylamidase but not cathepsin B or C (MIER and VAN DEN HURK 1975). FRÄKI and HOPSU-HAVU (1977) showed that chloroquine (1 m*M*) inhibited 87% of the effect of cathepsin B-like proteinase and approximately 50% of the effects of the alkaline trypsin-like, chymotrypsin-like and acid cathepsin D-like proteinases from human skin. All these enzymes were slightly inhibited by dapsone (1 m*M*) and sulphapyridine (1 m*M*).

Aprotinin (Trasylol) is a low molecular weight polypeptide with a broad spectrum inhibition of proteolytic enzymes, including trypsin, chymotrypsin and kallikrein. THOMSON et al. (1978) demonstrated by an indirect immunoperoxidase technique that aprotinin is bound to human peripheral blood lymphocytes and polymorphonuclear leucocytes.

K. Conclusion

The inflammatory process is fundamentally a defence mechanism against injurious stimuli. The local tissue structures, both cellular and extracellular, exudating plasma factors as well as invading inflammatory cells are involved. All of these contain proteolytic enzymes or their readily activatable zymogens which contribute to and modify the inflammatory response. The extracellular deposition of intracellular proteases can be brought about by a number of mechanisms, e.g. cell death, secretion, release during migration, regurgitation during phagocytosis and reversed endocytosis. The early phase is characterised by increased vasodilatation and the entry of intravascular proteins into the extracellular space. In this stage a number of plasma-derived proteolytic systems are activated. This leads to formation of biologically active peptides which further augment the vasodilatation directly, and also by causing formation and liberation of vaso-active amines and prostaglandins. At the same time chemotactic factors are produced by a direct effect of proteases on complement components as well as by antigen–antibody reactions. The white cells phagocytose micro-organisms or foreign bodies during which process the granules discharge their enzymes. Cathepsin G, elastase and collagenase in leucocytes are supposed to be most important for the extracellular action of human leucocytes. These proteases have bactericidal and fungicidal properties, in addition to their capacity to hydrolyse immune complexes and a variety of tissue structures such as elastin, fibrin, collagen and proteoglycans. Proteolytic enzymes may also stimulate B- and T-lymphocytes in addition to providing a fibrinolytic system and stimulating platelet aggregation. The tissue damaged by the proteases may become foreign to the organism and then form the basis for an immunological reaction. During the repair phase, mononuclear cells and macrophages invade and clear the tissue of the debris. Increased vascular permeability and vasodilatation also lead to leakage of plasma-derived protease inhibitors into the inflamed tissue. Protease inhibitors may also be derived from local cells or invading inflammatory cells. The proteolytic activity is diminished due to inhibition and to dilution with plasma. The inflammatory oedema dilutes bacterial toxins and carries antibodies to the site of antigen. In conditions with generalised deficiency of one or other protease inhibitor the protease itself may cause disease.

The inflammatory response should be regarded as a general phenomenon which includes a complicated interplay of proteases and their inhibitors. In the skin it is modified by local skin factors such as epidermal and dermal proteases and locally formed protease inhibitors.

References

Allison AC (1969) Lysosomes. In: Bittar EE, Bittar N (eds) The biologic basis of medicine, vol 1. Academic, New York, pp 209–242

Allison AC, Paton GR (1965) Chromosome damage in human diploid cells following activation of lysosomal enzymes. Nature 207:1170–1173

Allison AC et al. (1966) The role of lysosomes and of cell membranes in photosensitization. Nature 209:874–878

Allison AC et al. (1968) An examination of the growth of cells in tissue culture. Proc Soc Exp Biol Med 126:112–114

Astrup T (1968) Blood coagulation and fibrinolysis in tissue culture and tissue repair. Biochem Pharmacol [Suppl] 17:241–257

Baggiolini M et al. (1978) Subcellular localization of granulocyte enzymes. In: Havemann K, Janoff A (eds) Neutral proteases of human polymorphonuclear leukocytes. Urban and Schwarzenberg, Baltimore, pp 3–17

Barrett AJ (1977) Cathepsin D and other carboxyl proteinases. In: Barrett AJ (ed) Proteinases in mammalian cells and tissues. North-Holland, Amsterdam, pp 209–248

Bentley C et al. (1976) In vitro synthesis of factor B of the alternative pathway of complement activation by mouse peritoneal macrophages. Eur J Immunol 6:393–398

Bleumink E, Doeglas HMG (1977) Deficiences of protease inhibitors in patients with skin diseases. Dermatologica 154:305

Bleumink E, Doeglas HMG (1980) Protease inhibitor deficiences in a patient with angioedema: results of family studies. Br J Dermatol 102:473

Bosman HB (1969) Glycoprotein degradation. Glycosidases in fibroblasts transformed by oncogenic viruses. Exp Cell Res 54:217–221

Bosman HB (1972) Elevated glycosidases and proteolytic enzymes in cells transformed by RNA tumor virus. Biochim Biophys Acta 264:339–343

Bosman HB et al. (1974) Surface biochemical changes accompanying primary infection with Rous sarcoma virus. II. Proteolytic and glycosidase activity and sublethal autolysis. Exp Cell Res 83:25–30

Braun-Falco O et al. (1977) Immunoelectronmicroscopical demonstration of in vivo bound complement C_3 in psoriatic lesions. Arch Dermatol Res 260:57–62

Burckhardt W (1941) Untersuchungen über die Photoaktivität einiger Sulfanilamide. Dermatologica 83:63–68

Burger M (1970) Proteolytic enzymes initiating cell division and escape from contact inhibition of growth. Nature 227:170–171

Christman JK et al. (1977) Plasminogen activators. In: Barrett AJ (ed) Proteases in mammalian cells and tissues. North Holland, Amsterdam, pp 91–149

Cochrane CG et al. (1959) The role of polymorphonuclear leukocytes in the initiation and cessation of the Arthus vasculitis. J Exp Med 110:481

Cormane RH et al. (1979) Immunologic implications of PUVA therapy in psoriasis vulgaris. Arch Dermatol Res 265:245–267

Davies P, Allison AC (1976) Secretion of macrophage enzymes in relation to the pathogenesis of chronic inflammation. In: Nelson DS (ed) Immunobiology of the macrophages. Academic, New York, p 427

Desai ID et al. (1964) Peroxidative and radiation damage to isolated lysosomes. Biochem Biophys Acta 86:277–285

Dingle JT et al. (1973) Immunoinhibition of intracellular protein digestion in macrophages. J Exp Med 137:1124–1141

Doeglas HMG, Bleumink E (1974) Familial cold urticaria: clinical findings. Arch Dermatol 110:382–388

Donaldson VH, Evans RR (1973) Biochemical abnormality in hereditary angioneurotic oedema: absence of serum inhibitor of C_1 esterase. Am J Med 35:37–44
Eftekhari N et al. (1980) Protease inhibitor profiles in urticaria and angioedema. Br J Dermatol 103:33–39
Fehr K et al. (1974) Digestion of autologous IgG by acid lysosomal protease (cathepsin D) and its role in immune complex formation and inflammation. Adv Clin Pharmacol 6:64
Fine RM et al. (1979) Accumulation of cancer-related glycoprotein, EDCI, in psoriasis. J Invest Dermatol 73:264–265
Fräki JE (1976) Human skin proteases. Separation and characterization of two acid proteases resembling cathepsin B_1 and cathepsin D and of an inhibitor to cathepsin B_1. Arch Dermatol Res 255:317
Fräki JE, Hopsu-Havu VK (1972) Human skin proteases. Differential extraction of proteases and of endogenous protease inhibitors. Arch Dermatol Forsch 242:329–342
Fräki JE, Hopsu-Havu VK (1975) Human skin proteases. Separation and characterization of two alkaline proteases, one splitting trypsin and the other chymotrypsin substrates. Arch Dermatol Res 253:261–276
Fräki JE, Hopsu-Havu VK (1976) Human skin proteases. Fractionation of psoriasis scale proteases and separation of a plasminogen activator and a histone hydrolysing protease. Arch Dermatol Res 256:113–126
Fräki JE, Hopsu-Havu VK (1977) Inhibition of human skin proteinases by chloroquine, dapsone and sulfapyridine. Arch Dermatol Res 259:113–115
Fräki JE et al. (1978) Plasminogen activators of psoriatic scale extracts. Separation of two plasminogen activators by isoelectric focusing. Arch Dermatol Res 261:259–266
Fräki JE et al. (1979) Proteinase binding and enhancing factor in human skin. Arch Dermatol Res 264:185–191
Fritz H (1980) Proteinase inhibitors in severe inflammatory processes (septic shock and experimental endotoxaemia): biochemical, pathophysiological and therapeutic aspects. Ciba Found Symp 75:351–379
Fruton JS (1946) On the proteolytic enzymes of animal tissues. V. Peptidases of skin, lung and serum. J Biol Chem 166:721–738
Gadek JE et al. (1980) Replacement of therapy in hereditary angioedema: successful treatment of acute episodes of angioedema with partly purified C_1 inhibitor. N Engl J Med 302:542–546
Ginsburg I et al. (1977) The role played by leukocyte extracts and inflammatory exudates in the release of lipopolysaccharides from Gram-negative bacteria: relation to tissue damage induced during infection. In: Willoughby DA, Giroud JP, Velo GP (eds) Perspectives in inflammation. MTX Lancaster, England, pp 163–167
Glenner GG, Cohen LA (1960) Histochemical demonstration of a species specific trypsin-like enzyme in mast cells. Nature 185:846–847
Glenner GG et al. (1962) Histochemical demonstration of a trypsin-like esterase activity in mast cells. J Histochem Cytochem 10:109–110
Goldstein IM (1975) Effect of steroids on lysosomes. Transplant Proc 7:21–24
Goldstein IM, Weissmann G (1974) Generation of C_5-derived lysosomal enzyme releasing activity (C5a) by lysates of leukocyte lysosomes. J Immunol 113:1583–1588
Gordon S (1976) Macrophage neutral proteinases and chronic inflammation. Ann NY Acad Sci 278:176–189
Greenbaum LM (1979) Kininogenases in blood cells. In: Erdös ED (ed) Bradykinin, kallidin and kallikrein. Springer, Berlin Heidelberg New York, pp 91–102 (Handbook of experimental pharmacology, vol 25, supplement)
Hällgren R, Venge P (1976) Cationic proteins of human granulocytes: enhancement of phagocytosis by staphylococcus protein A-IgG complexes. Inflammation 1:237–246
Hammarström S et al. (1975) Increased concentrations of non-esterified arachidonic acid 12:-hydroxy-5,8,10,14-eicosatetraenoic acid, prostaglandin E_2 and prostaglandin F_2 in epidermis of psoriasis. Proc Natl Acad Sci USA 72:5130–5134
Hayashi H (1975) The intracellular neutral SH-dependent protease associated with inflammatory reactions. Int Rev Cytol 40:101–151

Hayashi H et al. (1975) The nature of a mediator of leukocyte chemotaxis in inflammation. Antibiot Chemother 19:296–332
Hibino T et al. (1980) In vitro and in vivo inhibition of rat liver cathepsin L by epidermal proteinase inhibitor. Biochem Biophys Res Commun 93:440–447
Hönigsmann H et al. (1974) Epidermal lysosomes and ultraviolet light. J Invest Dermatol 63:337–342
Hopsu-Havu VK, Ekfors TO (1969) Identification in the rat skin and subcutaneous granuloma of an enzyme liberating N-terminal glycyl-proline from peptides. Arch Klin Exp Dermatol 235:301–307
Hopsu-Havu VK, Fräki JE (1980) Proteinases and their inhibitors in skin diseases. Int J Dermatol 20:159–163
Hopsu-Havu VK, Glenner GG (1966) A new dipeptide naphthylamidase hydrolyzing glycyl-prolyl-beta-naphthylamide. Histochemie 7:197–201
Hopsu-Havu VK, Jansén CT (1968) Biochemical demonstration of an aminopeptidase in rat skin specific for N-terminal basic amino acids (aminopeptidase B). Arch Klin Exp Dermatol 233:1–10
Hopsu-Havu VK et al. (1970) Partial purification and characterization of an acid dipeptide naphthylamidase (carboxytripeptidase) of the rat skin. Arch Klin Exp Dermatol 236:282–296
Hopsu-Havu VK et al. (1977) Proteolytic enzymes in the skin. In: Barrett AJ (ed) Proteinases in mammalian cells and tissues. Elsevier, North-Holland, Amsterdam New York Oxford, pp 545–581
Houck JC, Patel YM (1965) Proposed mode of action of corticosteroids on the connective tissue. Nature 206:158–160
Houck JC et al. (1967) The effects of anti-inflammatory drugs upon the chemistry and enzymology of rat skin. Biochem Pharmacol 16:1099–1111
Houck JC et al. (1968) Induction of collagenolytic and proteolytic activities by anti-inflammatory drugs in the skin and fibroblast. Biochem Pharmacol 17:2081–2090
Hugon J, Borgers M (1965) The ultrastructural localization of acid phosphatase in the crypt epithelium of the irradiated mouse duodenum. J Histochem Cytochem 13:526–525
Izaki S et al. (1979) Fibrin deposition and clearance in chronic granulomatous inflammation: correlation with T-cell function and proteinase inhibitor activity in tissue. J Invest Dermatol 73:561–565
Jablonska S et al. (1979) Ist die Psoriasis eine autoimmune Erkrankung? Hautarzt 30:634–539
Järvinen M (1976) Purification and properties of two protease inhibitors from rat skin inhibiting papain and other SH-protease. Acta Chem Scand [B] 30:933–940
Järvinen M, Hopsu-Havu VK (1975) α-*N*-benzoylarginine-2-naphthyl-amide hydrolase (cathepsin B_1?) from rat skin. II. Purification of the enzyme and demonstration of two inhibitors in the skin. Acta Chem Scand [B] 29:772–780
Järvinen M et al. (1978) The low-molecular-weight SH-protease inhibitor in rat skin is epidermal. J Invest Dermatol 71:119–121
Janoff A (1970) Mediators of tissue damage in leukocyte lysosomes. X. Further studies on human granulocyte elastase. Lab Invest 22:228
Janoff A (1972) Neutrophil proteases in inflammation. Annu Rev Med 23:177–190
Janoff A, Zeligs JD (1968) Vascular injury and lysis of basement membrane in vitro by neutral proteinase of human leukocytes. Science 162:702–704
Junnila SK et al. (1971) Elastase and trypsin inhibitors of human skin and serum. Partial purification and characterization. Acta Derm Venereol (Stockh) 51:251–256
Kass EH, Wolff SM (1973) Bacterial lipopolysaccharides, chemistry, biology and clinical significance of endotoxins. University of Chicago Press, Chicago
Kocmierska-Grodzka D, Goutier R (1971) Investigation on the catheptic activity of liver in irradiated rats. Strahlentherapie 142:345–348
Koob TJ et al. (1974) Regulation of human skin collagenase activity by hydrocortisone and dexamethasone in organ culture. Biochem Biophys Res Commun 61:1083–1088
Koono M et al. (1968) Proteases associated with Arthus skin lesions: their purification and biological significance. Tohoku J Exp Med 94:231

Krogh HK (1970) The occurrence of antibodies to stratum corneum in man. Int Arch Allergy Appl Immunol 37:649–659
Kueppers F, Bearn AG (1966) A possible experimental approach to the association of hereditary alpha-1-antitrypsin deficiency and pulmonary emphysema. Proc Soc Exp Biol Med 121:1207–1209
Lazarus GS (1974) The role of neutral proteinase and cathepsin D in turpentine induced inflammation. J Invest Dermatol 62:367–371
Lazarus GS et al. (1968) Degradation of collagen by a human granulocyte collagenolytic system. J Clin Invest 47:2622–2629
Lazarus GS et al. (1975) Lysosomes and the skin. J Invest Dermatol 65:259–271
Lazarus GS et al. (1977) Polymorphonuclear leukocytes: possible mechanism of accumulation in psoriasis. Science 198:1162–1163
Lie SO, Schofield B (1973) Inactivation of lysosomal functions in normal cultured human fibroblasts by chloroquine. Biochem Pharmacol 22:3109–3114
Lin T-Y, Williams HR (1975) Inhibition of the local haemorrhagic Shwartzman reaction by an acid proteinase. Experientia 31(2):209–212
Lycke AWJ et al. (1967) Effects of cortisol released cutaneous protease upon the permeability of the micro-circulation. J Invest Dermatol 48:318–325
Mathews KP et al. (1980) Familial carboxypeptidase N deficiency. Ann Intern Med 93:443–445
McClelland DBL, Van Furth R (1976) In vitro synthesis of B_1C/B_1A globulin (the C_3 component of complement) by tissues and leukocytes of mice. Immunology 31:855
Mier PD, Van den Hurk JJMA (1975) Inhibition of lysosomal enzymes by dapsone. Br J Dermatol 93:471–472
Mier PD, Van den Hurk JJMA (1976a) Acid hydrolases in psoriatic epidermis. Br J Dermatol 94:219–220
Mier PD, Van den Hurk JJMA (1976b) Lysosomal hydrolases of the epidermis. 4. Overall profile in comparison with dermis and other tissues. Br J Dermatol 94:443–446
Minta JO, Movat HZ (1979) The complement system and inflammation. In: Movat HZ (ed) Inflammatory reaction. Curr Top Pathol 68:135–178
Miyamoto M, Terayama H (1973) Serum factors affecting cathepsin release from lysosomes. Biochem Biophys Res Commun 64:617–624
Movat HZ (1979a) The kinin system and its relations to other systems. Curr Top Pathol 68:111–134
Movat HZ (1979b) The plasma kallikrein-kinin system and its interrelationship with other components of blood. In: Erdös EG (ed) Bradykinin, kallidin and kallikrein. Springer, Berlin Heidelberg New York, pp 1–89 (Handbook of experimental pharmacology, vol 25, Supplement)
Murphy G et al. (1977) Collagenase is a component of the specific granules of human neutrophil leukocytes. Biochem J 162:195–197
Neufahrt A, Henckel S (1974) Fraktionierte Zentrifugation und Gelfiltration zur Trennung von Proteinasen aus Psoriasisschuppen. Arch Dermatol Forsch 250:127–136
Odeberg H et al. (1975) Cationic proteins of human granulocytes. IV. Esterase activity. Lab Invest 32(1):86–90
Ohlsson K, Ohlsson I (1973) The neutral proteases of human granulocyte. Isolation and partial characterization of two granulocyte collagenases. Eur J Biochem 36:473–481
Ohyama T (1974) A neutral inflammatory protease associated with Arthus skin lesion in guinea pig. Tohoku J Exp Med 113:231
Olson RL, Nordquist RE (1966) Ultramicroscopic localization of acid phosphatase in human epidermis. J Invest Dermatol 46:431
Ossowski L et al. (1973) An enzymatic function associated with transformation of fibroblasts by oncogenic viruses. II. Mammalian fibroblast cultures transformed by DNA and RNA tumor viruses. J Exp Med 137:112–126
Page RC et al. (1974) Participation of mononuclear phagocytes in chronic inflammatory diseases. J Reticuloendothel Soc 15:413–438
Paris JE et al. (1969) Distribution and properties of lysosomal enzymes in untreated and in irradiated mouse mammary-gland carcinomas. JNCI 42:383–398

Peavy DL et al. (1972) Selective effect of bacterial endotoxins on various subpopulations of lymphoreticular cells. J Infect Dis 128 (Suppl):83
Postlethwaite AE et al. (1976) The chemotactic attraction of human fibroblasts to a lymphocyte-derived factor. J Exp Med 144:1188–1203
René AA, Evans AS (1970) Correlation of radiation-induced ultrastructural changes in mouse hepatocytes with alterations in plasma concentration of protein-bound neutral hexoses. Radiat Res 44:224–236
René AA et al. (1971) Radiation-induced ultrastructural and biochemical changes in lysosomes. Lab Invest 25:230–239
Ruddy S et al. (1972) The complement system of man. I. N Engl J Med 287:489–495
Ryan GB, Majno G (1977) Acute inflammation. Am J Pathol 86:185–276
Ryan TJ et al. (1971) Epithelial-endothelial interaction in the control of inflammation through fibrinolysis. Br J Dermatol 84:501–515
Ryan WL, Cardin C (1967) Lysosomal stimulation and inhibition of the growth of cells in tissue cultures. Proc Soc Exp Biol Med 126:112–114
Sasaki M et al. (1977) A new serum component which specifically inhibits thiol proteinases. Biochem Biophys Res Commun 76:917–924
Schmidt W, Havemann K (1974) Isolation of elastase-like and chymotrypsin-like neutral proteases from human granulocytes. Hoppe-Seylers Z Physiol Chem 355:1077–1082
Schnebli HP, Burger M (1972) Selective inhibition of growth of transformed cells by protease inhibitors. Proc Natl Acad Sci USA 69:3825–3827
Schorlemmer HU, Allison AC (1976) Effects of activated complement components on enzyme secretion by macrophages. Immunology 31:781
Schorlemmer HU et al. (1976) Ability of activated complement components to induce lysosomal enzyme release from macrophages. Nature 261:48
Schorlemmer HU et al. (1977) Interactions of macrophages and complement components in the pathogenesis of chronic inflammation. In: Willoughby DA, Giroud JP, Velo GP (eds) Perspectives in inflammation. MTX, Lancaster, England, pp 191–206
Schulz KH (1974) Hereditäres Quicke-Ödem. Neuere Wege der Therapie. Hautarzt 25:12–16
Schwartze G, Hanson H (1971) Zur proteolytischen Aktivität in Psoriasisschuppen. Dermatol Monatsschr 157:315–325
Seltzer JL et al. (1978) A component of normal human serum which enhances the activity of vertebrate collagenases. Biochem Biophys Res Commun 80:637–645
Seppä HJ (1978) Rat skin main neutral protease: immunohistochemical localization. J Invest Dermatol 71:311–315
Smolen JE, Weissmann G (1978) The granulocyte: metabolic properties and mechanisms of lysosomal enzyme release. In: Havemann K, Janoff A (eds) Neutral proteases of human polymorphonuclear leukocytes. Urban and Schwarzenberg, Munich, pp 56–76
Starkey PM (1977) Elastase and cathepsin G: the serine proteinases of human neutraphil leucocytes and spleen. In: Barrett AJ (ed) Proteinases in mammalian cells and tissues. North-Holland, Amsterdam, pp 57–89
Starkey PM, Barrett AJ (1973) Human cathepsin B_1. Inhibition by α_2-macroglobulin and other serum proteins. Biochem J 131:823–831
Starkey PM, Barrett AJ (1976a) Human lysosomal elastase. Catalytic and immunological properties. Biochem J 155:265–271
Starkey PM, Barrett AJ (1976b) Human cathepsin G. Catalytic and immunological properties. Biochem J 155:273–278
Steigleder GK, Blomayer U (1964) Zur Biochemie der krankhaften Verhornung, im besonderen bei Psoriasis. Arch Klin Exp Dermatol 218:461–468
Stetson CA Jr (1951) Similarities in the mechanisms determining the Arthus and Shwartzman phenomenon. J Exp Med 94:347
Stricklin GP et al. (1977) Human skin collagenase: isolation of precursor and active forms from both fibroblast and organ cultures. Biochemistry 16:1607–1615
Stüttgen G, Schilling T (1959) On the influenceability of proteolysis of human skin by prednisolone, kallikrein inactivator and roentgen rays. Z Gesamte Exp Med 132:51–63

Stüttgen G, Würdemann I (1959) Zur Darstellung der kateptischen Endopeptidase in der normalen menschlichen Haut und bei Dermatosen. Arch Klin Exp Dermatol 208:192–203
Tagami H, Ofuji S (1977) Characterization of a leukotactic factor derived from psoriatic scale. Br J Dermatol 97:509–518
Thomas CA et al. (1977) Cellular serine proteinase induced chemotaxis by complement activation. Nature 269:521–522
Thomson AW et al. (1978) Effects of the antiprotease Transylol on peripheral blood leukocytes. Experientia 34:528–530
Thorne KJ et al. (1976) Lysis and killing of bacteria by lysosomal proteinases. Infect Immun 14:555–563
Tokaji G (1971) The chemical pathology of thermal injury, with special reference to burns SH-dependent protease and its inhibitor. Kumamoto Med J 24:68–86
Toki N, Yamura T (1978) Plasma-kinin-forming enzyme in human skin. I. Extraction and column chromatographic of plasma-kinin-forming enzyme and its inhibitor. Acta Derm Venereol (Stockh) 58(5):395–399
Unkeless JC et al. (1973) An enzymatic function associated with transformation of fibroblasts by oncogenic virus. I. Chick embryo fibroblast cultures transformed by avian RNA tumor viruses. J Exp Med 137:85–111
Unkeless JC et al. (1974) Fibrinolysis associated with oncogenic transformation. Partial purification and characterization of the factor, a plasminogen activator. J Biol Chem 249:4295–4305
Vassalli JD et al. (1976) Macrophage plasminogen activator: modulation of enzyme production by anti-inflammatory steroids, mitotic inhibitors and cyclic nucleotides. Cell 8:271–281
Venge P, Olsson I (1975) Cationic proteins of human granulocytes. VI. Effects on the complement system and mediation of chemotactic activity. J Immunol 115:1505–1508
Vischer TL et al. (1976) In vitro stimulation of lymphocytes by neutral proteinases from human polymorphonuclear leukocyte granules. J Exp Med 144:863–872
Volden G (1978) Acid hydrolases in blister fluid. 4. Influence of ultraviolet radiation. Br J Dermatol 99:53–60
Volden G, Thune PO (1979a) Acid hydrolases in blister fluid. 5. Influence of ultraviolet radiation in patients with polymorphic light eruption. Br J Dermatol 100:277–282
Volden G, Thune PO (1979b) Acid hydrolases in blister fluid. 6. Photocontact and contact allergic skin reactions. Br J Dermatol 100:283–289
Volden G, Thune PO (1979c) Photosensitivity and cutaneous acid hydrolases in porphyria cutanea tarda. Ann Clin Res 11:129–132
Volden G et al. (1980) Release of intracellular enzymes from cutaneous cells after non-necrotizing damage by ultraviolet light. Arch Dermatol Res 268:225–230
Wahl LM et al. (1974) Collagenase production by endotoxin-activated macrophages. Proc Natl Acad Sci USA 71:3598–3601
Wakulchik SD et al. (1980) Photolysis of protoporphyrin-treated human fibroblasts in vitro: studies on the mechanism. J Lab Clin Med 96:158–167
Ward PA, Cochrane CG (1965) Bound complement and immunologic injury of blood vessels. J Exp Med 121:215–234
Ward PA, Hill JH (1970) C_5 chemotactic fragments produced by an enzyme in lysosomal granules of neutrophils. J Immunol 104:535–543
Ward PA et al. (1972) Leukotactic factors elaborated by virus-infected tissue. J Exp Med 135:1095–1103
Ward PA et al. (1978) The modulation by neutral proteases and other factors from neutrophils. In: Havemann K, Janoff A (eds) Neutral proteases of human polymorphonuclear leukocytes. Urban and Schwarzenberg, Munich, pp 276–286
Weissmann G, Dingle J (1961) Release of lysosomal protease by ultraviolet irradiation and inhibition by hydrocortisone. Exp Cell Res 25:207–210
Weissmann G, Fell HB (1962) The effect of hydrocortisone of the response of fetal rat skin in culture to ultraviolet irradiation. J Exp Med 116:365–380

Werb Z, Gordon S (1975a) Secretion of a specific collagenase by stimulated macrophages. Z Exp Med 142:346–360
Werb Z, Gordon S (1975b) Elastase secretion by stimulated macrophages. Characterization and regulation. J Exp Med 142:361–377
Wibo M, Poole B (1974) Protein degradation in cultured cells. II. The uptake of chloroquine by rat fibroblasts and the inhibition of cellular protein degradation and cathepsin B_1. J Cell Biol 63:430–440
Wilkinson PC (1974) Chemotaxis and inflammation. Churchill Livingstone, London
Willis I, Kligman AM (1968) The mechanism of photoallergic contact dermatitis. J Invest Dermatol 51:378
Willoughby DA (1978) Inflammation. Endeavour 2:57–65
Yu RJ et al. (1972) Inhibition of keratinases by α_2-macroglobulin. Experientia 28:886

CHAPTER 28

The Inflammatory Response – A Review

C. J. DUNN and D. A. WILLOUGHBY

A. Introduction

The capacity for survival of an organism residues in part within its ability to recognise foreign or mechanically inflicted trauma. It is this "recognition" that triggers a series of events culminating in a host defence or inflammatory reaction which prompted John Hunter to describe inflammation as "a salutary process".

Acute inflammation is characterised by a distinct vascular response consisting of transient vasoconstriction followed by a dilatation. This is accompanied by enhanced vascular permeability, giving rise to extravasation of plasma proteins, salts and water, leading to oedema formation. These changes account for the observed redness, heat, swelling and pain which all types of inflammation exhibit. In extreme circumstances, loss of function of the inflamed tissue may occur. At the same time leucocytes adhere to the endothelial cells lining local blood vessel walls, migrating into the extravascular connective tissue. Polymorphonuclear (PMN) leucocytes predominate in the early stages (acute phase), giving way to mononuclear leucocytes as the reaction progresses. Whether or not the reaction subsides depends very much on the nature of the irritant, removal of which usually leads to resolution and healing. Failure to do so results in a continuous influx of leucocytes, particularly mononuclear (monocytes, lymphocytes), in an attempt to remove the offending agent. The participation of lymphoid cells is not obligatory, being dependent on the antigenicity of the initiating stimulus. The increasing army of mononuclear phagocytes may be aided by local division of these cells as well as differentiation into cells more specialised for the task in hand (e.g. secretory epithelioid and multinucleate giant cells). Activated fibroblasts undergo mitosis and lay down collagen encapsulating the lesion, which now exhibits the features of a typical chronic inflammatory reaction.

The complex and *multifactorial* nature of the mechanisms underlying these processes is described below.

B. Mediators of Vascular Changes

The early responses to a wide range of noxious stimuli are related to changes in the microvasculature resulting from endogenous formation and release of a variety of mediators. Mast cells lining the blood vessel walls are extremely sensitive to tissue damage, undergoing degranulation and release of histamine and serotonin. Both substances induce increased permeability of venules to plasma proteins, resulting in oedema formation (BLOOM 1922, ROWLEY and BENDITT

1956). Vasoconstriction induced by serotonin may be accompanied by a similar though transient effect of histamine which leads to subsequent vasodilatation (DALE and RICHARDS 1918; RAPPORT et al. 1948). In contrast, plasma-derived kininogen is enzymatically cleaved by kallikrein, liberating vaso-active polypeptides of the kinin family which cause vasodilatation and increased vascular permeability (SPECTOR 1951; DUNN et al. 1976). The sequential release of histamine, serotonin and kinins during acute inflammatory reactions suggests a significant role for these substances in the mediation of early oedema formation (DIROSA et al. 1971).

The complement system is of potential significance in the mechanisms of vaso-active mediator release. Depletion of complement leads to significant suppression of acute inflammatory oedema in both immune and non-immune inflammation (WILLOUGHBY et al. 1968, 1969; CAPASSO et al. 1975; YAMAMOTO et al. 1975). WILLOUGHBY and DIROSA (1970) suggested that the complement system might be triggered by altered tissue proteins following lysosomal enzyme activity. Complement components C3 and C5 are activated to form anaphylatoxins which liberate mast cell histamine and serotonin (JOHNSTON and STROUD 1977), whereas cleavage of C4 and C2 by activated C1 esterase gives rise to kinin-like peptides (MAYER 1973).

Hageman factor may also be activated by injured blood vessels and traumatised tissue, converting pre-kallikrein to the active enzyme, kallikrein, by which kinins are generated (SCHULTZ 1978). In turn, kallikrein is able to activate complement C1, emphasising the potential interactions between coagulation–kinin–complement cascade pathways in acute inflammatory reactions.

Prostaglandins (PGs) and related fatty acid metabolites have been widely implicated in the inflammatory response. Direct effects include vasodilatation induced by prostacyclin (PGI_2) and PGE_2 (MORLEY 1981). Increased vascular permeability and pain elicited by histamine and bradykinin is potentiated by PGs (FERREIRA 1972; WILLIAMS and MORLEY 1973). The lipoxygenase-derived products leukotrienes C and D, which are related to SRS-A, have been shown to enhance vascular permeability and diminish blood flow in acute allergic inflammatory responses (WILLIAMS and PIPER 1980; SAMUELSON 1981). Inflammatory leucocytes are able to synthesise PGs, leukotrienes and various hydroxy-eicosatetraenoic acids via the cyclo-oxygenase and lipoxygenase pathways (BORGEAT et al. 1977; WEDMORE and WILLIAMS 1981), providing a local source of these mediators by secretion (WAHL et al. 1977; WEISSMAN et al. 1979).

The role of PGs and other lipid metabolites as mediators of cellular functions will be discussed in greater detail below.

C. Mediators of Cellular Responses

I. Adhesion

The vascular changes described above are accompanied by accumulation of phagocytic leucocytes in the extravascular inflamed connective tissue. Adhesion of leucocytes to vascular endothelium is an essential prerequisite and has been the subject of numerous "in vivo" studies (COHNHEIM 1867; CLARK et al. 1936;

ATHERTON and BORN 1972). Each observation is consistent in that, following diverse inflammatory stimuli, leucocytes begin to roll along the vessel wall, adhering to endothelium in the locality of the inflamed area. Migration between endothelial cells (diapedesis) into the extravascular connective tissue inevitably follows.

Calcium ions are essential for leucocyte–endothelial cell adhesion (LEA), possibly acting as a bridge to overcome the net negative membrane charge of both cell types involved (ALLISON and LANCASTER 1960; ATHERTON and BORN 1972; HOOVER et al. 1978). Activation of complement components C3a, C5a, and C5a-des-Arg stimulates PMN adhesion "in vitro" and "in vivo" (O'FLAHERTY et al. 1978; GALLIN J. I. et al. 1980; JACOB et al. 1980). GALLIN E. K. et al. (1975) showed that C5a modifies the leucocyte membrane potential, which may well facilitate LEA as described above.

Adhesion of PMN leucocytes to foreign surfaces and endothelium can occur independently of blood elements (MACGREGOR et al. 1978 a) although enhancement may be achieved by plasma (PEARSON et al. 1979). Enhanced PMN adhesion in vitro and in vivo has been observed for these cells under inflammatory conditions, suggesting both local and systemic control (PERILLIE et al. 1962; LENTNEK et al. 1976; MACKAY et al. 1985). With respect to systemic control, plasma from acute inflammatory conditions significantly enhances PMN adhesion compared with normal plasma (LENTNEK et al. 1976). Complement is required as a co-factor to a heat-stable factor derived from the clotting system. In this respect it is of interest that factors derived from the clotting and kinin systems induce LEA in vivo (GRAHAM et al. 1965).

Additional evidence suggests the presence of a plasma inhibitory factor (MACGREGOR 1977; MACGREGOR et al. 1978 b) which may decrease LEA, resulting in termination of the inflammatory response. The origin and precise nature of this factor awaits clarification.

Chemotactic factors, prostaglandins and leukotriene B_4 enhance LEA in vitro (HOOVER et al. 1978; PEARSON et al. 1979; HOOVER et al. 1984), suggesting a role for these inflammatory mediators in the local control of LEA following their liberation during the course of an inflammatory reaction.

As observed for other PMN functions (SMITH and IGNARRO 1975), intracellular cyclic nucleotide levels affect the adhesive properties of PMN leucocytes (BRYANT and SUTCLIFFE 1974; MACGREGOR 1976). Opposing effects between cyclic AMP (inhibitory) and cyclic GMP (stimulatory) have been observed for PMN adherence, being associated with decreased and increased intracellular calcium flux respectively.

In vivo studies strongly suggest that it is the endothelial cell (EC) which primarily undergoes certain specific changes following injury (CLARK and CLARK 1935, 1936; ALLISON et al. 1955), becoming "sticky" and attractive to circulating leucocytes. In support of these observations, recent studies have shown that LEA may be achieved through stimuli which are specific for ECs. Thus, interleukin-1 (IL-1) induced ECs to become hyperadhesive for PMNs and monocytes (DUNN and FLEMING 1984); adhesion was protein synthesis dependent and probably mediated by EC membrane surface changes rather than by the secretion of humoral substances. Increased adhesiveness was not observed upon exposure of other cell types (PMNs, monocytes, fibroblasts) to IL-1, emphasising the specificity for ECs (DUNN and FLEMING 1985b). Similar effects were obtained using bacterial lipopolysaccharide (DUNN and FLEMING 1984) and tumor necrosis factor (POHLMAN et al. 1986). IL-1 has also been shown to induce vascular pathology in vivo characterised by extensive subendothelial cell leucocyte migration, de-endothelialisation and formation of leucocyte–platelet aggregates (DUNN et al. 1987). The mechanisms mediating IL-1/lipopolysaccharide-induced EC hyperadhesiveness are unknown although these substances stimulate EC synthesis and membrane surface expression of tissue thromboplastin (BEVILACQUA et al. 1984) and the secretion of an intercellular matrix proteoglycan (MONTESANO et al. 1984). Whether these events are related to increased EC–leucocyte adhesion remains to be determined. The 1- to 2-h latency period required for protein synthesis and expression of EC hyperadhesiveness agrees well with observations in vivo where significant leucocyte accumulation does not occur until 1–2 h after injury (HURLEY 1964). It could, therefore, be envisaged that the initial response to injury is the release of IL-1 from local connective tissue cells, such as histiocytes, which in turn stimulates the vascular endothelium of the adjacent microvasculature to become hyperadhesive for passing leucocytes.

II. Chemotaxis and Leucocyte Migration Inhibition

Having successfully adhered to the inflamed vessel walls, leucocytes exhibit enhanced motility (kinesis) and directional movement (taxis). Despite difficulties experienced in in vivo demonstration, overwhelming evidence supports the contention that a variety of chemical substances (chemotaxins) are responsible for such behaviour, being liberated at the site of inflammation. PMN leucocytes are able to sense the chemical gradient (ZIGMOND 1977) which directs the cell towards the focal area of injury.

Among the naturally occurring chemotaxins C5a, kallikrein (GALLIN and GALLIN 1977; GALLIN et al. 1979) and fibrinopeptides (RIDDLE et al. 1967; BARNHART 1968) have been identified. PMN leucocytes possess specific receptors for C5a (CHENOWETH and HUGLI 1978), a PMN-derived glycoprotein (SPILBERG and MEHTA 1979) and bacterial-related synthetic peptides (ASWANIKUMAR et al. 1977; WILLIAMS et al. 1977). Furthermore, a degree of specificity has been shown for chemotaxis of particular cell types. Thus, lymphokines secreted by sensitised lymphocytes following antigen contact are chemotactic for monocytes and macrophages (ALTMAN et al. 1973), providing a plausible explanation for the predominance of these cells at sites of inflammation induced by delayed hypersensitivity.

Interleukin-1 has recently been shown to be chemotactic for PMN and mononuclear leucocytes (LUGER et al. 1983). This interesting peptide, which is secreted from a variety of cell types, including epidermal cells (LUGER et al. 1983), probably plays a significant role in inflammation as indicated by both its variety of biological properties (KAMPSCHMIDT 1984) and its presence in inflammatory exudates (OPPENHEIM et al. 1982; WOOD et al. 1983; GOTO et al. 1984). Tumor necrosis factor α (TNFα), a 17000-dalton peptide derived from macrophages, also has a wide spectrum of pro-inflammatory activities similar to those of IL-1 (SHA-

LABY et al. 1986) and has been implicated as a potential mediator of inflammatory diseases (ISSEKUTZ and MEGYERI 1987; COTRAN et al. 1986).

Other chemotactic factors relate to the type of inflammatory stimulus and include tetrapeptides specific for eosinophils (ECF-A), in addition to a peptide specific for neutrophils liberated during acute allergic anaphylactic reactions (GOETZL and AUSTEN 1975; WASSERMAN et al. 1977; SOTER and AUSTEN 1977) and late phase allergic responses (METZGER et al. 1986). Eosinophils are also specifically receptive to histamine, as shown by CLARK et al. (1975). PMN phagocytosis of urate crystals liberates a factor thought to be responsible for the intense accumulation of PMN leucocytes in acute gouty arthritis (SPILBERG et al. 1976).

PMN chemokinetic and chemotactic activity is also derived from lipoxygenation of polyenoic fatty acids (e.g. arachidonate). Leukotriene B_4 (LTB_4) is the most potent, its activity in vitro approximating that of C5a (FORD-HUTCHINSON et al. 1980; SMITH et al. 1980). However, the role of LTB_4 in vivo is more controversial (FORD-HUTCHINSON et al. 1984; STEELE et al. 1984). Since chemotactic PGs and lipoxygenase metabolites are liberated from stimulated PMN and mononuclear leucocytes (GORDON et al. 1976; WEISSMAN et al. 1979; BACH et al. 1980), these cells provide a positive feedback mechanism for further accumulation of inflammatory cells. Thus, not only the potential for chemotactic factor generation but also the specificity exists whereby the required type of leucocyte may be selectively localised at the focal site of an inflammatory lesion.

Stimulation of PMN and mononuclear leucocytes by chemo-attractants is characterised by changes in membrane potential (GALLIN J. I. et al. 1975; GALLIN and GALLIN 1977; SELIGMANN et al. 1977). It is thought that transduction of this signal leads to intracellular biochemical changes, particularly protein and lipid transmethylation which appears to be essential for chemotactic responsiveness (O'DEA et al. 1978; PIKE et al. 1978; SNYDERMAN and GOETZL 1981). Organisation of microfilaments and microtubules is crucial for motility, in which intracellular calcium ions and cyclic nucleotide levels play a role (ESTENSEN et al. 1973; CLARK et al. 1977).

Having discussed how leucocytes might reach the focal area of injury, the means by which these cells are retained there deserve consideration. The ideal solution would be to inhibit the highly mobile leucocytes at the appropriate site, where they might then set about the task of localisation and removal of the respective irritant. Numerous studies suggest this is so, providing evidence for the existence of leucocyte migration inhibitory factors (LIF) which may be liberated by stimulation of a variety of both tissue and haematogenous cells (PICK 1976; YOSHIDA et al. 1975). Thus, LIFs are released following antigenic stimulation of specifically sensitised lymphocytes (DUMONDE et al. 1969). However, it should be emphasised that immunological specificity is not essential since non-sensitised lymphocytes (PICK 1976) and PMN leucocytes (STASTNY and ZIFF 1970) are able to release LIFs. Moreover, different types of LIF have been demonstrated in immune complex (Arthus), delayed hypersensitivity and non-immune inflammatory exudates (YAMAMOTO et al. 1976a, b; DUNN and WILLOUGHBY 1981), reinforcing the widespread distribution of the leucocyte migration inhibition phenomenon.

It is probable that a variety of substances initiate leucocyte immobilisation (e.g. endotoxin, lymphokines, C3b, C5a chemotactic factors). However, it has been proposed by DUNN and WILLOUGHBY (1981) that these may act via a common mechanism, namely stimulation of leucocyte tissue thromboplastin/thromboplastin-like enzyme, the release of which converts exudate fibrinogen to a dense fibrin meshwork, erecting a physical barrier against further leucocyte migration.

III. Phagocytosis and Release of Inflammatory Mediators

Once established at the site of tissue injury it is the natural function of leucocytes to eliminate the offending agent. This is achieved by phagocytosis of opsonised particles (i.e. coated with immunoglobulin or complement components) by PMN leucocytes and macrophages, a process which is facilitated by C3b and IgG surface membrane receptors (HENSON 1969; HENSON et al. 1972; SHEVACH et al. 1973).

Contact between inflammagen and phagocyte membrane alone, as well as phagocytosis, causes membrane perturbation leading to a series of reactions which play a major part in tissue injury. The primary reaction consists of a respiratory burst of oxidative metabolism (BABIOR et al. 1973; WEISSMANN et al. 1979) and generation of free oxygen radicals which are both microbicidal and capable of local tissue damage (DORMANDY 1978; FANTONE and WARD 1982). Invagination of inflammagen is followed by fusion with lysosomes and subsequent digestion by lysosomal enzymes, the leakage or secretion of which may also cause local tissue disruption (ALLISON and DAVIES 1974; WEISSMANN et al. 1979). Of the lysosomal enzymes neutral proteinases, such as collagenase and elastase, are released by PMN leucocytes and macrophages (LAZARUS 1973; WERB and GORDON 1975a, b), contributing to connective tissue collagen breakdown which threatens the structural integrity of the surrounding tissue, an undesirable effect of the inflammatory response. The digestion and activation by elastase of a wide range of plasma proteins (fibrinogen, kininogen, complement, immunoglobulins) as well as structural connective tissue proteins (proteoglycans, collagen, elastin) may exacerbate tissue injury (PLOW and EDGINGTON 1975; FRITZ 1978; HUGLI et al. 1976; SOLOMON 1978; JANOFF 1972; STARKEY et al. 1977). The experimental induction of chronic arthritis by intra-articular injection of PMN lysosomal enzymes emphasises only too clearly their inflammatory and destructive potential (WEISSMANN et al. 1969).

Lysosomal acid hydrolases may be selectively secreted by PMN leucocytes (COCHRANE and AIKEN 1966) or by macrophages (DAVIES et al. 1974a, b; CARDELLA et al. 1974; ALLISON and DAVIES 1975), depending on the nature of the irritant. Macrophages synthesise and secrete complement components C3a and C3b, thus perpetuating the inflammatory process, which may develop into chronicity. The situation may be aggravated by concomitant stimulation of leucocyte membrane phospholipid metabolism, resulting in secretion of substantial amounts of PGs and lipoxygenase products from both PMN leucocytes and macrophages (GORDON et al. 1976; GEMSA et al. 1977; PASWELL et al. 1979; WEISSMAN et al. 1979; BRUNE et al. 1978, 1984).

More recently, considerable attention has been focused on the pathogenic potential of IL-1, a polypeptide secreted by a variety of cells, including monocytes and macrophages. Current evidence suggests that IL-1 stimulates destruction of bone (THOMPSON et al. 1985) and cartilage in vitro (see ZIFF 1983; WOOD et al. 1983; KRAKAUER et al. 1985) and in vivo (PETTIPHER et al. 1986) in connective tissue diseases and excessive proliferation of fibroblasts in fibrotic conditions (for review see KAMPSCHMIDT 1984). Recombinant IL-1 has recently been shown to induce chronic granulomatous inflammation in the skin of mice implanted with a slow-release polymer containing this monokine (DUNN and GIBBONS 1987). Experiments in the rabbit cornea indicate the specific induction of angiogenesis by IL-1 (PRENDERGAST et al. 1987). Similar pathological properties have also been attributed to TNFα (SAKLATVALA 1986). Both IL-1 and TNFα stimulate intravascular coagulation via induction of endothelial cell procoagulant activity, tissue plasminogen activator inhibitor and decreased protein C activation in vitro and in vivo (NAWROTH and STERN 1986; NAWROTH et al. 1986; EMEIS and KOOISTRA 1986), which may contribute to inflammatory diseases, such as glomerular nephritis and rheumatoid arthritis, where intravascular and extravascular coagulation is a prominent feature.

Inflammation may, however, be minimised assuming there is efficient intracellular elimination of inflammagen. In the case of micro-organisms, leucocytes are equipped with lysosomal microbicidal enzymes myeloperoxidase (KLEBANOFF 1968; KLEBANOFF and HAMON 1972) and lysozyme together with cationic proteins. Toxic free radicals (e.g. superoxide anion) generated during phagocytosis also represent an important microbicidal mechanism. Their potentially harmful effects on local tissue may be prevented by caeruloplasmin present in extracellular fluids (DORMANDY 1978; GOLDSTEIN et al. 1979), while intracellular superoxide dismutase prevents leucocyte autolysis by free radicals leaking into the cytosol (SALIN and MCCORD 1974). Similarly, the damaging effects of secreted lysosomal enzymes are controlled by inhibitors such as β_2-macroglobulin, β_1-antitrypsin and C1 esterase inhibitor which are synthesised in the liver and accumulate at sites of inflammation (GLENN et al. 1968; BILLINGHAM and GORDON 1976; HUDIG and SELL 1978). These examples demonstrate some of the feedback mechanisms which serve to limit the injurious side-effects of an inflammatory reaction.

Finally, it should be borne in mind that not all destructive lesions are dependent on a "classical" inflammatory response. A good example is pemphigus vulgaris, an auto-immune disease of the skin and mucous membranes. IgG autoantibody specifically induces synthesis and secretion of plasminogen activator by epidermal cells which mediates tissue damage via activation of extracellular plasmin in the absence of inflammatory leucocytes (HASHIMOTO et al. 1983). Recent evidence for direct interaction of sensitised T-lymphocytes with neural tissue in vitro, resulting in demyelination, also suggests that the inflammatory response may be less critical in the pathogenesis of such diseases as multiple sclerosis (LYMAN et al. 1986).

References

Allison AC, Davies P (1974) Mechanisms underlying chronic inflammation. In: Velo GP, Willoughby DA, Giroud JP (eds) Future trends in inflammation. Piccin, Padua, pp 449–452

Allison AC, Davies P (1975) Increased biochemical and biological activities of mononuclear phagocytes exposed to various stimuli, with special reference to secretion of lysosomal enzymes. In: VanFurth R (ed) Mononuclear phagocytes in immunity, infection, and pathology. Blackwell, Oxford, pp 487–506
Allison F, Lancaster MG (1960) Studies on the pathogenesis of acute inflammation. II. The relationship of fibrinogen and fibrin to the leucocytic sticking reaction in ear chambers of rabbits injured by heat. J Exp Med 111:45–64
Allison F, Smith MR, Wood WB (1955) Studies on the pathogenesis of acute inflammation. I. The inflammatory reaction to thermal injury as observed in the rabbit ear chamber. J Exp Med 102:655–667
Altman LC, Snyderman R, Oppenheim JJ, Mergenhagen SE (1973) A human mononuclear leucocyte chemotactic factor: characterization, specificity and kinetics of production by homologous leukocytes. J Immunol 110:801–810
Aswanikumar S, Corcoran B, Schiffmann E, Day AR, Freer RJ, Showell HJ, Becker EL, Pert CB (1977) Demonstration of a receptor on rabbit neutrophils for chemotactic peptides. Biochem Biophys Res Commun 74:807–810
Atherton A, Born GVR (1972) Quantitative investigations of the adhesiveness of circulating polymorphonuclear leucocytes to blood vessel walls. J Physiol (Lond) 222:447–474
Babior BM, Kipnes RS, Curnutte JT (1973) Biological defense mechanisms. The production by leukocytes of superoxide, a potential bactericidal agent. J Clin Invest 52:741–744
Bach MK, Brashler JR, Hammarstrom S, Samuelsson B (1980) Identification of leukotriene C-1 as a major component of slow-reacting substance from rat mononuclear cells. J Immunol 125:115–117
Barnhart MI (1968) Role of blood coagulation in acute inflammation. Biochem Pharmacol 16 (Suppl):206–219
Bevilacqua MP, Pober JS, Majeau GR, Cotran RS, Gimbrone MA (1984) Interleukin 1 (IL-1) induces biosynthesis and cell surface expression of procoagulant activity in human vascular endothelial cells. J Exp Med 160:618–623
Billingham MEJ, Gordon AH (1976) The role of the acute phase reaction in inflammation. Agents Actions 6:195–200
Bloom W (1922) Histamine as an inflammatory agent. Bull Johns Hopkins Hosp 33:185–188
Borgeat P, Hamberg M, Samuelsson B (1977) Transformation of arachidonic acid and homo-γ-linolenic acid by rabbit polymorphonuclear leukocytes, monohydroxy acids from novel lipoxygenases. J Biol Chem 252:8772–8775
Brune K, Glatt M, Kalin H, Peskar BA (1978) Pharmalogical control of prostaglandin and thromboxane release from macrophages. Nature 274:261–263
Brune K, Aehringhaus U, Peskar BA (1984) Pharmacological control of leukotriene and prostaglandin production from mouse peritoneal macrophages. Agents Actions 14(5/6):729–734
Bryant RE, Sutcliffe MC (1974) The effect of 3′,5′-adenosine monophosphate on granulocyte adhesion. J Clin Invest 54:1241–1244
Capasso F, Dunn CJ, Yamamoto S, Deporter DA, Giroud JP, Willoughby DA (1975) Pharmacological mediators of various immunological and non-immunological inflammatory reactions produced in the pleural cavity. Agents Actions 5:528–533
Cardella CJ, Davies P, Allison AC (1974) Immune-compleyes induce selective release of lysosomal hydrolases from macrophages. Nature 247:46–48
Chenoweth DE, Hugli TE (1978) Demonstration of specific C5A receptor on intact human polymorphonuclear leukocytes. Proc Natl Acad Sci USA 75:3943–3947
Clark ER, Clark EL (1935) Observations on changes in blood vascular endothelium in the living animal. Am J Anat 57:385–405
Clark ER, Clark EL, Rex RO (1936) Observations on polymorphonuclear leukocytes in the living animal. Am J Anat 59:123–135
Clark RAF, Gallin JI, Kaplan AP (1975) The selective eosinophil chemotactic activity of histamine. J Exp Med 142:1462–1476

Clark RAF, Sandler JA, Gallin JI, Kaplan AP (1977) Histamine modulation of eosinophil migration. J Immunol 118:137–145
Cochrane CG, Aiken BS (1966) Polymorphonuclear leukocytes in immunologic reactions. The destruction of vascular basement membrane in vivo and in vitro. J Exp Med 124:733–752
Cohnheim J (1867) Über Entzündung und Eiterung. Arch Pathol Anat 40:1–49
Cotran RS, Gimbrone MA, Bevilacqua MP, Mendrick DL, Pober JS (1986) In situ induction and detection of a human-endothelial cell activation antigen in immunological inflammation. Fed Proc 45:379 Abstr no 1309
Dale HH, Richards AN (1918) The vasodilator action of histamine and of some other substances. J Physiol 52:110–165
Davies P, Allison AC, Ackerman J, Butterfield A, Williams S (1974 a) Asbestos induces release of lysosomal enzymes from mononuclear phagocytes. Nature 251:423–425
Davies P, Page RC, Allison AC (1974 b) Changes in cellular enzyme levels and extracellular release of lysosomal acid hydrolases in macrophages exposed to group A streptococcal cell wall substance. J Exp Med 139:1262–1282
DiRosa M, Giroud JP, Willoughby DA (1971) Studies of the mediators of the acute inflammatory response induced in rats in different sites by carrageenan and turpentine. J Pathol 104:15–29
Dormandy TL (1978) Free-radical oxidation and antioxidants. Lancet I:647–650
Dumonde DC, Wolstencroft RA, Panayi GS, Matthew M, Morley J, Howson WT (1969) Lymphokines: non-antibody mediators of cellular immunity generated by lymphocyte activation. Nature 224:38–42
Dunn CJ, Fleming WE (1984) Increased adhesion of polymorphonuclear leukocytes to vascular endothelium by specific interaction of endogenous (interleukin-1) and exogenous (lipopolysaccharide) substances with endothelial cells in vitro. Eur J Rheumatol Inflamm 7:80–86
Dunn CJ, Fleming WE (1985) The role of interleukin-1 in the inflammatory response with particular reference to endothelial cell–leukocyte adhesion. In: Kluger MJ, Oppenheim JJ, Powanda MC (eds) Progress in leukocyte biology. Liss, New York, pp 45–54
Dunn CJ, Gibbons AJ (1987) Human recombinant interleukin-1 induces chronic granulomatous inflammation. J Leukocyte Biol 42:615
Dunn CJ, Willoughby DA (1981) Leukocyte and macrophage migration inhibitory activities in inflammatory exudates – involvement of the coagulation system. Lymphokines 4:231–269
Dunn CJ, Giroud JP, Willoughby DA (1976) Cellular pharmacokinetics in inflammation. In: Huskisson EC, Velo GP (eds) Inflammatory arthropathies. Excerpta Medica, Amsterdam, pp 23–30
Dunn CJ, Schaub RG, Fleming WE, Gibbons AJ (1987) Vascular changes induced by interleukin 1 in vivo: scanning electron microscope studies. In: Movat HZ (ed) Leukocyte emigration and its sequelae. Karger, Basel, pp 58–61
Emeis JJ, Kooistra T (1986) Interleukin 1 and lipopolysaccharide induce an inhibitor of tissue-type plasminogen activator in vivo and in cultured endothelial cells. J Exp Med 163:1260–1266
Estensen RD, Hill HR, Quie PG, Gogan N, Goldberg ND (1973) Cyclic GMP and cell movement. Nature 254:458–460
Fantone JC, Ward PA (1982) Role of oxygen-derived free radicals and metabolites in leukocyte-dependent inflammatory reactions. Am J Pathol 107(3):396–418
Ferreira SH (1972) Prostaglandins, aspirin-like drugs and analgesia. Nature 240:200–203
Ford-Hutchinson AW, Bray MA, Doig MV, Shipley ME, Smith MJH (1980) Leukotriene B, a potent chemokinetic and aggregating substance released from polymorphonuclear leukocytes. Nature 286:264–265
Ford-Hutchinson AW, Brunet G, Savard P, Charleson S (1984) Leukotriene B_4, polymorphonuclear leukocytes and inflammatory exudates in the rat. Prostaglandins 28(1):13–27
Fritz H (1978) Necessity of a critical consideration of the homogeneity PMH proteases applied to biological assay systems: failure to detect intrinsic kininogenase activity in

PMH elastase. In: Havemann K, Janoff A (eds) Neutral proteases of human polymorphonuclear leukocytes – biochemistry, physiology, and clinical significance. Urban and Schwarzenberg, Baltimore, pp 261–263
Gallin EK, Gallin JI (1977) Electrophysiology of human macrophage interaction with chemotactic factors. J Cell Biol 75:277–289
Gallin EK, Wiederhold ML, Lipsky PE, Rosenthal AS (1975) Spontaneous and induced membrane hyperpolarizations in macrophages. J Cell Physiol 86:653–661
Gallin EK, Seligmann BE, Gallin JI (1980) Alteration of macrophage and monocyte membrane potential by chemotactic factors. In: VanFurth R (ed) Mononuclear phagocytes – functional aspects. Martinus Nijhoff, The Hague, pp 505–526
Gallin JI, Durocher JR, Kaplan AP (1975) Interaction of leukocyte chemotactic factors with cell surface. 1. Chemotactic factor-induced changes in human granulocyte surface charge. J Clin Invest 55:967–974
Gallin JI, Gallin EK, Shiffmann E (1979) Mechanism of leukocyte chemotaxis. In: Paoletti R, Samuelson B, Weissmann G (eds) Advances in inflammation research. Raven, New York, pp 123–138
Gallin JI, Wright DG, Malech HL, Davis JM, Klempner MS, Kirkpatrick CH (1980) Disorders of phagocyte chemotaxis. Ann Intern Med 92:520–538
Gemsa D, Steggemann L, Till G, Resch K (1977) Enhancement of PGE_1 response of macrophages by concanavalin-A and colchicine. J Immunol 119:524–529
Glenn EM, Bowman BJ, Koslowske TC (1968) The systemic response to inflammation. Biochem Pharmacol 16 (Suppl):27–49
Goetzl EJ, Austen KF (1975) Purification and synthesis of eosinophilotactic tetrapeptides of human lung tissue: identification as eosinophil chemotactic factor of anaphylaxis. Proc Natl Acad Sci USA 72:4123–4127
Goldstein IM, Kaplan HB, Edelson HS, Weissmann G (1979) Ceruloplasmin. A scavenger of superoxide anion radicals. J Biol Chem 254:4040–4045
Gordon D, Bray MA, Morley J (1976) Control of lymphokine secretion by prostaglandins. Nature 262:401–402
Goto K, Nakamura S, Goto F, Yoshinaga M (1984) Generation of an interleukin-I-like lymphocyte-stimulating factor at inflammatory sites: correlation with the infiltration of polymorphonuclear leucocytes. Br J Exp Pathol 65:521–532
Hashimoto K, Shafran KM, Webber PS, Lazarus GS, Singer KJ (1983) Anti-cell surface pemphigus autoantibody stimulates plasminogen activator activity of human epidermal cells. A mechanism for the loss of epidermal cohesion. J Exp Med 157:259–272
Henson PM (1969) The adherence of leucocytes and platelets induced by fixed IgG antibody or complement. Immunology 16:107–121
Henson PM, Johnson HB, Spiegelberg HL (1972) The release of granule enzymes from human neutrophils stimulated by aggregated immunoglobulins of different classes and subclasses. J Immunol 109:1182–1192
Hoover RL, Briggs RT, Karnovsky MJ (1978) The adhesive interaction between polymorphonuclear leucocytes and endothelial cells in vitro. Cell 14:423–428
Hoover RL, Karnovsky MJ, Austen KF, Corey EJ, Lewis RA (1984) Leukotriene B_4 action on endothelium mediates augmented neutrophil/endothelial adhesion. Proc Natl Acad Sci USA 81:2191–2193
Hudig D, Sell S (1978) Serum concentrations of alpha-macro fetoprotein (acute-phase alpha-2-macroglobulin), a proteinase inhibitor, in pregnant and neonatal rats and in rats with acute inflammation. Inflammation 3(2):137–148
Hugli TE, Taylor JC, Crawford JP (1976) Selective cleavage of human C3 by human leukocyte elastase (HLE). J Immunol 116:1737–1739
Hunter J (1794) In: A treatise on the blood, inflammation and gunshot wounds. Richardson, London
Hurley JV (1964) Substances promoting leukocyte emigration. Ann NY Acad Sci 116:918–935
Issekutz AC, Megyeri P (1987) Induction of leukocyte infiltration by endotoxin. In: Movat HZ (ed) Leukocyte emigration. Karger, Basel, pp 24–37
Jacob HS, Craddock PR, Hammerschmidt DE, Moldow CF (1980) Complement induction of granulocyte aggregation: an unsuspected mechanism of disease. N Engl J Med 302:789–794

Janoff A (1972) Human granulocyte elastase. Further delineation of its role in connective tissue damage. Am J Pathol 68:579–592
Johnson RL, Ziff M (1976) Lymphokine stimulation of collagen accumulation. J Clin Invest 58(1):240–252
Johnston RB Jr, Stroud RM (1977) Complement and host defense against infection. J Pediatr 90:169–179
Kampschmidt RF (1984) The numerous postulated biological manifestations of interleukin-1. J Leukocyte Biol 36:341–355
Klebanoff SJ (1968) Myeloperoxidase-halide-hydrogen peroxide antibacterial system. J Bacteriol 95:2131–2138
Klebanoff SJ, Hamon CB (1972) Role of myeloperoxidase-mediated antimicrobial systems in intact leukocytes. J Reticuloendothel Soc 12:170–196
Krakauer T, Oppenheim JJ, Jasin HE (1985) Human interleukin 1 intermediates cartilage matrix degradation. Cell Immunol 91(1):92–99
Lazarus GS (1973) Studies on the degradation of collagen by collagenases. In: Dingle JT (ed) Lysosomes in biology and pathology, vol 3. North-Holland, Amsterdam, pp 338–364
Lentnek AL, Schreiber AD, MacGregor RR (1976) The induction of augmented granulocyte adherence by inflammation. Mediation by a plasma factor. J Clin Invest 57:1098–1103
Luger TA, Charon JA, Colot M, Micksche M, Oppenheim JJ (1983) Chemotactic properties of partially purified human epidermal cell-derived thymocyte-activating factor (ETAF) for polymorphonuclear and mononuclear cells. J Immunol 131(2):816–820
Lyman WD, Roth GA, Chui FC, Brosnan CF, Bornstein MB, Raine CS (1986) Antigen specific T cells can mediate demyelination in organotypic central nervous system cultures. Cell Immunol 102:217–226
MacGregor RR (1976) Cyclic nucleotide induction as a mechanism for modification of granulocyte adherence (GA) by plasma factors. Clin Res 24:348A
MacGregor RR (1977) Granulocyte adherence changes induced by hemodialysis, endotoxin epinephrine, and glucocorticoids. Ann Intern Med 86:35–39
MacGregor RR, Macarak EJ, Kefalides NA (1978a) Comparative adherence of granulocytes to endothelial monolayers and nylon fiber. J Clin Invest 61:697–702
MacGregor RR, Negendank WB, Schreiber AD (1978b) Impaired granulocyte adherence in multiple myeloma: relationship to complement system, granulocyte delivery, and infection. Blood 51:591–599
Mackay AR, Sedgwick AD, Dunn CJ, Fleming WE, Willoughby DA (1985) The transition from acute to chronic inflammation. Br J Dermatol 113 (Suppl 28):34–48
Mayer MM (1973) The complement system. Sci Am 229:54–66
Metzger WJ, Richerson HB, Waserman SI (1986) Generation and partial characterization of eosinophil chemotactic activity and neutrophil chemotactic activity during early and late-phase asthmatic responses. J Allergy Clin Immunol 78:282–290
Montesano R, Mossaz A, Ryser JE, Orci L, Vassalli P (1984) Leukocyte interleukin induced cultured endothelial cells to produce a highly organized, glycosaminoglycan-rich pericellular matrix. J Cell Biol 99:1706–1715
Morley J (1981) Role of prostaglandins secreted by macrophages in the inflammatory process. Lymphokines 4:377–394
Nawroth PP, Stern DM (1986a) Modulation of endothelial cell hemostatic properties by tumor necrosis factor. J Exp Med 163:740–746
Nawroth PP, Handley DA, Esmon CT, Stern DM (1986b) Interleukin 1 induces endothelial cell procoagulant while suppressing cell-surface anticoagulant activity. Proc Natl Acad Sci USA 83:3460–3464
O'Dea R, Viveros OH, Axelrod J, Awanikaumar S, Schiffmann E, Corcoran BA (1978) Rapid stimulation of protein carboxymethylation in leukocytes by a chemotactic peptide. Nature 272:462–465
O'Flaherty JT, Craddock PR, Jacob HS (1978) Effect of intravascular complement activation on granulocyte adhesiveness and distribution. Blood 51:731–733
Oppenheim JJ, Charon JA, Luger TA (1982) Evidence for an in vivo inflammatory role of interleukin 1 (IL-1). Transplant Res 14:553–557

Passwell JH, Dayer JM, Merler E (1979) Increased prostaglandin production by human monocytes after membrane receptor activation. J Immunol 123:115–117
Pearson JD, Carleton JS, Beesley JE, Hutchings A, Gordon JL (1979) Granulocyte adhesion to endothelium in culture. J Cell Sci 38:225–231
Perillie PE, Nolan JP, Finch SC (1962) Studies of the resistance to infection in diabetes mellitus: local exudate cellular response. J Lab Clin Med 59:1008–1012
Pettipher ER, Higgs GA, Henderson BA (1986) Interleukin 1 induces leukocyte infiltration and cartilage proteoglycan degradation in the synovial joint. Proc Natl Acad Sci USA 83:8749–8752
Pick E (1976) The mechanism of action of macrophage migration inhibitory factor (MIF). A personal view. Curr Titles Immunol 4:565–568
Pike MC, Kredich NM, Snyderman R (1978) Requirement of *S*-adenosyl-L-methionine-mediated methylation for human monocyte chemotaxis. Proc Natl Acad Sci USA 75:3928–3930
Plow EF, Edgington TS (1975) An alternative pathway for fibrinolysis. I. The cleavage of fibrinogen leukocyte proteases at physiologic pH. J Clin Invest 56:30–35
Pohlman TH, Stanness KA, Beatty PG, Ochs HD, Harlan JM (1986) An endothelial cell surface factor(s) induced in vitro by lipopolysaccharide, interleukin 1, and tumor necrosis factor-alpha increases neutrophil adherence by a CDw18-dependent mechanism. J Immunol 136:4548–4551
Prendergast RA, Lutty GA, Dinarello CA (1987) Interleukin-1 induces corneal neovascularization. Fed Proc 46:1200
Rapport MM, Green AA, Page IH (1948) Crystalline serotonin. Science 108:329–330
Riddle JM, Odle NA, Bluhm GB, Barnhart MI (1967) Leucocyte–fibrin interactions. Thromb Diath Haemorrh 18:302–303
Rowley DA, Benditt EP (1956) 5-Hydroxytryptamine and histamine as mediators of the vascular injury produced by agents which damage mast cells in rats. J Exp Med 103:399–411
Saklatvala J (1986) Tumor necrosis factor alpha stimulates resorption and inhibits synthesis of proteoglycan in cartilage. Nature 32:547–550
Salin ML, McCord JM (1974) Superoxide dismutases in polymorphonuclear leukocytes. J Clin Invest 54:1005–1009
Samuelsson B (1981) Leukotrienes: mediators of allergic reactions and inflammation. Int Arch Allergy Appl Immunol 66(1):98–101
Schultz DR (1978) Mediators of the inflammatory process. In: Talbott JH (ed) Fundamental and clinical aspects of internal medicine. Elsevier, Amsterdam, pp 27–35
Seligmann B, Gallin EK, Martin DL, Shain W, Gallin JI (1977) Evidence of membrane potential changes in human polymorphonuclear leukocytes during exposure to the chemotactic factor formylemethionylleucylphenyl alanine as measured with the fluorescent dye D1 pentyloxa carbo cyanine. J Cell Biol 75:103a
Shalaby MR, Pennica D, Palladino MA Jr (1986) An overview of the history and biologic properties of tumor necrosis factor. Springer Semin Immunopathol 9:33–38
Shevach EM, Jaffe ES, Green I (1973) Receptors for complement and immunoglobulin on human and animal lymphoid cells. Transplant Rev 16:3–11
Smith MJ, Ford-Hutchinson AW, Bray MA (1980) Leukotriene B, a potential mediator of inflammation. J Pharm Pharmacol 32:517–521
Smith RJ, Ignarro LJ (1975) Bioregulation of lysosomal enzyme secretion from human neutrophils: roles of guanosine 3′,5′-monophosphate and calcium in stimulus-secretion coupling. Proc Natl Acad Sci USA 72(1):108–111
Snyderman R, Goetzl EJ (1981) Molecular and cellular mechanisms of leukocyte chemotaxis. Science 213:830–833
Solomon A (1978) Possible role of PMN proteases in immunoglobulin degradation and amyloid formation. In: Havemann K, Janoff A (eds) Neutral proteases of human polymorphonuclear leukocytes. Urban and Schwarzenberg, Baltimore, pp 423–436
Soter NA, Austen FK (1977) Urticaria, angioedema, and mediator release in humans in response to physical environmental stimuli. Fed Proc 36:1736–1739
Spector WG (1951) The role of some higher peptides in inflammation. J Pathol Bacteriol 63:93–97

Spilberg I, Mehta J (1979) Demonstration of a specific neutrophil receptor for a cell-derived chemotactic factor. J Clin Invest 63:85–90
Spilberg I, Gallacher J, Mehta JM, Mandell B (1976) Urate crystal-induced chemotactic factor: isolation and partial characterization. J Clin Invest 58:815–820
Starkey PM, Barrett AJ, Burleigh MC (1977) The degradation of articular collagen by neutrophil proteinases. Biochim Biophys Acta 483:386–389
Stastny P, Ziff M (1970) Inhibitor of macrophage migration produced by polymorphonuclear leucocytes. J Reticuloendothel Soc 7:140–145
Steele L, Hunneyball IM, Mason CG (1984) Contribution of LTB_4 to leukocyte migration in inflammatory lesions. J Pharm Pharmacol 36:644–647
Thomson B, Saklatvala J, Chambers TJ (1985) The effect of interleukin 1 (Catabolin) on bone resorption by disaggregated osteoclasts. Lymphokine Res 4(1):75
Wahl LM, Ohlsen CE, Sandberg AL, Mergenhagen SE (1977) Prostaglandin regulation of macrophage collagenase production. Proc Natl Acad Sci USA 74:4955–4958
Wasserman SI, Soter NA, Center DM, Austen KF (1977) Cold urticaria. Recognition and characterization of a neutrophil chemotactic factor which appears in serum during experimental cold challenge. J Clin Invest 60:189–194
Wedmore CV, Williams TJ (1981) Control of vascular permeability by polymorphonuclear leukocytes in inflammation. Nature 289:646–648
Weissmann G, Korchak HM, Perez HD, Smolen JE, Goldstein IM, Hoffstein ST (1979) The secretory code of the neutrophil. J Reticuloendothel Soc 26:687–691
Werb Z, Gordon S (1975a) Secretion of a specific collagenase by stimulated macrophages. J Exp Med 142:346–350
Werb Z, Gordon S (1975b) Elastase secretion by stimulated macrophages. Characterization and regulation. J Exp Med 142:361–367
Williams LT, Snyderman R, Pike MC, Lefkowitz RJ (1977) Specific receptor sites for chemotactic peptides on human polymorphonuclear leukocytes. Proc Natl Acad Sci USA 74:1204–1207
Williams TJ, Morley J (1973) Prostaglandins as potentiators of increased vascular permeability in inflammation. Nature 246:215–218
Williams TJ, Piper PJ (1980) The action of chemically pure SRS-A on the microcirculation in vivo. Prostaglandins 19:779–782
Willoughby DA, DiRosa M (1970) A unifying concept for inflammation: a new appraisal of some old mediators. Excerpta Medica Int Congr Ser 229:28–38
Willoughby DA, Polak L, Turk JL (1968) Suppression of contact hypersensitivity and acute inflammation by anti-complement serum. Nature 219:192–194
Willoughby DA, Coote E, Turk JL (1969) Complement in acute inflammation. J Pathol 97:295–301
Wood DD, Ihrie EJ, Dinarello CA, Cohen PL (1983) Isolation of an interleukin-1-like factor from human joint effusions. Arthritis Rheum 26(8):975–978
Yamamoto S, Dunn CJ, Deporter DA, Capasso F, Willoughby DA, Huskinsson EC (1975) A model for the quantitative study of Arthus (immunologic) hypersensitivity in rats. Agents Actions 5:374–377
Yamamoto S, Dunn CJ, Willoughby DA (1976a) Studies on delayed hypersensitivity pleural exudates in guinea pigs. I. Demonstration of substances in the cell-free exudate which cause inhibition of mononuclear cell migration in vitro. Immunology 30:505–511
Yamamoto S, Dunn CJ, Willoughby DA (1976b) Studies on delayed hypersensitivity pleural exudates in guinea pigs. II. The interrelationship of monocytic and lymphocytic cells with respect to migration activity. Immunology 30:513–519
Yoshida T, Bigazzi PE, Cohen S (1975) Biologic and antigenic similarity of virus-induced migration inhibition factor in conventional, lymphocyte-derived migration inhibition factor. Proc Natl Acad Sci USA 72:1641–1643
Ziff M (1983) Factors involved in cartilage injury. J Rheumatol (Suppl) 11:13
Zigmond SH (1977) The ability of polymorphonuclear leukocytes to orient in gradients of chemotactic factor. J Cell Biol 75:606–610

CHAPTER 29

Specific Acute Inflammatory Responses

M. W. GREAVES and F. LAWLOR

A. Introduction

Although inflammation is the major component of some dermatological disorders, it is also an important though subsidiary component of many more. The skin shows striking clinical and histopathological specificity, compared with other organs and tissues, in its reponse to injury. It is capable of responding to different forms of injury in widely varying ways, ranging from erytheme and wealing to blister formation, reactions seen in no other organ or tissue surfaces. Itching is also a symptom of inflammation unique to the skin. At a microscopic level the skin also manifests a striking cellular reaction, seen in no other tissue, called spongiosis, i.e. intercellular oedema. Although the pharmacological events occurring in inflamed skin and in other tissues show points of similarity, it is not surprising that there are also major differences. It is imortant to attempt to understand these differences, not only in order to appreciate the pathophysiology of inflammatory skin disorders but also for the sake of future progress in developing innovative drug treatment for these conditions.

Acute inflammatory responses of the skin to injury differ from chronic inflammation principally in the dynamics of the dermal cellular population. In acute inflammation turnover of these cells is rapid, either because the cells are in transit through the injured tissue or because they are destroyed locally. In chronic inflammation the immigrating cells, mainly macrophages and lymphocytes, take up permanent residence in the skin, the most extreme example being granuloma formation. This account deals with the more important types of acute inflammatory response which are characteristic of skin, and their pharmacological heterogeneity. Acute inflammatory responses can be conveniently subdivided into immediate short-lived weal and flare responses, and delayed acute inflammation.

B. Immediate Weal and Flare Responses

Immediate weal and flare responses are, as originally indicated by LEWIS (1927), closely similar in appearance, symptoms and time course to the responses of human skin to intradermal injection of histamine. Thus, the main features are redness, weal formation, itching and pain. Histological changes, which are confined to the dermis, are usually minimal and include vasodilatation and dermal oedema.

It is usually possible to demonstrate that histamine is released into the affected skin, and that it plays a part in the production of the observed changes in the skin. But present evidence suggests histamine is unlikely to be the sole mediator in some instances.

I. Responses to Intradermal Antigen Injection

The association between histamine, mast cells and immediate hypersensitivity reactions in human skin is well established. LEWIS (1927) found that high concentrations of morphine injected intradermally elicited a triple response reaction which closely mimicked that due to histamine. He concluded that morphine caused release in the skin of histamine or H-substance. As early as 1941 direct evidence supporting this view was reported by KATZ. He used cantharidin to raise an intra-epidermal bulla in allergic subjects, and was able to demonstrate release of histamine into the bulla cavity following introduction of specific antigen into the blister fluid. Subsequently KATZ's findings have received ample confirmation (SONDERGAARD and GREAVES 1971; DUNSKY and ZWEIMAN 1978; TING et al. 1980) using both skin perfusion and skin chamber methods. In SONDERGAARD and GREAVES' perfusion studies perfusates from weal and flare reactions in seven of nine subjects yielded histamine ranging in concentration from 5 to 56 ng/ml perfusate. The conclusion that the observed histamine is derived from skin mast cells is supported by the observation of marked cytological changes in mast cells of the involved skin as detected by electron microscopy. Onset of these changes preceded the recovery of histamine from the antigen-challenged skin (TING et al. 1980). Direct evidence of involvement of other mediators, whether derived from mast cells or not, is at present lacking. Even if histamine is the sole primary mediator, the histamine weal is likely to be associated with secondary formation or release of other pharmacological activity (COOK and SHUSTER 1980). Efforts to identify leukotrienes C_4, D_4, and E_4 (the constituents of slow-reacting substance of anaphylaxis, SRS-A), prostaglandins, bradykinin and neuropeptides (substance P, VIP) are needed.

II. The Urticarias

Pharmacological aspects of the urticarias are dealt with in detail elsewhere (Chap. 18, 23, 27, Vol. I and Chap. 18, 31, Vol. II) so only a very brief outline will be given here. Two broad categories of urticaria need to be mentioned: the physical urticarias and chronic "idiopathic" urticaria.

1. Physical Urticarias

Most information has been gained from study of cold contact urticaria. Cooling of the skin is achieved with an ice cube or, more quantitatively, by a refrigerated probe which can be cooled to a pre-determined temperature. Application for 5 min leads to an immediate weal and flare reaction associated with itching. That this reaction is accompanied by histamine release from the cold-challenged skin has been confirmed by direct evidence from two different methods: use of a suction bulla technique, which enables recovery of pharmacological activity directly from involved skin, and analysis of venous effluent draining the cold challenged hand and forearm (KEAHEY and GREAVES 1980; BLACK et al. 1981; MISCH et al. 1982). These studies confirmed by bioassay the release of histamine into affected skin, causing itching, but suggested that other mediators, including prostaglandin E_2, played a significant part in the genesis of the vascular changes. Similar findings have been obtained in other physical urticarias (GREAVES and SONDERGAARD 1970a; LAWRENCE et al. 1981; SIBBALD et al. 1981).

2. Chronic Idiopathic Urticaria and Angio-Oedema

Chronic idiopathic urticaria may be conveniently defined as the occurrence of daily or almost daily weals and/or short-lived subcutaneous swellings for at least 3 months without any obvious precipitating cause.

Induction of suction bullae in involved skin of patients with chronic idiopathic urticaria led to the detection of greatly elevated levels of histamine in the affected lesions (KAPLAN et al. 1978). PHANUPHAK et al. (1982) have also claimed that tissue extracts from skin biopsy samples contain increased histamine activity compared with comparable samples from healthy skin. Since most of the skin histamine is stored in mast cells, this must imply either increased biosynthesis or reduced histamine clearance (or both) in affected skin. As both of these studies relied upon the radio-enzyme method for identification and assay of histamine, the possibility of additional mediators being involved remains open. Individual weals of chronic urticaria generally persist for up to 24 h, which argues against straightforward release of histamine from skin mast cells as the sole mechanism. The triggering factors in histamine release of chronic urticaria are unknown. However, the releasability of pharmacological mediators from peripheral blood leucocytes appears to be defective in chronic urticaria (GREAVES et al. 1974; KERN and LICHTENSTEIN 1976; CZARNETZKI et al. 1976).

C. Delayed Acute Inflammation

Delayed acute inflammation may be defined as an acute sustained inflammatory response with an onset one or more hours after injury and with a total duration of 24 h or more. The types of injury which cause this response are diverse and include ultraviolet radiation, thermal injury, lymphocyte-mediated allergic reactions and chemical irritancy. They also frequently involve cellular infiltration of the affected tissue. Long-continued inflammation leads to organisation of the cellular infiltrate as well as reactive changes in the connective tissue of the dermis and in the epidermis. In some instances granuloma formation may occur. These chronic inflammatory changes are likely to involve very complex pharmacological factors and are not considered here.

I. Ultraviolet Inflammation

Much of our present knowledge on pharmacological mechanisms of skin inflammation has been gained by studying the effects of ultraviolet (UV) irradiation.

This is because ultraviolet erythema is easily reproducible, quantifiable and acceptable to volunteer subjects. Although much work has been done in animals, there is a substantial body of information in man, to which this discussion will be confined. Ultraviolet radiation is conveniently divided into three fractions, depending on the wavelength: UV-C (100–290 nm), UV-B (290–320 nm), and UV-A (320–400 nm). Skin inflammation is readily produced by UV-C or UV-B but high dosage of UV-A is required to cause visible changes in the skin. However, patients previously treated by the photosensitising drug 8-methoxypsoralen become highly sensitive to the erythemogenic effects of UV-A. This combination has been used in the treatment of psoriasis (photochemotherapy; PUVA treatment).

1. Ultraviolet B Radiation

UV-B causes both erythema and tanning of human skin, and the ability of sunlight to produce these effects is mainly attributable to UV-B. Irradiation of human untanned skin with 3 times the minimal erythema dose of UV-B causes visible erythema which appears at 2 h after the irradiation, is maximal at 24 h and at 48 h, and gradually fades thereafter. Lewis originally proposed that UV erythema was caused by local release of a diffusible mediator (LEWIS 1927). He suggested the involvement of "H-substance", but anti-histamines do not inhibit the human UV reaction (PARTINGTON 1954) so a role for histamine is unlikely. Of the vaso-active agents investigated, the prostaglandins and related compounds most closely fulfil the essential criteria for putative mediators of UV inflammation (DALE 1934). These state that the properties of the mediator should be appropriate to its proposed pro-inflammatory role, that its identity should be fully established, that it should be present in increased concentrations in inflamed tissue but in lesser concentrations or absent in non-inflamed tissue, that anti-inflammatory drugs should antagonise or inhibit the putative mediator and that mechanisms should exist in the skin for its formation and release, and removal. The properties of the prostaglandins and their congeners are discussed elsewhere (Chap. 21) and will not be considered further.

The possibility that prostaglandin-like activity might be involved in the pathogenesis of UV-B erythema arose from the skin perfusion studies in UV-irradiated human skin. Perfusion of the inflamed skin revealed the presence in the perfusate of increased concentrations of smooth muscle contracting activity which resembled prostaglandin E (SONDERGAARD and GREAVES 1970). Subsequently more detailed studies of UV-B erythema were carried out using a suction bulla technique to recover inflammatory exudate (BLACK et al. 1980). Clinically normal human abdominal skin was irradiated with 3 times the minimum erythema dose of UV-B. Accompanying the developing erythema, increased arachidonic acid and prostaglandin E_2, $F_{2\alpha}$, and D_2 concentrations were found in the exudate as measured by gas chromatography–mass spectrometry (GC-MS). The maximum increase in these agents occurred at the height of the erythema at 24 h. At 48 h, when the erythemal reaction was still maximally intense, arachidonic acid and PGE_2 levels had returned to near normal (Fig. 1). In further studies (BLACK et al. 1982a) PGI_2 (measured by GC-MS as its stable metabolite, 6-keto-$PGF_{1\alpha}$) was

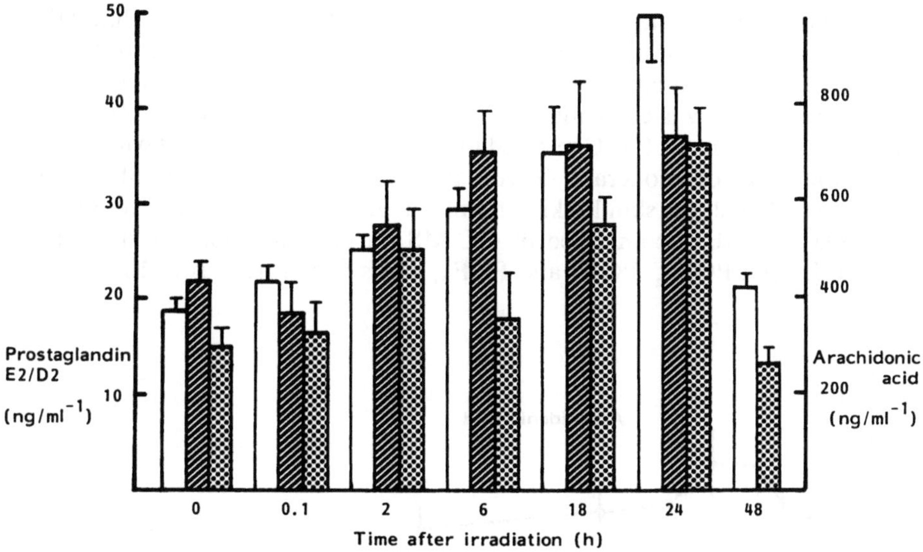

Fig. 1. The recoverable levels (mean ± SEM) of PGE_2 (*open rectangles*), PGD_2 (*shaded rectangles*) and arachidonic acid (*stippled rectangles*) measured by GC-MS in exudates from control skin and at various time intervals after 3 times the minimum erythema dose of UV-B irradiation

shown to be present in increased concentrations during the first 24 h after UV-B irradiation. Thus, in human skin, concentrations of arachidonic acid and its cyclo-oxygenase products parallel the development of the erythema during the first 24 h. The origin of the increased prostaglandin activity is uncertain. The presence of increased levels of PGI_2 suggests a contribution from dermal vascular sources. The effects of indomethacin on both UV-B erythema and the corresponding pharmacological changes are of special interest. Oral indomethacin only partially reduced the erythema, but totally suppressed the evoked increases in PGE_2 and $PGF_{2\alpha}$ concentrations (Black et al. 1978). Topical indomethacin had similar effects. These findings suggest that increased prostaglandin formation is involved in the pathogenesis of the earlier stages of UV-B erythema but that other mediators, formed by indomethacin-insensitive pathways, play a part in the later stages of the reaction. Alternative mediators have been proposed. Johnson and Daniels (1969) suggested that lysosomal enzyme release might be the key molecular event in inflammation. Rather unsatisfactory evidence for transitory release of kinin-like material in UV-irradiated skin was reported by Epstein and Winkelmann (1967). Since prostaglandins are known to enhance the vaso-active actions of other mediators, including histamine and bradykinin (Williams and Morley 1973), it is possible that synergism exists between prostaglandins and other mediators in UV inflammation.

2. UV-C Irradiation

The inflammatory response to UV-C radiation differs from that to UV-B irradiation. The former is less intense, develops and fades more rapidly and does not cause melanin pigmentation. VAN DER LEUN (1966) considered that UV-C erythema was due to a direct action of UV-C on the vessels, independently of a diffusible mediator, in contrast to UV-B erythema, which involved one or more mediators. This view is challenged by the findings of CAMP et al. (1978). Irradiation of human skin by UV-C resulted in erythema which appeared within 3 h of irradiation, was of moderate degree at 6 h and was maximal at 12–24 h. Thereafter it faded, the skin looking normal at 72 h. Exudate, obtained by a suction bulla method, was analysed by GC-MS for concentrations of arachidonic acid, PGE_2 and $PGF_{2\alpha}$. PGE_2 and $PGF_{2\alpha}$ levels were raised at 6, 18, and 24 h

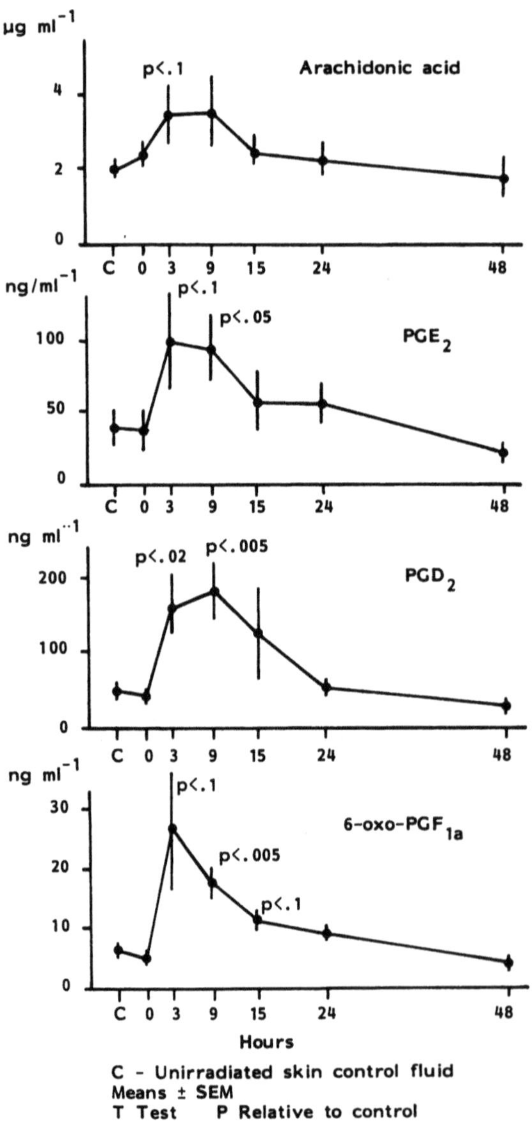

Fig. 2. Concentrations (means ± SEM) of arachidonic acid and prostaglandins in human skin suction bulla fluid after UV-A irradiation; *t*-test; probability values relative to controls: *C*, unirradiated skin control fluid

after UV-C irradiation and the concentration of PGE_2 was still elevated at 48 h. Arachidonic acid levels were also elevated at 6, 18, and 24 h, but not at 48 h. Since less than 2% UV-C penetrates to the dermis (EVERETT et al. 1966), an epidermal origin of the observed increase in prostaglandin activity seems probable. Although EAGLESTEIN and MARSICO (1975) were unable to suppress UV-C erythema by administration of the cyclo-oxygenase inhibitor indomethacin, BLACK et al. (1978) were able completely to suppress the increases in PGE_2 and $PGF_{2\alpha}$ measured by GC-MS in suction bulla fluid from UV-C irradiated human skin. Erythema was also reduced, but not completely inhibited, by indomethacin. These results are therefore compatible with the view that cyclo-oxygenase products contribute to the observed erythemal response.

3. UV-A Irradiation

Progress in investigating the molecular effects of UV-A radiation in human skin has been held back by lack of UV-emitting equipment with a high UV-A output and minimal UV-B contamination. The latter is important because even high energy sources of UV-A require long exposure times to produce visible erythema and even a small amount of stray UV-B could, over a prolonged period, cause a significant biological effect. UV-A produces erythema with certain specific characteristics (KAIDBY and KLIGMAN 1979). It appears during irradiation, fades over the succeeding few hours and then increases to maximum intensity and persists for hours or days thereafter. The doses required to produce this reaction are about 1000 times greater than those required to produce a comparable UV-B or UV-C reaction. HAWK et al. (1983) obtained suction blister exudate at 0, 5, 9, 15, 24 and 48 h after UV-A irradiation as well as from a contralateral control site. Increased concentrations, as measured by GC-MS, or arachidonic acid and prostaglandins D_2, E_2, and 6-oxo-$PGF_{1\alpha}$ were found between 3 and 9 h after irradiation, before the onset of maximal erythema (Fig. 2). Elevated histamine concentrations, measured by radioenzyme assay, were also found between 3 and 9 h after irradiation. At 24 h, when the erythema was still maximal, all values had returned to near control levels. Thus a causal role for these mediators early on in the UV-A response seems likely, though other mechanisms are probably involved in the later stages.

II. Photochemotherapy (PUVA) Erythema

Photochemotherapy involves oral administration of the furocoumarin drug 8-methoxypsoralen (8-MOP), followed by irradiation with long wavelength ultraviolet (UV-A). It is used principally for the treatment of psoriasis, the administration of 8-MOP greatly lowering the minimum erythema dose (MED) of UV-A. Irradiation with 3 MEDs of UV-A 2 h after administration of a single oral dose of 8-MOP results in erythema which appears at 18–24 h, reaches a maximum at 24–48 h and gradually fades thereafter. Exudate collected from irradiated uninvolved skin of psoriatic volunteers 24, 48, and 72 h after irradiation showed no elevation of PGE_2 or $PGF_{2\alpha}$ concentrations, measured by GC-MS (PGE_2) and radioimmunoassay ($PGF_{2\alpha}$), when compared with unirradiated (control) skin of the same patients (PLUMMER et al. 1978). In contrast, arachidonic acid concentrations (measured by GC-MS) were significantly higher in psoriatic uninvolved skin than in healthy non-psoriatic subjects, whether the skin had been irradiated or not. These results do not support the view that prostaglandins are involved in the pathogenesis of PUVA erythema, but they do raise the possibility of an abnormality of the arachidonic acid cascade in psoriasis. It is of interest that HEILIGSTADT et al. (1978), in a study of the biosynthesis of prostaglandins in vitro in biopsy material from irradiated skin of nine patients with psoriasis, found no differences in rates of biosynthesis of PGE_2 or $PGF_{2\alpha}$ compared with unirradiated skin of the same patients. The turnover of arachidonic acid was not estimated in that study. These findings are also supported by the failure of indomethacin to inhibit PUVA erythema (GSCHNAIT and PEHAMBERGER 1977).

III. Heat-Induced Inflammation

Considering the frequency of skin injury due to burns, surprisingly little work has been done on the pharmacological aspects of thermal injury. LEWIS originally described histamine-like activity in burn blister fluid (LEWIS and GRANT 1924). However, the identity of this activity was not confirmed. GOODWIN et al. (1963) described the presence of kinin-like activity in the urine and blister fluid of burnt subjects. In 1960 JOHANSSON claimed the presence of serotonin (5-hydroxytryptamine) in human burn exudate in a concentration of 0.1 μg/ml. That prostaglandins are regularly associated with thermal injury to skin was more recently suggested by ARTURSON et al. (1973). Burn blister fluid obtained from injured patients was extracted and the acidic lipid extracts subjected to thin-layer and silicic acid chromatography. Biological activity was found which co-chromatographed with prostaglandins E and F and the presence of prostaglandin E_2 was confirmed and quantified by GC-MS. Studies using non-steroidal anti-inflammatory drugs in thermal burns would be of interest, especially if combined with simultaneous pharmacological studies.

IV. Trafuril (Tetrahydrofurfurylnicotinate)

Trafuril is an ester of nicotinic acid which, as a 5% cream, is still used as a rubefacient in inflammatory joint disease. Application to the skin of a normal subject results in erythema, heat and oedema lasting at least 1–2 h, but little or no itching or pain. Early studies proposed that histamine played a significant role as a mediator, although the reaction could not be inhibited by anti-histamine administration (STREHLER 1949). In 1959 TRUELOVE and DUTHIE made the important observation that administration of aspirin inhibits the normal response to Trafuril. However, topical Trafuril application, apart from causing the cutaneous changes outlined above, also induces a reactive state in the treated skin whereby light stroking of the treated skin induces a dermographic response (WINKELMANN et al. 1965). Both the initial inflammatory reaction and the dermographic response were inhibited by aspirin but not by the anti-histamine diphenhydramine, by procaine or by pre-injection of the test site with the histamine-depleting agent com-

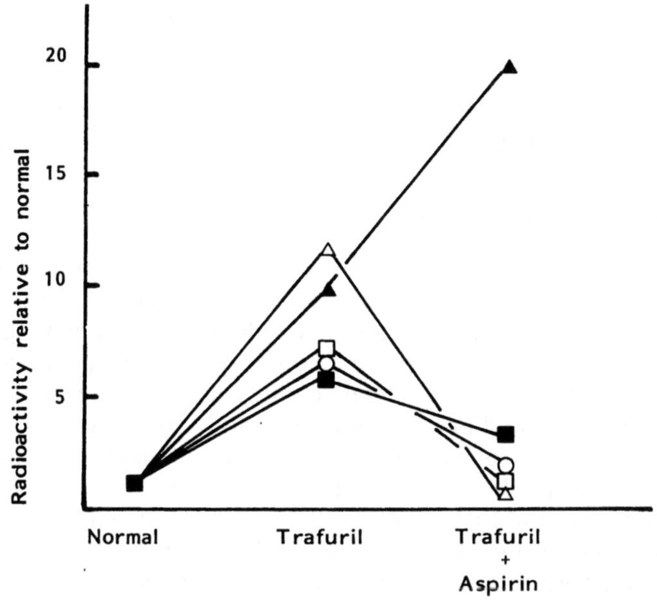

Fig. 3. Radioactivity in eluates from a Lipidex 5000 gel-partition column after extraction of exudates from Trafuril-treated skin before and after aspirin treatment, relative to normal skin. Peaks co-chromatograph as follows: ▲, arachidonic acid; ○, PGE; □, $PGF_{2\alpha}$; ■, PGD_2; △, 15-keto-PGE_2

pound 48/80. WINKELMANN carried out skin perfusion studies in Trafuril-treated skin, the perfusate being examined by bioassay. No histamine was found, but kinin-like smooth muscle contracting activity was detected in some subjects both in the initial Trafuril-induced erythema and in dermographism of Trafuril-treated skin. Since this kinin-like material was not produced in the presence of vascular occlusion of the treated area, it seems probable that the activity derived not from the skin but from the blood. A more comprehensive study of the Trafuril reaction before and after aspirin treatment was carried out by PLUMMER et al. (1977) using a suction bulla method to obtain exudate from inflamed and unaffected (control) skin. Examination of exudate from Trafuril-inflamed skin by bioassay using the cascade superfusion method (VANE 1964) revealed little or no histamine or bradykinin-like activity before or after application of Trafuril. However, prostaglandin-like activity was detected in normal skin which increased in Trafuril-treated skin. Gel partition chromatography of the extracted exudates after conversion to ^{14}C methyl esters revealed activity co-chromatographing with arachidonic acid, PGE_2, PGD_2, $PGF_{2\alpha}$ and several prostaglandin metabolites. Confirmation of the identity of PGE_2 and $PGF_{2\alpha}$ was obtained by gas chromatography. Of special interest was the finding that prior administration of aspirin caused suppression of radioactivity co-chromatographing with PGE_2, PGD_2 and $PGF_{2\alpha}$, but elevation of levels of arachidonic acid-like activity (Fig. 3). These results support the view (VANE 1971) that aspirin owes its anti-inflammatory activity to inhibition of prostaglandin biosynthesis. In contrast, the same workers (BLACK et al. 1982 b) have now shown that the glucocorticoid prednisolone, administered systemically, suppresses both arachidonic acid and prostaglandin levels in Trafuril-treated skin near to values obtained in untreated skin. However, Trafuril-induced erythema was only partially inhibited, suggesting that other mediators may be involved.

V. Contact Allergic Dermatitis

Although contact allergic dermatitis has for many years been the subject of intense research by immunologists, comparatively little interest has been shown in the soluble mediators presumed to cause the redness, oedema, itching, pain and vesiculation which characterise this common eruption. The presence of blood basophil polymorphonuclears in allergic contact dermatitis reactions has long been recognised in man and in the guinea-pig (WOLF-JURGENSON 1966; DVORAK and MIHM 1972). Since human basophils contain histamine (GRAHAM et al. 1955) it seems conceivable that histamine, released from basophils within the antigen-challenged skin, may fulfil a mediator role or possibly regulate T-lymphocyte function via lymphocyte membrane histamine H_2-receptors (ROCKLIN 1976). The lymphokines comprise a large group of soluble pharmacologically active substances derived from stimulated lymphocytes. Two of these, lymph node permeability factor (WILLOUGHBY et al. 1965) and skin reactive factor (SRF), are highly vaso-active. Most interest has centred on SRF (PICK et al. 1970), whose vascular actions are probably mediated through kinin formation. However, direct evidence of the involvement of SRF in allergic contact dermatitis is lacking. The possibility that prostaglandins and related compounds are involved in the genesis of allergic contact dermatitis was first suggested by the skin perfusion studies of GREAVES and SONDERGAARD (1970 b). These authors obtained ethyl acetate-soluble smooth muscle contracting activity in perfusates from involved skin of 15 out of 22 patients with allergic contact dermatitis. Subsequent studies (GREAVES et al. 1971) showed that this activity co-chromatographed with prostaglandins E and F. Histamine activity was found in only three of the subjects.

Independent support for these findings was subsequently obtained from GOLDYNE et al. (1973), who reported immunoreactive prostaglandin activity in increased amounts in perfusates from inflamed skin of allergic contact dermatitis. However, the cross-reactivity of their antibody did not enable them to distinguish between the different prostaglandins. Recently the role of arachidonate metabolites has been reinvestigated using a suction bulla technique to obtain exudate from the involved skin. Arachidonic acid, PGE_2, 12-HETE and 5-HETE were assayed by GC-MS and leukotriene B_4 by combined high performance liquid chromatography and bioassay (MISCH et al. 1982). MISCH found significantly elevated levels of arachidonic acid and PGE_2 in involved skin. 5-HETE was also raised, and leukotriene B_4 was found in 8 of 14 subjects but only in 2 of 14 control (uninvolved) sites. These data confirm previous findings of cyclo-oxygenase products in contact allergic dermatitis and also indicate involvement of chemo-attractant lipoxygenase products. These may mediate the complex cellular infiltrate which is a feature of allergic contact dermatitis.

VI. Primary Irritant Dermatitis

Despite the fact that primary irritant dermatitis still causes about 50% of occupational dermatoses, knowledge of the pharmacological substances involved in its production is extremely scanty. Increased histamine levels have been identified by CRAPS and INDERBITZIN (1957) in involved skin of primary irritant dermatitis.

More recently SONDERGAARD and GREAVES (1974) examined perfusates from 13 subjects in whom primary irritant dermatitis had been induced by application of benzalkonium chloride. Perfusates from involved skin contained histamine in only three of the subjects and in none of 22 control perfusates. Elevated levels of PGE- and PGF-like activity were obtained from inflamed skin in 7 of the 13 subjects as demonstrated by ethyl acetate extraction followed by thin layer chromatography. Our more recent studies in benzalkonium primary irritant dermatitis using a suction bulla technique and combined GC-MS analysis have confirmed the above findings (MISCH et al. 1982), but have not yielded any evidence of lipoxygenase products in this model.

VII. Atopic Dermatitis

Most research into the molecular mechanisms of atopic dermatitis has centred on its immunological aspects and particularly on the importance of IgE-mediated and T-lymphocyte-mediated immunity in the disorder. However, affected skin shows several consequences of pharmacological disorder which contribute, to a greater or lesser degree, to the visible inflammatory changes.

Most interest has centred on the role of histamine. That elevated levels of histamine are characteristic of atopic dermatitic skin has long been recognised (JOHNSON et al. 1960), and this increase may correlate with an increased mast cell population density in atopic eczema (MIKHAIL and MILLER-MILINSKA 1964) although a more recent study (MIHM et al. 1976) failed to demonstrate any change in the skin mast cell population in atopic dermatitis. WINKELMANN (1966a, b) reported finding elevated histamine concentrations in dermal perfusates from involved skin of atopic eczema. Other abnormal pharmacological findings include altered vascular responses of atopic eczema skin to histamine and kallikrein (MICHAELSSON 1970) and abnormal sweat responses to propranolol (HEMELS 1970). The latter findings lend support to SZENTIVANYI's β-adrenergic blockade theory of atopy (SZENTIVANYI 1979). Defective mast cell β-adrenergic responses may allow unchecked release of mast cell mediators in atopic eczema skin through a disorder of the adenylate cyclase–cyclic AMP system of these cells. In support of this notion, RING et al. (1980) reported increased evoked in vitro histamine from leucocytes of atopic subjects compared with controls. This may be due to enhanced leucocyte phosphodiesterase activity in atopic subjects, which leads to lowered cellular cyclic adenosine monophosphate levels and consequent disturbed regulation of histamine release (CHAN et al. 1981).

VIII. Psoriasis

Psoriasis is a hereditary inflammatory and epidermoproliferative disorder. There is a voluminous literature on the possible pharmacological control mechanisms involved in epidermal cell growth and differentiation, but comparatively little interest has been shown in the molecular basis of the inflammatory component of the disorder. One of the most prominent features of psoriasis is the accumulation of neutrophil polymorphonuclear leucocytes within the affected epidermis. Recently a number of attempts have been made to elucidate the pharmacological mechanisms underlying this process. TAGAMI and OFUJI (1976, 1977) have reported finding leucocytic chemotactic activity in psoriasis scale which they suggested might be due to C3a. LAZARUS et al. (1977) reported that psoriatic plaque contained a complement-dependent chemo-attractant factor, characterisation of which suggested it was a protease. Leukotriene B_4 is a recently described product of arachidonic acid transformation (BORGEAT and SAMUELSSON 1979) which is at least as potent as C5a (CAMP et al. 1982). Using a skin chamber method to recover exudate in vivo from involved and uninvolved skin of psoriatics, BRAIN et al. (1982) have shown that involved but not uninvolved skin of psoriatics contains ethyl acetate extractable polar acidic chemokinetic material which co-chromatographs on both straight- (Fig. 4) and reverse-phase high performance liquid chromatography (HPLC) with authentic leukotriene B_4. On straight-phase HPLC a second peak was identified which co-chromatographed with monohydroxy eicosatetraenoic acids, including 12-HETE and 5-HETE. Subsequent analysis by GC-MS has confirmed the identity of these compounds. Psoriatic epidermis contains aggregations of neutrophil polymorphonuclears and these, when activated, secrete leukotriene B_4. This possibility is reinforced by the recent finding that topical application of nanomolar amounts of leukotriene B_4 to intact human skin leads to development of a delayed erythematous and oedematous papule. Histo-

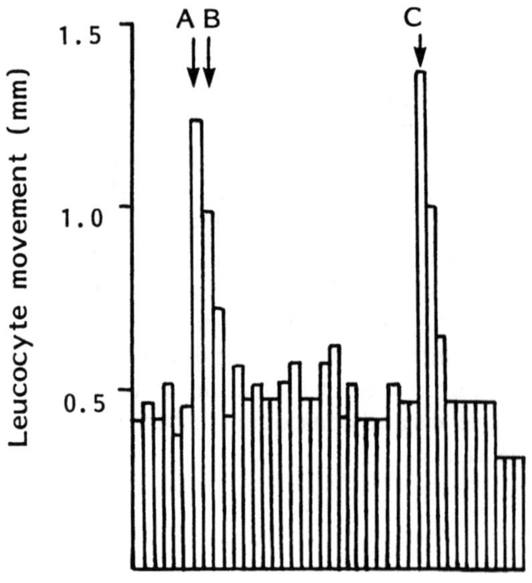

Fig. 4. Chemotactic activity seen after straight-phase HPLC of a lesional skin extract. Activity is expressed as the distance moved by leucocytes. Mean of duplicate observations is given for each fraction. The positions of standard 12-HETE (*A*), 5-HETE (*B*) and leukotriene B_4 (*C*) are shown

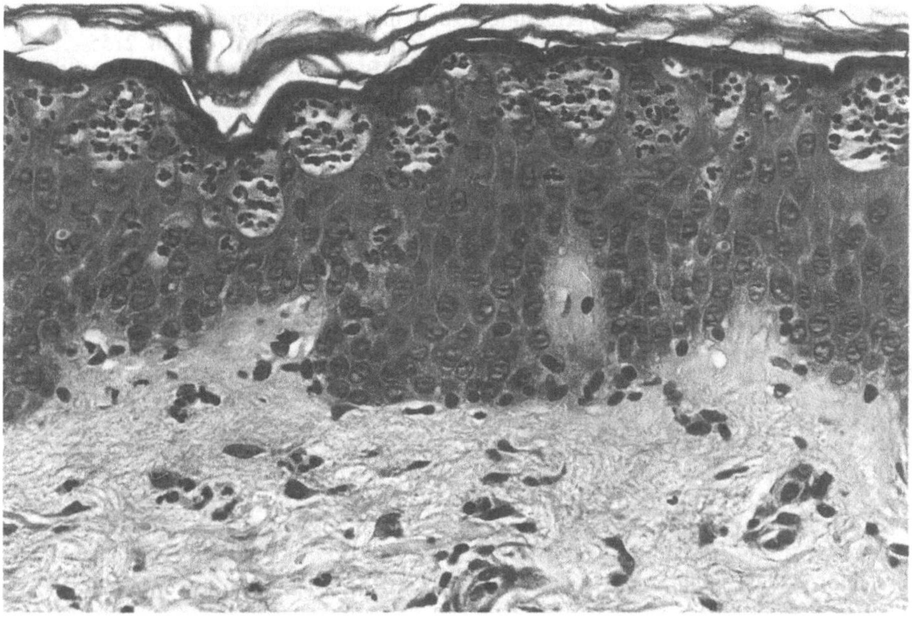

Fig. 5. Histological appearance of the 24-h leukotriene B_4 reaction. The upper epidermis is occupied by several vacuoles containing numerous intact neutrophils. H & E, × 320

logical examination at 24 h reveals a striking picture. Well-defined intra-epidermal micro-abscesses are present (Fig. 5) which are populated exclusively by neutrophil polymorphonuclears (CAMP et al. 1983). Although these changes differ in detail compared with clinical spongiform pustules of psoriasis, it is conceivable that increased levels of leukotriene B_4 found in psoriatic skin play a role in the pathogenesis of the psoriatic lesion.

References

Arturson G, Hamberg M, Jonson CE (1973) Prostaglandins in human burn blister fluid. Acta Physiol Scand 87:270–276

Black AK, Greaves MW, Hensby CN, Plummer NA, Warin AP (1978) The effects of indomethacin on arachidonic acid and prostaglandins E_2 and $F_{2\alpha}$ levels in human skin 24 h after UV-B and UV-C irradiation. Br J Clin Pharmacol 6:261–266

Black AK, Fincham N, Greaves MW, Hensby CN (1980) Time course changes in levels of arachidonic acid and prostaglandins D_2, E_2, $F_{2\alpha}$ in human skin following ultraviolet B irradiation. Br J Clin Pharmacol 10:453–457

Black AK, Keahey TM, Eady RAJ, Greaves MW (1981) Dissociation of histamine release and clinical improvement following treatment of acquired cold urticaria by prednisone. Br J Clin Pharmacol 12:327–331

Black AK, Hensby CN, Greaves MW (1982a) Increased levels of 6-oxo-$PGF_{1\alpha}$ in human skin following ultraviolet B irradiation. Br J Clin Pharmacol 13:351–354

Black AK, Greaves MW, Hensby CN (1982b) The effect of systemic prednisolone on arachidonic acid and prostaglandin E_2 and $F_{2\alpha}$ levels in human cutaneous inflammation. Br J Clin Pharmacol 14:391–394

Borgeat P, Samuelsson B (1979) Arachidonic acid metabolism in polymorphonuclear leucocytes: effects of ionophore A23187. Proc Natl Acad Sci USA 76:2148–2155
Brain SD, Camp RDR, Dowd PM, Greaves MW, Black AK, Mallet AI, Greaves MW (1982) Psoriasis and leukotriene B_4. Lancet II:762
Camp RD, Greaves MW, Hensby CN, Plummer NA, Warin AP (1978) Irradiation of human skin by short wavelength ultraviolet radiation (100–290 nm) (UV-C): increased concentrations of arachidonic acid and prostaglandins E_2 and $F_{2\alpha}$. Br J Clin Pharmacol 6:145–148
Camp RDR, Coutts AA, Greaves MW, Kay AB, Walport MJ (1982) Responses of human skin to intradermal injection of leukotrienes C_4, D_4, and B_4. Br J Pharmacol 80:497–502
Camp RDR, Russel-Jones R, Brain S, Wollard P, Greaves MW (1984) Production of intraepidermal microabscesses by topical application of leucotriene B_4. J Invest Dermatol 82:202–204
Chan SC, Greive SR, Sagko MJ, Hanigin JM (1981) Enhanced leukocyte phosphodiesterase activity in atopic dermatitis and after histamine-mediated desensitisation. Clin Res 29:597 (Abstract)
Cook J, Shuster S (1980) Histamine weal formation and absorption in man. Br J Pharmacol 69:579–585
Craps L, Inderbitzin T (1957) Histamine cutanee et reactions inflammatories. Arch Belges Dermatol 13:1–19
Czarnetzki B, Kern F, Lichtenstein LM (1976) Defective release of eosinophil chemotactic factor from peripheral leucocytes in patients with chronic urticaria. J Invest Dermatol 67:276–278
Dale HH (1934) Progress in autopharmacology. A survey of present knowledge of the chemical regulation of certain functions by natural constituents of the tissues. I. Introduction. The action of histamine and evidence restricting its possible significance. Evidence for other natural vasodilators. Bull Johns Hopkins Hosp 53:297–307
Dunsky EH, Zweiman B (1978) The direct demonstration of histamine release in allergic reactions in the skin using a skin chamber technique. J Allergy Clin Immunol 62:127–130
Dvorak HF, Mihm JC (1972) Basophil leukocytes in allergic contact dermatitis. J Exp Med 135:235–254
Eaglestein WH, Marsico AR (1975) Dichotomy in response to indomethacin in UV-C and UV-B induced ultraviolet light inflammation. J Invest Dermatol 65:238–240
Epstein JH, Winkelmann RK (1967) Ultraviolet light induced kinin formation in human skin. Arch Dermatol 95:532–535
Everett MA, Yeargers E, Sayre RM, Olson RC (1966) Penetration of epidermis by ultraviolet rays. Photochem Photobiol 5:533–542
Goldyne ME, Winkelmann RK, Ryan RJ (1973) Prostaglandin activity in human cutaneous inflammation: detection by radioimmunoassay. Prostaglandins 4:737–749
Goodwin LG, Jones CR, Richards WHG, Kohn J (1963) Pharmacologically active substances in the urine of burned patients. Br J Exp Pathol 44:551–560
Graham HT, Lowry OH, Wheelwright F, Lenz M, Parish HH (1955) Distribution of histamine among leukocytes and platelets. Blood 10:467–475
Greaves MW, Søndergaard J (1970a) Urticaria pigmentosa and factitious urticaria: direct evidence for release of histamine and other smooth muscle contracting agents in dermographic skin. Arch Dermatol 101:418–425
Greaves MW, Søndergaard J (1970b) A new pharmacological finding in human allergic contact eczema. Arch Dermatol 101:659–661
Greaves MW, Søndergaard J, McDonald-Gibson W (1971) Recovery of prostaglandins in human cutaneous inflammation. Br Med J [Clin Res] 2:258–260
Greaves MW, Plummer VM, McLaughlan P, Stanworth DR (1974) Serum and cell-bound IgE in chronic urticaria. Clin Allergy 4:265–271
Gschnait F, Pehamberger H (1977) Indomethacin does not affect PUVA induced erythema. Arch Dermatol Res 259:109–111

Hawk JLM, Black AK, Jaenicke KF, Barr RM, Soter N, Mallet AI, Gilchrest BA, Hensby CN, Parish JA, Greaves MW (1983) Increased concentrations of arachidonic acid, prostaglandins E_2, D_2, 6-oxo-$F_{1\alpha}$ and histamine in human skin following ultraviolet A irradiation. J Invest Dermatol 80:496–499

Heiligstadt H, Kassis V, Wiesmann K, Sondergaard J (1978) Prostaglandins in PUVA-treated psoriasis. Acta Derm Venereol (Stockh) 58:213–216

Hemels HG (1970) The effect of propranolol on the acetyl choline-induced sweat response in atopic and non-atopic subjects. Br J Dermatol 83:312–315

Johansson SA (1960) 5-Hydroxytryptamine in burns. Acta Physiol Scand 48:126–131

Johnson BE, Daniels F (1969) Lysosomes and reactions of the skin to ultraviolet radiation. J Invest Dermatol 53:85–89

Johnson HH, De Oreo GA, Lascheid WB, Mitchell F (1960) Skin histamine levels in chronic atopic dermatitis. J Invest Dermatol 34:237–241

Kaidby KH, Kligman AM (1979) The acute effects of long wavelength ultraviolet on human skin. J Invest Dermatol 72:253–256

Kaplan AP, Horakova Z, Katz SI (1978) Assessment of tissue fluid histamine levels in patients with urticaria. J Allergy Clin Immunol 61:350–354

Katz G, Cohen S (1941) Experimental evidence for histamine release in allergy. JAMA 117:1782–1783

Keahey TM, Greaves MW (1980) Cold urticaria: disassociation of cold-evoked histamine release and urticaria following cold challenge. Arch Dermatol 116:174–177

Kern F, Lichtenstein LM (1976) Defective histamine release in chronic urticaria. J Clin Invest 57:1369–1377

Lawrence CM, Jorizzo JL, Black AK, Coutts A, Greaves MW (1981) Cholinergic urticaria with associated angio-oedema. Br J Dermatol 105:543–550

Lazarus GS, Yost FJ, Thomas CA (1977) Polymorphonuclear leucocytes: possible mechanism of accumulation in psoriasis. Science 198:1162–1163

Lewis T (1927) Blood vessels of the human skin and their responses. Shaw, London

Lewis T, Grant RT (1924) Vascular reactions of the skin to injury. II. The liberation of histamine-like substance in injured skin: the underlying cause of factitious urticaria and of weals produced by burning; and the observations upon the nervous control of certain skin reactions. Heart 2:209–265

Michaelsson G (1970) Decreased cutaneous reactions to kallikrein in patients with atopic dermatitis and psoriasis. Acta Derm Venereol (Stockh) 50:37–44

Mihm MC, Soter NA, Dvorak HF, Austen KF (1976) The structure of normal skin and the morphology of atopic eczema. J Invest Dermatol 67:305–312

Mikhail GR, Miller-Milinska A (1964) Mast cell population in human skin. J Invest Dermatol 43:249–253

Misch KJ, Black AK, Barr R, Hensby CN, Mallet AI, Greaves MW (1982) Histamine and non-histamine pharmacological activity in cold urticaria. J Invest Dermatol 78:329 (Abstract)

Misch KJ, Black AK, Barr RM, Greaves MW (in preparation) Arachidonic acid lipoxygenase products in contact allergic dermatitis

Partington MW (1954) The vascular response of skin to ultraviolet light. Clin Sci 13:425–430

Phanuphak P, Schocket AL, Arroyave CM, Kohler PF (1980) Skin histamine in chronic urticaria. J Allergy Clin Immunol 65:371–375

Pick E, Krejci J, Turk JL (1969) Interaction between sensitised lymphocytes and antigen in vitro. I. The release of a skin reactive factor. Immunology 17:741–767

Plummer NA, Hensby CN, Black AK, Greaves MW (1977) Prostaglandin activity in sustained inflammation of human skin before and after aspirin. Clin Sci Mol Med 52:615–620

Plummer NA, Hensby CN, Warin AP, Camp RDR, Greaves MW (1978) Prostaglandins E_2, F_2 and arachidonic acid levels in irradiated and unirradiated skin of psoriatic patients receiving PUVA treatment. Clin Exp Dermatol 3:367–369

Ring J, Allen DH, Mathison DA, Spiegelberg HL (1980) In vitro releasibility of histamine and serotonin: studies of atopic patients. J Clin Immunol 3:85–91

Rocklin RE (1976) Modulation of cellular immune responses in vivo and in vitro by histamine receptor-bearing lymphocytes. J Clin Invest 57:1051–1057
Sibbald RG, Black AK, Eady RAJ, James M, Greaves MW (1981) Aquagenic urticaria: evidence of cholinergic and histaminergic basis. Br J Dermatol 105:297–301
Søndergaard J, Greaves MW (1970) Pharmacological studies in inflammation due to ultraviolet radiation. J Pathol 101:93–97
Søndergaard JS, Greaves MW (1971) Direct recovery of histamine from cutaneous anaphylaxis in man. Acta Derm Venereol (Stockh) 51:98–100
Søndergaard JS, Greaves MW (1974) Recovery of prostaglandins in human primary irritant dermatitis. Arch Dermatol 110:556–558
Strehler E (1949) Über die Wirkungsweise eines neuen Hauthyperämiemittels Trafuril (Nikotinsäuretetrahydrofurfurylester). Schweiz Med Wochenschr 79:144–152
Szentivanyi A (1979) The conformational flexibility of adrenoreceptors and the constitutional basis of atopy. Triangle 18:109–115
Tagami H, Ofuji S (1976) Leucotactic properties of soluble substances in psoriasis scale. Br J Dermatol 95:1–8
Tagami H, Ofuji S (1977) Characterisation of a leucotactic factor derived from psoriatic scale. Br J Dermatol 97:509–518
Ting S, Dunsky EH, Lavker RM, Zweiman B (1980) Patterns of mast cell alterations and in vivo mediator release in human allergic skin reactions. J Allergy 66:417–423
Truelove LH, Duthie JJ (1959) Effect of aspirin on the cutaneous response to local application of an ester of nicotinic acid. Ann Rheum Dis 18:137–141
Van der Leun JC (1966) Ultraviolet erythema: a study on diffusion processes in human skin. PhD thesis, Utrecht, Netherlands
Vane JR (1964) The use of isolated organs for detecting active substances in the circulating blood. Br J Pharmacol 23:360–373
Vane JR (1971) Inhibition of prostaglandin synthesis as a mechanism of action for aspirin-like drugs. Nature New Biol 231:232–235
Williams TJ, Morley J (1973) Prostaglandins as potientiators of increased vascular permeability. Nature 246:215–217
Willoughby DA, Walter MN, Spector WG (1965) Lymph node permeability factor in dinitrochlorbenzene skin hypersensitivity reaction in guinea pigs. Immunology 8:578–584
Winkelmann RK (1966a) Technique of dermal perfusion. J Invest Dermatol 46:220–284
Winkelmann RK (1966b) Non-allergic factors in atopic dermatitis. J Allergy 37:29–37
Winkelmann RK, Wilhelmj CM, Horner FA (1965) Experimental studies on dermographism. Arch Dermatol 92:436–442
Wolf-Jurgensen P (1966) Basophil leucocytes in delayed hypersensitivity: experimental studies in man using the skin window technique. Munksgaard, Copenhagen

Subject Index

MIX
Papier aus verantwortungsvollen Quellen
Paper from responsible sources
FSC® C105338

If you have any concerns about our products,
you can contact us on
ProductSafety@springernature.com

In case Publisher is established outside the EU,
the EU authorized representative is:
Springer Nature Customer Service Center GmbH
Europaplatz 3, 69115 Heidelberg, Germany

Printed by Libri Plureos GmbH
in Hamburg, Germany